TEXTS AND READINGS IN MATHEMATICS **27**

Algebraic Topology
A Primer

Second Edition

Algebraic Topology
A Primer
Second Edition

Satya Deo
Harish-Chandra Research Institute
Allahabad

HINDUSTAN
BOOK AGENCY

Published by

Hindustan Book Agency (India)
P 19 Green Park Extension
New Delhi 110 016
India

Email: info@hindbook.com
www.hindbook.com

ISBN 978-93-86279-67-5

Preface

Algebraic Topology is an important branch of topology having several connections with many areas of modern mathematics. Its growth and influence, particularly since the early forties of the twentieth century, has been remarkably high. Presently, it is being taught in many universities at the M.A./M.Sc. or beginning Ph.D. level as a compulsory or as an elective course. It is best suited for those who have already had an introductory course in topology as well as in algebra. There are several excellent books, starting with the first monograph 'Foundations of Algebraic Topology' by S. Eilenberg and N.E. Steenrod, which can be prescribed as a textbook for a first course on algebraic topology by making proper selections. However, there is no general agreement on what should be the 'first course' in this subject. Experience suggests that a comprehensive coverage of the topology of simplicial complexes, simplicial homology of polyhedra, fundamental groups, covering spaces and some of their classical applications like invariance of dimension of Euclidean spaces, Brouwer's Fixed Point Theorem, etc. are the essential minimum which must find a place in a beginning course on algebraic topology. Having learnt these basic concepts and their powerful techniques, one can then go on in any direction of the subject at an advanced level depending on one's interest and requirement.

This book is designed to serve as a textbook for a first course in algebraic topology described above. In order to lay down the real foundation of algebraic topology, we have also included a brief introduction to singular homology and cohomology. Approach to the contents has been dictated by several advanced courses taught by the author on different topics of algebraic topology at many universities. It is necessary to mention that the subject appears a bit abstract to begin with, but after a while it presents concrete topological as well as geometrical results of great insight and depth. The beginning student is expected to have some patience before appreciating the depth of these results. Maximum care has been taken to emphasize the subtle points of the concepts and results in a lucid manner. Examples are given at every stage to illustrate and bring out the underlying concepts and the results. It is hoped that the detailed explanations will help the student to grasp the results correctly and to have a sound understanding of the subject with confidence. The prerequisites for the study of this book are very little, and these have been briefly discussed in Chapter 1 and the Appendix to make it self-contained. These are included to maintain continuity between what the student has already learnt and what he is going to learn. It may be either quickly reviewed or even skipped.

We introduce and study the fundamental groups and its properties in Chapter 2. Starting with the concept of pointed spaces we show that the fundamental groups are topological invariants of path-connected spaces. After computing the fundamental group of the circle, we show how it can be used to compute fundamental groups of other spaces by geometric methods. In Chapter 3, we explain the topology of simplicial complexes, introduce the notion of barycentric subdivision and then prove the simplicial approximation theorem. In Chapter 4, we introduce the first classical homology theory, viz., the simplicial homology of a simplicial complex and then proceed to define the simplicial homology of a compact polyhedron. We provide detailed discussion of how a continuous map between compact polyhedra induces a homomorphism in their simplicial homology, and then prove the topological invariance of these groups. A few classical applications of the techniques of homology groups include the proof of the fact that the Euclidean spaces of different dimensions are not homeomorphic, Brouwer's Fixed Point Theorem, Lefschetz Fixed Point Theorem and the Borsuk-Ulam Theorem, etc. In Chapter 5, we study the theory of covering projections and its relation with fundamental groups. Important results on lifting of a map, classification of covering projections and the universal coverings are discussed at length.

The final topic dealing with the singular homology and cohomology has been discussed in Chapter 6. We have presented the important properties of singular homology including proofs of the homotopy axiom and the excision axiom. All of this is done in a way so that singular homology gets established as a 'homology theory' on the category of all topological pairs in the sense of Eilenberg-Steenrod. Then we go on to formally explain the definitions of an abstract homology as well as an abstract cohomology theory. Apart from the singular homology, we show that simplicial homology is also a homology theory on the category of all compact polyhedral pairs. Related topics such as homology and cohomology with coefficients, the Universal Coefficient Theorem, the Künneth Formula, the Mayer-Vietoris Sequence and the cohomology algebra of a space etc., have been discussed and adequately illustrated by examples. The student can test his understanding of the subject by working out the exercises given in every chapter. All of the basic material covered in this book should make the subject fascinating for beginners and should enable them to pursue advanced courses in any branch of algebraic topology.

Except for a few remarks, we have refrained from giving a detailed history of the subject which is, of course, always very enlightening and interesting.

However, we feel that it can be appreciated adequately when the student has learnt the subject at a more advanced level than what is presented here. The material presented here is contained in most of the books on algebraic topology. My contribution is basically in organizing and presenting it afresh. Definitions, propositions, corollaries etc. are numbered by 3 digits, i.e., Theorem 4.5.1 means Theorem 1 of Section 5 of Chapter 4. A suggested guideline for teaching a two-semester (or one academic year) course from this book is as follows: in the first semester one can cover Fundamental Group (Sections 2.1 to 2.6), Simplicial Complexes (Sections 3.1 to 3.4), Simplicial Homology (Sections 4.1 to 4.8) and Covering Projection (Sections 5.1 to 5.6). The chapter on Covering Projection can be taught immediately after the chapter on Fundamental Group. Proofs of some of the technical theorems given in Chapters 4 and 5 can be omitted. The remaining section of Chapter 4 on Simplicial Homology, viz. Section 4.9 and the whole of Chapter 6 on Singular Homology can be easily covered in the second semester. It is hoped that the material prescribed for the first semester is quite adequate for acquainting the student with the basic concepts of algebraic topology. The material for the second semester, on the other hand, will allow him to take a deeper plunge in any of the advanced topics of modern algebraic topology.

The author would like to express his thanks to many of his friends and colleagues, specially to G.A. Swarup, A.R. Shastri, Ravi Kulkarni and K. Varadarajan, all of whom have read the manuscript and made important suggestions for improvement of the book. The author appreciates the facilities made available to him by the Department of Mathematics, University of Arkansas, where the first draft of the whole book was completed. Rajendra Bhatia, the Managing Editor of the series 'Texts and Readings in Mathematics' of the Hindustan Book Agency is thanked for his keen interest in the editorial work. I thank my wife Prema Tripathi for her insistence that the job of completing the book should get preference over administrative duties. Finally, J.K. Maitra, one of my students and a colleague, deserves special thanks for his constant assistance in typesetting the book on LaTeX.

Satya Deo
July 1, 2003 Jabalpur

Preface to the Second Edition

In this edition of 'Algebraic Topology, A Primer', some necessary changes have been made in Chapters 3, 4 and 6. In the original edition, a simplicial complex was assumed, for simplicity, to be finite, which meant that all the polyhedra were assumed to be compact. This was indeed a sweeping restriction, especially when one is defining singular homology on the category of all topological pairs. Therefore, Chapter 3, has now been expanded by including another section on general simplicial complexes. As a consequence of this, basic results on the topology of a simplicial complex have also been included. Then, Chapter 4, on simplicial homology has been expanded to include simplicial homology of an arbitrary polyhedral pair. This allows us to state the important fact later in Chapter 6 that the simplicial homology is a homology theory on the category of all polyhedral pairs and their maps in the sense of Eilenberg-Steenrod.

Earlier, the Chapter 6 on singular homology was really somewhat sketchy. It has now been expanded with more details. In order that the book remains only elementary and within size, we have not included the proof of theorems such as the Acyclic Model Theorem, Eilenberg-Zilber Theorem nor the properties of the cup products in singular cohomology. However, the classical applications like the Jordan-Brouwer Separation Theorem, Jordan Curve Theorem and the Invariance of Domain Theorem, etc. have now been included in reasonable detail.

Many colleagues and students have pointed out typos and even a few mathematical errors in the original edition. The author is thankful to all of those, especially to H.K. Mukherjee, Krishnendu Gangopadhyay and Snigdha Bharati Choudhury, who read the book carefully and made many comments. All the corrections have now been made to the best of my knowledge. The numbering of definitions, theorems, corollaries, propositions, lemmas, examples and remarks have now been made in one sequence, i.e., Proposition 3.2 will follow Lemma 3.1 and Theorem 6.4 will follow Proposition 6.3 etc. Some exercises have been changed and the remaining ones have been properly ordered. The author is thankful to all those who followed this book as a text for a course on the subject and offered their valuable suggestions for improving the book. The 'Primer Character' of the book, however, has not been changed. My student Dr V.V. Awasthi of VNIT, Nagpur deserves thanks for helping me in the typesetting of the text. Finally, the author expresses his sincere thanks to the authorities of the Harish-Chandra Research Institute, Allahabad for the wonderful facilities and to the National Academy of Sciences, India for the Platinum Jubilee Senior Scientist Fellowship.

Satya Deo
Harish-Chandra Research Institute
Allahabad.
May 29, 2017

Contents

Chapter 1

Basic Topology: a review

1.1 Introduction

In this chapter, we briefly recall and collect some of the basic definitions and results of point set topology which will be needed later for explaining the concepts of algebraic topology. Important definitions are followed by some quick examples and, if necessary, by some comments. The results are mostly stated without proofs, but occasionally well-known proofs are indicated merely to highlight some important routine arguments. Emphasis has been laid mostly on those topics of set topology which appear necessary for a correct understanding of subsequent chapters. We assume that the student is somewhat already familiar with the basic notions of set topology, particularly with quotient spaces, connectedness and compactness, etc., **which are used here freely regardless of the order in which they occur in these preliminaries.** Definitions and results can be referred back in the book while reading the material of algebraic topology as and when they are required.

1.2 Euclidean Spaces and their Subspaces

Let $\mathbb{R}^n$ denote the n-fold cartesian product of the real line $\mathbb{R}$. Then an arbitrary point $x = (x_1, \ldots, x_n)$ of $\mathbb{R}^n$ will have n real components $x_1, x_2, \ldots, x_n$. The set $\mathbb{R}^n$ has several kinds of interesting and useful mathematical structures. It is an n-dimensional vector space over the field $\mathbb{R}$ of reals; it has the Euclidean inner product with respect to which it is a Hilbert space. The set of all $n \times n$ matrices over reals corresponds to the set of all linear transformations on the vector space $\mathbb{R}^n$. In this section, we recall the usual topology on $\mathbb{R}^n$. By Euclidean metric on $\mathbb{R}^n$, we mean the distance function d defined by the following formula for any pair of points $x = (x_1, \ldots, x_n)$, $y = (y_1, \ldots, y_n)$ in $\mathbb{R}^n$:

$$d(x, y) = \{\textstyle\sum_{i=1}^{n}(x_i - y_i)^2\}^{1/2}.$$

This is easily seen to be a metric on $\mathbb{R}^n$. For any $x \in \mathbb{R}^n$, the distance of x from the origin, denoted by $|x|$, is called the **Euclidean norm** of x. For any $x \in \mathbb{R}^n$ and for any $r \geq 0$, the subset $B = \{y \in \mathbb{R}^n \mid d(x, y) < r\}$ is called the open ball (or ball) of radius r with center x. Using these balls, we define a topology τ_d on $\mathbb{R}^n$ as follows: A subset U of $\mathbb{R}^n$ is a member of τ_d or is an open set if for each point $x \in U$, there exists a ball of some radius $r > 0$ such that $B(x, r) \subseteq U$. It can be easily verified that τ_d is indeed a topology on $\mathbb{R}^n$. This topology is called the **usual topology** or **Euclidean topology** on $\mathbb{R}^n$. More generally, for any metric space (X, ρ), the same procedure defines a topology τ_ρ on the set X and the resulting topology is called the topology on X induced by the metric ρ. It is in this way that a metric space is always regarded as a topological space.

For any subset Y of $\mathbb{R}^n$, we have a metric on Y induced by the Euclidean metric of $\mathbb{R}^n$. On the other hand, the Euclidean topology of $\mathbb{R}^n$ induces the subspace topology on Y. It can be easily verified that the topology on Y, induced by the metric of Y, is identical with the above subspace topology. In the sequel, whenever we refer to a subspace of $\mathbb{R}^n$, which we will frequently do, it will be assumed to have the induced topology from $\mathbb{R}^n$, unless explicitly stated otherwise.

Note that the subset $\mathbb{Z}$ of integers gets the discrete topology from $\mathbb{R}$. On the other hand, the set $\mathbb{Q}$ of rational numbers gets a topology from $\mathbb{R}$ which is totally disconnected but is not discrete. The unit interval $I = [0, 1]$ gets the induced topology, which is compact as well as connected. The subspace

$$\mathbb{S}^n = \{x \in \mathbb{R}^{n+1} \mid |x| = 1\}$$

is called the standard n-**sphere**.

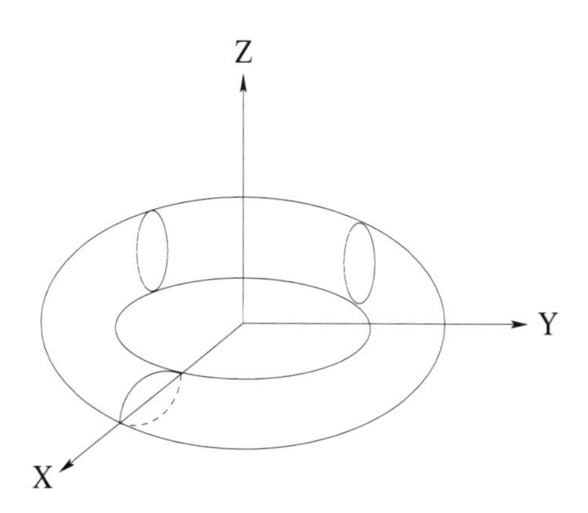

Fig. 1.1: The torus in Euclidean 3-space

Again, it is a compact, connected subset of the Euclidean space $\mathbb{R}^{n+1}$. Note that $\mathbb{S}^0 = \{1, -1\} \subset \mathbb{R}$ has just two points and is discrete whereas

$\mathbb{S}^1 = \{x \in \mathbb{R}^2 \mid |x| = 1\}$ is the set of all unimodular complex numbers. It is a topological group with respect to multiplication of complex numbers and is called the **circle group**. This group is evidently abelian; its k-fold product $T^k = \prod_1^k \mathbb{S}^1$ is called the **k-dimensional torus group**, which too is a compact, connected, abelian topological group. It is canonically a subspace of $\mathbb{R}^{2n} = \mathbb{C}^n$, where $\mathbb{C}$ stands for the set of complex numbers. We are identifying $\mathbb{R}^{2n}$ as $\mathbb{C}^n$ by saying that every pair (x_1, x_2) of real numbers can be regarded as a complex number $z = x_1 + ix_2$ and, therefore, every $2n$-tuple $(x_1, \ldots, x_{2n})$ can be thought of as an n-tuple $(z_1, \ldots, z_n)$ of complex numbers where $z_i = (x_{2i-1}, x_{2i}), i = 1, \ldots, n$. It may be mentioned here that the set $\mathbb{C}$ of complex numbers has an additional structure, viz., multiplication of complex numbers defined on the set of points of the plane $\mathbb{R}^2$.

For each $n \geq 1$, the compact, connected subset $\mathbb{D}^n = \{(x_1, \ldots, x_n) \in \mathbb{R}^n \mid \sum_1^n x_i^2 \leq 1\}$ of $\mathbb{R}^n$ is called an **n-disk** or an **n-cell**. Its boundary is clearly $\mathbb{S}^{n-1}$. In the 3-dimensional Euclidean space, let us consider the surface S obtained by rotating the circle $\{(y-3)^2 + z^2 = 1, x = 0\}$ about the z-axis. This surface, which is a compact, connected subset of $\mathbb{R}^3$, is homeomorphic to the 2-dimensional torus $\mathbb{S}^1 \times \mathbb{S}^1$ mentioned earlier. To see why these are homeomorphic one has to observe that both of these spaces are compact Hausdorff spaces and there is a continuous bijection (Exercise) from S onto $\mathbb{S}^1 \times \mathbb{S}^1$. Furthermore, we will prove later (see the Section 1.7) that **any continuous bijection from a compact space onto a Hausdorff space is always a homeomorphism.** Our assertion now follows immediately from these facts. We will frequently use this type of argument in the sequel.

Simplexes

Let $v_0, v_1, \ldots, v_n$ be $n+1$ points of $\mathbb{R}^n, n \geq 1$, so that the set of vectors $\{v_1 - v_0, v_2 - v_0, \ldots, v_n - v_0\}$ is linearly independent. Then the subset σ^n of $\mathbb{R}^n$ defined by

$$\sigma^n = \{x \in \mathbb{R}^n \mid x = \textstyle\sum_0^n \alpha_i v_i,\ 0 \leq \alpha_i \leq 1 \text{ and } \sum_0^n \alpha_i = 1\},$$

is called an **n-simplex**.

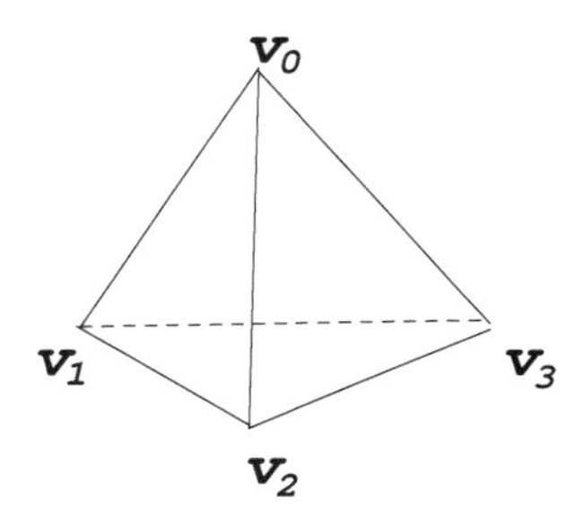

Fig. 1.1(a): A 3-simplex

Note that the above procedure cannot be used to define a $(n+1)$-simplex in $\mathbb{R}^n$ because we cannot find $n+1$ linearly independent vectors in $\mathbb{R}^n$. However, a k-simplex σ^k can be defined in $\mathbb{R}^n$ for all $k = 0, 1, \ldots, n$. A 0-simplex is just a point of $\mathbb{R}^n$. Also, notice that a 1-simplex is just a line segment joining two points whereas a 2-simplex is a triangle (with interior) obtained by taking the convex linear combinations of three points which are not on the same straight line. A 3-simplex is a tetrahedron, etc. An **n-simplex** σ^n is a compact, connected subset of the Euclidean space which carries a good geometric structure. When several simplexes, possibly of different dimensions, are suitably combined together, we get objects called **simplicial complexes**. These objects carry a very nice geometric structure and possess a topology which retains the Euclidean topology of every simplex forming the simplicial complex. A separate chapter has been devoted to the study of these simplicial complexes and maps between them, called **simplicial maps**. Here is a question which one may like to answer: Is there a space X that admits n-simplexes for all $n \geq 0$? In other words, what is that space X in which all n-simplexes, $n \geq 0$, are embedded, with each n-simplex embedded in the next $(n+1)$-simplex for all $n \geq 0$? The answer will become clear from the discussion of the concept of product spaces.

1.3 Continuous Maps and Product Spaces

A map $f: X \to Y$ between two topological spaces X and Y is said to be **continuous** if the inverse image $f^{-1}(V)$ of each open set V of Y is open in X. This is equivalent to saying that the inverse image of every closed subset of Y is closed in X. Continuity of a map is indeed a local property. To understand this, let $x \in X$ and W be a neighbourhood of the point $f(x)$ in Y. This means there is an open set V of Y such that $f(x) \in V \subset W$. Now, continuity of f implies that $f^{-1}(V)$ is an open set of X containing the point x. Hence $N = f^{-1}(V)$ is a neighbourhood of x and has the property that $f(N) = ff^{-1}(V) \subseteq V \subseteq W$. Thus, what we have proved is that $f: X \to Y$ *continuous means for every $x \in X$ and for every neighbourhood W of $f(x)$, there exists a neighbourhood N of x in X such that $f(N) \subseteq W$*. This condition is taken as the definition of continuity at the point $x \in X$. In other words, we say that f is continuous at $x \in X$ if for each neighbourhood W of $f(x)$, there exists a neighbourhood N of x such that $f(N) \subseteq W$. This is quite valid because if, according to this definition, $f: X \to Y$ is continuous at each point $x \in X$, then f is easily seen to be continuous. To prove this last statement, let V be an open set of Y. Suppose $x \in f^{-1}(V)$. This means $f(x) \in V$ and V is a neighbourhood of $f(x)$ in Y. By the given condition, there exists a neighbourhood N of x in X such that $f(N) \subseteq V$. This means $N \subseteq f^{-1}(V)$ and so $f^{-1}(V)$ is also a neighbourhood of x. Since $f^{-1}(V)$ has just been shown to be a neighbourhood of each of its points, it is an open set of X. Hence f is continuous.

In order to verify that a particular map $f: X \to Y$ is continuous, it is frequently easy and convenient to check that f is continuous at each point $x \in X$. It must now be mentioned here that the classical (ϵ, δ)-definition of continuity of a function in analysis is just another way of expressing the local definition of continuity in the context of metric spaces. It says that a map $f: (X, d) \to (Y, d')$ is continuous at $x = x_0 \in X$ if for all $\epsilon > 0$, $\exists\ \delta > 0$ such that $d(x, x_0) < \delta$ implies that $d'(f(x), f(x_0)) < \epsilon$. Interpreted in terms of neighbourhoods, this definition says that for each ϵ-ball around $f(x_0)$, there exists a δ-ball around x_0 in X such that $f(B(x_0, \delta)) \subseteq B(f(x_0), \epsilon)$. This is precisely the definition of local continuity of the map f at $x_0 \in X$.

There are several standard methods of constructing new topological spaces from the given ones. Defining product spaces is one such method. Let $\{X_\alpha \mid \alpha \in I\}$ be a family of topological spaces, and consider the cartesian product $\prod X_\alpha, \alpha \in I$ of sets X_α. For each α, let $p_\alpha: \prod X_\alpha \to X_\alpha$ denote the canonical projection. Then the topology on $\prod X_\alpha$ generated by subsets

$$\{p_\alpha^{-1}(U_\alpha) \mid U_\alpha \text{ is open in } X_\alpha, \alpha \in I\},$$

is called the **product topology or the Tychonoff topology** on $\prod X_\alpha$. It is important to understand the basic open sets of the above product topology – these are, by definition, just finite intersections of the sets of the above type. Thus, a basic open set is

$$p_{\alpha_1}^{-1}(U_{\alpha_1}) \cap \cdots \cap p_{\alpha_n}^{-1}(U_{\alpha_n}),$$

where U_{α_i} is an open set of $X_{\alpha_i}, i = 1, 2, \ldots, n$. It is necessary to realize that the above basic open set is just $\prod_{\alpha \in I} A_\alpha$, where $A_{\alpha_i} = U_{\alpha_i}$ for $i = 1, 2, \ldots, n$ and $A_\alpha = X_\alpha$ for all $\alpha \neq \alpha_1, \ldots, \alpha_n$. This basic open set is usually denoted by $\langle U_{\alpha_1}, \ldots, U_{\alpha_n} \rangle$. Using these basic open sets, it is easy to see that projection maps $p_\alpha: \prod X_\alpha \to X_\alpha$ are continuous, open surjections. The product topology can also be characterized as the smallest topology on the set $\prod X_\alpha$ making all projection maps continuous. It must be pointed out here that the projection maps need not be closed. For instance, consider the projection from the plane $\mathbb{R}^2$ to either of the factors. The set $F = \{(x, \frac{1}{x}) \mid x \neq 0\}$ is a closed subset of $\mathbb{R}^2$ but its projection on either factor is $\mathbb{R} - \{0\}$, which is not closed in $\mathbb{R}$.

Another important observation which can be inferred from the description of basic open sets is: An infinite product of discrete spaces $\{X_\alpha \mid \alpha \in I\}$, each space having more than one point, is never discrete. Note that if the product space was discrete, then each singleton $\{x_\alpha\}$ of the product must be open and so must contain a basic open set $\langle U_\alpha \rangle$, which is impossible!

A space X is embedded in the product space $X \times Y$ under the inclusion map $x \mapsto (x, y_0)$, where $y_0 \in Y$ is a fixed point. The restriction of the projection map $X \times Y \to X$ to the subspace $X \times \{y_0\}$ is the continuous inverse of this inclusion. If $Y = \{y_0\}$ consists of a single point, then X is homeomorphic to

$X \times Y$ and we don't get anything new. In the case of Euclidean spaces, we note that $\mathbb{R}^n$ is embedded in $\mathbb{R}^{n+1}$ under the map $x \mapsto (x, 0)$, where the last coordinate is zero. It is in this sense that we sometimes write:

$$\mathbb{R} \subset \mathbb{R}^2 \subset \mathbb{R}^3 \subset \cdots \subset \mathbb{R}^n \subset \cdots,$$

and their union is denoted by $\mathbb{R}^\infty = \cup \mathbb{R}^n$. It is also true that $\mathbb{R}^n$ is a vector subspace of $\mathbb{R}^{n+1}$ for any $n \geq 1$. The product $\mathbb{R}^\omega = \prod \mathbb{R}$ of countably infinite number of real lines is a space which contains $\mathbb{R}^n$ as a subspace for all n and so it contains $\mathbb{R}^\infty$. This means $\mathbb{R}^\infty$ has the subspace topology induced from $\mathbb{R}^\omega$. We can also topologize the set $\mathbb{R}^\infty$ by taking the weak topology (see later at the end of Section 1.5) induced by the inclusion maps $\mathbb{R}^n \to \mathbb{R}^\infty, \ \ n \geq 1$. Here we remark that the two topologies on $\mathbb{R}^\infty$ are not the same (Exercises later on). It now follows that either of the two spaces $\mathbb{R}^\omega$ or $\mathbb{R}^\infty$ (with subspace topology) can be taken as a universal space for n-simplices for all $n \geq 0$.

A nice result, which is used frequently in showing that a map going into a product space is continuous, and can be easily proved, is the following:

A map $f\colon Y \to \prod X_\alpha$ is continuous if and only if the composites $p_\alpha \circ f$ are continuous for each projection map $p_\alpha\colon \prod X_\alpha \to X_\alpha$.

1.4 Homeomorphisms and Examples

The following simple result about continuity of maps will be frequently used while dealing with paths and homotopies in the chapter on fundamental groups.

Lemma 1.4.1. (Continuity Lemma). *Let a space X be written as $X = F_1 \cup F_2$, where F_1, F_2 are closed subsets of X. Suppose $f\colon X \to Y$ is a map such that $f|_{F_1}, f|_{F_2}$ are continuous. Then the map f itself is continuous.*

Proof. Let $B \subset Y$ be a closed set in Y. Then $f^{-1}(B) = (f|_{F_1})^{-1}(B) \cup (f|_{F_2})^{-1}(B)$. Since $(f|_{F_1})^{-1}(B)$ is closed in F_1 and F_1 is closed in X, we find that $(f|_{F_1})^{-1}(B)$ is closed in X. Similarly, $(f|_{F_2})^{-1}(B)$ is also closed in X. Therefore, $f^{-1}(B)$, being union of two closed sets, is closed. This proves the lemma. ■

One can easily generalize the above Lemma when $X = \cup_{i=1}^n F_i$ is a finite union of closed sets. We must point out that if the sets F_1, F_2 are not closed, then the lemma is false. For example, consider the characteristic function $\chi_{\mathbb{Q}}\colon \mathbb{R} \to I$ of rationals $\mathbb{Q}$ defined by $\chi_{\mathbb{Q}}(x) = 1, x \in \mathbb{Q}$, and 0 otherwise; being constant on rationals as well as on irrationals, its restriction on each one of them is continuous, but the function $\chi_{\mathbb{Q}}$ itself is discontinuous as a function from $\mathbb{R} \to I$.

Recall that a map $f\colon X \to Y$ is said to be a **homeomorphism** if f is a bijection and both f, f^{-1} are continuous. A space X is called homeomorphic to a space Y if there exists a homeomorphism from one to the other and, in topology, two homeomorphic spaces are considered to be topologically same or equivalent. Note that the composite of two homeomorphisms is a homeomorphism, inverse of a homeomorphism is again a homeomorphism. It follows that for any space X, the set of all self-homeomorphisms of X forms a group with respect to composition of maps. It is an interesting problem to determine the group Homeo(X) of all homeomorphisms of a given space X.

Example 1.4.2. Any two open intervals (finite or infinite) of the real line $\mathbb{R}$ are homeomorphic. To see this, note that the map $f\colon (a,b) \to (0,1), a < b$, defined by $f(x) = (x-a)/(b-a)$ is a homeomorphism. Moreover, the map $g\colon \mathbb{R} \to (-1,1)$ defined by $g(x) = x/(1+|x|)$ is a homeomorphism, whereas the map $h\colon (0,\pi/2) \to (0,\infty)$ defined by $h(x) = \tan x$ is also a homeomorphism. Combining all these homeomorphisms one can easily prove the result stated above. One can, similarly, show that any two closed intervals are homeomorphic.

Example 1.4.3. For each $n \geq 1$, the boundary $B(\sigma^n)$ of the n-simplex σ^n is homeomorphic to the sphere $\mathbb{S}^{n-1}$. To see this, first we observe that if an object in an Euclidean space is moved under some translation, rotation or scaling (downward or upward) by a nonzero scalar, then the object changes only up to homeomorphism. Hence, without any loss of generality, we can assume that the sphere $\mathbb{S}^{n-1}$ is placed inside the n-simplex σ^n(imagine the case $n = 3$ as shown in the Fig.1.2).

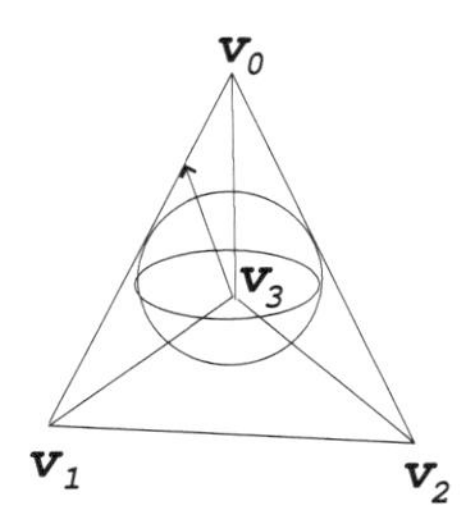

Fig. 1.2: The radial map

Take a point P in the interior of σ^n. If we draw a ray starting from P along any direction then the ray will intersect each of the sphere and the simplex in question, at exactly one point, say, R_1 and R_2, respectively. We define a map $h: \mathbb{S}^{n-1} \to B(\sigma^n)$ simply by taking R_1 to R_2. It is now evident that this map is a bijective continuous map having a continuous inverse and, therefore, h is indeed a homeomorphism. The map h defined above is called the **radial map**.

An interesting generalization of the foregoing example is discussed in Proposition 3.2.1.

Example 1.4.4. Let $\mathbb{D}^n$ be an n-disk, I^n be an n-cube and σ^n be an n-simplex. We will now show that any two of these three spaces is homeomorphic to each other. For simplicity of explanation, let us take the case of a 2-simplex and a square. As remarked in the previous example, we can assume that the simplex is inside the square (see Fig. 1.3).

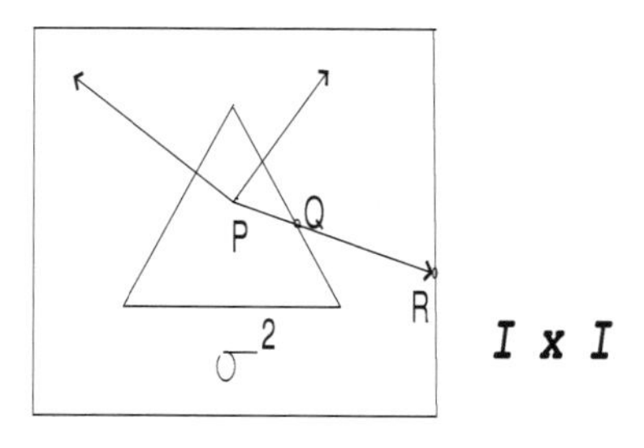

Fig. 1.3

Again, take a point P inside the simplex and consider the radial map emanating from P. In this case, every ray defines a unique line segment PQ of the simplex and a unique line segment PR of the square. We map the segment PQ to the segment PR by a linear homeomorphism. It is now plain to see that when these linear homeomorphisms along all possible directions are combined together, we get a homeomorphism mapping the 2-simplex to the square containing the simplex. The argument is clearly valid for all dimensions. The same argument applies to the case of an n-disk and an n-cube. It may be noted that the above homeomorphism restricted to the boundary of a disk, a simplex or a cube gives us the radial map of the previous example.

Example 1.4.5 (Stereographic Projection). Let $X = \mathbb{S}^2 - \{(0,0,1)\}$, where $\mathbb{S}^2$ is the standard 2-sphere in $\mathbb{R}^3$. Define a map $f\colon X \to \mathbb{R}^2$ by the formula:

$$f(x_1, x_2, x_3) = (x_1/(1-x_3), x_2/(1-x_3), 0),$$

where $\mathbb{R}^2$ is embedded in $\mathbb{R}^3$ by taking the third coordinate to be zero. It can be verified that f is a continuous bijection which has a continuous inverse. Therefore, f is a homeomorphism. This particular map is called **stereographic projection** (Fig. 1.4). The same map can be generalized to give a stereographic projection from $\mathbb{S}^n - \{(0,0,\ldots,0,1)\} \to \mathbb{R}^n$, which again will be a homeomorphism. *This fact is sometimes expressed by saying that an n-sphere, when punctured at a point, yields the Euclidean space $\mathbb{R}^n$.*

Now, one can see something more: Consider $\mathbb{S}^2$ punctured at two points, say, at the north pole and at the south pole (they don't make any difference though). Then the resulting space is homeomorphic to the hyperboloid of one sheet as well as to a right circular cylinder having infinite length on both sides. Again, here the length is not important – just as a finite open interval is

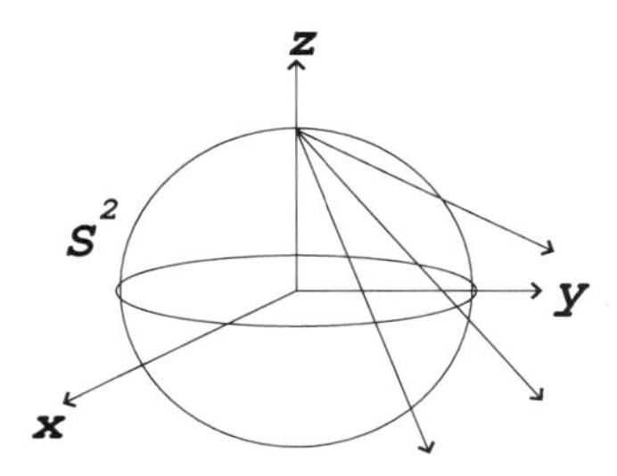

Fig. 1.4: Stereographic Projection

homeomorphic to the whole real line, an infinite cylinder is homeomorphic to a cylinder of finite height provided its boundary circles are removed; the same thing applies to the hyperboloid of one sheet also. A similar consideration shows that an open cylinder is homeomorphic to an open annulus in the plane. It also follows that the plane punctured at one point is homeomorphic to an open cylinder (Fig. 1.5).

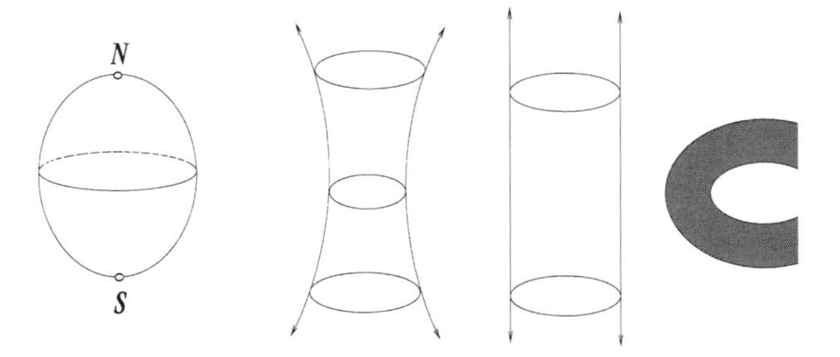

Fig. 1.5: Homeomorphic spaces

Exercises

1. Prove that if we remove a point from the ellipse $4x^2 + 9y^2 = 1$, then the remaining space is homeomorphic to the open interval $(0, 1)$.

2. Use the (ϵ, δ)-definition of a continuous map to show that the map $f\colon \mathbb{R} \to \mathbb{R}$ defined by

$$f(x) = \begin{cases} \sin \frac{1}{x} & x \neq 0 \\ 0 & x = 0 \end{cases}$$

is not continuous, but the map $g\colon \mathbb{R} \to \mathbb{R}$ defined by

$$g(x) = \begin{cases} x \cdot \sin \frac{1}{x} & x \neq 0 \\ 0 & x = 0 \end{cases}$$

is continuous.

3. Give a continuous map from the interval $[-1,1]$ onto the boundary of the sqare $I \times I$, which makes exactly five rounds of the boundary of the square.

4. A family $\{A_\alpha\}$ of subsets of a space X is called **locally finite** if each point x of X has a neighbourhood N which meets only finitely many members of $\{A_\alpha\}$. Let $\{F_\alpha\}$ be a locally finite family of closed sets such that $X = \cup F_\alpha$. If $f\colon X \to Y$ is a map such that $f|_{F_\alpha}$ is continuous for each α, then show that f is continuous.

5. Show, by an example, that an injective continuous map from the half-open interval $[0,1)$ into the Euclidean plane need not be a homeomorphism.

6. Show that the surface of a cone (end-circle included) is homeomorphic to the closed 2-disk $\mathbb{D}^2$, and the solid cone itself is homeomorphic to the closed 3-disk $\mathbb{D}^3$.

7. Prove that the ascending union $\mathbb{R}^\infty$ of Euclidean spaces $\mathbb{R}^n, n \geq 1$ with weak topology is not homeomorphic to the infinite product $\mathbb{R}^\omega$ of real lines. (**Hint**: Look at the metrizability property)

1.5 Quotient Spaces

Quotient spaces are of fundamental importance in topology. The intuitive idea of "identifying" or "pasting" is made rigorous by the concept of a quotient space. To make the things clear, let X be a topological space and R be an equivalence relation in X. It is well-known that R decomposes the set X into mutually disjoint equivalence classes. For each $x \in X$, let C_x be the equivalence class containing the element x, i.e., $C_x = \{y \in X \mid yRx\}$. Then for any two $x, y \in X, C_x = C_y$ or $C_x \cap C_y = \emptyset$. There is a natural surjective map $\nu\colon X \to X/R$, where X/R denotes the set of all distinct equivalence classes of X, defined by $\nu(x) = C_x$. The collection τ of all subsets $V \subset X/R$ such that $\nu^{-1}(V)$ is an open set of the space X is easily seen to be a topology on X/R. This topology on X/R evidently makes the quotient map $\nu\colon X \to X/R$ a continuous map. Indeed, τ is the largest topology on X/R which makes the map ν continuous. The set X/R with the topology τ defined above is called a **quotient space** of the space X. Changing equivalence relations on the set X produces all quotient spaces of X.

Take the unit interval $I = [0,1]$ and consider the equivalence relation R on I for which the equivalence classes are: $C_0 = \{0,1\}$ and $C_x = \{x\}$ for all x, $0 < x < 1$. Then the quotient space I/R can be easily seen to be homeomorphic to the circle $\mathbb{S}^1$. In such a case, we normally say that $\mathbb{S}^1$ is obtained from the unit interval by identifying the two end points 0 and 1 of the interval I.

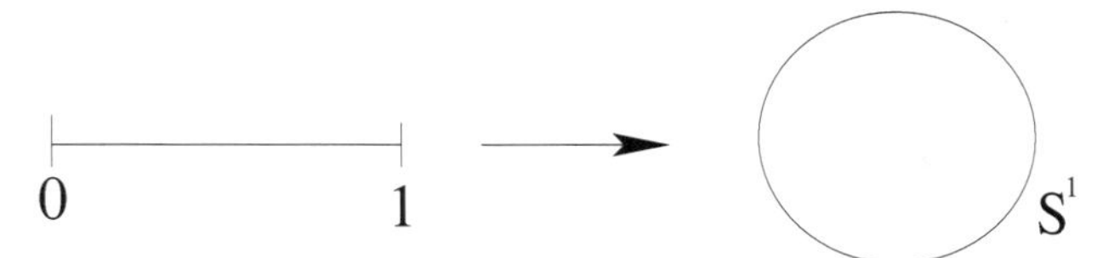

Fig. 1.6: Circle as the quotient space of unit interval

Interesting quotient spaces can be obtained by above kind of identifications. As another example, consider the unit square $I \times I$ and identify the points $(0, t)$ with $(1, t)$ for all $t \in I$. Then the resulting quotient space is homeomorphic to an Euclidean cylinder whose boundary consists of two disjoint circles.

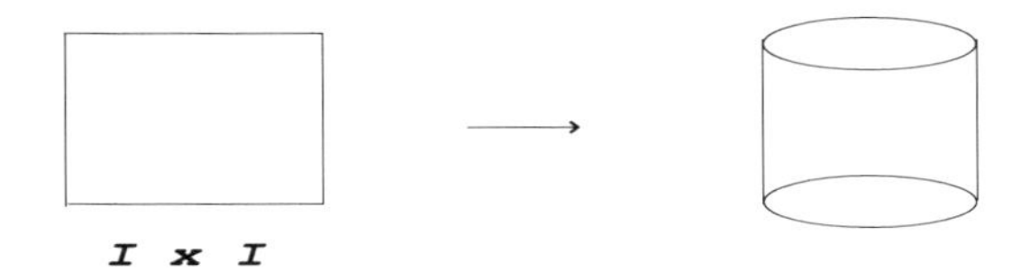

Fig. 1.7: Cylinder as the quotient space of the unit square

In the same square $I \times I$, let us identify the points $(0, t)$ with $(1, t)$ for all $t \in I$ and also identify $(t, 0)$ with $(t, 1)$ for every $t \in I$. Then the resulting quotient space, this time, is homeomorphic to the **torus** embedded in the Euclidean space $\mathbb{R}^3$ (Fig.1.8).

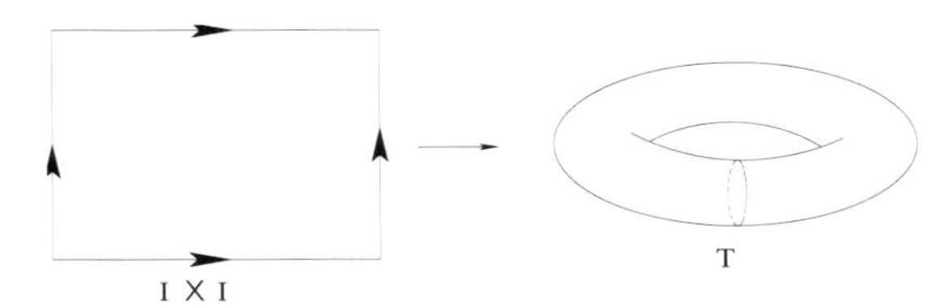

Fig. 1.8: Torus as the quotient space of unit square

As yet another example, let us again start with the square $I \times I$ and identify the point $(0, t)$ with $(1, 1 - t)$ for each $t \in I$. Then the resulting quotient space is homeomorphic to the Möbius band M (or strip) contained in $\mathbb{R}^3$ (Fig.1.9).

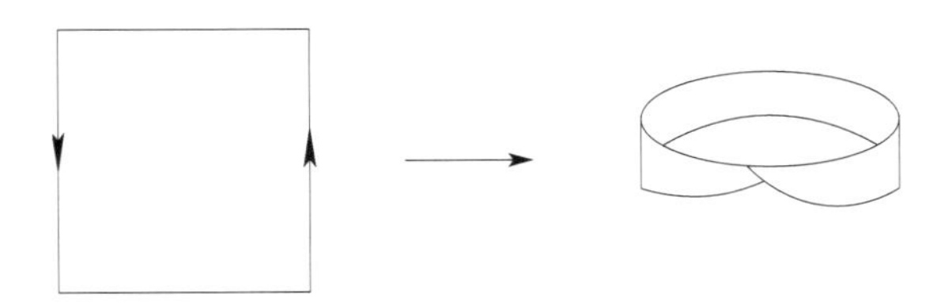

Fig. 1.9: Möbius band as the quotient space of square

The following procedure of the same kind yields a totally different topological space: Start with $I \times I$ and for each $t \in I$, identify $(t, 0)$ with $(t, 1)$ and $(0, t)$ with $(1, 1-t)$. Then the resulting quotient space, being continuous image of $I \times I$, is a compact, connected space. It is called the **Klein bottle**. Unfortunately, this space is not homeomorphic to any subspace of $\mathbb{R}^3$, i.e., the Klein bottle K cannot be embedded in $\mathbb{R}^3$ (Why?). However, it is an interesting topological space which we will consider again later (Fig. 1.9a).

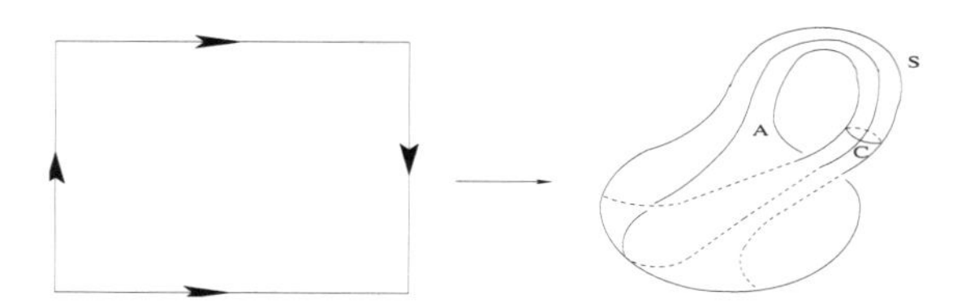

Fig. 1.9a: The Klein bottle as the quotient space of a square

As the final example, we have the **projective plane** as a quotient space of the 2-sphere $\mathbb{S}^2$. Here, we identify a point (x, y, z) of $\mathbb{S}^2$ with its antipodal point, viz., $(-x, -y, -z)$. This quotient space too, denoted by $\mathbb{RP}^2$, cannot be embedded in $\mathbb{R}^3$ (Why?), and hence cannot be easily visualized.

The following result is quite useful in showing that the maps from the quotient spaces are continuous.

Proposition 1.5.1. *Suppose Y is a quotient space of X with quotient map $\nu \colon X \to Y$ and $g \colon Y \to Z$ is a map to some space Z. Then g is continuous if and only if $g \circ \nu \colon X \to Z$ is continuous.*

Let us have some more examples of quotient spaces generalizing the idea of projective plane $\mathbb{RP}^2$, which are very important in topology as well as in geometry.

Example 1.5.2. (Real Projective Space) Let us consider the standard n-sphere $\mathbb{S}^n, n \geq 1$. For any $x \in \mathbb{S}^n$, let us identify x with its antipodal point $-x \in \mathbb{S}^n$. Then the resulting quotient space $\mathbb{S}^n/\sim$, denoted by $\mathbb{RP}^n$, is called the n-dimensional **real projective space**. It's easily seen that the real projective space is a compact, connected, Hausdorff space. To see the Hausdorff property, let us remark that the quotient map $\mathbb{S}^n \to \mathbb{RP}^n$ is closed. This is so because if F is a closed subset of $\mathbb{S}^n$ and $f \colon \mathbb{S}^n \to \mathbb{S}^n$ is the antipodal map, then $f^{-1}f(F) = F \cup f(F)$. Since f is a homeomorphism, this saturation of F is closed in S^n.

There is another way to define the same real projective space $\mathbb{RP}^n, n \geq 1$. If we consider the upper hemi-sphere $\mathbb{D}^n_+ \subset \mathbb{S}^n$, and define an equivalence relation $\sim$ in the hemi-sphere by identifying the antipodal points lying on the boundary of the hemi-sphere. Then there is a continuous canonical bijection from the resulting quotient space $\mathbb{D}^n_+/\sim$ and the preceding quotient space $\mathbb{S}^n/\sim$. Since the first space is compact and the latter is Hausdorff, the canonical map is a homeomorphism.

There is yet another way of defining the real projective space in practice. Consider the set $\mathbb{R}^n - \{0\}$, identify all points on a line passing through the origin and let X be the quotient space. Note that every such line intersects $\mathbb{S}^n \subset \mathbb{R}^{n+1} - \{0\}$ in exactly two points, which are antipodal. Now it is obvious that there is again a canonical bijection from the space X onto $\mathbb{S}^n/\sim$, which is bicontinuous and hence a homeomorphism.

Example 1.5.3. (Complex Projective Space) Consider the n-dimensional complex sphere, viz. $\{(z_0, \dots, z_n) \in \mathbb{C}^n \mid \sum |z_i|^2 = 1\}$. It is evident that the above set is, in fact the real $(2n+1)$-dimensional sphere $\mathbb{S}^{2n+1}$. Identify two points $(z_0, \dots, z_n)$, $(w_0, \dots, w_n)$ if there exists a $\lambda \in \mathbb{S}^1 (\neq 0)$ such that

$$(w_0, \dots, w_n) = \lambda(z_0, \dots, z_n).$$

The resulting quotient space $\mathbb{S}^{2n+1}/\sim$, denoted by $\mathbb{CP}^n$, is called n-dimensional **complex projective space.** This space is again a compact, connected, Hausdorff space.

To show that it is Hausdorff, it is enough to show that the quotient map $\nu \colon \mathbb{S}^{2n+1} \to \mathbb{CP}^n$ is a closed map. For this observe that if F is a closed set of $\mathbb{S}^{2n+1}$, then its saturation $\nu^{-1}(\nu(F)) = \mathbb{S}^1 \cdot F$, where $\mathbb{S}^1$ is the unit circle. Because $\mathbb{S}^1$, F both are compact, the product $\mathbb{S}^1 \cdot F$ under the map $\mathbb{S}^1 \times \mathbb{S}^{2n+1} \to \mathbb{S}^{2n+1}$ is the image of $\mathbb{S}^1 \times F$, which is compact, and so is closed.

Analogous to the real projective space there is another way to define $\mathbb{CP}^n$, $n \geq 1$. Let us consider the set $\mathbb{C}^n \setminus \{0\}$ and identify all complex lines passing through origin. Then it is easily seen that the resulting quotient space is homeomorphic to $\mathbb{CP}^n$.

The above projective spaces are not only compact, connected, Hausdorff spaces, but they have additional properties viz. these are n-dimensional manifolds having very nice cellular structures, and it is essentially these properties which make projective spaces extremely useful in algebraic topology.

Note that a quotient map $\nu \colon X \to X/R$, which is continuous and surjective, need not be an open map or a closed map. However, if $f \colon X \to Y$ is a closed (respectively, open) surjective continuous map, then the topology of Y is the quotient topology of X. The desired equivalence relation is obtained by identifying each point of $f^{-1}(y)$ with the single point y of Y. Also, some caution is required when we consider the subspaces of quotient spaces or the quotient spaces of subspaces : Let Y be a subspace of X. Then an equivalence relation R in X restricts to an equivalence relation S in Y. Then the quotient space Y/S has two topologies on it, one as the "quotient topology" coming from Y, and the other as the "subspace topology" coming from X/R. In general, the two topologies may be different from each other. To see an example of this situation, consider the square $X = I \times I \subset \mathbb{R}^2$. Take the subspace $Y = \{(t, 0) : 0 \leq t < 1\}$ of X and let R be the equivalence relation in X which produces the cylinder, i.e., identify each $(0, t)$ with the point $(1, t), t \in I$. Note that Y, as a subspace of X, is homeomorphic to the half-open interval $[0, 1)$ and the induced equivalence relation S on Y is the identity relation. This means the quotient space Y/S is again homeomorphic to $[0, 1)$. However, if we consider the image of Y under the quotient map $X \to X/R$ and take the subspace topology on Y/S (bottom circle) induced from the cylinder X/R, then the induced subspace topology on Y/S is homeomorphic to the Euclidean circle $\mathbb{S}^1$. The assertion made earlier now follows because the two spaces $[0, 1)$ and $\mathbb{S}^1$ are clearly not homeomorphic. It may be mentioned that, in general, the quotient topology on Y/S is stronger than the subspace topology induced from X/R because the following square is commutative:

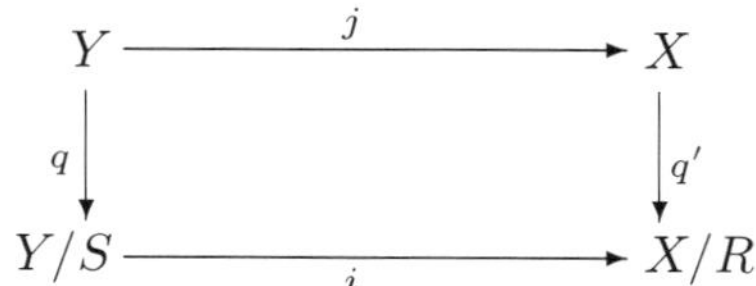

Since $q'j = iq$ is continuous and Y/S has the quotient topology, the map i is also continuous. This at once implies that the quotient topology on Y/S is stronger than the subspace topology on Y/S. The following statement can be easily proved: *Let Y be a closed (respectively, open) subspace of X and let $X \to X/R$ be a closed map. Then the subspace topology on Y/S coincides with the quotient topology induced from the subspace Y of X.* Similarly, if R and S are equivalence relations in spaces X and Y, respectively, then $R \times S$ is an equivalence relation in $X \times Y$. There are examples to show that the quotient space $(X/R) \times (Y/S)$ need not be homeomorphic to the quotient space $(X \times Y)/(R \times S)$ (see Dugundji [5], pp 130-131).

A quotient space of a Hausdorff space need not be Hausdorff. To see this, take the unit interval $[0, 1]$. Identify all the points of the set $A = [0, 1)$ to one equivalence class C_0. Then this quotient space has two points C_0, C_1 and the space $\{C_0, C_1\}$ is not even a T_1-space, since the point $\{C_0\}$ is not closed.

Recall that an equivalence relation R is just a subset of $X \times X$. The relation R is said to be **closed** if it is a closed subset of $X \times X$. In this terminology, note that X/R is Hausdorff implies that R is closed. In the case of a compact space X, one can prove that X/R is Hausdorff if and only if R is closed.

For any space X, if we identify the top copy $X \times \{1\}$ of X in $X \times I$ to a point, then the quotient space $X \times I/R$ is called **cone** over X. It is denoted by $C(X)$. In the product space $X \times [-1, 1]$, if we identify top copy

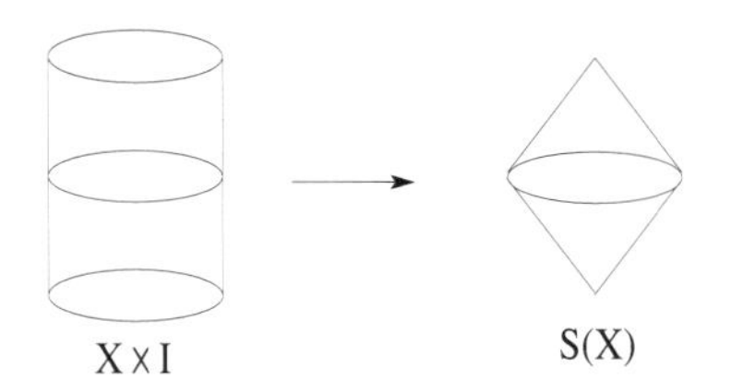

Fig. 1.10: Suspension of a space

$X \times \{1\}$ and bottom copy $X \times \{-1\}$ to two different points, then the quotient space, denoted by $S(X)$, is called **suspension** of X. It must be noticed that constructions of cone and suspension are natural in X, i.e., any continuous map $f: X \to Y$ induces continuous maps from $C(X)$ to $C(Y)$ as well as from $S(X)$ to $S(Y)$.

The concept of a mapping cylinder is also useful. Let $f: X \to Y$ be any continuous map. Consider the product $X \times I$ with unit interval $I = [0, 1]$ and, in the disjoint union $(X \times I) \cup Y$, identify for each $x \in X$, the points $(x, 0)$ with $f(x) \in Y$. Then the quotient space $M_f = (X \times I) \cup Y/ \sim$ obtained in this way is called the **mapping cylinder** of f. The space Y is embedded in M_f under the map $y \to [y]$. Note that whenever $y = f(x)$ for some $x \in X$, then $[y] = [f(x)]$. The space X is also embedded in M_f in several ways but we normally identify X with $X \times \{1\} \subset M_f$. The Figure 1.11 depicts an idea of the mapping cylinder M_f.

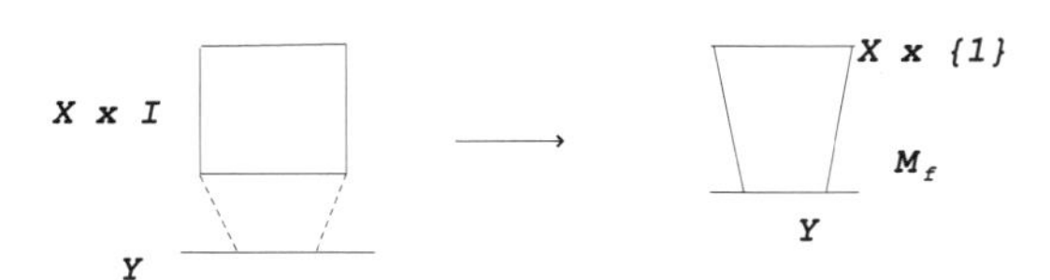

Fig. 1.11: Mapping cylinder of a map

Exercises

1. If X is a regular space and A is a closed subset of X, then show that the quotient space X/A is Hausdorff.

2. In the real line $\mathbb{R}$, let us identify all rational numbers to a single point. Then show that the quotient space is an indiscrete space.

3. Let $X = [-1, 1] \times \{0\} \cup \{(0, 1)\}$ be the subset of the plane $\mathbb{R}^2$. Define neighbourhoods of points $(x, 0), -1 \leq x \leq 1$ to be those induced from $\mathbb{R}^2$, but the neighbourhoods of the point $(0, 1)$ are sets of the type $[(-\epsilon, \epsilon) \times \{0\} - (0, 0)] \cup (0, 1)$. Show hat X is locally homeomorphic to real line, but is not Hausdorff (this is an example of a 1-manifold, which is not Hausdorff!)

4. Let R be an equivalnce relation in a compact Hausdorff space X. Show that the quotient space X/R is Hausdorff if and only if R is closed in $X \times X$.

5. Let $\mathbb{S}^n$ be the standard n-sphere, $n \geq 0$. Prove that the suspension $S(\mathbb{S}^n)$ of $\mathbb{S}^n$ is homeomorphic to the $(n+1)$-sphere $\mathbb{S}^{n+1}$.

6. Let $\mathbb{D}^2$ be the closed 2-disk in the plane $\mathbb{R}^2$. Prove that the quotient space $\mathbb{R}^2/\mathbb{D}^2$, obtained by collapsing the disk $\mathbb{D}^2$ to a point, is again homeomorphic to the plane $\mathbb{R}^2$.

7. Prove that when two copies of the disk $\mathbb{D}^n$ are identified along their boundaries, the resulting quotient space is homeomorphic to the n-sphere $\mathbb{S}^n$.

8. The Möbius band is obtained by identifying an edge of a square with its opposite edge after rotating the first edge by π radians. What will be the quotient space when the edge is rotated by 2π instead of π ?

Separation Axioms

Recall that a space X is said to be **Hausdorff** or a T_2-space if given any two distinct points $x_1, x_2 \in X$, we can find two disjoint open sets U_1 and U_2 of X such that $x_1 \in U_1$ and $x_2 \in U_2$. This is usually expressed by saying that x_1, x_2 can be separated in X by disjoint open sets. A condition weaker than T_2-axiom is T_1-axiom, viz., a space X is said to be a T_1-space if given any two distinct points x_1, x_2 of X, we can find two open sets U_1 and U_2 (not necessarily disjoint) such that one open set contains one point but not the other. It is easy to see that a space X is a T_1-space if and only if every point of X is closed in X. A T_1-space X is said to be T_3-space (or regular) if a point $x \in X$ and a closed set F of X, not containing x, can be separated by disjoint open sets. A stronger property than T_3-property is $T_{3\frac{1}{2}}$-property (or completely regular) which says that a point x and a closed set F, not containing x, can be separated by real-valued continuous functions, i.e., there exists a continuous function $f\colon X \to [0,1]$ such that $f(x) = 0$ and $f(F) = 1$. Finally, a T_1-space X is said to be a T_4-space (or normal) if any two disjoint closed sets of X can be separated by disjoint open sets, i.e., if F_1, F_2 are two disjoint closed subsets of X, then there exist disjoint open sets U_1, U_2 of X such that $F_1 \subseteq U_1$ and $F_2 \subseteq U_2$. It is well known that for each $i = 2, 3, 4$, a T_i-space is a T_{i-1}-space, but there are examples to show that the converse of none of these is true. There is an example (this one is not easy) of a T_3-space which is not $T_{3\frac{1}{2}}$. One of the most important results regarding these separation axioms is the famous **Urysohn Lemma** (see Dugundji [5], for all the results on separation axioms), which asserts that in a normal space X, any pair of disjoint closed sets can be separated by real-valued continuous functions. This lemma, in particular, implies that a normal space is completely regular, i.e., a T_4-space is also a $T_{3\frac{1}{2}}$-space whereas a $T_{3\frac{1}{2}}$-space is easily seen to be a T_3-space. It may also be mentioned here that on a $T_{3\frac{1}{2}}$-space X, there exists a large family of real-valued continuous functions; and the study of algebraic properties of the family of continuous functions is a beautiful topic of topology.

Weak Topology

Let X be a set and $\{A_\alpha \mid \alpha \in I\}$ be a family of subsets of X such that $X = \cup_{\alpha \in I} A_\alpha$. Suppose each A_α is a topological space in its own right. Can we define a topology on X so that, as a subspace of X, A_α gets its own topology? In order that the answer is in affirmative, it is clearly necessary that whenever A_α, A_β intersect, the set $A_\alpha \cap A_\beta$ must get the same topology from each of the spaces A_α and A_β. Suppose this is the case and also assume that for each $\alpha, A_\alpha \cap A_\beta$ is closed in A_β for all β. Now, we define a subset U of X to be open if $U \cap A_\alpha$ is closed in A_α for each α. Then the collection of all such subsets of X forms a topology on X and this induces the same topology on A_α as its original topology (Exercise). This topology on X is called the **weak topology** or the **coherent topology** on X induced by the given topologies of A_α for $\alpha \in I$. Note that with this topology on X, the inclusion maps $i_\alpha \colon A_\alpha \to X$ are continuous. In fact, the weak topology on X is the largest topology on X which makes all inclusion maps $i_\alpha \colon A_\alpha \to X$ continuous.

One interesting aspect of the weak topology on X defined by the family $\{A_\alpha\}$ is the following: Let $f \colon X \to Y$ be a map from the space X to any space Y and suppose the restriction $f|_{A_\alpha} : A_\alpha \to Y$ is continuous for each α. Then the map f itself is continuous. Important objects of algebraic topology require spaces like simplicial complexes and CW-complexes. The topology of these spaces is the weak topology defined by the Euclidean topology of simplexes and cells, respectively.

1.6 Connected and Path-Connected Spaces

A topological space X is said to be **disconnected** if it can be expressed as a union of two nonempty disjoint open sets. This is equivalent to saying that X can be expressed as a union of two nonempty disjoint closed sets. Another way of saying that X is disconnected is to say that there exists a subset $U(\neq \emptyset, X)$ of X which is both open and closed (clopen). A space X is said to be **connected** if it is not disconnected.

It follows that a discrete space having more than one point is disconnected. The space $\mathbb{Q}$ of rationals is disconnected; the real line $\mathbb{R}$ (with Euclidean topology) is connected. In fact, the following result identifies all connected subsets of the real line:

Example 1.6.1. A subset A of $\mathbb{R}$ is connected if and only if A is an interval (it may be finite, infinite, open, closed, or half open). To see this, note that if A is not an interval, then there is point $p \in \mathbb{R} - A$ such that A has points of $\mathbb{R}$

on the left as well as right of p. This means $(-\infty, p) \cap A$, $(p, \infty) \cap A$ are two nonempty disjoint open sets of A whose union is A and so A is disconnected. Conversely, suppose $A = [a, b], a < b$ is an interval. If A is not connected, then there is a clopen set U of A ($\emptyset \neq U \neq A$). Let $x_0 \in A - U$. Then there are points of U which are either on the left of x_0 or on the right of x_0. Suppose there are points of U on the left of x_0 and consider the set $L = \{x \in U \mid x < x_0\}$. Let $p = \sup L$. Then $p \in A$, since $\bar{L} \subset \bar{U}$, and since U is closed, $p \in U$. But U is also open means there is an $\epsilon > 0$ such that $[p, p + \epsilon) \subset U$. This contradicts the supremum property of p and so A is connected. We can similarly verify the result for other intervals.

We have mentioned the above result just because it answers the question "what subsets of real line are connected?", in a most simple way. Any result which classifies a class of mathematical objects in easily detectable terms is considered a beautiful theorem and there are very few results of this kind. Heine-Borel Theorem is a result of this type because it identifies all compact subsets of the Euclidean space $\mathbb{R}^n$ simply as "closed and bounded" subsets of $\mathbb{R}^n$. There is yet another result of this kind which has been termed as a "theorem par excellence". That theorem classifies all compact, connected 2-manifolds (viz., surfaces) in terms of a 2-sphere, a connected sum of tori or a connected sum of projective planes (see W.Massey [13] p.10 and p.33 for details).

The following results are quite well-known:

Proposition 1.6.2. *A continuous image of a connected space is connected.*

Theorem 1.6.3. *The product of an arbitrary family of connected spaces is connected.*

It follows from the latter result that the infinite cube I^∞, where I is the unit interval, is connected and so is the product $\mathbb{R}^\omega = \prod_1^\infty \mathbb{R}$ of real lines. Evidently, a subset of a connected space is hardly connected. But the following is a useful result:

Theorem 1.6.4. *Let A be a connected subset of a space X. Then $\bar{A}$ is also connected. More generally, if B is any subset of X such that $A \subseteq B \subseteq \bar{A}$, then B is connected.*

Proof. Since the closure of A in the subspace B is obviously B, the general result follows from the special one when we take B as X. To prove the first result, it suffices to show that any continuous map $f \colon \bar{A} \to \{0, 1\}$, where $\{0, 1\}$ has the discrete topology, is a constant map. For such a map, the restriction $f|_A$ of f to A will be continuous and, since A is connected, $f|_A$ is constant. But then, since A is dense in $\bar{A}$, f itself must be constant. ∎

Example 1.6.5 (Topologist's Sine Curve). This example shows that certain subsets of plane which visually do not appear to be connected, may indeed be connected (see Fig.1.12).

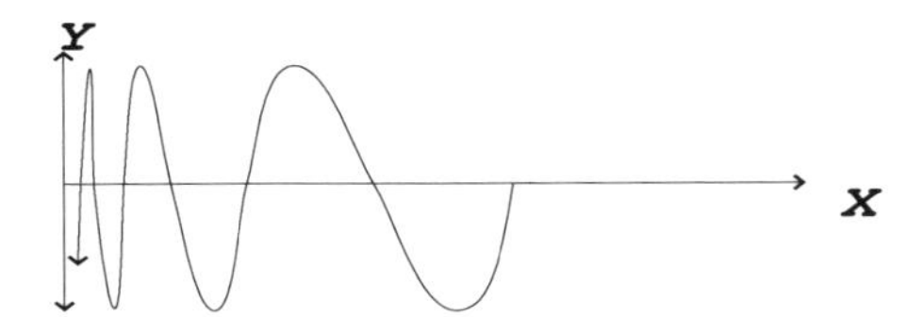

Fig. 1.12: Topologist's Sine Curve

Consider the subspace $G = \{(x,y) \in \mathbb{R}^2 \mid y = \sin\frac{1}{x}, 0 < x \leq \frac{1}{\pi}\}$. Since the interval $(0, \pi]$ is connected and G is its image under the continuous map $x \to (x, \sin\frac{1}{x})$, G is a connected set. Hence, by earlier result, its closure $T = \bar{G}$, i.e.,

$$T = G \cup \{(0,y) \mid |y| \leq 1\}$$

must be connected. This space T is popularly known as Topologist's Sine Curve and it has the property that any subset B satisfying $G \subseteq B \subseteq T$ is connected. In particular, the subset B obtained by adjoining an additional point, say, $(0, 1/2)$ or even a part of y-axis lying between the points $(0, -1)$ and $(0, 1)$, to G is connected. Note that the set B does not appear visually to be connected; it appears to have two pieces.

Next, we consider conditions when the union of connected subsets of a space X is connected. We have

Proposition 1.6.6. *Let $\{A_\alpha \mid \alpha \in I\}$ be a family of connected subsets of a space X such that $\bigcap A_\alpha \neq \emptyset$. Then the union $\bigcup_{\alpha \in I} A_\alpha$ is connected.*

The above result shows, in particular, that the plane $\mathbb{R}^2$ (or, more generally, even $\mathbb{R}^n, n \geq 2$) is connected. In fact, the plane $\mathbb{R}^2$ is the union of all straight lines in $\mathbb{R}^2$ passing through origin and every line is evidently connected. Indeed, the union of any number (finite or infinite) of lines passing through a fixed point of the plane is connected; in particular, a star-shaped subset of the Euclidean space $\mathbb{R}^n (n \geq 2)$ is connected. It seems appropriate to ask: what are the connected subsets of the plane $\mathbb{R}^2$? Unfortunately, there is no good answer to this question.

Example 1.6.7. Deleting a single point p from the real line $\mathbb{R}$ makes the remaining space $\mathbb{R} - \{p\}$ disconnected. However, if we remove one point from the plane $\mathbb{R}^2$ or even higher dimensional space $\mathbb{R}^n$, the remaining space is still connected. In fact, we can delete even a countable number of points $A = \{p_1, p_2, \ldots, p_n, \ldots\}$ and yet the remaining set $\mathbb{R}^n - A$ is connected. To see this, let $a \in \mathbb{R}^n - A$. It suffices to show that any other point $z \in \mathbb{R}^n - A$ and the point a lie in some connected subset of $\mathbb{R}^n - A$. For this, consider any bisector line L of the line segment az and for each $y \in L$, let $l_y = ay \cup yz$. Then, for each $y \in L, l_y$ is a connected set and any two of them meet only in a and z. We claim that there exists at least one y on L such that the set l_y lies completely in the set $\mathbb{R}^n - A$. If that is not the case then each l_y will contain at least one point of A, which means the set A will have at least as many points as the line L. Since L has uncountable number of points, this is a contradiction. Therefore, $\mathbb{R}^n - A$, being union of connected sets, all having the point a common, must be connected.

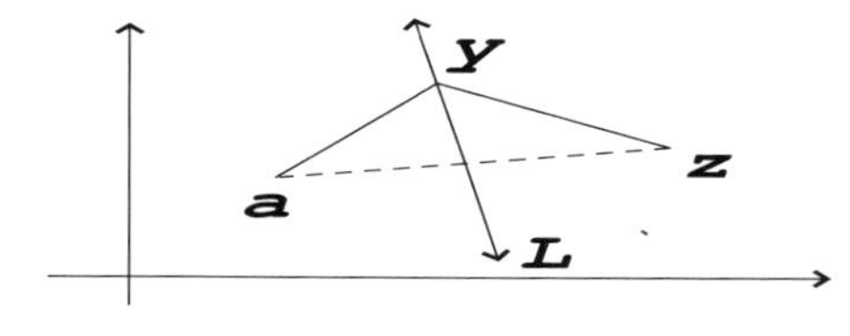

Fig. 1.13

Non-homeomorphic Spaces

Example 1.6.8. We have already seen that any two open (or closed) intervals are homeomorphic. However, an open interval $(a, b), a < b$ is not homeomorphic to a closed interval $[c, d], c < d$. The question is: How to prove this? The answer is very easy: The closed interval is compact, but the open interval is not compact. Hence, they cannot be homeomorphic because compactness is a topological property and so if they were homeomorphic, then one being compact will imply that the other must be compact, which is not the case. Now, how to prove that the open interval (a, b) is not homeomorphic to the half-open interval (c,d]? As both are noncompact, the argument involving compactness will not work. Here, however, we can use connectedness as follows: Let $f: X \to Y$ be a homeomorphism. If a subset A of X is removed from X and its image $f(A)$ is removed from Y, then the remaining spaces $X - A$ and $Y - f(A)$ must be homeomorphic (Just consider restrictions of f and f^{-1} to $X - A$ and $Y - f(A)$, respectively). Now, it is straight to see that $(c, d]$ cannot be homeomorphic to (a, b) simply because removing the single point d from $(c, d]$ leaves it connected, whereas removing any one point from the open interval (a, b) leaves the remaining space disconnected. The connectedness argument can also be applied to show that (a, b) is not homeomorphic to $[c, d]$: removing the two end points c and d from $[c, d]$ leaves it connected whereas removing any two points from (a, b) renders it disconnected.

The above argument can be applied to prove that the real line $\mathbb{R}$ is not homeomorphic to the plane $\mathbb{R}^2$: removal of one point from $\mathbb{R}$ makes it disconnected whereas $\mathbb{R}^2$ remains connected even after removing hundreds of points

(Recall the earlier example). Now, here is an interesting problem: Can we apply connectedness argument or some other argument involving a topological property to prove that the Euclidean spaces $\mathbb{R}^m, \mathbb{R}^n (m \neq n)$ are not homeomorphic? An answer (affirmative or negative) will provide an answer to the same problem for spheres $\mathbb{S}^m$ and $\mathbb{S}^n (m \neq n)$ since these are one-point compactifications of $\mathbb{R}^m$ and $\mathbb{R}^n$, respectively. We can ask other questions in the same spirit: With any of these methods, can we decide whether or not (i) The projective plane $\mathbb{P}^2$ is homeomorphic to the torus $T^2 = \mathbb{S}^1 \times \mathbb{S}^1$?, (ii) Klein bottle is homeomorphic to the torus? **Results of algebraic topology, as we will see in due course of time, can quickly answer these and numerous other questions of this type.** These and a few others were historically the first applications of algebraic topology.

Components

If X is a disconnected space, then the maximal connected subsets of X, called **components**, are occasionally quite useful. For any point x of such a space X, consider the union C_x of all connected subsets of X which contain the point x. Then, since union of any family of connected subsets having a nonempty intersection is connected, C_x is obviously connected and is a maximal connected subset of X containing the point x. The set of all components of X forms a **partition** of the set X. Since closure of a connected set is connected, every component C_x must be closed in X. The components of a discrete space are singleton subsets. A space X is said to be **totally disconnected** if the components of X are singleton sets. In the space $\mathbb{Q}$ of rationals also, the components are singleton sets and so $\mathbb{Q}$ is totally disconnected but, as we know, $\mathbb{Q}$ is not discrete. It may be mentioned that such spaces are not of much use in algebraic topology except for constructing pathological examples.

Paths

The concept of a **path** is of basic importance in algebraic topology. By a path ω in a space X, we mean a continuous map $\omega\colon I \to X$, where $I = [0, 1]$ is the unit interval. The point $\omega(0)$ is called the **initial point** and $\omega(1)$ is called the **terminal point** of the path ω. Sometimes, we identify the path ω with its image set $\omega(I)$, though the two things are quite different. Since I is compact as well as connected, the image set $\omega(I)$ is a compact and connected subset of the space X. It must be pointed out that strange things can happen with the image of a general path – even if ω is a 1-1 map, the image set $\omega(I)$ need not be homeomorphic to I as, for example, when the space X is indiscrete. If X is the Euclidean space $\mathbb{R}^n$ (or, more generally, any Hausdorff space), then a 1-1 path ω is indeed a homeomorphism from I to $\omega(I)$ (why?). It all depends on the space X where the path is defined. Peano gave an example of a path ω in $\mathbb{R}^2$ where the image $\omega(I)$ filled the whole square $I \times I$ (the space-filling curve) – of course, this path cannot be 1-1.

A topological space X is said to be **path-connected** if any two points of X can be joined by a path. Note that if ω_1 and ω_2 are two paths in a space X such that $\omega_1(1) = \omega_2(0)$, then we can define a new path $\omega_1 * \omega_2$, called the **product** of ω_1 and ω_2, in X as follows:

$$(\omega_1 * \omega_2)(t) = \begin{cases} \omega_1(2t) & 0 \le t \le 1/2 \\ \omega_2(2t-1) & 1/2 \le t \le 1 \end{cases}$$

The continuity of the path $\omega_1 * \omega_2$ follows from the Pasting Lemma. Also, the initial point of $\omega_1 * \omega_2$ is the initial point of ω_1 and the terminal point of the new path is the terminal point of ω_2. Note that the image set of the path $\omega_1 * \omega_2$ is just the union of the image sets of ω_1 and ω_2. The inverse of a path ω is the path ω^{-1} defined by $\omega^{-1}(t) = \omega(1-t)$. Observe that the inverse path ω^{-1} travels in the direction opposite to that of ω. Using these ideas regarding paths, one can easily prove that **a space X is path-connected if and only if there is a point x_0 of X which can be joined by a path to any other point $x \in X$.**

A path connected space X is connected. To see this, let $x_0 \in X$. For any other point $x \in X$, let $\omega_x \colon I \to X$ be the path joining x_0 to x. Then, $\cup_{x \in X} \omega_x(I) = X$. Since $\omega_x(I)$ is connected for all x and since $\cap_{x \in X} \omega_x = \{x_0\}$, it follows that X is connected. We can define path-components of a space X analogous to components as the maximal path connected subsets of X. Though the components are closed subsets of X, path-components need not be closed. The following example shows this and also the fact that *a connected space need not be path-connected.*

Example 1.6.9. Consider the Topologist's Sine Curve $T = S \cup G$ where $S = \{(0, y) \mid |y| \le 1\}$ and $G = \{(x, y) \in \mathbb{R}^2 \mid y = \sin\frac{1}{x}, 0 < x \le \frac{1}{\pi}\}$. We have already seen (Example 1.6.5) that T is connected. We show now that T is not path connected. In fact, we claim that the point $(0, 0) \in T$ cannot be joined to $(\frac{1}{\pi}, 0) \in T$ by a path in T. If possible, let $\omega \colon I \to T \subset \mathbb{R}^2$ be a path in T so that $\omega(0) = (0, 0)$ and $\omega(1) = (\frac{1}{\pi}, 0)$. Since S is a closed subset of T, $\omega^{-1}(S)$ is a closed subset of I such that $0 \in \omega^{-1}(S)$ but $1 \notin \omega^{-1}(S)$. This means $\omega^{-1}(S)$ has a maximal element, say a, $0 < a < 1$. Now the restriction $\omega|_{[a,1]}$ is a path joining a point of S and the point $(\frac{1}{\pi}, 0)$ having the property that $\omega(t) \in G$, $\forall\ t > a$, For convenience, let us assume $a = 0$ and write $\omega(t) = (x(t), y(t))$. Observe that $\forall\ t > 0$, $x(t)$ runs over the interval $(0, \frac{1}{\pi})$ and $y(t)$ runs along the sine curve $y(t) = \sin\frac{1}{t}$. Now we show that there is a sequence $\{t_n\}$ of points in I such that $t_n \to 0$, and $y(t_n) = (-1)^n$, which will give a contradiction to the continuity of ω. For each given n choose u such that $0 < u < x(\frac{1}{n})$ and $\sin\frac{1}{u} = (-1)^n$. Then apply the Intermediate value theorem to find an element t_n, $0 < t_n < \frac{1}{n}$ such that $x(t_n) = u$.

A space X is said to be **locally connected** (respectively, locally path-connected) if every neighbourhood N of $x \in X$ contains a connected (respectively, path-connected) neighbourhood. It can be easily seen that in a locally connected space, the components are not only closed but are open also. The following is an interesting result.

Proposition 1.6.10. *A connected, locally path-connected space is path-connected.*

Proof. Let X be a connected space which is also locally path connected and $x_0 \in X$. Then we claim that the subset $U = \{x \in X \mid x, x_0 \text{ can be joined by a path}\}$ of X must be an open set. This follows at once because $y \in U$ implies that there is a neighbourhood $N(y)$ which is path connected. But that means each point of $N(y)$ can be joined to x_0 by a path and hence, $N \subseteq U$. Thus, U is open. On the other hand, if $z \notin U$, then z has a path connected neighbourhood $N(z)$ so that no point of $N(z)$ can be in U. This means $X - U$ is an open set. Since X is connected and $U \neq \emptyset, U = X$, i.e., X is path-connected. ∎

It can be easily seen that *a continuous image of a path-connected space is path-connected*; *a product of any family of path-connected spaces is path-connected.*

Remark 1.6.11. It must be pointed out that path-components of a space X, as we will see later on in Chapter 6, are closely related to the zero-dimensional singular homology groups of X. On the other hand, the components of a space are tied up with the zero-dimensional Čech cohomology groups of a space X. In fact, these homology and cohomology groups are the basic objects of study in algebraic topology.

1.7 Compact Spaces and Locally Compact Spaces

We recall that a topological space X is said to be **compact** if every open cover of X has a finite subcover. What this says is that given any open cover $\mathcal{U} = \{U_\alpha \mid \alpha \in J\}$ of X, there are finite number of members, say, $U_{\alpha_1}, \ldots, U_{\alpha_n}$ of $\mathcal{U}$ which cover the space X. Thus, compactness is a kind of finiteness condition on the topology of X.

Note that the unit interval $I = [0, 1]$ is compact whereas the real line $\mathbb{R}$ itself is not compact. Any n-sphere $\mathbb{S}^n, n \geq 0$, is a compact subset of $\mathbb{R}^{n+1}$. The famous Heine-Borel Theorem asserts that **a subset A of the Euclidean space $\mathbb{R}^n$ is compact if and only if it is closed and bounded** – this yields a very easy way of verifying whether or not a subset of $\mathbb{R}^n$ is compact. Thus, for instance, the ellipsoid, viz., $\{(x, y, z) \in \mathbb{R}^3 \mid x^2/a^2 + y^2/b^2 + z^2/c^2 = 1\}$, where a, b, c are any three given positive constants, being closed as well as bounded subset of $\mathbb{R}^3$, is compact. On the other hand, the hyperboloid $\{(x, y, z) \in \mathbb{R}^3 \mid x^2/a^2 + y^2/b^2 - z^2/c^2 = 1\}$, being unbounded, is not a compact space. The graph G of the sine curve $y = \sin \frac{1}{x}, 0 < x \leq \pi$, in the plane $\mathbb{R}^2$ is not compact since it is not closed – the point $(0, a) \in \mathbb{R}^2, |a| \leq 1$ are limit points of G, but do not belong to the graph. But the topologist's sine curve $T = G \cup \{(0, a) \mid |a| \leq 1\}$ is a compact space. The torus $\mathbb{S}^1 \times \mathbb{S}^1 \subseteq \mathbb{R}^3$ is compact but a punctured torus is not compact.

It can be directly proved from the definition that *a continuous image of a compact space is compact; a closed subspace of a compact space is compact.* Compactness, when combined with Hausdorff separation axiom, yields some interesting results : (i) a compact subset of a Hausdorff space X is closed in X, (ii) a compact Hausdorff space is normal. The first part at once implies that a continuous map $f\colon X \to Y$, where X is compact and Y is Hausdorff, is always closed. For, let F be a closed subspace of X. Then, since X is compact, F must be compact and since a continuous image of a compact space is compact, $f(F)$ is compact. Finally, since a compact subset of a Hausdorff space is closed, $f(F)$ must be closed in Y. With this last result, the following fact is immediate: **Any continuous bijection $f\colon X \to Y$ from a compact space X to a Hausdorff space Y is a homeomorphism.** This result is of common use in identifying quotient spaces with more familiar spaces. To see an example, take a disk $\mathbb{D}^2$ and identify all the boundary points to a single point. Then it is asserted that the result is homeomorphic to $\mathbb{S}^2$. How? Note that $\mathbb{D}^2 - \mathbb{S}^1$ is homeomorphic to $\mathbb{S}^2 - \{N\}$, where N is the north pole of the sphere. Now, if the homeomorphism is extended to a map $f\colon \mathbb{D}^2 \to \mathbb{S}^2$ by mapping whole of $\mathbb{S}^1$ to the point N, then f is continuous and onto. Hence, f induces a continuous map $h\colon \mathbb{D}^2/\mathbb{S}^1 \to \mathbb{S}^2$, which is evidently a continuous bijection. As $\mathbb{D}^2/\mathbb{S}^1$ is compact and $\mathbb{S}^2$ is Hausdorff, the above result says that h is a homeomorphism.

One of the most important theorems of set topology, which has beautiful consequences in other branches of mathematics also, is the following:

Theorem 1.7.1. (Tychonoff). *Let $\{X_\alpha \mid \alpha \in I\}$ be any (finite or infinite) family of compact spaces. Then the product space $\prod_{\alpha \in I} X_\alpha$ is compact.*

While dealing with covering projections, we will frequently require the following result called Lebesgue's Covering Lemma:

Lemma 1.7.2. (Lebesgue). *Let (X, d) be a compact metric space and $\mathcal{U} = \{U_\alpha\}$ be an open covering of X. Then there exists a positive real number δ (called Lebesgue number for the given covering) such that any ball $B(x, \delta)$ around any point $x \in X$ is contained in some member of $\mathcal{U}$.*

Proof. For every $x \in X$, first let us choose a positive number $r(x)$ such that $B(x, r(x))$ is contained in some member of $\mathcal{U}$. Then $\{B(x, r(x)/2) \mid x \in X\}$ is an open covering of X. Since X is compact, there exists a finite subcover, say, $B(x_1, r(x_1)/2), \ldots, B(x_n, r(x_n)/2)$ which covers X. Let $\delta = \min\{r(x_1)/2, \ldots, r(x_n)/2\}$. Then, $\delta > 0$ is the required Lebesgue number. To see this, let us consider a ball $B(x, \delta)$ for some $x \in X$. Then there is an $i = 1, \ldots, n$ such that $x \in B(x_i, r(x_i)/2)$. Now, if $y \in B(x, \delta)$, then

$$d(y, x_i) \leq d(y, x) + d(x, x_i) \leq \delta + r(x_i)/2 \leq r(x_i).$$

Hence, $B(x, \delta) \subseteq B(x_i, r(x_i)) \subseteq U$ for some $U \in \mathcal{U}$. ■

Recall that a map $f\colon (X, d) \to (Y, d')$ is said to be **uniformly continuous** if given $\epsilon > 0, \exists$ a $\delta > 0$ (independent of the points $x, x' \in X$) such that $d(x, x') < \delta$ implies $d'(f(x), f(x')) < \epsilon$. A continuous map need not be uniformly continuous. However, for a compact metric space X, we have the following useful result:

Proposition 1.7.3. *Let X be a compact metric space and $f\colon (X, d) \to (Y, d')$ be a continuous map. Then f is uniformly continuous, i.e., a continuous map from a compact metric space to another metric space is uniformly continuous.*

Proof. Consider the balls $\{B(y, \epsilon/2) \mid y \in Y\}$ and let $\delta > 0$ be a Lebesgue number for the open covering $\{f^{-1}(B(y, \epsilon/2)) \mid y \in Y\}$ of the compact space X. Since each $B(x, \delta)$ lies in one of these sets, $f(B(x, \delta)) \subseteq B(y, \epsilon/2)$ for some y, and since $f(x) \in B(y, \epsilon/2)$, we find that for any $z \in B(x, \delta)$,

$$d(f(z), f(x)) \leq d(f(z), y) + d(y, f(x)) < \epsilon/2 + \epsilon/2 = \epsilon,$$

which says that $f(B(x, \delta)) \subseteq B(f(x), \epsilon)$. ■

A generalization of the concept of compactness is local compactness. A Hausdorff space X is said to be **locally compact** if every point $x \in X$ has a neighbourhood $N(x)$ in X which is compact. It can be easily proved that this, in fact, implies that any neighbourhood N of x will contain a compact neighbourhood of x. The real line $\mathbb{R}$ is obviously locally compact whereas the space $\mathbb{Q}$ of rational numbers is not locally compact. To see the latter, suppose $N(q)$ is a compact neighbourhood of $q \in \mathbb{Q}$. Then $N(q)$ must contain a closed ϵ-neighbourhood, say, $[q - \epsilon, q + \epsilon]$ of q in $\mathbb{Q}$ whose end points are irrational numbers and which contains all rationals lying between $q - \epsilon$ and $q + \epsilon$. Being the closed subset of a compact space $N(q)$, the rational interval $[q - \epsilon, q + \epsilon] \cap \mathbb{Q}$ must be compact. But this contradicts the Heine-Borel Theorem because it is not closed in $\mathbb{R}$. This example also shows that a subspace of a locally compact space need not be locally compact. However, the following is an interesting result.

Proposition 1.7.4. *A subset Y of a locally compact space X, which is the intersection $Y = F \cap U$ of a closed subset F and an open subset U of X, is locally compact. Conversely, if Y is locally compact, then there exists a closed set F of X and an open set U of X such that $Y = F \cap U$.*

Discrete spaces are evidently locally compact. Since every topological space X is a continuous image of a discrete space, it follows that a continuous image of a locally compact space need not be locally compact. However, the image under a continuous open map of a locally compact space is locally compact. The infinite product $\mathbb{R}^\omega$ of real lines is not locally compact - any neighbourhood of origin must contain a product neighbourhood $\prod_1^\infty U_i$ where all, except finitely many U_i, must be real lines but that can never be compact. In fact, an infinite product $\prod_{i \in I} X_i$ of locally compact spaces is locally compact if and only if all except finitely many are compact.

An important feature of a locally compact Hausdorff space X, which is not already compact, is that it can always be compactified by just adding one more point in the space X and by enlarging the topology of X. This idea is due to P.S. Alexandroff and is known as **one-point compactification of** X. To see this, let ∞ be a point which is not in the given locally compact space X. Consider the set $\hat{X} = X \cup \{\infty\}$. We define a topology on X by defining the neighbourhoods of every point. Declare neighbourhoods of all points of X to be the same which were in the space X, but define neighbourhoods of ∞ to be those subsets U of $\hat{X}$ which contain the point ∞ and the complement of some nonempty compact subset C of the space X, i.e., $U = \{\infty\} \cup (X - C)$, where C is a nonempty compact subset of X. Then one can easily verify that these neighbourhoods define a topology on $\hat{X}$. It can also be proved that the resulting topology on $\hat{X}$ is compact, Hausdorff and the original space X is an open dense subset of $\hat{X}$. The new space $\hat{X}$ is called the one-point compactification of X. Here, it is necessary to point out that the topology on $\hat{X}$ defined above is, in fact, unique, i.e., if $\hat{X}$ is a compact, Hausdorff space such that X is an open dense subset of $\hat{X}$, then the topology of $\hat{X}$ must be exactly the way it has been defined above by Alexndroff. Using this uniqueness property, the following two examples are obvious:

Example 1.7.5. The one-point compactification of the discrete space of natural numbers $\mathbb{N}$ is homeomorphic to the subspace $\{0\} \cup \{\frac{1}{n} \mid n \in \mathbb{N}\}$ of $\mathbb{R}$.

Example 1.7.6. For each $n \geq 1$, the one point compactification of $\mathbb{R}^n$ is homeomorphic to the n-sphere $\mathbb{S}^n$ (see Example 1.4.5).

A locally compact space X is also completely regular. For, just take the one-point compactification $\hat{X}$ of X. Then, being compact Hausdorff, $\hat{X}$ is normal, which means, $\hat{X}$ is completely regular and so, X being a subspace of $\hat{X}$, must be completely regular. Indeed, given a compact set K and a closed subset F of a locally compact space (F is disjoint from K), there exists a continuous map $f\colon X \to \mathbb{R}$ such that $f(F) = 0$ and $f(K) = 1$. Again, this can be proved by going to the one-point compactification of X.

Example 1.7.7 (Long Line)**.** A locally compact space X need not be a metric space. An interesting example of such a space is the long line L. The space L is defined by considering the set $\Omega = \{0, 1, 2, \ldots, \omega, \omega + 1, \ldots\}$ of ordinals which are less than the first uncountable ordinal ω_1. Between any two consecutive ordinals α and $\alpha + 1$, we insert a copy of the open interval $(0, 1)$ so that α

and $\alpha + 1$ become the end points. The resulting set L_+ is linearly ordered and, therefore, we can give it the order topology. Take another copy L_- of L_+ by giving it the opposite linear order. Then their union $L = L_+ \cup L_-$, where the ordinal 0 of the two spaces L_+ and L_- are identical, i.e., $L_+ \cap L_- = \{0\}$, equipped with order topology, is called the **long line**. It is nonmetrizable and locally compact. It is also countably compact. In fact, L is a nonmetrizable one-dimensional manifold.

1.8 Compact Surfaces

The following concept is quite useful, especially in the study of surfaces.

Definition 1.8.1 (connected sum). *Let us take two tori T_1, T_2 (see Fig.1.15) and choose small closed disks D_1, D_2 in T_1 and T_2, respectively. Note that the boundary of each of these disks is homeomorphic to the circle $\mathbb{S}^1$. Let us remove the interiors of the two disks from T_1 and T_2 and choose a homeomorphism h from the boundary of the first disk to the boundary of the second. Let T_1' and T_2' be the remaining portions of the two tori when the interiors of the disks have been removed. Now, consider the disjoint union $X = T_1' \cup T_2'$ and identify the boundary point $x \in T_1'$ with the boundary point $h(x) \in T_2'$ for all x in the boundary of D_1. The identification space or quotient space X/R so obtained, denoted by $T_1 \# T_2$, is called the* **connected sum** *of T_1 and T_2.*

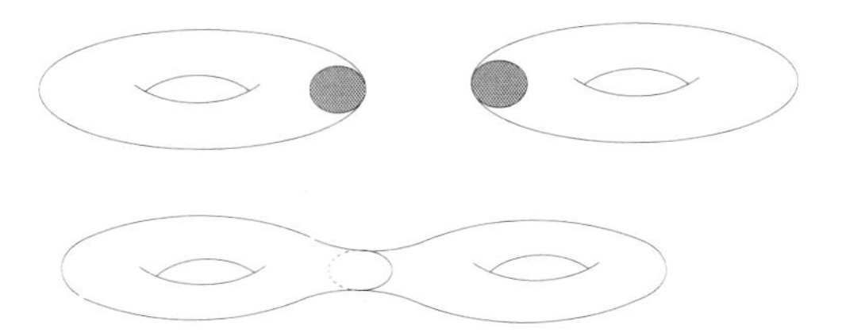

Fig. 1.15: The connected sum of two tori

It can be shown that the connected sum $T_1 \# T_2$ is independent (up to homeomorphism) of the choice of the disks D_1, D_2 and the homeomorphism h. Note that the connected sum of two tori is a new surface. We can similarly define the connected sum of two projective planes and, in fact, we can define the connected sum of any two spaces X and Y each of which has two disks of the same dimension embedded in it, e.g., we can form the connected sum of any two n-dimensional manifolds, $n \geq 1$.

Classification of Compact Surfaces

We have already mentioned a few compact surfaces like a 2-sphere, a torus, a projective plane and a Klein bottle (we are leaving out Möbius band and cylinder, etc., because they are surfaces with boundary which we are not considering at the moment). A Hausdorff topological space X, which has the property that each point of X has an open neighbourhood homeomorphic to the Euclidean space $\mathbb{R}^n$ or to an open ball $\mathbb{B}^n$, is called a **topological n-manifold**. The 2-dimensional manifolds are called **surfaces**. To avoid pathological situations, we consider only those n-manifolds which are embedded in some Euclidean space $\mathbb{R}^n$. This is equivalent to saying that they have a countable basis. It is a deep result of low-dimensional topology that every 2-manifold is triangulable. Here, our objective is to know all possible compact surfaces. The ones like a 2-sphere, etc., mentioned earlier, are all compact surfaces. The question is that besides these, what else do we have? The answer is known and it is very interesting: They are either a connected sum of a finite number of tori (T_i) or they are the connected sum of a finite number of projective planes (P_i)! This result, called the **Classification Theorem for compact surfaces**, is one of the most exciting theorems of the whole of topology. It says that any compact surface is homeomorphic to a 2-sphere or to a connected sum $T_1\#T_2\cdots\#T_n$ or to a connected sum $P_1\#P_2\cdots\#P_n$ for some $n \geq 1$. The theorem also asserts that each of the above surfaces is distinct, i.e., no two of them are homeomorphic to each other. The other fascinating result regarding these compact surfaces is that each one of these surfaces can be obtained as a quotient of an ordinary polygon. To be specific, let us recall how a torus was obtained from a polygon having four sides by identifying the opposite sides of the square or how the Klein bottle was obtained from the same 4-gon by identifying sides somewhat differently. In the general case, we start with a n-gon and then identify various edges of this polygon, without disturbing the interior of the polygon, in a suitable manner so as to obtain any given compact surface (the detailed statements as well as the proofs of these results with enough geometric insight can be found in W.Massey [13]).

Paracompact Spaces

This concept is due to J. Dieudonné. Recall that a family $\mathcal{A} = \{A_\alpha : \alpha \in I\}$ of subsets of a space X is said to be **locally finite** if each point $x \in X$ has a neighbourhood which meets only finitely many members of $\mathcal{A}$. A topological space X is said to be paracompact if every open cover of X has a locally finite open refinement covering the space X. Every compact space is trivially paracompact but a locally compact space need not be paracompact, e.g., the long line mentioned earlier is locally compact, but not paracompact. It can be proved that a paracomact, Hausdorff space is normal and so a paracompact Hausdorff space satisfies all the separation axioms. It is easy to see that a closed (not arbitrary) subset of a paracompact space is paracompact, and (this one is not easy!) a closed continuous image of a paracompact space is paracompact. One of the most important properties of a paracompact spaces is the following: *a paracompact space X has the partition of unity subordinated to every open covering of X*. What it means is that given any open covering $\mathcal{U}$ of X, there exists a locally finite refinement $\mathcal{V}$ of $\mathcal{U}$ and a family of continuous functions $f_V: X \to \mathbb{R}, V \in \mathcal{V}$ such that the closure of the set $\{x \in X : f_V(x) \neq 0\} \subset V$ for every $V \in \mathcal{V}$ and $\Sigma_{V \in \mathcal{V}} f_V = 1$.

Another deep result about paracompact spaces is the following theorem of A.H.Stone: *Every metric space is paracompact* (see Dugundji [5] for all details). The class of paracompact spaces has a very special status in algebraic topology simply because *every CW-complex is paracompact* which implies, in particular, that every simplicial complex (finite ones are already compact) is paracompact. In fact, most of the important cohomology theories, e.g., the Čech cohomology, singular cohomology, de Rham cohomology, etc., coincide on a paracompact manifold; the general n-manifolds acquire interesting properties only when they are paracompact, not otherwise. We have the following important result in this direction:

Theorem 1.8.2. *Let X be an n-manifold. Then the following are equivalent:* (1) *X is paracompact* (2) *X is second countable* (3) *X is Lindeloff* (4) *X is metrizable.*

Exercises

1. Let X be a Hausdorff space. Show that any two disjoint compact subsets of X can be separated by two disjoint open sets of X.

2. Let us consider the following subspace X of the Euclidean plane: X consists of all closed line segments joining the origin and the points $(1, 1/n), n \in \mathbb{N}$ and the line segment $\{(x, 0) : 0 \leq x \leq 1\}$. Prove that the weak topology on X defined by these line segments is strictly finer than the topology of X.

3. Let $p: X \times Y \to X$ be the projection on the first factor and Y be a compact space. Show that p is a closed map, i.e., the projection parallel to a compact factor is a closed map.

4. Let A be any subset of X and Y be a compact space. Prove that any neighbourhood U of the subset $A \times Y$ contains a tubular neighbourhood, i.e., there exists a neighbourhood V of A in Y such that $V \times Y \subset U$.

5. Show that an arbitrary product of path-connected spaces is again path-connected.

6. **(Hawaiian earring)** For each $n \geq 1$, let C_n be the circle in plane of radius $1/n$ with center $(1/n, 0)$. Let $X = \bigcup_{n \geq 1} C_n$ and consider the following two topologies on X: one, the subspace topology from $\mathbb{R}^2$, and the other, the weak topology defined by the family of circles C_n. Show that the two topologies on X are distinct. (The space X with subspace topology is called the Hawaiian earring.)

7. Consider the ascending union $\mathbb{R}^\infty = \cup_{n\geq 1}\mathbb{R}^n$. This set has two topologies on it: one, the weak topology defined by the spaces $\mathbb{R}^n, n \geq 1$, and the other is the subspace topology coming from $\mathbb{R}^\omega = \prod_1^\infty \mathbb{R}$. Prove that the weak topology is strictly stronger than the subspace topology.

8. Consider the ascending union $\mathbb{S}^\infty = \bigcup_{n\geq 1} \mathbb{S}^n$ of n-spheres and give it the weak topology defined by the family of Euclidean spheres. Show that $\mathbb{S}^\infty$ is not locally compact.

9. Suppose X is a locally compact space and Y is a locally compact subspace of X. Show that there exists a closed subset F and an open subset U of X such that $Y = F \cap U$.

10. A subset Y of a space X is said to be **locally closed** if for each $y \in Y$ there exists a neighbourhood $N(y)$ of y in X such that $Y \cap N(y)$ is closed in $N(y)$. Prove that the following conditions are equivalent: (i) Y is locally closed (ii) Y is open in $\bar{Y}$, the closure of Y. (iii) Y is the intersection of an open and a closed subset of X.

11. Suppose Y is a quotient space of X and Z is a locally compact space. Show that $Y \times Z$ is a quotient space of $X \times Z$. (This exercise indicates that the product of two quotient spaces need not be a quotient space).

12. Prove that the connected sum $\mathbb{R}\#\mathbb{R}$ is homeomorphic to a hyperbola, and $S^1\#S^1$ is homeomorphic to S^1.

13. Let X, Y, Z be arbitrary surfaces (2-manifolds). Then prove that there are the following homeomorphisms: (i) $X\#Y \cong Y\#X$. (ii) $X\#\mathbb{S}^2 \cong X$. (iii) $(X\#Y)\#Z \cong X\#(Y\#Z)$.

1.9 What is Algebraic Topology?

There are many questions in point-set topology which are quite easy to ask. But it is surprising that answering them is a lot difficult. To take a simple example, let us ask whether or not the two Euclidean spaces $\mathbb{R}^3$ and $\mathbb{R}^4$ are homeomorphic. On intuitive basis, one immediately tends to assert that these are not homeomorphic. The focal point of the problem is: what is a proof of this assertion? Let us observe that the same question for the Euclidean spaces $\mathbb{R}$ and $\mathbb{R}^4$ is answered very easily. These are not homeomorphic because if they were homeomorphic, then removing one point from both of these spaces must leave homeomorphic spaces. However, removing one point from $\mathbb{R}$ leaves the remainder space as disconnected whereas removing one point from $\mathbb{R}^4$ leaves the remainder as a connected space and so they cannot be homeomorphic. This is a proof of our assertion using a method of point-set topology. As another example, it appears on the same intuitive grounds that the two Euclidean spheres $\mathbb{S}^3$ and $\mathbb{S}^4$ are not homeomorphic; likewise, the 2-sphere $\mathbb{S}^2$ and the torus $\mathbb{S}^1 \times \mathbb{S}^1$ are apparently not homeomorphic; the Klein bottle and the torus also appear to be non-homeomorphic. However, using only the methods of point-set topology, it is not easy to give a proof of these facts except in the case of a 2-sphere and torus. In this case, one can argue that there is a nontrivial circle on torus, which when removed, leaves the torus connected, whereas removing any circle from $\mathbb{S}^2$ leaves it disconnected. The general question whether or not two given spaces X and Y are homeomorphic is known as the "Homeomorphism Problem". Methods of algebraic topology using the concepts of **homology groups** and **homotopy groups** of the spaces X and Y, as we will see later, are frequently very powerful for resolving the homeomorphism problems stated above.

The French mathematician C. Jordan (1858-1922) was the first to ask for a proof of the following somewhat obvious statement:

Jordan Curve Theorem: A simple closed curve C in the Euclidean plane $\mathbb{R}^2$ separates the plane into two open connected sets with C as their common boundary. Exactly one of these two open sets is unbounded.

In the first instance, the above statement seems obvious on intuitive grounds, but Jordan asserted that intuition is not a proof. We just cannot accept this as a theorem unless it has a rigorous mathematical proof. It was Oswald Veblen (1880-1960) who used the method of algebraic topology to give a correct mathematical proof of the above statement in 1905. This result is now well-known as the Jordan Curve Theorem.

Another type of question which we come across in point-set topology is the following **Extension Problem**. Let $A \subset X$ be a subspace and $f\colon A \to Y$ be a continuous map. Does there exist a continuous map $F\colon X \to Y$ such that $F|_A = f$? In other words, can we extend $f\colon A \to Y$ to whole of X continuously?

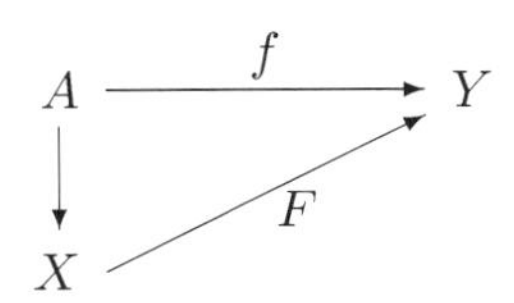

In the special case, when X is a normal space, $A \subset X$ is a closed subspace of X and $Y = \mathbb{R}$ is the real line, the Tietze's Extension Theorem says that any continuous map $f\colon A \to \mathbb{R}$ can be extended continuously to whole of X. However, we are interested in the general Extension Problem for arbitrary spaces X, Y, A and an arbitrary continuous map $f\colon A \to Y$. The methods of general topology may sometimes be successful to answer an extension problem, but only rarely. For instance, can the sine curve function $f\colon (0, \infty) \to [-1, 1]$ defined by $f(x) = \sin 1/x$ be extended to $(-\infty, \infty)$? The answer is "No" because it is easy to see by continuity considerations that setting $f(0)$ to be any point in $[-1, 1]$ can never make the extended function continuous. Now, let us consider an interesting question of the same kind. The n-sphere $\mathbb{S}^n$ is the boundary of the $(n+1)$ disk $\mathbb{D}^{n+1} \subset \mathbb{R}^{n+1}$. Let $f\colon \mathbb{S}^n \to \mathbb{S}^n$ be the identity map. Can we extend this map continuously to $\mathbb{D}^{n+1}$? Here, the answer could be guessed on geometric grounds to be in negative but the proof is difficult. In fact, once again, as we will see later, algebraic topology provides a quick proof of this result for the case $n > 0$. The case $n = 0$ is clear from only continuity considerations.

Yet another kind of question which we confront in point-set topology is the important **Lifting Problem**. Suppose $p\colon \tilde{X} \to X$ is an onto continuous map. Let $f\colon Y \to X$ be a given continuous map. Does there exist a continuous map $g\colon Y \to \tilde{X}$ such that $p \circ g = f$?

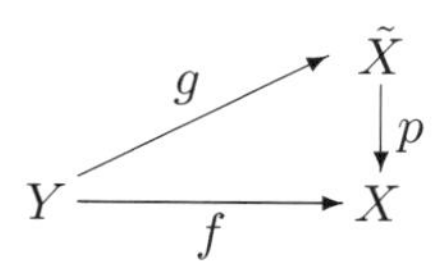

Unlike the Tietze's Extension Theorem, we have no result in point-set topology which answers this Lifting Problem even in a significant special case. However, as we will see later in this book, the methods of algebraic topology provide a complete solution to this Lifting Problem in terms of a concept called **fundamental group** of a topological space.

The concepts of fundamental group and homology group mentioned above were the first basic notions of algebraic topology which were introduced by the famous French mathematician H. Poincaré during 1895-1901. It must be mentioned here that the subject of point-set topology did not even exist when the concepts of algebraic topology mentioned above were created and used in solving problems of other branches of mathematics. The point-set topology grew out of the works done by M. Fréchet in 1906 on general metric spaces and developed further in 1914 by the classic work of F. Hausdorff.

Given a topological space X, the basic aim of algebraic topology is to associate some algebraic object, say, a group $H(X)$ to the space X (it could be an abelian group or, more generally, a module or even a sequence of modules, etc.). The algebraic object so associated depends on the topology of X in such a manner that if $f\colon X \to Y$ is a continuous map between X and Y, then it naturally induces a homomorphism $H(f)$ between the groups $H(X)$ and $H(Y)$. We emphasise that the ingenuity of creating the concept lies in capturing the topology of the space X in defining the associated algebraic object. Furthermore, the induced homomorphisms have the following two properties:

(i) If $f\colon X \to Y$ and $g\colon Y \to Z$ are continuous maps, then the induced homomorphisms $H(f), H(g)$ and $H(g \circ f)$ are related as follows:

$$H(g \circ f) = H(g) \circ H(f).$$

(ii) If $I_X : X \to X$ is the identity map of the space X, then $H(I_X) = I_{H(X)}$ is the identity map of the group $H(X)$.

We note that H is a covariant functor from the category of topological spaces and continuous maps to the category of groups and homomorphisms. These two properties of the associated groups and induced homomorphisms immediately yield the important fact that the group $H(X)$ is a **topological invariant** of the space X, i.e., if X and Y are homeomorphic spaces, then the associated groups $H(X)$ and $H(Y)$ must be isomorphic (prove it as an easy exercise). Thus, suppose we have two spaces X and Y for which the homeomorphism problem is to be solved. If we find that the associated groups $H(X)$ and $H(Y)$ are not isomorphic, then as a consequence of the above topological invariance of associated groups, the spaces X and Y cannot be homeomorphic. This is a general method of algebraic topology which solves the Homeomorphism Problem to a large extent. We must point out that this solution of the Homeomorphism Problem is only one way, i.e., it is successful only when X and Y are not homeomorphic. The method does not assert that X and Y are homeomorphic if the associated groups are isomorphic, except in a few rare cases.

The strategy for solving the Extension Problem is now quite straightforward. Let $f\colon A \to Y$ be a continuous map and $A \subset X$ be a subspace. If this f can be extended to a continuous map $F\colon X \to Y$, then the triangle

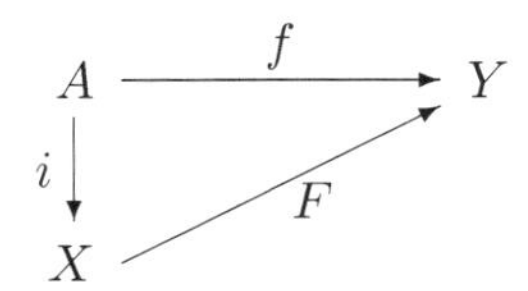

must be commutative, where $i\colon A \to X$ is the inclusion map. Let us consider the following diagram of associated groups and induced homomorphisms.

$$\begin{array}{ccc} H(A) & \xrightarrow{H(f)} & H(Y) \\ {\scriptstyle H(i)}\downarrow & \nearrow{\scriptstyle H(F)} & \\ H(X) & & \end{array}$$

By the properties of associated groups and induced homomorphisms stated earlier, this new diagram of groups and homomorphisms must also be commutative, i.e., we must have

$$H(f) = H(F) \circ H(i).$$

However, if for some reason we find that this last equation is false, then we conclude that the map F cannot exist and so the map f cannot be extended. This solves the Extension Problem. We remark that in case the above equation is valid, we cannot assert that the map f can be extended. Thus, the solution of the Extension Problem using the above method is also one way. A similar strategy is also used to solve the Lifting Problem. This will be discussed in detail in the chapter on covering spaces because, for Lifting Problem, the technique of algebraic topology is surprisingly successful both ways rather than only one way as was the case with the Homeomorphism Problem or the Extension Problem. In fact, we obtain a necessary and sufficient condition in terms of the associated fundamental groups and induced homomorphism for the existence of a map $g\colon Y \to \tilde{X}$ lifting the given map $f\colon Y \to X$.

The subject, called **Algebraic Topology**, basically aims at creating and defining various kinds of algebraic objects associated to a topological space and homomorphisms induced by a continuous map. It is true that a particular kind of algebraic object may be better suited for a particular problem in topology. More often than not, computing these associated algebraic objects for various spaces and identifying the induced homomorphisms between the associated algebraic objects is an interesting question in its own right. The subject has grown into a vast field as a result of developing new tools, proving theorems for computational problems and applying them fruitfully elsewhere in various branches of mathematics.

Chapter 2

The Fundamental Group

2.1 Introduction

Let X be a topological space. Often we associate with X an object that depends on X as well as on a point x of X. The point x is called a **base point** and the pair (X, x) is called a **pointed space**. If (X, x) and (Y, y) are two pointed spaces, then a continuous map $f\colon X \to Y$ such that $f(x) = y$ is called a map between pointed spaces. Let $f\colon X \to Y$ be a homeomorphism and x be a point of X. Then f is a homeomorphism between pointed spaces (X, x) and $(Y, f(x))$. The composite of two maps between pointed spaces is again a map between pointed spaces and the identity map $I_{(X,x)}\colon (X, x) \to (X, x)$ is always a homeomorphism of pointed spaces for each $x \in X$.

In this chapter, we show how to each pointed space (X, x), we can associate a group $\pi_1(X, x)$, called **the fundamental group** of the space X at x. Each map $f\colon (X, x) \to (Y, y)$ between pointed spaces (X, x) and (Y, y) then induces a homomorphism (denoted by $f_{\#}$ also)

$$f_*\colon \pi_1(X, x) \to \pi_1(Y, y)$$

between groups $\pi_1(X, x)$ and $\pi_1(Y, y)$ such that the following two conditions are satisfied:

(i) If $f\colon (X, x) \to (Y, y)$ and $g\colon (Y, y) \to (Z, z)$ are two maps of pointed spaces, then
$$(g \circ f)_* = g_* \circ f_*\colon \pi_1(X, x) \to \pi_1(Z, z).$$

(ii) If $I_{(X,x)}\colon (X, x) \to (X, x)$ is the identity map, then the induced group homomorphism
$$I_{(X,x)_*}\colon \pi_1(X, x) \to \pi_1(X, x)$$
is also the identity map.

We will also compute fundamental group of several interesting spaces and exhibit some of their uses. The two properties of the induced homomorphism stated above are known as *functorial properties*, which at once yield the following important consequence : **the fundamental group** $\pi_1(X, x)$ **is a topological invariant of the pointed space** (X, x). The detailed meaning of this statement is given below:

Proposition 2.1.1. *If* (X, x) *and* (Y, y) *are two pointed spaces which are homeomorphic, then their fundamental groups* $\pi_1(X, x)$ *and* $\pi_1(Y, y)$ *are isomorphic.*

Proof. Suppose $f\colon (X, x) \to (Y, y)$ is a homeomorphism of pointed spaces. Then the inverse map $f^{-1}\colon (Y, y) \to (X, x)$ is evidently a map of pointed spaces and has the property that

$$f^{-1} \circ f = I_{(X,x)}, \; f \circ f^{-1} = I_{(Y,y)}.$$

Let us consider the induced homomorphisms

$$f_*\colon \pi_1(X, x) \to \pi_1(Y, y), \; f_*^{-1}\colon \pi_1(Y, y) \to \pi_1(X, x)$$

in the fundamental groups. By functorial property (i), we find that

$$(f^{-1} \circ f)_* = f_*^{-1} \circ f_*,$$

and, by (ii), we see that $I_{(X,x)_*}$ is the identity map on $\pi_1(X, x)$. Thus, we conclude that $f_*^{-1} \circ f_*$ is the identity map on $\pi_1(X, x)$. Similarly, we can see that $f_* \circ f_*^{-1}$ is the identity map on $\pi_1(Y, y)$. Therefore, f_* and f_*^{-1} are inverses of each other, i.e., $f_*\colon \pi_1(X, x) \to \pi_1(Y, y)$ is an isomorphism. ■

The fact that the fundamental group $\pi_1(X, x)$ of a pointed space (X, x) is a topological invariant is an interesting and a very useful result. It says that *if* X *and* Y *are two spaces such that for some* $x_0 \in X$, $\pi_1(X, x_0)$ *is not isomorphic to any of* $\pi_1(Y, y)$, $y \in Y$, *then the spaces* X *and* Y *cannot be homeomorphic.* For, suppose X and Y are homeomorphic and let $f\colon (X, x) \to (Y, y)$ be a homeomorphism. Then $f\colon (X, x_0) \to (Y, f(x_0)$ is a homeomorphism of pointed spaces and so, by the above proposition, the induced group homomorphism $f_*\colon \pi_1(X, x_0) \to \pi_1(Y, f(x_0))$ must be an isomorphism, i.e., $\pi_1(X, x_0)$ is isomorphic to $\pi_1(Y, f(x_0))$, a contradiction.

One of the most important problems in topology, known as the *classification problem* in a given class of topological spaces, is to decide whether or not two given spaces of that class are homeomorphic. To prove that two spaces X and Y are indeed homeomorphic, the problem is really to find out some specific homeomorphism from X to Y, and invariably, the only method to do this is the knowledge of point-set topology. However, to prove that X and Y are *not* homeomorphic, one looks for some topological invariant possessed by one space and not by the other. For example, when we have to show that $\mathbb{R}^1$ is not homeomorphic to $\mathbb{R}^2$, we argue as follows : if we remove one point from both, then the remaining spaces, first being disconnected and the second being connected, are not homeomorphic and so $\mathbb{R}^1$ cannot be homeomorphic to $\mathbb{R}^2$. Similarly, the circle $\mathbb{S}^1$ cannot be homeomorphic to the figure of eight (two circles touching at a point) because if we remove the point of contact from the figure of eight, then the remaining space is disconnected whereas if we remove any point from $\mathbb{S}^1$, the remaining space remains connected. To prove that a closed interval $[0, 1]$ is not homeomorphic to an open interval $(0, 1)$, we say that one is compact whereas the other is not compact and so they cannot be homeomorphic. These methods are known as the methods of point set topology. Now, let us ask whether or not the 2-sphere $\mathbb{S}^2$ is homeomorphic to the 2-torus $\mathbb{T} = \mathbb{S}^1 \times \mathbb{S}^1$ (Fig. 2.1). This can also be resolved using point-set topology as follows : Take a circle C in $\mathbb{T}$ as shown in Fig. 2.1. If we remove C from $\mathbb{T}$, the remaining space is clearly connected. However, if we remove any circle from $\mathbb{S}^2$, the remaining space is evidently disconnected. This says that $\mathbb{T}$ and $\mathbb{S}^2$ can not be homeomorphic. Now, let us ask whether or not the 3-sphere $\mathbb{S}^3$ is homeomorphic to the 3-torus $\mathbb{T} = \mathbb{S}^1 \times \mathbb{S}^1 \times \mathbb{S}^1$.

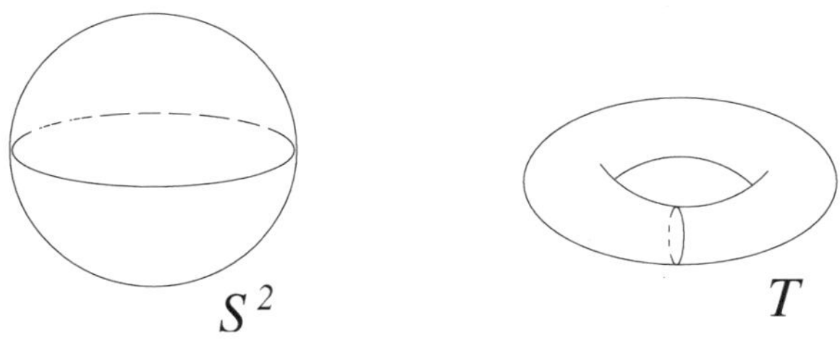

Fig. 2.1: A 2-sphere and a 2-torus

The reader is invited to discover some method of point-set topology to show that they are not homeomorphic (they are really not homeomorphic!) and see for himself that this can be extremely difficult. In such a case, however, the methods of algebraic topology sometimes work very well. We will, later on, prove that if X is a path-connected space then the fundamental group $\pi_1(X, x)$ is independent of the base point (up to isomorphism), and we denote it simply by $\pi_1(X)$. This fact, combined with the previous proposition, will say that if X and Y are two path-connected spaces such that $\pi_1(X)$ is not isomorphic to $\pi_1(Y)$ then X and Y cannot be homeomorphic. By popular belief based on experience, it is normally much easier to decide that two groups are not isomorphic than to decide that two given spaces are not homeomorphic. Now, granting that $\pi_1(\mathbb{S}^3) = 0$ and $\pi_1(\mathbb{S}^1 \times \mathbb{S}^1 \times \mathbb{S}^1) \cong \mathbb{Z} \times \mathbb{Z} \times \mathbb{Z}$ (we shall prove these facts later), we immediately conclude that $\mathbb{S}^3$ cannot be homeomorphic to the torus $\mathbb{S}^1 \times \mathbb{S}^1 \times \mathbb{S}^1$ as their fundamental groups are evidently not isomorphic. This is just one of the several methods of algebraic topology in proving that two spaces are not homeomorphic. The crucial point is the result that the fundamental group $\pi_1(X)$ is a topological invariant of pathr-connected spaces. Several objects such as homology groups, Euler characteristics, etc., are other important invariants of topological spaces. We shall come to them later.

We have only indicated that fundamental group $\pi_1(X, x)$ of a pointed space (X, x) is a topological invariant. This is a good result, but by no means the best result. More general and interesting results, including the best ones about fundamental groups, will be studied only after we have defined them. The definition requires the important concept of *homotopy* and the generalization of results will require the concept of *homotopy type* in the class of topological spaces. We take up these notions in the next section.

2.2 Homotopy

We now proceed to define fundamental group $\pi_1(X, x)$ of a given pointed space (X, x). It will take some time to do so, but that is true of most of the topological invariants in algebraic topology. The fundamental group is, of course, the first such invariant we are going to deal with. There are several important concepts which will be introduced on our way to the definition of $\pi_1(X, x)$. The first one is the concept of **homotopy**. We have

Definition 2.2.1. *Let X, Y be two spaces and $f, g\colon X \to Y$ be two continuous maps. We say that f is* **homotopic** *to g (and denote it by writing $f \cong g$) if there exists a continuous map $F\colon X \times I \to Y$ such that $F(x, 0) = f(x)$ and $F(x, 1) = g(x)$ for all $x \in X$. The map F is called a* **homotopy** *from f to g.*

We know that for each $t \in I$, the map $i_t\colon X \to X \times I$ defined by $i_t(x) = (x, t)$ is an embedding. So, $f_t = F \circ i_t\colon X \to Y$ is a family of continuous maps from X to Y, where t runs over the interval I. By the definition of homotopy F, f_t is the map f for $t = 0$ and for $t = 1$ it is the map g. Thus, a homotopy F is simply a family of continuous maps from X to Y which starts from f, changes continuously with respect to t and terminates into the map g. In other words, f gets continuously transformed by means of the homotopy F and finally changes or deforms itself into the map g. See Fig. 2.2.

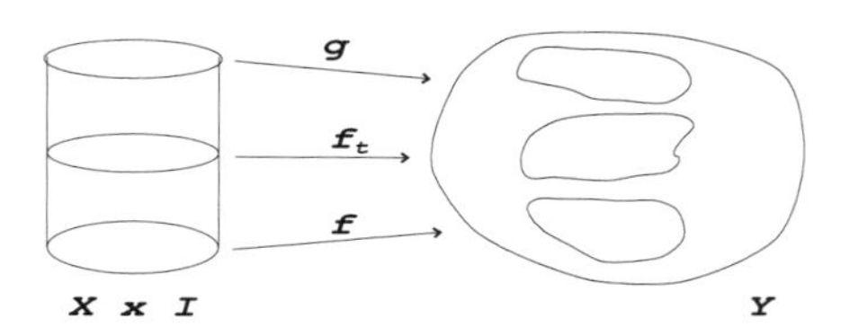

Fig. 2.2: F is a Homotopy from $f = F \circ i_0$ to $g = F \circ i_1$

Example 2.2.2. Let $X = \mathbb{R}^n = Y$ be the Euclidean spaces and let $f, g\colon X \to Y$ be defined by $f(x) = x$ and $g(x) = 0$, $x \in X$. Define the map $H\colon \mathbb{R}^n \times I \to \mathbb{R}^n$ by

$$H(x, t) = (1 - t)x.$$

Then, clearly, H is continuous and for all $x \in X$, $H(x,0) = x = f(x)$, $H(x,1) = g(x)$. Thus, H is a homotopy from f to g and so f is homotopic to g.

Note that if we define $F\colon \mathbb{R}^n \times I \to \mathbb{R}^n$ by $F(x,t) = (1-t^2)x$, then F is also a homotopy from f to g. In other words, there can be several ways of deforming a map f into a given map g.

Example 2.2.3. This is a generalization of Example 2.2.2 above. Let X be any topological space and Y be a convex subset of $\mathbb{R}^n$, i.e., Y has the property that whenever $y_1, y_2 \in Y$, the line segment joining y_1 to y_2 is completely contained in Y. Let $f, g\colon X \to Y$ be any two continuous maps. Then f is homotopic to g. To see this, let us define the map $H\colon X \times I \to Y$ by

$$H(x,t) = tg(x) + (1-t)f(x).$$

Then we see at once that H is well-defined, continuous, it starts with f and terminates into g. A homotopy of this kind is called a **straight-line homotopy**.

Example 2.2.4. Let $\mathbb{S}^1 = \{z \in C : |z| = 1\}$ be the unit circle. We know that we can also write $\mathbb{S}^1 = \{e^{i\theta} : 0 \le \theta \le 2\pi\}$. Define two maps $f, g\colon \mathbb{S}^1 \to \mathbb{S}^1$ by $f(z) = z$ and $g(z) = -z$, $z \in \mathbb{S}^1$. Then f is homotopic to g and the map $F\colon \mathbb{S}^1 \times I \to \mathbb{S}^1$ defined by

$$F(e^{i\theta}, t) = e^{i(\theta + t\pi)}$$

is a homotopy from f to g. Note that F is continuous because it is the composition of maps

$$\begin{gathered} \mathbb{S}^1 \times I \to \mathbb{S}^1 \times \mathbb{S}^1 \to \mathbb{S}^1 \\ (e^{i\theta}, t) \to (e^{i\theta}, e^{it\pi}) \to e^{i(\theta+t\pi)}, \end{gathered}$$

where the second map is multiplication of complex numbers. *Note that in this example, the family of maps $\{f_t\colon \mathbb{S}^1 \to \mathbb{S}^1\}$ is just the family of rotations by the angle $t\pi, 0 \le t \le 1$.*

The next result implies that the set of all maps from a space X to a space Y can be decomposed into disjoint equivalence classes.

Theorem 2.2.5. *Let X, Y be fixed topological spaces and $C(X,Y)$ denote the set of all continuous maps from X to Y. Then the relation of "being homotopic to" is an equivalence relation in the set $C(X,Y)$.*

Proof. Note that each continuous map $f\colon X \to Y$ is homotopic to itself because $H\colon X \times I \to Y$ defined by $H(x,t) = f(x)$ is a homotopy from f to itself. Next, suppose $H\colon f \cong g$. Then the map $H'\colon X \times I \to Y$ defined by

$$H'(x,t) = H(x,(1-t))$$

is a homotopy from g to f; to see this, note that $H'(x,0) = H(x,1) = g(x)$ and $H'(x,1) = H(x,0) = f(x)$ for all $x \in X$. Moreover, H' is continuous because H' is simply the composite of continuous maps

$$X \times I \to X \times I \to Y,$$

where the first map is the map $(x,t) \mapsto (x, 1-t)$ and the second is H. The first map itself is continuous because its composite with the two projection maps, viz., $(x,t) \to x$ and $(x,t) \to (1-t)$ is continuous. Thus, the relation is symmetric. Finally, suppose $H_1 \colon f \cong g$ and $H_2 \colon g \cong h$. Define a map $H \colon X \times I \to Y$ by

$$H(x,t) = \begin{cases} H_1(x, 2t), & 0 \le t \le 1/2, \\ H_2(x, 2t-1), & 1/2 \le t \le 1. \end{cases}$$

Then H is continuous by the continuity lemma and $H(x,0) = f$ and $H(x,1) = h$. Hence $H \colon f \cong h$, proving the relation to be transitive. ■

The relation of homotopy in the set $C(X,Y)$ of all continuous maps, therefore, decomposes this set into mutually disjoint equivalent classes. The equivalence classes are called the **homotopy classes** of maps from X to Y and the set of all **homotopy classes** is denoted by $[X,Y]$. We will need the following result very often.

Theorem 2.2.6. *Let $f_1, g_1 \colon X \to Y$ be homotopic and $f_2, g_2 \colon Y \to Z$ be also homotopic. Then the composite maps $f_2 \circ f_1$, $g_2 \circ g_1 \colon X \to Z$ are homotopic too, i.e., composites of homotopic maps are homotopic.*

Proof. Let $H_1 \colon f_1 \cong g_1$ and $H_2 \colon f_2 \cong g_2$. Then, clearly, $f_2 \circ H_1 \colon X \times I \to Z$ is a homotopy from $f_2 \circ f_1$ to $f_2 \circ g_1$. Next, define a map $H \colon X \times I \to Z$ by $H(x,t) = H_2(g_1(x), t)$, i.e., the map H is simply the following composite:

$$\begin{gathered} X \times I \to Y \times I \xrightarrow{H_2} Z \\ (x,t) \to (g_1(x), t) \to H_2(g_1(x), t) \end{gathered}$$

Then H is continuous and $H(x,0) = H_2(g_1(x),0) = f_2(g_1(x))$, $H(x,1) = H_2(g_1(x),1) = g_2(g_1(x))$, i.e., $H \colon f_2 \circ g_1 \cong g_2 \circ g_1$. Now, because $f_2 \circ f_1$ is homotopic to $f_2 \circ g_1$ and $f_2 \circ g_1$ is homotopic to $g_2 \circ g_1$, it follows by the transitive property of homotopy relation that $f_2 \circ f_1$ is homotopic to $g_2 \circ g_1$. ■

Exercises

1. Let X be a topological space and $Y \subset \mathbb{S}^2$ be the open upper hemisphere. Prove that any two maps $f, g \colon X \to Y$ are homotopic.
2. Let $P = \{p\}$ be a point space and X be a topological space. Show that X is path connected if and only if the set $[P, X]$ of homotopy classes of maps is a singleton.

3. Let X be a discrete space. Show that if a map $f: X \to X$ is homotopic to the identity map $I_X: X \to X$, then $f = I_X$. (**Hint:** The given condition implies that there is a path joining x and $f(x)$.)

4. Suppose X is a connected space and Y is a discrete space. Prove that the two maps $f, g: X \to Y$ are homotopic if and only if $f = g$.

5. Let $\mathbb{S}^1$ be the unit circle of the complex plane and $f, g: \mathbb{S}^1 \to \mathbb{S}^1$ be two maps defined by $f(z) = z$ and $g(z) = z^2$. What is wrong in saying that the map $F: \mathbb{S}^1 \times I \to \mathbb{S}^1$ defined by $F(z, t) = z^{t+1}$ is a homotopy from f to g?

6. Let X be a locally compact Hausdorff space and the set $C(X, Y)$ of all continuous maps from X to Y be given the compact open topology. Prove that two maps $f, g \in C(X, Y)$ are homotopic if and only if these can be joined by a path in the space $C(X, Y)$.
(**Hint:** Use the exponential correspondence theorem. (Spanier [15] p. 6))

2.3 Contractible Spaces and Homotopy Type

The notion of a contractible space is very important and the definition itself has some geometric appeal, as we shall see later. Recall that a map $f: X \to Y$ is said to be a **constant map** provided each point of X is mapped by f to some fixed point $y_0 \in Y$. If this is the case, then it is convenient to denote such a constant map by the symbol C_{y_0} i.e., $C_{y_0}(x) = y_0$, for every $x \in X$. We have

Definition 2.3.1. *A topological space X is said to be a contractible space if the identity map $I_X: X \to X$ is homotopic to some constant map $C_x: X \to X$, where, of course, $x \in X$. Any homotopy from I_X to C_x is called a* **contraction** *of the space to the point $x \in X$.*

There are numerous examples of contractible spaces. For instance, we note that *any convex subset of an Euclidean space $\mathbb{R}^n$ is contractible.* For, let S be a convex subset of $\mathbb{R}^n$. This means for any two points $x, y \in S$, the point $tx + (1-t)y$ is also in S for all t, $0 \le t \le 1$. Now let $x_0 \in S$. Define a map $H: S \times I \to S$ by

$$H(x, t) = (1-t)x + tx_0$$

Then it is clear that H is a homotopy from the identity map on S to the constant map $C_{x_0}: S \to S$. Hence, H is a contraction and so S is contractible. In particular, the Euclidean space $\mathbb{R}^n$, the disk $\mathbb{D}^n$ are contractible spaces. More generally, a subspace X of $\mathbb{R}^n$ is said to be **star-shaped** if there exists a point $x_0 \in X$ such that the line segment joining any point of X to x_0 lies completely in X. For example, the subset $X \subset \mathbb{R}^2$ (Fig. 2.3) is star-shaped.

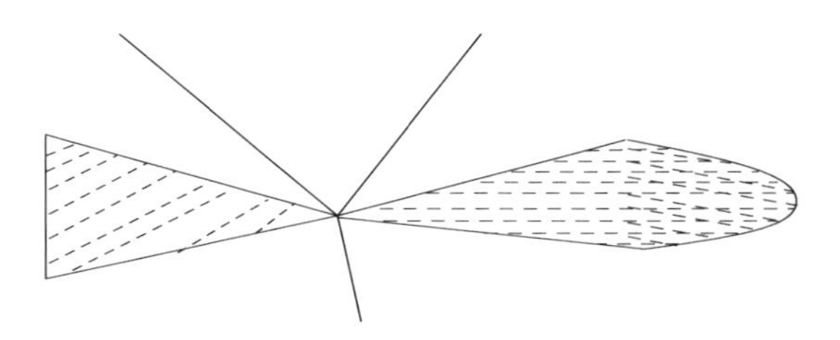

Fig. 2.3: A star-shaped region

Then one can easily see that X is contractible to the point x_0 and the same contraction, as defined above, works in this case also. Now, we ask the following question: *Determine whether or not the n-sphere* $\mathbb{S}^n$, $n \geq 1$ *is contractible*? It is obviously not star-shaped. That does not mean, however, that X is not contractible (see Example 2.3.10). The answer to this question is "No"; it will take quite sometime before we can prove it. The following concept is again extremely basic.

Definition 2.3.2. *Let* $f\colon X \to Y$ *be a continuous map. We say that* f *is a* **homotopy equivalence** *if there exists a continuous map* $g\colon Y \to X$ *such that* $g \circ f$ *is homotopic to the identity map* I_X *on* X *and* $f \circ g$ *is homotopic to the identity map* I_Y *on* Y. *Two spaces* X *and* Y *are said to be* **homotopically equivalent** *or of the same* **homotopy type** *if there exists a homotopy equivalence from one to the other.*

We must observe that *two homeomorphic spaces are of the same homotopy type*. For, suppose X and Y are homeomorphic and let $f\colon X \to Y$ be a homeomorphism. Then the inverse map $f^{-1}\colon Y \to X$ is continuous and satisfies the condition that $f^{-1} \circ f \cong I_X$ and $f \circ f^{-1} \cong I_Y$. This means f is a homotopy equivalence, i.e., X and Y are of the same homotopy type. The converse is not true. We have

Example 2.3.3. Consider the unit disk $\mathbb{D}^2$ (open or closed) and a point $x_0 \in \mathbb{D}^2$. Let $i\colon P = \{x_0\} \to \mathbb{D}^2$ be the inclusion map and $C_{x_0}\colon \mathbb{D}^2 \to P$ be the constant map. Then evidently $C_{x_0} \circ i = I_P$. On the other hand, the map $H\colon \mathbb{D}^2 \times I \to \mathbb{D}^2$ defined by

$$H(x,t) = (1-t)x + tx_0$$

is a homotopy from $I_{\mathbb{D}^2}$ to $i \circ C_{x_0}$. Thus, $\mathbb{D}^2$ is of the same homotopy type as a point space P and these are clearly not homeomorphic.

One can easily verify that the relation of "homotopy equivalence" in the class of all topological spaces is an equivalence relation. Also, in view of the above example, the relation of homotopy equivalence is strictly weaker than the relation of "homeomorphism". It is also clear from the above example that if a space X is compact then a space Y, which is homotopically equivalent to X, need not be compact, i.e., the compactness is not a homotopy invariant. Similarly, the topological dimension is not a homotopy invariant because dimension of the plane $\mathbb{R}^2$ is 2 whereas a point has dimension zero. These and several other examples show that topological invariants are, in general, not "homotopy" invariants and so the homotopy classification of spaces is quite a weak classification. However, it is still very important because we will later define some homotopy invariants which would be evidently topological invariants. These invariants would also be computed for a large class of topological spaces. The moment we notice that if any of these invariants is not the "same" for any two given spaces X and Y, *we can immediately assert that X and Y are not of the same homotopy type and therefore cannot be homeomorphic.* This is a well established strategy of algebraic topology for proving that two given spaces are *not* homeomorphic. Example 2.3.3 is a trivial case of the following:

Theorem 2.3.4. *A topological space X is contractible if and only if X is of the same homotopy type as a point space $P = \{p\}$.*

Proof. Suppose X is contractible. Let $H: X \times I \to X$ be a homotopy from the identity map I_X to the constant map $C_{x_0}: X \to X$. Define maps $i: P \to X$ and $C: X \to P$ by $i(p) = x_0$ and $C(x) = p$, $x \in X$. Then, clearly, $c \circ i = I_P$. Also, the map H is a homotopy from I_X to $i \circ C$ because $H(x, 0) = x$ and

$$H(x, 1) = C_{x_0}(x) = x_0 = i \circ C(x)$$

for each $x \in X$. Hence X and P are of the same homotopy type.

Conversely, suppose there are maps $f: X \to P$ and $g: P \to X$ such that $g \circ f \cong I_X$ and $f \circ g \cong I_P$. Let $g(p) = x_0$ and $H: X \times I \to X$ be a homotopy from I_X to $g \circ f$. Then because $g \circ f(x) = g(f(x)) = g(p) = x_0$, for all $x \in X$, $g \circ f$ is the constant map $C_{x_0}: X \to X$. Thus, I_X is homotopic to the constant map C_{x_0} and so the space X is contractible. ∎

Thus, the contractible spaces are precisely those spaces which are homotopically equivalent to a point space. The intuitive picture of a contractible space X is quite interesting. A homotopy H which starts from the identity map on X and terminates into a constant map C_{x_0}, $x_0 \in X$, should be thought of as a continuous deformation of the space X which finally shrinks the whole space X into the point x_0. In other words, if we imagine the unit interval I as a time interval then at the time $t = 0$, every point $x \in X$ is at its original place; as t varies from 0 to 1 continuously, x moves continuously and approaches the point x_0; even x_0 moves accordingly and comes back to itself. Furthermore, all of these points move simultaneously in such a way that their relative positions do not change abruptly. Thus, if we follow the movement of an arbitrary point $x \in X$, we note that it describes a path in X starting from x which terminates at x_0. In particular, we intuitively see that X is path connected. We make this statement precise:

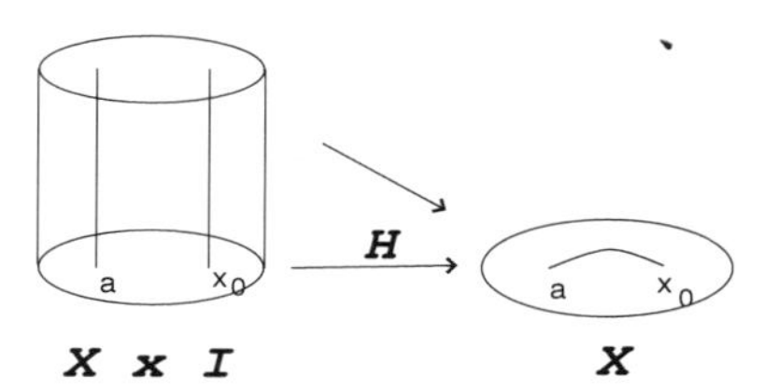

Fig. 2.4: Contractible space is path connected

Proposition 2.3.5. *If X is a contractible space, then X is path connected.*

Proof. Suppose X is contractible to a point x_0 and $H: X \times I \to X$ is a homotopy from I_X to C_{x_0}. Let $a \in X$. It suffices to show that a can be joined to x_0 by a path in X. Note that H maps whole of bottom to itself whereas the entire top to the point x_0. Define a map $f: I \to X \times I$ by $f(t) = (a, t)$ and note that it is continuous. Then $\omega = H \circ f$ is a path (Fig. 2.4) in X such that

$$\omega(0) = H(f(0)) = H(a, 0) = a,$$

and

$$\omega(1) = H(f(1)) = H(a, 1) = x_0.$$

■

Now suppose X is contractible. This means X can be contracted to some point $x \in X$. Can we then contract X to an arbitrary point $x_0 \in X$? The answer to this question is "yes". We will now explain this. Let us prove

Proposition 2.3.6. *A topological space X is contractible if and only if an arbitrary map $f: T \to X$ from any space T to X is homotopic to a constant map.*

Proof. Suppose X is contractible. This means the identity map $I_X: X \to X$ is homotopic to some constant map, say $C_{x_0}: X \to X$. Now let $f: T \to X$ be any map. By Theorem 2.2.6, we find that $I_X \circ f$ is homotopic to $C_{x_0} \circ f$. But $I_X \circ f = f$ and $C_{x_0} \circ f: T \to X$ is the constant map.

For the converse, take $T = X$ and the map $f: T \to X$ to be the identity map. Then, by the given condition, we find that $I_X: X \to X$ is homotopic to a constant map, i.e., X is contractible. ■

It now follows from above that if X is a contractible space, then any map $f: X \to X$ is homotopic to a constant map $C_{x_0}: X \to X$. In particular, for any $x \in X$, the constant map C_x and the identity map $I_X: X \to X$ both are homotopic to C_{x_0}, i.e., C_x is homotopic to I_X for all $x \in X$.

Corollary 2.3.7. *If X is a contractible space, then the identity map $I_X: X \to X$ is homotopic to a constant map $C_x: X \to X$ for all $x \in X$. In particular, X can be contracted to any arbitrary point of X.*

Once again, let X be a contractible space. We ask now a slightly stronger question. *Can we contract X to some point $x_0 \in X$ so that the point x_0 does not move at all?* The answer to this question is "No". We will give an example of this later on (See Example 2.3.10). This question leads to the concept of relative homotopy which is stronger than the homotopy defined earlier.

Definition 2.3.8. *Let $A \subset X$ be an arbitrary subset and $f, g\colon X \to Y$ be two continuous maps. We will say that f is homotopic to g "relative to A" if there exists a continuous map $F\colon X \times I \to Y$ such that*

$$F(x,0) = f(x), F(x,1) = g(x), \text{ for all } x \in X, \text{ and}$$
$$F(a,t) = f(a) = g(a), \text{ for all } a \in A.$$

Note that if we take A to be null set $\emptyset$, then the concept of relative homotopy reduces to that of homotopy. It is also to be noted that if $f, g\colon X \to Y$ are to be homotopic relative to some subset A of X, then f and g must agree on A to start with. The map f will change into the map g by a family of continuous maps $h_t\colon X \to Y$, $t \in I$, but the points of A will remain unchanged under h_t when t varies from 0 to 1.

If $A \subseteq X$ is a fixed subset and $C(X,Y)$ denotes the set of all continuous maps $f\colon X \to Y$, then following the proof of Theorem 2.2.5, one can prove that the relation of being "relatively homotopic to" with respect to A is an equivalence relation in the set $C(X,Y)$.

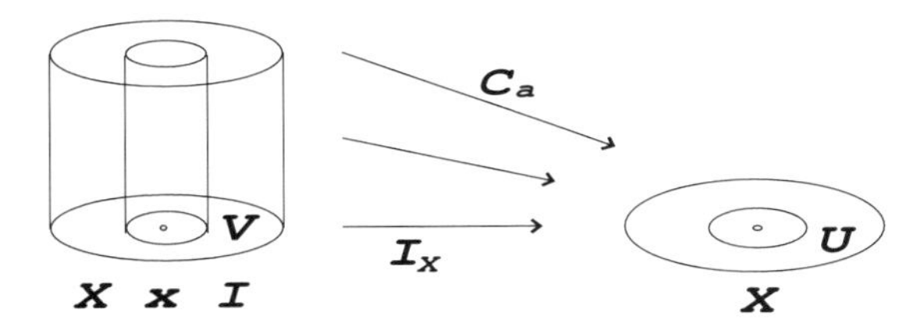

Fig. 2.5: Strongly contractible at a is semi-locally path connected at a.

Let us now assume that a space X is contractible to a point $a \in X$ relative to the subset $A = \{a\}$, i.e., X can be contracted to the point a in such a manner that all points of X move and finally get changed into the point a, but the point a itself does not undergo any change. In other words, we have a homotopy $F\colon X \times I \to X$ which starts with the identity map I_X and terminates into the constant map C_a relative to the subset $\{a\}$. This means, (see Fig. 2.5) under the continuous map F, the line $\{a\} \times I$ is mapped to the point $a \in X$. If we take any neighbourhood U of a, the continuity of F will give for each $t \in I$, neighbourhoods $V_t(a)$ of a in X and $W(t)$ of t in I such that $F(V_t(a) \times W(t)) \subset U$. The compactness of I means that the open covering $\{W(t) : t \in I\}$ of I will have a finite subcover, say, $W(t_1), \ldots, W(t_n)$, such that

$F(V_{t_i}(a) \times W(t_i)) \subset U$, for all $i = 1, \ldots, n$. Therefore, $V(a) = \bigcap_{i=1}^{n} V_{t_i}(a)$ is a neighbourhood of a in X such that $F(V(a) \times I) \subset U$. Now, if $b \in V(a)$, then considering the image $F(V(a) \times I)$ in U, we find that b can be joined to a by a path which lies in U. This completes the proof of the following:

Theorem 2.3.9. *If a space X is contractible to a point $a \in X$ relative to the subset $\{a\}$, then for each neighbourhood U of a in X, there exists a neighbourhood V of a contained in U such that any point of V can be joined to a by a path lying completely inside U, i.e. X is semi-locally path connected.*

Let us now consider the famous

Example 2.3.10. (Comb Space). We consider the following subset C (Fig. 2.6) of the plane (shown only partly).

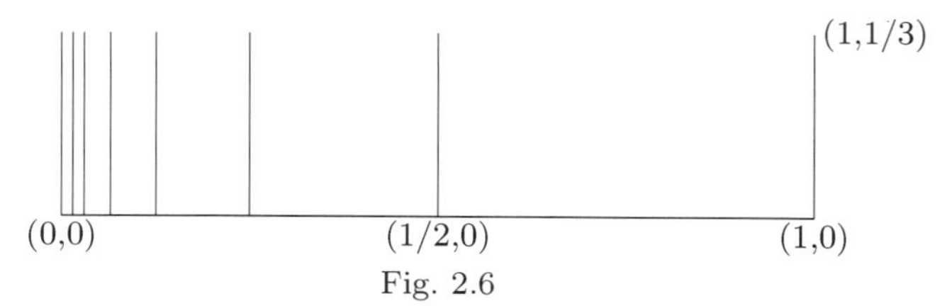

Fig. 2.6

It consists of the horizontal line segment joining (0,0) to (1,0) and vertical unit closed line segments standing on points $(1/n, 0)$ for each $n = 1, 2, \ldots$ together with the line segment joining (0,0) with (0,1).

The comb space C is contractible. The projection map $p\colon C \to L$, where L is the line segment joining (0,0) with (1,0), is a homotopy equivalence. For, if $i\colon L \to C$ is the inclusion map, then $p \circ i = I_L$ and the map $F\colon C \times I \to C$ defined by

$$F((x, y), t) = (x, (1 - t)y).$$

is a homotopy between I_C and $i \circ p$. We already know that L is homeomorphic to the unit interval which is of the same homotopy type as a point space and so by Theorem 2.3.1, C is contractible.

C is not contractible relative to $\{(0,1)\}$. Note that any small neighbourhood V of (0,1) has infinite number of path components. So, if we take the neighbourhood $U = D \cap C$ of (0,1) in C, where D is the open disk around (0,1) of radius 1/2, then U cannot have any neighbourhood V each of whose points can be joined to (0,1) by a path lying in U. This observation combined with Theorem 2.3.9, shows that C is not contractible relative to $\{(0,1)\}$.

Remark 2.3.11. Note that if $f, g\colon X \to Y$ are two continuous maps which are homotopic relative to some subset A of X, then obviously f is homotopic to g. The converse, however, is not true, i.e., there are maps $f, g\colon X \to Y$ which are homotopic, and even agree on a subset A of X, yet they need not be homotopic relative to A. For example, consider the identity map $I_X : X \to X$ of the comb space $X = C$ and the constant map $C_{(0,1)} : X \to X$. Then, obviously, I_X and $C_{(0,1)}$ agree on (0,1) and are also homotopic by Corollary 2.3.1, since X is contractible. However, we have just seen in the above example that I_X and $C_{(0,1)}$ are not homotopic relative to (0,1). This means the concept of relative homotopy is definitely stronger than that of homotopy.

There are some fundamental concepts related to "contractible" spaces. It is appropriate to discuss them now.

Definition 2.3.12. *Let $A \subset X$. We say that A is a* **retract** *of X if there exists a continuous map $r : X \to A$ such that $r(a) = a$, for all $a \in A$. The map r is called the* **retraction map**.

If $i : A \to X$ is the inclusion map, then the condition $r(a) = a$, for all $a \in A$ is equivalent to saying that $r \circ i = I_A$. Thus, A is a retract of X if and only if the inclusion map i has left inverse. As an example, note that every single point $x_0 \in X$ of an arbitrary topological space X is a retract of X and the constant map C_{x_0} is the retraction map. Let $X = [0, 1]$, the unit interval and $A = \{0, 1\}$, the boundary of X. Then A cannot be a retract of X because X is connected whereas A is disconnected. Let us ask a general question : Let $X = \mathbb{D}^{n+1}$ be the $(n + 1)$-disk and $\mathbb{S}^n$ be its boundary. Determine whether or not $\mathbb{S}^n$ is a retract of $\mathbb{D}^{n+1}$. The answer is again "No" and this fact is known as *Brouwer's No Retraction Theorem.* The proof, however, will be given later on.

Definition 2.3.13. *A topological space X is said to be* **deformable** *into a subspace $A \subset X$ if there is a map $f : X \to A$ which is right homotopy inverse of the inclusion map $i : A \to X$, i.e., the identity map I_X is homotopic to $i \circ f : X \to X$.*

The above definition asserts that there is a homotopy, say, $D : X \times I \to X$ such that

$$D(x, 0) = x, D(x, 1) = i(f(x)) = f(x).$$

Any such homotopy D is called a **deformation** of X into A and we say that X is **deformable** into A. It must be observed that the homotopy D, which starts with identity map $I_X : X \to X$, simply moves each point of X continuously, including the points of A and finally pushes every point of X into a point of A. In particular, if a space X is deformable into a point $a \in X$, then X is contractible and vice versa. If we can find a deformation D which deforms X into A but the points of A do not move at all, then the homotopy D will be "relative to A" and we say that X is **strongly deformable** into A. In such a case, note that for each $a \in A$, $D(a, 1) = f(a) = a$ and so the map $f : X \to A$ is automatically a retraction of X onto A. We have

Definition 2.3.14. *A space X is said to be* **strongly deformable** *into a subspace A if there is a continuous map $f : X \to A$ which is the right homotopy inverse of the inclusion map $i : A \to X$ relative to A, i.e., the identity map $I_X : X \to X$ is homotopic to $i \circ f : X \to X$ relative to A.*

Clearly, if X is strongly deformable into a subspace A, then it is also deformable into A. But the converse is not true even if the map $f : X \to A$ is onto: the comb space C (Example 2.3.10) is deformable into the point $\{(0,1)\}$ because it is contractible and so the identity map I_X is homotopic to the map $i \circ C_{(0,1)}$ where $C_{(0,1)}$ is the constant map $X \to X$. However, we have already seen that I_X is not homotopic to $i \circ C_{(0,1)}$ relative to the point $\{(0,1)\}$. Finally, we have

Definition 2.3.15. *A subspace A of a topological space X is said to be a* **deformation retract** *of X if X is deformable into A so that the final map is a retraction of X onto A.*

Definition 2.3.16. *A subspace A of a topological space X is said to be a* **strong deformation retract** *of X if X is deformable into A strongly* (A is then automatically a retract of X).

It follows from the above definitions that if A is a deformation retract of X, then the inclusion map $i : A \to X$ has a two-sidedhomotopy inverse, i.e., it is a homotopy equivalence, and consequently A and X are of the same homotopy type. On the other hand, if A is a strong deformation retract of X, then A is homotopically equivalent to X and something more is true, viz., X can be deformed into A without moving the points of A at all. For instance, the point (0,1) of the comb space X is a deformation retract of X but is not a strong deformation retract of X. Quite often we will use the following:

Example 2.3.17. For $n \geq 1$, $\mathbb{S}^n \subset \mathbb{R}^{n+1} - \{(0, \ldots, 0)\} = X$ is a strong deformation retract of X. Fig. 2.7 depicts the case $n = 1$.

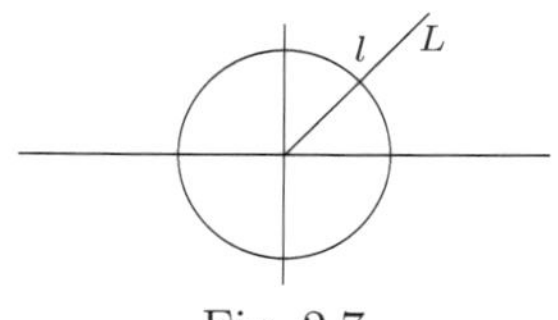

Fig. 2.7

Each infinite line segment L starting from its origin intersects the circle $\mathbb{S}^1$ at exactly one point, say, l. Since origin is not a point of $\mathbb{R}^2 - \{(0,0)\}$, these lines are disjoint and their union is $\mathbb{R}^2 - \{(0,0)\}$. We define a map $r : \mathbb{R}^2 - \{(0,0)\} \to \mathbb{S}^1$ by $r(x) = l$ for all points $x \in L$. Then, clearly r is continuous and $\mathbb{S}^1$ becomes a retract of $\mathbb{R}^2 - \{(0,0)\}$. Let us define a deformation $D : (\mathbb{R}^2 - \{(0,0)\} \times I \to \mathbb{R}^2 - \{(0,0)\}$ by

$$D(x,t) = (1-t)x + t\frac{x}{\|x\|}.$$

Then D is clearly a strong deformation retraction of $\mathbb{R}^2 - \{(0,0)\}$ relative to $\mathbb{S}^1$ into $\mathbb{S}^1$. A similar argument shows that $\mathbb{S}^n$ is a strong deformation retract of $\mathbb{R}^{n+1} - \{(0,0)\}$.

Exercises

1. Let $A \subset X$ be a retract of X where X is Hausdorff. Then prove that A must be closed in X. (This implies that an open interval (0,1) can never be a retract of any closed subset of the real line.)
2. Let X be a connected space and $x_0, x_1 \in X$ be two points of X which have disjoint open neighbourhoods in X. Show that $A = \{x_0, x_1\}$ can never be a retract of X.
3. Prove that a space X is contractible if and only if every map $f : X \to T$ to any space T is null-homotopic.
4. Show that if A is a strong deformation retract of X and B is a strong deformation retract of A, then B is a strong deformation retract of X.
5. Prove that an arbitrary product of contractible space is again contractible.
6. Prove that a retract of a contractible space is contractible.
7. For any space X consider the cylinder $X \times I$ over X and collapse the top $X \times 1$ of this cylinder to a point. The resulting quotient space, called **cone** over X, is denoted by $C(X)$. Prove that $C(X)$ is contractible for any space X.
8. Let $I^2 = [0,1] \times [0,1]$ be the unit square and $C \subset I^2$ be the comb space (Example 2.3.10). Prove that C is not a retract of I^2. (**Hint:** Given any open neighbourhood $U = B((0, \frac{1}{2}), \frac{1}{4}) \cap C$ in C, there exist a connected neighbourhood $V \subset U$ of $(0, \frac{1}{2})$ in $I \times I$ such that $r(V) \subset U$, but $r(V)$ is disconnected.)
9. Determine which of the following spaces are contractible:
   (i) Unit interval $I = [0,1]$.
   (ii) $\mathbb{S}^2 - \{p\}$, where $\mathbb{S}^2$ is a 2-sphere and p is any point of $\mathbb{S}^2$.
   (iii) Any solid or hollow cone in $\mathbb{R}^3$.
   (iv) The subspace $\{0\} \cup \{1/n : n \in N\}$ of real line.
10. Consider the following subspace X of plane $\mathbb{R}^2$; X consist of all closed line segments joining origin with points $(1, \frac{1}{n})$, $n \geq 1$ and the line $\{(x,0) \mid 0 \leq x \leq 1\}$. Prove that X is contractible, but none of the points $(x,0)$, $x \geq 0$, is a strong deformation retract of X.
11. Let I be the unit interval and X be any path connected space. Prove that the sets $[I, X]$ and $[X, I]$ each has only one element. (**Hint:** The space I is contractible)

12. Give an example of a space X which is of the same homotopy type as a discrete space $D = \{0, 1, 2, 3\}$, but is not homeomorphic to D.

13. Prove that a homotopy invariant is also a topological invariant. Give an example to show that a topological invariant need not be a homotopy invariant. (**Hint:** There are contractible spaces which are not compact, not locally compact, not locally conneted etc.)

2.4 Fundamental Group and its Properties

Recall that a **path** in a topological space X is just a continuous map $\alpha : I = [0, 1] \to X$; $\alpha(0)$ is called the **initial point** and $\alpha(1)$ is called the **terminal point** of the path α. If α, β are two paths in X such that $\alpha(1) = \beta(0)$, then we can define a new path (Fig. 2.8), called the **product** of α and β denoted by $\alpha * \beta$, as follows:

$$(\alpha * \beta)(t) = \begin{cases} \alpha(2t) & 0 \leq t \leq 1/2 \\ \beta(2t-1) & 1/2 \leq t \leq 1 \end{cases} \tag{4.1}$$

Note that $\alpha * \beta : I \to X$ is continuous by the continuity lemma, the initial point of $\alpha * \beta$ is the initial point of α and the terminal point of $\alpha * \beta$ is the terminal point of β.

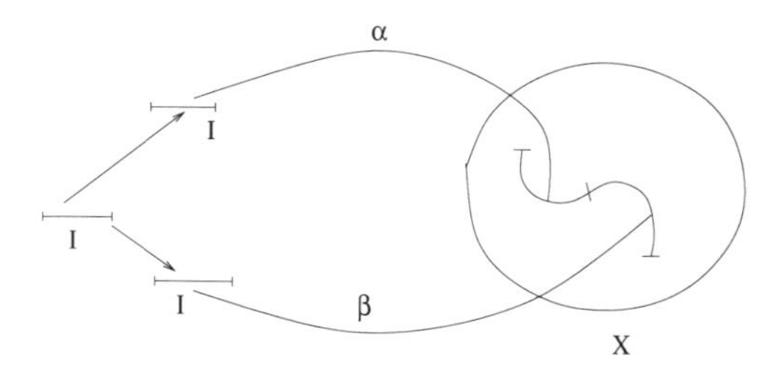

Fig. 2.8: Product of two paths

One can verify that if α, β, γ are three paths in X such that $\alpha(1) = \beta(0)$, $\beta(1) = \gamma(0)$, then $(\alpha * \beta) * \gamma$ and $\alpha * (\beta * \gamma)$ are paths in X which are not necessarily the same paths. To see this, just apply definition (2.4.1) and look at the image of the point $t = 1/2$ under both the paths. Therefore, the product of paths is not an associative operation. Let us now fix a point $x_0 \in X$ and consider the set of all closed paths at x_0, i.e., those paths whose initial and terminal points are x_0; such a path is also known as **loop** in X based at x_0. It is clear that the product of two loops based at x_0 is always defined. The difficulty that the product of loops based at x_0 need not be associative is still there. To surmount this difficulty and to finally get a group structure, we will introduce an equivalence relation in the set of all loops in X based at $x_0 \in X$. First, we have an important as well as general

Definition 2.4.1. *Let α, β be two paths in X with the same initial and terminal points, i.e., $\alpha(0) = \beta(0) = x_0$, $\alpha(1) = \beta(1) = x_1$. We will say that α is* **equivalent** *to β, and write it as $\alpha \sim_{(x_0,x_1)} \beta$, if there exists a homotopy between α and β relative to the subset $\{0,1\}$ of I. In other words, the homotopy keeps the end points fixed.*

Thus, the path α is equivalent to β if there exists a continuous map $H : I \times I \to X$ such that

$$H(s,0) = \alpha(s), H(s,1) = \beta(s)$$

$$H(0,t) = \alpha(0) = \beta(0), H(1,t) = \alpha(1) = \beta(1)$$

for all $s \in I$ and for all $t \in I$. In other words, the path α changes continuously and finally it becomes the path β, but during all this transformation the end points remain fixed (Fig. 2.9).

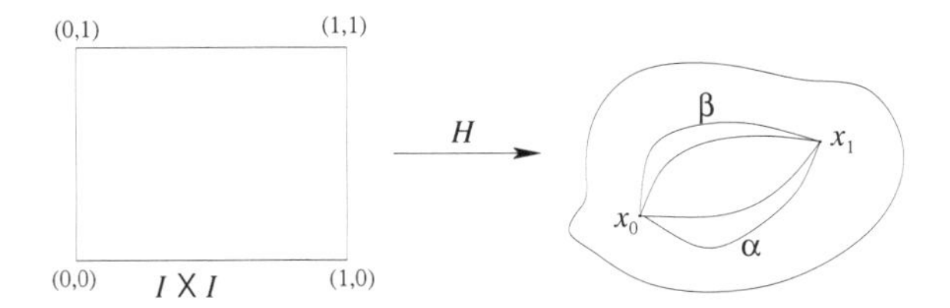

Fig. 2.9: H is a path-homotopy

The relative homotopy H, which keeps the end points fixed, is sometimes called a **path homotopy**, just for convenience. The next result, which is in fact true, more generally, for homotopy relative to any subset A of the domain space (cf. Definition 2.3.8), shows that the above relation is an equivalence relation.

Theorem 2.4.2. *Let $x_0, x_1 \in X$. Then the relation of "being equivalent" in the set of all paths starting from x_0 and terminating at x_1 is an equivalence relation.*

Proof. Let $\alpha : I \to X$ be any path with $\alpha(0) = x_0$, $\alpha(1) = x_1$. Then the map $H\colon I \times I \to X$ defined by $H(s,t) = \alpha(s)$ is a homotopy from α to itself relative to {0,1}. Thus, the relation is reflexive. Next, suppose α, β are two paths from x_0 to x_1 and $H\colon I \times I \to X$ is a continuous map such that

$$H(s,0) = \alpha(s), H(s,1) = \beta(s), \forall s \in I$$

$$H(0,t) = x_0, H(1,t) = x_1, \forall t \in I.$$

Define a map $H'\colon I \times I \to X$ by

$$H'(s,t) = H(s, 1-t).$$

Then H' is continuous and has the property that

$$H'(s,0) = H(s,1) = \beta(s),\ H'(s,1) = \alpha(s), s \in I$$

$$H'(0,t) = x_0,\ H'(1,t) = x_1, \forall t \in I.$$

Thus, $\alpha \sim \beta$ relative to {0,1} implies $\beta \sim \alpha$ relative to {0,1}, i.e., the relation is symmetric. To prove the transitivity of the relation, suppose $\alpha \sim \beta$ relative to {0,1}, $\beta \sim \gamma$ relative to {0,1}. Let H_1 and H_2 be two homotopies such that

$$H_1(s,0) = \alpha(s),\ H_1(s,1) = \beta(s), \forall s \in I$$

$$H_1(0,t) = x_0,\ H_1(1,t) = x_1, \forall t \in I$$

and

$$H_2(s,0) = \beta(s),\ H_2(s,1) = \gamma(s), \forall s \in I$$

$$H_2(0,t) = x_0, H_2(1.t) = x_1, \forall t \in I.$$

Define a map $H\colon I \times I \to X$ by

$$H(s,t) = \begin{cases} H_1(s,2t) & 0 \leq t \leq 1/2 \\ H_2(s,2t-1) & 1/2 \leq t \leq 1 \end{cases}$$

Then H is continuous by the continuity lemma and is indeed a homotopy relative to {0,1} from α to γ. This completes the proof of the theorem. ■

When $x_0 = x_1$, we conclude the following result.

Corollary 2.4.3. *The relation "being equivalent" in the set of all loops in X based at $x_0 \in X$ is an equivalence relation.*

Next, we are going to deal with only loops in X based at given point $x_0 \in X$. If α, β are two loops based at x_0 which are equivalent, then we will write this as $\alpha \sim_{x_0} \beta$. Also, the equivalence class of a loop α based at x_0 will be denoted by the symbol $[\alpha]$ and called the **homotopy class of the loop** α. It must be emphasized at this point that if α is treated as a map $\alpha\colon I \to X$ with $\alpha(0) = \alpha(1) = x_0$, then the *homotopy class of the map* α, according to Theorem 2.2.5 is different from the path homotopy class of loop α specified by Theorem 2.4.2. In fact, the former homotopy class is, in general, larger than the latter path-homotopy class. Let $\pi_1(X, x_0)$ denote the set of all homotopy classes of loops in X based at x_0 , i.e.,

$$\pi_1(X, x_0) = \{[\alpha] : \alpha \text{ is a loop in X based at } \mathrm{x_0}\}.$$

The next proposition implies that the product of loops induces a product in the set $\pi_1(X, x_0)$ of all homotopy classes of loops based at x_0. Recall that if α, β are two loops at x_0, then their product $\alpha * \beta$ is also a loop at $x_0 \in X$.

Proposition 2.4.4. *Suppose $\alpha, \beta, \alpha', \beta'$ are loops in X based at x_0. If $\alpha \sim_{x_0} \alpha'$, $\beta \sim_{x_0} \beta'$ then $\alpha * \beta \sim_{x_0} \alpha' * \beta'$.*

Proof. Let H_1 be a homotopy from α to α', H_2 be a homotopy from β to β', i.e., H_1, H_2 are maps from $I \times I \to X$ such that

$$H_1(s,0) = \alpha(s),\ H_1(s,1) = \alpha'(s), \forall s \in I$$
$$H_1(0,t) = x_0 = H_1(1,t), \forall t \in I$$

and

$$H_2(s,0) = \beta(s),\ H_2(s,1) = \beta'(s), \forall s \in I$$
$$H_2(0,t) = x_0 = H_2(1,t), \forall t \in I.$$

Define a map $H\colon I \times I \to X$ by

$$H(s,t) = \begin{cases} H_1(2s,t) & 0 \le s \le 1/2 \\ H_2(2s-1,t) & 1/2 \le s \le 1. \end{cases}$$

Then H is continuous by continuity lemma, and

$$\begin{aligned} H(s,0) &= \begin{cases} H_1(2s,0) & 0 \le s \le 1/2 \\ H_2(2s-1,0) & 1/2 \le s \le 1 \end{cases} \\ &= \begin{cases} \alpha(2s) & 0 \le s \le 1/2 \\ \beta(2s-1) & 1/2 \le s \le 1 \end{cases} \\ &= (\alpha * \beta)(s). \end{aligned}$$

By a similar calculation, $H(s,1) = (\alpha' * \beta')(s)$ and $H(0,t) = x_0 = H(1,t)$, $\forall\ t \in I$. Thus, $\alpha * \beta \sim_{x_0} \alpha' * \beta'$. ■

One can easily observe that the above proof yields the following general result, viz.,

Corollary 2.4.5. *Let α, α' be two path homotopic paths joining x_0 with x_1 and β, β' be two path homotopic paths joining x_1 to x_2. Then $\alpha*\beta$ is path-homotopic to $\alpha' * \beta'$ joining x_0 to x_2, and the path homotopy can be chosen so that the point x_1 remains fixed.*

Sometimes we will need to consider the path homotopy classes of paths joining x_0 to x_1. If α is a path joining x_0 with x_1, then $[\alpha]$ will also be used to denote the path homotopy class represented by α. The set of all path homotopy classes of paths joining x_0 to x_1 will be denoted by $\pi_1(X, x_0, x_1)$. In this terminology, we can define an operation

$$\circ\colon \pi_1(X, x_0, x_1) \times \pi_1(X, x_1, x_2) \to \pi_1(X, x_0, x_2)$$

by

$$[\alpha] \circ [\beta] = [\alpha * \beta].$$

The above corollary says that the map $\circ$ is well defined. In case x_0, x_1, x_2 are the same points, the map $\circ$ defines a binary operation in the set $\pi_1(X, x_0)$ of all homotopy classes of loops based at x_0. Consequently, we can state the next

Definition 2.4.6. *Let* $[\alpha], [\beta]$ *be any two elements of* $\pi_1(X, x_0)$. *Then we define their product* $[\alpha] \circ [\beta]$ *by*

$$[\alpha] \circ [\beta] = [\alpha * \beta].$$

The following basic result can now be proved.

Theorem 2.4.7. *The set* $\pi_1(X, x_0)$ *of all path homotopy classes of loops based at* x_0 *is a group with respect to the binary operation "* $\circ$ *" defined above.*

Proof. We must prove that the operation $\circ$ is associative, there exists identity element in $\pi_1(X, x_0)$ and each element of $\pi_1(X, x_0)$ has an inverse in $\pi_1(X, x_0)$. The proof of each of these statements is achieved by constructing a suitable path homotopy relative to {0,1} between appropriate paths and we discuss them below one by one:

The operation is associative: Let $[\alpha], [\beta], [\gamma]$ be three elements of $\pi_1(X, x_0)$. Since

$$([\alpha] \circ [\beta]) \circ [\gamma] = [(\alpha * \beta) * \gamma]$$

and

$$[\alpha] \circ ([\beta] \circ [\gamma]) = [\alpha * (\beta * \gamma)],$$

it is sufficient to show that $(\alpha * \beta) * \gamma \sim_{x_0} \alpha * (\beta * \gamma)$. By definition

$$((\alpha * \beta) * \gamma)(s) = \begin{cases} \alpha(4s) & 0 \leq s \leq 1/4 \\ \beta(4s-1) & 1/4 \leq s \leq 1/2 \\ \gamma(2s-1) & 1/2 \leq s \leq 1 \end{cases}$$

and

$$(\alpha * (\beta * \gamma))(s) = \begin{cases} \alpha(2s) & 0 \leq s \leq 1/2 \\ \beta(4s-2) & 1/2 \leq s \leq 3/4 \\ \gamma(4s-3) & 3/4 \leq s \leq 1. \end{cases}$$

Thus we should define a homotopy $H \colon I \times I \to X$ such that

$$\begin{cases} H(s,0) = ((\alpha * \beta) * \gamma)(s), \ H(s,1) = (\alpha * (\beta * \gamma))(s) \\ H(0,t) = x_0 = H(1,t), \forall \ s, t \in I \end{cases} \tag{2.6.1}$$

Such a homotopy is given by the formula

$$H(s,t) = \begin{cases} \alpha(4s/(t+1)) & 0 \leq s \leq (t+1)/4 \\ \beta(4s-1-t) & (t+1)/4 \leq s \leq (t+2)/4 \\ \gamma((4s-2-t)/(2-t)) & (t+2)/4 \leq s \leq 1 \end{cases}$$

The motivation for writing this homotopy comes from Fig. 2.10.

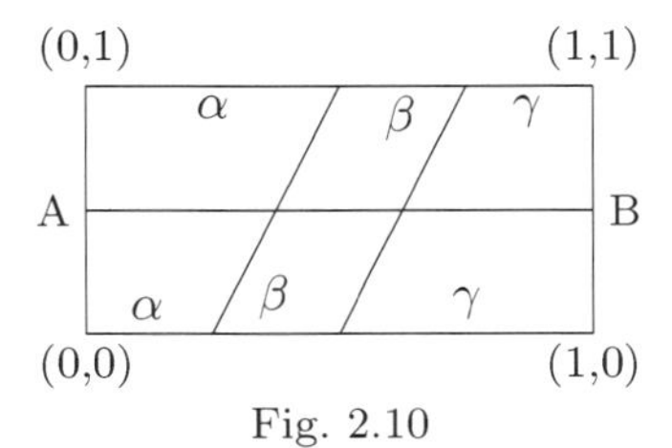

Fig. 2.10

We divide the square $I \times I$ into three quadrilaterals:

$Q_1 : 0 \leq s \leq (t+1)/4$
$Q_2 : (t+1)/4 \leq s \leq (t+2)/4$
$Q_3 : (t+2)/4 \leq s \leq 1, t \in I$

For instance, the equation of the line joining (1/4,0) and (1/2,1) would be $t = 4s - 1$ and so the equation of the region Q_1 would be

$$0 \leq s \leq (t+1)/4, \ 0 \leq t \leq 1.$$

For a fixed $t \in I$, a typical horizontal line AB would have three parts. When t moves from 0 to 1, these three parts also change their positions. For $t = 0$, we get a partition defining $(\alpha * \beta) * \gamma$ and for $t = 1$, we get a partition defining $\alpha * (\beta * \gamma)$. The map H is defined by α on Q_1, β on Q_2, γ on Q_3, each of which is continuous. On their common boundary, the two definitions match yielding a nice map H. Hence, by the continuity lemma, H is continuous. Moreover, conditions (2.6.1) are evidently satisfied. This completes the proof that $\circ$ is associative.

Remark 2.4.8. If α is a path joining x_0 with x_1, β is a path joining x_1 with x_2 and γ is a path joining x_2 with x_3, then the same proof as above says, more generally, that

$$\begin{aligned} ([\alpha] \circ [\beta]) \circ [\gamma] &= [(\alpha * \beta) * \gamma] \\ &= [\alpha * (\beta * \gamma)] \\ &= [\alpha] \circ ([\beta] \circ [\gamma]). \end{aligned}$$

There exists an identity element in $\pi_1(X, x_0)$: Let us consider the constant loop $C_{x_0} : I \to X$. We claim that the class $[C_{x_0}] \in \pi_1(X, x_0)$ is an identity element, i.e., for each loop α in X based at x_0, we show that $C_{x_0} * \alpha \sim_{x_0} \alpha$ and $\alpha * C_{x_0} * \sim_{x_0} \alpha$. For this let us consider Fig. 2.11.

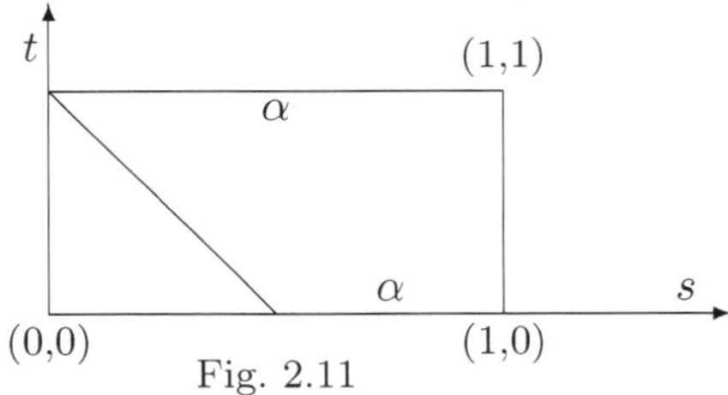

Fig. 2.11

Define a map $H: I \times I \to X$ by

$$H(s,t) = \begin{cases} x_0 & 0 \le s \le (1-t)/2 \\ \alpha((2s+t-1)/(t+1)) & (1-t)/2 \le s \le 1 \end{cases}$$

Note that the equation of the line joining (1/2,0) and (0,1) is $2s = 1-t$. H maps the whole triangle below this line to the point x_0 and the two definitions of H on the line $2s = 1-t$ match. The continuity of H follows from the continuity lemma. Furthermore,

$$\begin{aligned} H(s,0) &= \begin{cases} x_0 & 0 \le s \le 1/2 \\ \alpha(2s-1) & 1/2 \le s \le 1 \end{cases} \\ &= (C_{x_0} * \alpha)(s) \\ H(s,1) &= \alpha(s), \end{aligned}$$

and $H(0,t) = x_0 = H(1,t)$. This shows that $C_{x_0} * \alpha \sim_{x_0} \alpha$. To prove that $[C_{x_0}]$ is also right identity, we can write a suitable path homotopy by looking at Fig. 2.12.

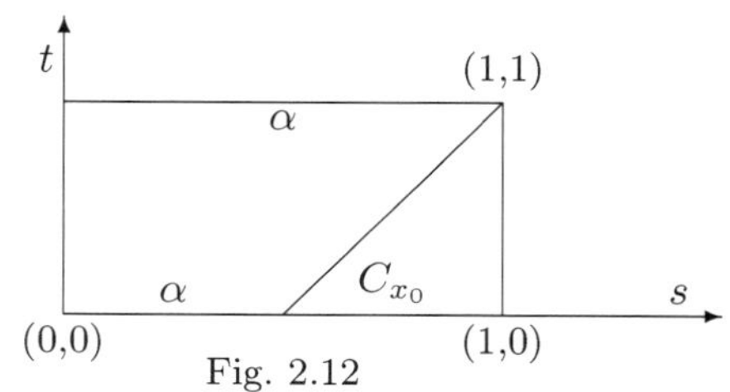

Fig. 2.12

This would be

$$H(s,t) = \begin{cases} \alpha(2s/(1+t)) & 0 \le s \le (t+1)/2 \\ x_0 & (1+t)/2 \le s \le 1 \end{cases}$$

More generally, we have

Remark 2.4.9. Let α be any path joining x_0 with x_1. Then the above proof implies

(1) $[C_{x_0}] \circ [\alpha] = [C_{x_0} * \alpha] = [\alpha]$,

(2) $[\alpha] \circ [C_{x_1}] = [\alpha * C_{x_1}] = [\alpha]$.

In other words, C_{x_0} serves as the left identity and C_{x_1} serves as the right identity for any $[\alpha]$.

Each element of $\pi_1(X, x_0)$ has an inverse: Let $[\alpha] \in \pi_1(X, x_0)$. We choose a representative, say α, of the homotopy class $[\alpha]$. For this α, we define a loop $\alpha': I \to X$ by

$$\alpha'(t) = \alpha(1-t).$$

Geometrically, α' simply describes the same path as α, but in reverse direction. We claim that the homotopy class $[\alpha'] \in \pi_1(X, x_0)$ does not depend on the choice of α from the class $[\alpha]$. For, suppose $\beta \sim_{x_0} \alpha$ by a homotopy H. Then we can define a homotopy $H': I \times I \to X$ by

$$H'(s,t) = H(1-s,t).$$

Then, obviously, H' is a homotopy from β' to α', i.e., $[\beta'] = [\alpha']$. Now we claim that

$$[\alpha'] \circ [\alpha] = [C_{x_0}] = [\alpha] \circ [\alpha'].$$

For this, we must construct a homotopy from $\alpha * \alpha'$ to C_{x_0} relative to $\{0,1\}$. We now consider Fig. 2.13.

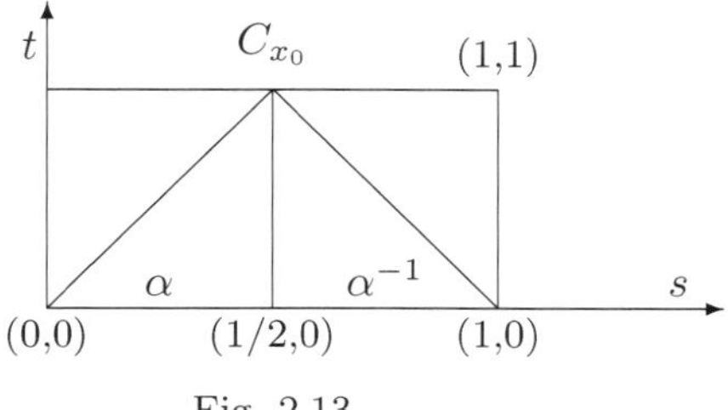

Fig. 2.13

Just as in earlier cases, the required homotopy $H\colon I \times I \to X$ is given by

$$H(s,t) = \begin{cases} x_0 & 0 \le s \le t/2 \\ \alpha(2s-t) & t/2 \le s \le 1/2 \\ \alpha(2-2s-t) & 1/2 \le s \le 1-t/2 \\ x_0 & 1-t/2 \le s \le 1. \end{cases}$$

It is easily verified that H is well defined, continuous and has the required properties. By exactly a similar argument we can show that

$$[\alpha'] \circ [\alpha] = [C_{x_0}].$$

Thus, $[\alpha']$ is an inverse of $[\alpha]$ in $\pi_1(X, x_0)$. ■

Remark 2.4.10. The same proof says that, more generally, if α is a path joining x_0 with x_1, then its inverse path α^{-1} has the following properties:

$$[\alpha] \circ [\alpha^{-1}] = [C_{x_0}],$$
$$[\alpha^{-1}] \circ [\alpha] = [C_{x_1}].$$

Remark 2.4.11. In the proof of the preceding theorem, it was enough to show that $\pi_1(X, x_0)$ has a left identity and each element of $\pi_1(X, x_0)$ has a left inverse – this is a result from elementary group theory. However, we have shown the existence of two-sided identity and two-sided inverse only to give more practice of writing path homotopies.

Remark 2.4.12. We also note that if $\alpha, \beta, \gamma, \delta$ are four loops based at $x_0 \in X$, then by the associative law proved above, we find that

$$(\alpha * \beta) * (\gamma * \delta) \sim_{x_0} \alpha * (\beta * (\gamma * \delta)) \sim_{x_0} ((\alpha * \beta) * \gamma) * \delta$$

and so the generalized associative law is valid in the sense that placing of parentheses does not make any difference in the homotopy class. Therefore, we can just ignore the parentheses and write the above loop only as $[\alpha * \beta * \gamma * \delta]$.

Definition 2.4.13. *Let X be a topological space and $x_0 \in X$. Then the group $\pi_1(X, x_0)$ obtained in Theorem 2.4.7 is called the* **fundamental group** *or the* **Poincaré group** *of the space X based at $x_0 \in X$.*

Remark 2.4.14. At this stage one would like to see examples of fundamental groups of some spaces. We will give several examples somewhat later, but before that it would be helpful to prove a few results on the behaviour of fundamental groups so that one can have some idea about the possibilities of the nature of fundamental group of a given space. We should also point out here that it is in the very nature of algebraic topology that computing associated algebraic objects is normally a long process and sometime can also be extremely difficult.

Having defined the fundamental group of a space X based at a point $x_0 \in X$, one would naturally like to ask: how important is the role of base point x_0 in the group $\pi_1(X, x_0)$? How are $\pi_1(X, x_0)$ and $\pi_1(X, x_1)$ related if $x_0 \neq x_1$? In fact, if X is arbitrary, then a loop at x_0, being itself path connected, will lie completely in the path component of x_0 and so if x_0, x_1 are in distinct path components of X, then $\pi_1(X, x_0)$, $\pi_1(X, x_1)$ are not related at all. However, if x_0, x_1 belong to the same path component of X, then $\pi_1(X, x_0)$ and $\pi_1(X, x_1)$ are indeed isomorphic. This follows from the next

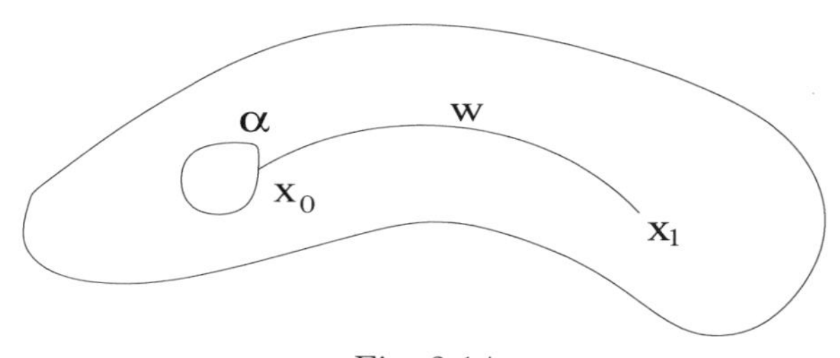

Fig. 2.14

Theorem 2.4.15. *Let X be a path connected space and x_0, x_1 be any two points of X. Then $\pi_1(X, x_0)$ and $\pi_1(X, x_1)$ are isomorphic. In fact, each path joining x_0 to x_1 defines an isomorphism from $\pi_1(X, x_0)$ to $\pi_1(X, x_1)$.*

Proof. Consider Fig. 2.14. Let $\omega\colon I \to X$ be a path joining x_0 to x_1 and suppose ω^{-1} is the inverse path of ω, i.e., $\omega^{-1}(t) = \omega(1-t)$ for each $t \in I$. If α is any loop based at x_0, then it is clear that $\omega^{-1} * \alpha * \omega$ is a loop based at x_1. Thus, we can define a map

$$P_\omega\colon \pi_1(X, x_0) \to \pi_1(X, x_1)$$

by

$$P_\omega[\alpha] = [\omega^{-1} * \alpha * \omega].$$

First, we check that P_ω is well defined. Suppose $\alpha \sim_{x_0} \beta$ and let $H\colon I \times I \to X$ be a homotopy relative to $\{0,1\}$ from α to β. Then by Corollary 2.4.2, $\omega^{-1} * \alpha * \omega \sim_{x_0} \omega^{-1} * \beta * \omega$, which means $P_\omega[\alpha] = P_\omega[\beta]$. Now, let us see the following computation:

$$\begin{aligned}
P_\omega([\alpha] \circ [\beta]) &= P_\omega[\alpha * \beta] \\
&= [\omega^{-1} * (\alpha * \beta) * \omega] \\
&= [\omega^{-1} * \alpha] \circ [\beta * \omega] \\
&= [\omega^{-1} * \alpha * C_{x_0}] \circ [\beta * \omega] \\
&= [\omega^{-1} * \alpha * \omega * \omega^{-1}] \circ [\beta * \omega] \\
&= [\omega^{-1} * \alpha * \omega] \circ [\omega^{-1} * \beta * \omega] \\
&= P_\omega[\alpha] \circ P_\omega[\beta].
\end{aligned}$$

If we use ω^{-1} instead of ω, then we get a homomorphism

$$P_{\omega^{-1}}\colon \pi_1(X, x_1) \to \pi_1(X, x_0).$$

Now, for each $[\alpha] \in \pi_1(X, x_0)$, we have

$$\begin{aligned}
P_{\omega^{-1}} \circ P_\omega[\alpha] &= P_{\omega^{-1}}[\omega^{-1} * \alpha * \omega] \\
&= [\omega * \omega^{-1} * \alpha * \omega * \omega^{-1}] \\
&= [\alpha],
\end{aligned}$$

which means $P_{\omega^{-1}} \circ P_\omega$ is identity on $\pi_1(X, x_0)$. By a similar argument, we see that $P_\omega \circ P_{\omega^{-1}}$ is identity on $\pi_1(X, x_1)$. It follows that P_ω is an isomorphism. ∎

Remark 2.4.16. It follows from the above theorem that for a path connected space X, the fundamental group $\pi_1(X, x)$ is independent of the base point x up to isomorphism of groups. Therefore, for a path connected space X, we can denote $\pi_1(X, x)$ simply by $\pi_1(X)$ ignoring the mention of the base point x, and call it the **fundamental group** of the space X.

Since the isomorphism P_ω depends on the path ω joining x_0 with x_1, we can examine the question as to how much P_ω depends on the path ω itself. We have,

Proposition 2.4.17. *If ω, ω' are two paths joining x_0 to x_1 which are path homotopic, then the induced isomorphisms P_ω and $P_{\omega'}$ are identical.*

Proof. If ω, ω' are path homotopic, then it is easily seen that ω^{-1} and $(\omega')^{-1}$ are also path homotopic. It follows that for any loop α based at x_0, $\omega^{-1} * \alpha$ is path homotopic to $(\omega')^{-1} * \alpha$ and therefore $\omega^{-1} * \alpha * \omega$ is path homotopic to $(\omega')^{-1} * \alpha * \omega'$. This means $P_\omega[\alpha] = P_{\omega'}[\alpha]$. ∎

Proposition 2.4.18. *Let X be a path connected space and $x_0, x_1 \in X$. Then $\pi_1(X, x_0)$ is abelian if and only if for each pair of paths ω, ω' from x_0 to x_1, $P_\omega = P_{\omega'}$.*

Proof. Assume that $\pi_1(X, x_0)$ is abelian. Since $\omega * (\omega')^{-1}$ is a loop based at x_0, we observe that for each $[\alpha] \in \pi_1(X, x_0)$,

$$[\omega * (\omega')^{-1}] \circ [\alpha] = [\alpha] \circ [\omega * (\omega')^{-1}],$$

which means

$$[(\omega')^{-1} * \alpha * \omega'] = [\omega^{-1} * \alpha * \omega].$$

Conversely, suppose $[\alpha], [\beta]$ are two elements of $\pi_1(X, x_0)$. Let ω be a path in X joining x_0 with x_1. Then $\beta * \omega$ is also a path joining x_0 with x_1. Hence, by the given condition, $P_{\beta*\omega}[\alpha] = P_\omega[\alpha]$. This means

$$[(\beta * \omega)^{-1} * \alpha * (\beta * \omega)] = [\omega^{-1} * \alpha * \omega].$$

Since $(\beta * \omega)^{-1} = \omega^{-1} * \beta^{-1}$, therefore, $[\beta^{-1} * \alpha * \beta] = [\alpha]$, i.e., $[\alpha][\beta] = [\beta][\alpha]$. ■

Note that to each pointed topological space (X, x), we have associated its fundamental group $\pi_1(X, x)$. Next, we show that for every continuous map $f : (X, x) \to (Y, y)$ of pointed spaces, there is an induced group homomorphism $f_\#$ between their fundamental groups.

Theorem 2.4.19. *Every continuous map $f : (X, x) \to (Y, y)$ of pointed spaces induces a group homomorphism $f_\# : \pi_1(X, x) \to \pi_1(Y, y)$.*

Proof. Let α be a loop in X based at x. Then $f \circ \alpha$ is a loop based at y. Moreover, if $\alpha \sim_{x_0} \alpha'$, say, by a homotopy H, then one can easily see that $f \circ \alpha \sim_y f \circ \alpha'$ by the homotopy $f \circ H$. Hence, we define a map $f_\# : \pi_1(X, x) \to \pi_1(Y, y)$ by

$$f_\#([\alpha]) = [f \circ \alpha].$$

To see that $f_\#$ is a homomorphism, observe that for any two loops α, β based at x, we have

$$\begin{aligned}(f \circ (\alpha * \beta))(t) &= f((\alpha * \beta)(t)) \\ &= \begin{cases} f(\alpha(2t)) & 0 \le t \le 1/2 \\ f(\beta(2t-1)) & 1/2 \le t \le 1 \end{cases} \\ &= \begin{cases} (f \circ \alpha)(2t) & 0 \le t \le 1/2 \\ (f \circ \beta)(2t-1) & 1/2 \le t \le 1 \end{cases} \\ &= ((f \circ \alpha) * (f \circ \beta))(t)\end{aligned}$$

for each $t \in I$, which means $f \circ (\alpha * \beta) = (f \circ \alpha) * (f \circ \beta)$. Hence

$$\begin{aligned}f_\#([\alpha] \circ [\beta]) &= f_\#[\alpha * \beta] \\ &= [f \circ (\alpha * \beta)] \\ &= [(f \circ \alpha) * (f \circ \beta)] \\ &= f_\#[\alpha] \circ f_\#[\beta].\end{aligned}$$

■

The two basic properties of the induced homomorphism $f_{\#}$ are given by the next result. In the language of categories and functors these two properties are sometimes expressed by declaring that the fundamental group π_1 is a **covariant functor** from the category of all pointed topological spaces and base point preserving continuous maps to the category of all groups and group homomorphisms.

Theorem 2.4.20. (i) *If $f\colon X \to X$ is the identity map, then $f_{\#} : \pi_1(X,x) \to \pi_1(X,x)$ is also the identity map for all $x \in X$.*

(ii) *If $f\colon (X,x) \to (Y,y)$ and $g : (Y,y) \to (Z,z)$ are two maps of pointed spaces, then*

$$(g \circ f)_{\#} = g_{\#} \circ f_{\#}.$$

Proof. (i) If f is the identity map on X, then for each loop α based at x, $f \circ \alpha = \alpha$ and so $f_{\#}[\alpha] = [f \circ \alpha] = [\alpha]$. Hence $f_{\#}$ is also identity on $\pi_1(X,x)$.
(ii) For any loop α based at x, we have

$$\begin{aligned}(g \circ f)_{\#}[\alpha] &= [(g \circ f) \circ \alpha] \\ &= [g \circ (f \circ \alpha)] \\ &= g_{\#}[f \circ \alpha] \\ &= g_{\#}(f_{\#}[\alpha]) \\ &= g_{\#} \circ f_{\#}[\alpha].\end{aligned}$$

Hence, $(g \circ f)_{\#} = g_{\#} \circ f_{\#}$. ■

Just before stating the above theorem, we pointed out that the above two properties of the induced homomorphisms are basic. The reason as to why they are important is that they imply a deep result, viz., *the fundamental group of a path connected topological space is a topological invariant.* In other words, the above properties yield the fact that if X and Y are two path connected topological spaces which are homeomorphic, then their fundamental groups must be isomorphic. We have already explained the proof in the introduction. However, we present the proof once more:

Theorem 2.4.21. *If X and Y are path connected spaces which are homeomorphic, then their fundamental groups are isomorphic.*

Proof. We have already proved in Theorem 2.4.15 that for path connected spaces, the fundamental group is independent of the base point. Let $f\colon X \to Y$ be a homeomorphism and $x \in X$. Then $f\colon (X,x) \to (Y,f(x))$ is a homeomorphism of pointed spaces and its inverse $f^{-1}\colon (Y,f(x)) \to (X,x)$ is a continuous map which has the property that

$$f^{-1} \circ f = I_X, f \circ f^{-1} = I_Y.$$

Considering the induced homomorphisms in fundamental groups, we see that

$$(f^{-1} \circ f)_\# = (I_X)_\#.$$

However, by (i) of Theorem 2.4.20, $(I_X)_\# = I_{\pi_1(X,x)}$ and by (ii) of the same theorem,

$$(f^{-1} \circ f)_\# = (f^{-1})_\# \circ f_\#.$$

Therefore,

$$(f_\#)^{-1} \circ f_\# = I_{\pi_1(X,x)}.$$

By a similar argument, we find that

$$f_\# \circ (f^{-1})_\# = I_{\pi_1(Y,f(x))}.$$

Combining the last two identities, we conclude that $f_\# \colon \pi_1(X,x) \to \pi_1(Y, f(x))$ is an isomorphism. ∎

Corollary 2.4.22. *Let $A \subset X$. Suppose $r \colon X \to A$ is a retraction and $i \colon A \to X$ is the inclusion. Then $r_\# \colon \pi_1(X,a) \to \pi_1(A,a)$ is onto and $i_\# \colon \pi_1(A,a) \to \pi_1(X,a)$ is 1-1 for each $a \in A$.*

Proof. For each point $a \in A$, the composite map $(A,a) \xrightarrow{i} (X,a) \xrightarrow{r} (A,a)$ is identity on (A,a). Consequently, by the two basic properties of the induced homomorphisms, the composite $\pi_1(A,a) \xrightarrow{i_\#} \pi_1(X,a) \xrightarrow{r_\#} \pi_1(A,a)$ is identity on the group $\pi_1(A,a)$. This implies at once that $i_\#$ is 1-1 and $r_\#$ is onto. ∎

Given two topological spaces X and Y, we are very often required to determine whether or not X and Y are homeomorphic. This is briefly known as the "homeomorphism problem". Similarly, to determine whether or not two given groups are isomorphic, is abbreviated as "isomorphism problem" for groups. Algebraic topologists agree that the latter problem is normally easier than the former. The fundamental group functor partially converts a homeomorphism problem into an isomorphism problem in the sense that if we can assert that $\pi_1(X)$ is not isomorphic to $\pi_1(Y)$, then it follows from topological invariance property of fundamental groups that X and Y cannot be homeomorphic. To quote another example, we will later on see that for the projective plane $\mathbb{RP}^2$, $\pi_1(\mathbb{RP}^2) \cong \mathbb{Z}_2$ and $\pi_1(\mathbb{S}^1 \times \mathbb{S}^1) \cong \mathbb{Z} \oplus \mathbb{Z}$. Hence, by our discussion above, it follows that the projective plane $\mathbb{RP}^2$ cannot be homeomorphic to the two dimensional torus $\mathbb{S}^1 \times \mathbb{S}^1$. Here one might recall that both these spaces are compact, connected, two dimensional and even locally homeomorphic. As a matter of fact, invariants of point set topology do not appear adequate to distinguish these two spaces. The homeomorphism problem is only partially converted into isomorphism problem because if the fundamental groups of X and Y are isomorphic, we cannot say that X and Y are homeomorphic. We will later on prove that the fundamental group of a contractible space X is isomorphic to the fundamental group of a point space

P, but X and P are never homeomorphic unless X consists of a single point.

We have some further properties of induced homomorphisms.

Theorem 2.4.23. *If $f, g\colon (X, x) \to (Y, y)$ are continuous maps of pointed spaces which are homotopic relative to $\{x\}$, then*

$$f_\# = g_\#\colon \pi_1(X, x) \to \pi_1(Y, y).$$

Proof. Let α be a loop in X based at x. Since f is homotopic to g and composite of homotopic maps is always homotopic, the loops $f \circ \alpha$ and $g \circ \alpha$ based at y are homotopic. This is not enough because they may not be homotopic relative to $\{0,1\}$. But, since f is homotopic to g relative to $\{x\}$, we see that $f \circ \alpha \sim_y g \circ \alpha$, i.e., $[f \circ \alpha] = [g \circ \alpha]$. This means

$$f_\#[\alpha] = g_\#[\alpha],\ \forall\ [\alpha] \in \pi_1(X, x). \qquad \blacksquare$$

Using Theorem 2.4.23, we can extend Theorem 2.4.20 as follows:

Theorem 2.4.24. *If $f\colon (X, x_0) \to (Y, y_0)$ is a homotopy equivalence in the relative sense, then $f_\#\colon \pi_1(X, x_0) \to \pi_1(Y, y_0)$ is an isomorphism of groups.*

Proof. Recall that two continuous maps $k, h\colon (X, x_0) \to (Y, y_0)$ of pointed spaces are said to be homotopic in the relative sense if there exists a homotopy $H\colon X \times I \to Y$ such that

$$H(x, 0) = k(x),\ H(x, 1) = h(x), \forall x \in X$$
$$H(x_0, t) = k(x_0) = h(x_0), \forall t \in I.$$

Thus, $f\colon (X, x_0) \to (Y, y_0)$ is a homotopy equivalence in the relative sense means there exists a map $g\colon (Y, y_0) \to (X, x_0)$ of pointed spaces such that $g \circ f \sim I_{(X, x_0)}$ and $f \circ g \sim I_{(Y, y_0)}$ in the relative sense. By the two basic properties of induced homomorphisms, we note that homomorphisms $f_\#\colon \pi_1(X, x_0) \to \pi_1(Y, y_0)$ and $g_\#\colon \pi_1(Y, y_0) \to \pi_1(X, x_0)$ are inverse of each other. Hence, $f_\#$ is an isomorphism. ∎

Corollary 2.4.25. *Let A be a strong deformation retract of a space X. Then for each point $a \in A$, $\pi_1(A, a)$ is isomorphic to $\pi_1(X, a)$.*

Proof. Since X is deformable into A strongly, we find that for each $a \in A$, the inclusion map $i\colon (A, a) \to (X, a)$ is a homotopy equivalence relative to $\{a\}$. Hence, by the above theorem, $i_\#\colon \pi_1(A, a) \to \pi_1(X, a)$ is an isomorphism. ∎

Remark 2.4.26. If the subspace A of X is a singleton, i.e., $A = \{a\}$ and A is strong deformation retract of X, then $\pi_1(X, a) \cong \pi_1(\{a\}, a)$. We usually write this as $\pi_1(a)$ ignoring the base point. We also note that if $\{a\}$ is a strong deformation retract of X, then X is contractible to $\{a\}$ and so is path connected. Therefore, $\pi_1(X, x) \cong \pi_1(X, a), \forall x \in X$. Thus, $\pi_1(X) \cong \pi_1(P) = \{e\}$, where P is a point space.

We now proceed to prove our final result, viz., *if two path-connected spaces X and Y have the same homotopy type, then their fundamental groups are isomorphic.* In particular, we show that the fundamental group of a contractible space X is trivial. First, we have

Theorem 2.4.27. *Let F be a homotopy between two maps $f, g\colon X \to Y$. Let $x_0 \in X$ and $\sigma\colon I \to Y$ be the path joining $f(x_0)$ and $g(x_0)$ defined by $\sigma(t) = F(x_0, t)$. Then the triangle*

$$\begin{array}{ccc} \pi_1(X, x_0) & \xrightarrow{f_\#} & \pi_1(Y, f(x_0)) \\ & \searrow^{g_\#} & \downarrow P_\sigma \\ & & \pi_1(Y, g(x_0)) \end{array}$$

of induced homomorphisms is commutative.

Proof. Let α be a loop in X based at x_0. Then we know that

$$P_\sigma \circ f_\#[\alpha] = P_\sigma[f \circ \alpha] = [\sigma^{-1} * (f \circ \alpha) * \sigma]$$

and

$$g_\#[\alpha] = [g \circ \alpha].$$

Hence, it suffices to show that the two loops $\sigma^{-1} * (f \circ \alpha) * \sigma$ and $g \circ \alpha$ based at $g(x_0)$ are equivalent. Define a continuous map $G\colon I \times I \to Y$ by $G(s,t) = F(\alpha(s), t)$.

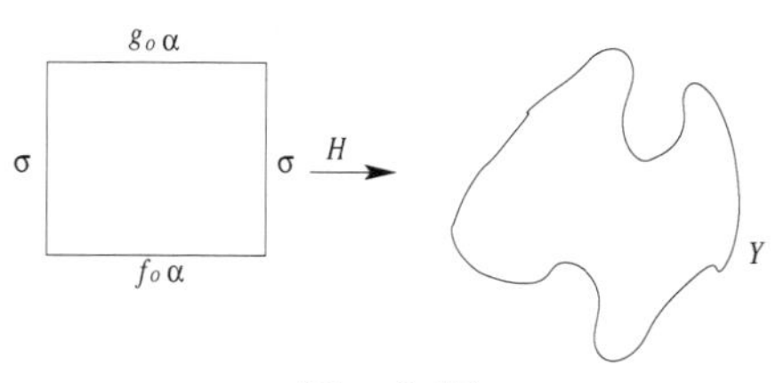

Fig. 2.15

Then (Fig. 2.15) for each $s \in I$, $t \in I$, we have

$$\begin{aligned} G(s,0) &= F(\alpha(s), 0) = f \circ \alpha(s), \\ G(s,1) &= F(\alpha(s), 1) = g \circ \alpha(s), \\ G(0,t) &= F(\alpha(0), t) = F(x_0, t) = \sigma(t), \\ G(1,t) &= F(\alpha(1), t) = F(x_0, t) = \sigma(t). \end{aligned}$$

Our objective is now to construct a homotopy $H\colon I \times I \to Y$ which starts with the loop $\sigma^{-1} * (f \circ \alpha) * \sigma$, terminates with the loop $g \circ \alpha$ and keeps the end points fixed.

The required homotopy H is defined by

$$H(s,t) = \begin{cases} \sigma^{-1}(2s) & 0 \le s \le (1-t)/2 \\ G((4s+2t-2)/(3t+1), t) & (1-t)/2 \le s \le (t+3)/4 \\ \sigma(4s-3) & (t+3)/4 \le s \le 1. \end{cases}$$

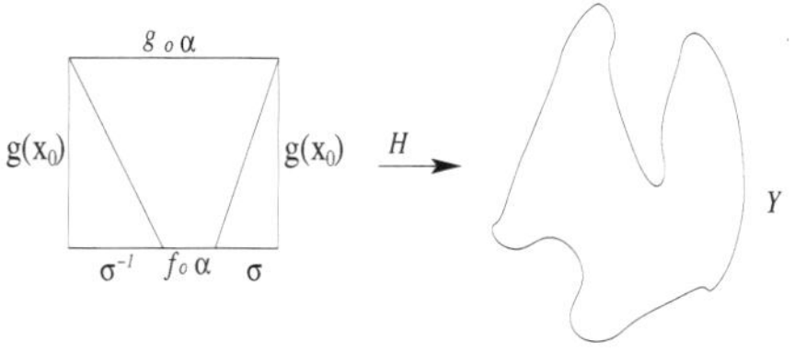

Fig. 2.16

It is easily verified that H is continuous and has the desired properties (Fig. 2.16). ■

Corollary 2.4.28. *If $f\colon X \to Y$ is homotopic to a constant map $C\colon X \to Y$, then the induced homomorphism $f_{\#} : \pi_1(X, x_0) \to \pi_1(Y, f(x_0))$ is the zero map.*

Proof. In the commutative triangle, notice that P_σ is an isomorphism and $C_{\#}\colon \pi_1(X, x_0) \to \pi_1(Y, C(x_0))$ is a trivial map. ■

In case $f(x_0) = g(x_0)$, the path σ is just a loop at $g(x_0) = f(x_0)$ and the automorphism $P_\sigma\colon \pi_1(Y, f(x_0)) \to \pi_1(Y, f(x_0))$ is simply the inner automorphism

$$[\alpha] \to [\sigma^{-1} * \alpha * \sigma] = [\sigma]^{-1} \circ [\alpha] \circ [\sigma].$$

Moreover, if the homotopy F is relative to the point subspace $\{x_0\}$, then σ is just the constant loop $C_{f(x_0)}$ and so P_σ is the identity map, i.e., $f_{\#} = g_{\#}$ and our result reduces to Theorem 2.4.23.

Remark 2.4.29. It is interesting to observe that if the space X is contractible to a point x_0 relative to $\{x_0\}$ by a homotopy $F\colon X \times I \to X$, i.e., $\{x_0\}$ is a strong deformation retract of X, then any loop α based at x_0 is obviously homotopic to the constant loop C_{x_0} by a homotopy $(s, t) \to F(\alpha(s), t)$ which does not move the point x_0 at all. However, if X is contractible to a point x_0, then any loop α based at x_0 will be homotopic to the constant loop C_{x_0}, but as in the case of comb space, the contracting homotopy mentioned above may not necessarily fix the point x_0 and so the loop α, it appears, may not represent the zero element of $\pi_1(X, x_0)$. The last statement, however, is not possible. In fact, we now show that if X is a contractible space, then for any given loop based at a point $x_0 \in X$, we can always find a homotopy which deforms the loop into the constant loop C_{x_0} without moving the point x_0 at all, i.e., $\pi_1(X, x_0) = 0$. We have

Corollary 2.4.30. *Let X be a contractible space. Then each loop α based at any point $x_0 \in X$ is equivalent to the constant loop C_{x_0} at x_0, i.e., $\pi_1(X, x_0) = 0$.*

Proof. Since X is contractible, the identity map I_X is homotopic to the constant map C_{x_0} by a homotopy F. Hence, if σ is the loop at x_0 defined by

$$\sigma(t) = F(x_0, t),$$

then by Theorem 2.4.27, the triangle

$$\begin{array}{ccc} \pi_1(X, x_0) & \xrightarrow{(I_X)_\#} & \pi_1(X, x_0) \\ & \searrow^{(C_{x_0})_\#} & \downarrow P_\sigma \\ & & \pi_1(X, x_0) \end{array}$$

is commutative, i.e.,

$$P_\sigma \circ (I_X)_\#[\alpha] = (C_{x_0})_\#[\alpha],$$

for each $[\alpha] \in \pi_1(X, x_0)$. But this means $[\sigma^{-1} * \alpha * \sigma] = [C_{x_0}]$, i.e., $\sigma^{-1} * \alpha * \sigma \sim_{x_0} C_{x_0}$. This implies at once that

$$\alpha \sim_{x_0} \sigma * C_{x_0} * \sigma^{-1} \sim_{x_0} C_{x_0}. \qquad \blacksquare$$

Now, we have our main

Theorem 2.4.31. *Let X, Y be two path-connected spaces which are of the same homotopy type. Then their fundamental groups are isomorphic.*

Proof. The fundamental groups of both the spaces X and Y are independent of the base points because each of them is path connected. Since X and Y are of the same homotopy type, there exist continuous maps $f\colon X \to Y$ and $g\colon Y \to X$ such that $g \circ f \sim I_X$ by a homotopy, say, F and $f \circ g \sim I_Y$ by some homotopy, say G. Let $x_0 \in X$ be a base point. Let

$$f_\#\colon \pi_1(X, x_0) \to \pi_1(Y, f(x_0))$$

and

$$g_\#\colon \pi_1(Y, f(x_0)) \to \pi_1(X, g(f(x_0)))$$

be the induced homomorphisms. It is evidently enough to prove that $f_\#$ is a bijective map. Let σ be the path joining x_0 to $gf(x_0)$ defined by the homotopy F. Then by Theorem 2.4.27, the triangle

$$\begin{array}{ccc} \pi_1(X, x_0) & \xrightarrow{(I_X)_\#} & \pi_1(X, x_0) \\ & \searrow^{(g \circ f)_\#} & \downarrow P_\sigma \\ & & \pi_1(X, gf(x_0)) \end{array}$$

is commutative, i.e., $(g \circ f)_\# = P_\sigma \circ (I_X)_\#$ and so $g_\# \circ f_\# = P_\sigma$. Since P_σ is an isomorphism, we find that $f_\#$ is a 1-1 map. Next, let τ be the path joining $f(x_0)$ to $(f \circ g)(f(x_0))$ defined by the homotopy G. Then again the triangle

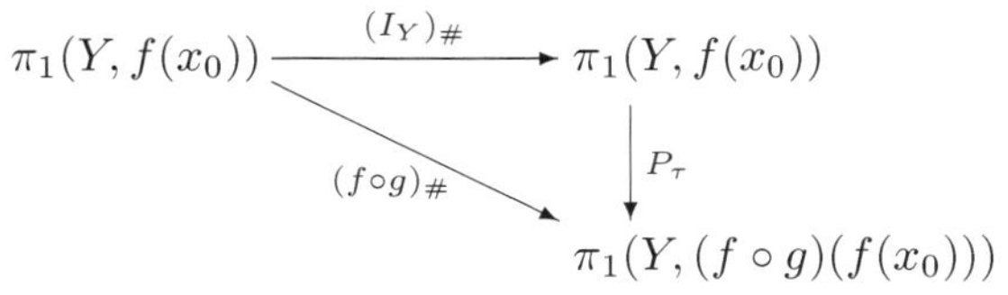

$$
\begin{array}{ccc}
\pi_1(Y, f(x_0)) & \xrightarrow{(I_Y)_\#} & \pi_1(Y, f(x_0)) \\
& \searrow^{(f \circ g)_\#} & \downarrow P_\tau \\
& & \pi_1(Y, (f \circ g)(f(x_0)))
\end{array}
$$

is commutative, i.e., $f_\# \circ g_\# = P_\tau$. Since P_τ is an isomorphism, we conclude that $f_\#$ is also onto. ■

Corollary 2.4.32. *Let X, Y be two path-connected spaces and $f\colon X \to Y$ be a homotopy equivalence. Then $f_\#\colon \pi_1(X) \to \pi_1(Y)$ is an isomorphism.*

2.5 Simply-Connected Spaces

We have already seen that if X is a contractible space then $\pi_1(X)$ is always trivial. In fact, those spaces whose fundamental groups are trivial play very important role in complex analysis as well as in algebraic topology. We have

Definition 2.5.1. *A space X is said to be* **simply connected** (*or* 1*-connected*) *if it is path connected and $\pi_1(X) = 0$.*

Notice that X being path connected, the fundamental group of X is independent of the base point. Consequently, X is simply connected if and only if it is path connected and its fundamental group vanishes at some point of X (hence it vanishes at every point). Corollary 2.4.30 asserts that a contractible space is simply connected. Later we will give example to show that the converse is not true. However, let us have

Proposition 2.5.2. *A path-connected space X is simply connected if and only if any two paths in X having the same endpoints are path homotopic.*

Proof. Suppose α, β are two paths in X, joining x_0 with x_1. Notice that $\alpha * \beta^{-1}$ is a loop based at x_0. Since X is simply connected, this loop must be null homotopic relative to x_0, i.e.,

$$[\alpha * \beta^{-1}] = [C_{x_0}].$$

Now, by Remark 2.4.12,

$$[(\alpha * \beta^{-1}) * \beta] = [C_{x_0} * \beta] = [\beta]$$

and

$$[\alpha * (\beta^{-1} * \beta)] = [\alpha * C_{x_1}] = [\alpha].$$

Hence, by the associative law of product of path homotopy classes, we find that $[\alpha] = [\beta]$, i.e., α is path homotopic to β.

The converse part is trivial. ■

If a space X is expressed as the union of two simply-connected spaces, say, X_1, X_2, then can we say that X itself is simply connected? The answer is "no": we shall see later that the unit circle $\mathbb{S}^1$ in the complex plane is not simply connected, though $\mathbb{S}^1$ can be expressed as the union of two contractible arcs. However, the next result is an important one.

Theorem 2.5.3. *Let $\{V_i : i \in \Lambda\}$ be an open covering of a space X, where each V_i is simply connected. Then X itself is simply connected provided*

(1) $\bigcap V_i \neq \phi$

(2) *for each $i \neq j$, $V_i \cap V_j$ is path connected.*

Proof. Since each of the open sets V_i is path connected and their intersection is non-empty, the space X itself is path connected. Hence it suffices to show that $\pi_1(X, x)$ is trivial for some $x \in X$. We choose a point $x_0 \in \cap_i V_i$.

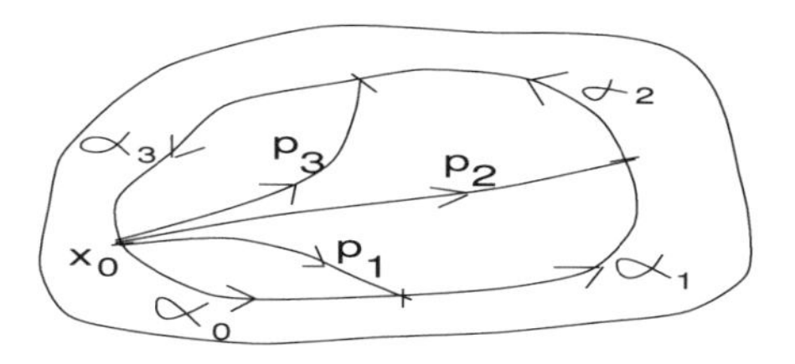

Fig. 2.17: The path α is null-homotopic (case $n = 3$)

Let $\alpha : I \to X$ be a loop based at x_0. We will show that $[\alpha]$ is the zero element of $\pi_1(X, x_0)$. Note that $\{\alpha^{-1}(V_i)\}$ is an open cover of I. Since I is a compact metric space, this open cover will have a Lebesgue number, say, $\epsilon > 0$. This means there will be a partition

$$0 = t_0 < t_1 < t_2 < \cdots < t_n = 1$$

of I such that for $0 \leq j \leq n-1$, $\alpha([t_j, t_{j+1}])$ will be contained in some V_i. For convenience, let us rename the open sets V_i so that $\alpha([t_j, t_{j+1}]) \subset V_j$, $0 \leq j \leq n-1$. For each j, we can define a path α_j in X by

$$\alpha_j(s) = \alpha((1-s)t_j + st_{j+1}).$$

Then, $\alpha_j(I)$ is contained in the simply-connected open set V_j for each j, and

$$[\alpha] = [\alpha_0 * \alpha_1 * \cdots * \alpha_{n-1}].$$

Look at the Fig. 2.17. Notice that $\alpha(t_1) \in V_0 \cap V_1$ and $V_0 \cap V_1$ is path connected containing the base point x_0. Hence we can find a path $\rho_1 : I \to X$ from x_0 to $\alpha(t_1)$ such that $\rho_1(I) \subset V_0 \cap V_1$. By a similar argument, we can find a path ρ_j from x_0 to $\alpha(t_j)$ lying completely in $V_{j-1} \cap V_j$ for $j = 1, 2, \ldots, n-1$. If ρ_j^{-1} denotes the reverse path of ρ_j, then we have

$$[\alpha] = [\alpha_0 * \rho_1^{-1} * \rho_1 * \alpha_1 * \rho_2^{-1} * \rho_2 * \alpha_2 * \cdots * \rho_{n-1} * \alpha_{n-1}].$$

Note that $\rho_1^{-1} * \rho_1$ is path homotopic to the constant loop based at $\alpha(t_1)$, etc. It follows that

$$[\alpha] = [\alpha_0 * \rho_1^{-1}] \circ [\rho_1 * \alpha_1 * \rho_2^{-1}] \circ \cdots \circ [\rho_{n-1} * \alpha_{n-1}].$$

The first term in the right hand side above is a loop based at x_0 and lying completely in the simply-connected space V_0. Similarly, the second term is a loop lying completely in the simply connected space V_1, and so on. Consequently, each term is actually null homotopic in V_j for some j and hence in X. Therefore, $[\alpha]$ represents the zero element of $\pi_1(X, x_0)$. ■

Example 2.5.4. The standard n-sphere $\mathbb{S}^n$, $n \geq 2$, is simply connected. For simplicity, consider the case of a 2-sphere $\mathbb{S}^2 = \{(x, y, z) \in \mathbb{R}^3 \mid x^2+y^2+z^2 = 1\}$. Let $N = (0, 0, 1)$ be the north pole of $\mathbb{S}^2$, $M = (0, 0, -1)$ be the south pole of $\mathbb{S}^2$. Then, $U = \mathbb{S}^2 - \{N\}$, $V = \mathbb{S}^2 - \{M\}$ are two open sets of $\mathbb{S}^2$ which cover the sphere $\mathbb{S}^2$. From the results of point-set topology, one can see that

(i) each of U and V are homeomorphic to the Euclidean plane $\mathbb{R}^2$, and so is contractible. Hence these are simply connected.

(ii) $U \cap V$ is path connected; it follows from the fact that even if a countable number of points are removed from $\mathbb{S}^2$ (here only two points have been removed), the remaining space is still path connected.

Hence, it follows from the previous theorem that $\mathbb{S}^2$ is simply connected. Exactly the same procedure shows that $\mathbb{S}^n$ is simply connected for each $n \geq 2$.

We will see later on that the 1-sphere $\mathbb{S}^1$ is not simply connected. In fact, we will prove that $\pi_1(\mathbb{S}^1) \cong \mathbb{Z}$, the additive group of integers.

Exercises

1. Show that union of two contractible spaces, having nonempty path-connected intersection, need not be contractible.

2. If P is a path component of X and $x_0 \in P$, then the inclusion map $i \colon P \to X$ can be regarded as a map of pointed spaces $(P, x_0) \to (X, x_0)$. Prove that the induced homomorphism

$$i_{\#} \colon \pi_1(P, x_0) \to \pi_1(X, x_0)$$

is an isomorphism.

3. Prove that a cone $C(X)$ over a topological space X is simply connected. If v_i is the vertex of a cone $C(X_i)$, $i \in \Lambda$, then show that their one-point union, viz., their disjoint union with all vertices v_i identified to one point, is simply connected.

4. Let α, β be two paths in X with initial point x_1 and terminal point x_2. Prove that $\alpha \sim \beta$ relative to {0,1} if and only if the loop $\alpha\beta^{-1} \sim C_{x_1}$ relative to {0,1}.

5. Let $A \subset X$ be a deformation retract. Show that the inclusion map $i\colon A \to X$ induces an isomorphism

$$i_{\#}\colon \pi_1(A, a) \to \pi_1(X, a)$$

for each $a \in A$.

6. Consider two spheres $\mathbb{S}^m$ and $\mathbb{S}^n$, $m, n \geq 2$ imbedded in some Euclidean space $\mathbb{R}^k$, k sufficiently large. If $\mathbb{S}^m$ intersects $\mathbb{S}^n$ tangentially in a single point, then show that their union is simply connected.

2.6 Results for Computing Fundamental Groups

In this section, we prove a few results which would be of help in computing the fundamental groups of some familiar subspaces of the Euclidean space $\mathbb{R}^n$, etc. Most importantly, we will prove in the next section that the fundamental group $\pi_1(\mathbb{S}^1)$ of the circle $\mathbb{S}^1$ is isomorphic to the additive group $\mathbb{Z}$ of integers. Then, combining that result with those of this section, we will determine the fundamental group of a number of familiar spaces.

Example 2.6.1. Let $\mathbb{D}^n (n \geq 1)$ be the n-dimensional unit disk in $\mathbb{R}^n$. Since $\mathbb{D}^n$ is path connected, we can compute its fundamental group at any point. For simplicity, let us choose origin as its base point. Note that the homotopy $H\colon \mathbb{D}^n \times I \to \mathbb{D}^n$ defined by

$$H(x, t) = (1 - t)x$$

shrinks every point of $\mathbb{D}^n$ to the origin without moving the origin at all. Therefore, each loop in $\mathbb{D}^n$ based at origin is path homotopic to the constant loop. Whence, $\pi_1(\mathbb{D}^n) = 0$.

More generally, if S is any convex subset of $\mathbb{R}^n$ and $s_0 \in S$ is a fixed point, then the homotopy D defined by

$$D(x, t) = (1 - t)x + ts_0$$

shrinks each point of S to the point s_0 without moving the point s_0. Therefore, each loop in S based at s_0 is null homotopic, and $\pi_1(S) = \{e\}$. Because the fundamental group of a contractible space is trivial, it follows directly that the fundamental group of any star-like subspace of $\mathbb{R}^n$ (convex subsets, in particular) is trivial.

Next, we proceed to give examples of spaces which have non-trivial fundamental groups. However, before we do so, let us prove the next useful result.

Theorem 2.6.2. *Let X, Y be two spaces with base points $x_0 \in X$ and $y_0 \in Y$, respectively. Then*

$$\pi_1(X \times Y, (x_0, y_0)) \cong \pi_1(X, x_0) \times \pi_1(Y, y_0),$$

i.e., the fundamental group of the product space is isomorphic to the product of fundamental groups of factors.

Proof. Suppose $\alpha\colon I \to X \times Y$ is a loop based at (x_0, y_0) and $p_1\colon (X \times Y) \to X$, $p_2\colon (X \times Y) \to Y$ are the canonical projections. Then p_1, p_2 are obviously maps of pointed spaces and hence induce group homomorphisms

$$p_{1\#}: \pi_1(X \times Y, (x_0, y_0)) \to \pi_1(X, x_0)$$

and

$$p_{2\#}: \pi_1(X \times Y, (x_0, y_0)) \to \pi_1(Y, y_0).$$

We recall that, by definitions, $p_{1\#}[\alpha] = [p_1 \circ \alpha]$ and $p_{2\#}[\alpha] = [p_2 \circ \alpha]$. Therefore, from the definition of direct product of groups, we get a group homomorphism

$$\rho\colon \pi_1(X \times Y, (x_0, y_0)) \to \pi_1(X, x_0) \times \pi_1(Y, y_0)$$

defined by

$$\rho([\alpha]) = ([\rho_1 \circ \alpha], [\rho_2 \circ \alpha]).$$

Now, it suffices to show that ρ is a bijection. For this, first note that given a pair of continuous maps $f\colon I \to X$, $g : I \to Y$, there is, by definition of product topology, a continuous map $(f, g)\colon I \to X \times Y$ defined by $(f, g)(t) = (f(t), g(t))$. Conversely, any continuous map $h\colon I \to X \times Y$ defines a pair of continuous maps $\rho_1 \circ h\colon I \to X$, $p_2 \circ h\colon I \to Y$. In fact, it is evident that this correspondence between the set of all continuous maps $h\colon I \to X \times Y$ and the set of pairs $(p_1 \circ h, p_2 \circ h)$ of continuous maps is a 1-1 correspondence. Now, let $([\alpha], [\beta]) \in \pi_1(X, x_0) \times \pi_1(Y, y_0)$. Then the pair (α, β) of loops based at x_0 and y_0 respectively corresponds to the loop $(\alpha, \beta)\colon t \to (\alpha(t), \beta(t))$ in $X \times Y$ based at (x_0, y_0) and defines a map

$$\eta\colon \pi_1(X, x_0) \times \pi_1(Y, y_0) \to \pi_1(X \times Y, (x_0, y_0))$$

given by

$$\eta([\alpha], [\beta]) = [(\alpha, \beta)].$$

To see that η is well defined, note that if $F_1\colon \alpha_1 \sim_{x_0} \alpha_2$, $F_2\colon \beta_1 \sim_{y_0} \beta_2$ are homotopies, then the map $F\colon I \times I \to X \times Y$ defined by

$$F(s, t) = (F_1(s, t), F_2(s, t))$$

is a homotopy from (α_1, β_1) to (α_2, β_2) relative to $\{0,1\}$. Now, it is clear from definitions of ρ and η that they are inverses of each other which means ρ is a bijection. ■

Note: Let $\{X_\alpha : \alpha \in \Lambda\}$ be a family of spaces, and for each $\alpha \in \Lambda$, let $x_\alpha \in X_\alpha$ be a base point. Then the above proof easily generalizes to show that

$$\pi_1(\prod X_\alpha, (x_\alpha)) \cong \prod \pi_1(X_\alpha, x_\alpha).$$

The Fundamental Group of Circle

Recall that the unit circle $\mathbb{S}^1 = \{(x, y) \in \mathbb{R}^2 \mid x^2 + y^2 = 1\}$ is a subspace of $\mathbb{R}^2$. It can also be described in terms of complex numbers, viz., $\mathbb{S}^1 = \{z \in \mathbb{C} : |z| = 1\}$. The description in terms of complex numbers at once implies that $\mathbb{S}^1$ is also a topological group under the multiplication induced by multiplication of complex numbers. We will choose $(1, 0) \in \mathbb{S}^1$ as the base point and denote this point, hence onwards, by p_0. As remarked earlier, we will finally prove that the fundamental group $\pi_1(\mathbb{S}^1, p_0)$ is isomorphic to the additive group $\mathbb{Z}$ of integers. A brief sketch of the proof runs as follows: We use a map $p : \mathbb{R} \to \mathbb{S}^1$, called the **exponential map**, defined by $p(t) = e^{2\pi it}$, $t \in \mathbb{R}$. This map has the nice property that each path $\alpha\colon I \to \mathbb{S}^1$ with initial point $p_0 \in \mathbb{S}^1$ can be lifted to a unique path $\alpha'\colon I \to \mathbb{R}$ such that $\alpha'(0) = 0 \in \mathbb{R}$. If α is a loop at p_0 then $\alpha'(1)$ must be an integer and this integer is called the **degree** of the loop α. We show that if $\alpha \sim_{p_0} \beta$, then deg α = deg β. This yields a map $f\colon \pi_1(\mathbb{S}^1, p_0) \to \mathbb{Z}$ defined by $f([\alpha])$ = deg α. We then complete the proof of our main result by showing that f is indeed an isomorphism.

Each of the steps outlined in the above sketch will be an interesting result in its own right and we now proceed to furnish their complete details. The exponential map $p\colon \mathbb{R} \to \mathbb{S}^1$ defined by $p(t) = e^{2\pi it}$, $t \in \mathbb{R}$ is in fact a continuous onto map which simply wraps the real line $\mathbb{R}$ onto the circle $\mathbb{S}^1$ infinite number of times (Fig. 2.18).

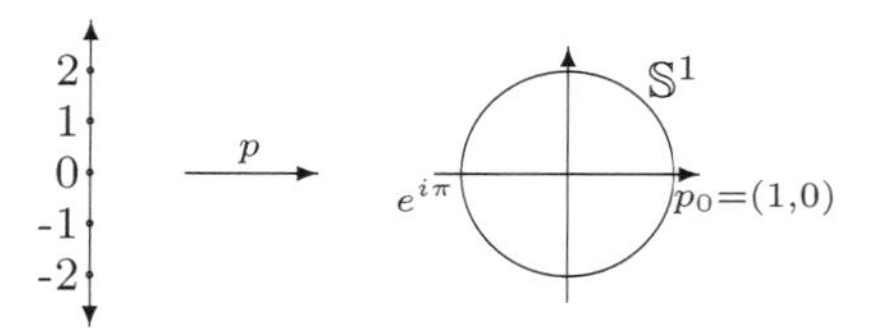

Fig. 2.18

In fact, the interval [0,1] gets wrapped counterclockwise around the circle once with $p(0) = p(1) = p_0$ and the interval $[-1, 0]$ also gets wrapped once but clockwise, and so on. The map, $p|_{(-1/2,1/2)}$ is a homeomorphism from the open interval $(-1/2, 1/2)$ onto $\mathbb{S}^1 - \{e^{i\pi}\}$. We let $\ln\colon \mathbb{S}^1 - \{e^{i\pi}\} \to (-1/2, 1/2)$ be the inverse of this homeomorphism. Also, note that $p(t_1 + t_2) = p(t_1) \cdot p(t_2)$, $t_1, t_2 \in \mathbb{R}$ and $p(t_1) = p(t_2)$ if and only if $t_1 - t_2$ is an integer. It follows that $p^{-1}(p_0) = \mathbb{Z}$.

Definition 2.6.3. *A continuous map $f'\colon X \to \mathbb{R}$ is called a* **lift** *or* **covering** *of a given continuous map $f\colon X \to \mathbb{S}^1$ provided the triangle is commutative, i.e., $p \circ f' = f$.*

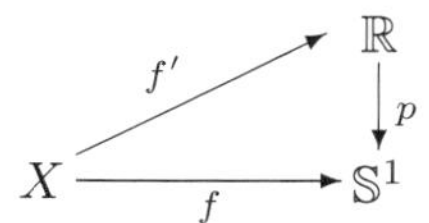

The following lemma says that any two lifts of a given map $f: X \to \mathbb{S}^1$, which agree at a single point of the connected space X, agree everywhere.

Lemma 2.6.4. *Let X be a connected space and $f', f'': X \to \mathbb{R}$ be continuous maps such that $p \circ f' = p \circ f''$. If f', f'' agree at one point $x_0 \in X$, then $f' = f''$.*

Proof. Since $\mathbb{R}$ has the additive group structure, we can define a map $g: X \to \mathbb{R}$ by $g = f' - f''$. Because $p \circ f' = p \circ f''$, we observe that for each $x \in X$,

$$\begin{aligned} p \circ g(x) &= e^{2\pi i g(x)} \\ &= e^{2\pi i (f'(x) - f''(x))} \\ &= e^{2\pi i f'(x)}/e^{2\pi i f''(x)} \\ &= p \circ f'(x)/p \circ f''(x) \\ &= 1. \end{aligned}$$

This means $g(x) \in p^{-1}(1) = \mathbb{Z}$. Since X is connected and g is continuous, the image set $g(X)$ must be a singleton. But $g(x_0) = 0$ implies that $g(x) = 0$, for all $x \in X$, i.e., $f'(x) = f''(x)$, for all $x \in X$. ■

The next two theorems are extremely basic.

Theorem 2.6.5. (The Path Lifting Property). *Let $p: \mathbb{R} \to \mathbb{S}^1$ be the exponential map and $\alpha: I \to \mathbb{S}^1$ be any path such that $\alpha(0) = p_0$. Then there exists a unique lift α' of α such that $\alpha'(0) = 0$.*

Theorem 2.6.6. (The Homotopy Lifting Theorem). *Let $p: \mathbb{R} \to \mathbb{S}^1$ be the exponential map. Suppose $F: I \times I \to \mathbb{S}^1$ is a homotopy such that $F(0,0) = p_0$. Then there exists a unique lift $F': I \times I \to \mathbb{R}$ of F such that $F'(0,0) = 0$.*

Theorem 2.6.5 says that any path in $\mathbb{S}^1$ starting at p_0 can be lifted to a unique path in $\mathbb{R}$ starting at the origin of $\mathbb{R}$. Similarly, Theorem 2.6.6 asserts that any homotopy between two given paths in $\mathbb{S}^1$ starting at p_0, can be lifted to a unique homotopy between the two lifted paths starting at the origin of $\mathbb{R}$. We are now going to prove a general result of which the above two theorems are obvious corollaries.

Theorem 2.6.7. *Let X be a convex compact subset of the Euclidean space $\mathbb{R}^n$ and $x_0 \in X$. Given any continuous map $f: X \to \mathbb{S}^1$ and any $t_0 \in \mathbb{R}$ such that $p(t_0) = f(x_0)$, there exists a unique continuous map $f': X \to \mathbb{R}$ such that $f'(x_0) = t_0$ and $p \circ f' = f$.*

Proof. Note that a convex subset of $\mathbb{R}^n$ is always connected and so, if f', f'' are two lifts of f which agree at the point $x_0 \in X$, then it follows from Lemma 2.6.4 that $f' = f''$. This proves the uniqueness of the lift f' satisfying the condition that $f'(x_0) = t_0$.

To prove the existence of a lift f' of f satisfying the condition that $f'(x_0) = t_0$, we point out that the point x_0 can be taken to be the origin of $\mathbb{R}^n$ without any loss of generality. For, if x_0 is not the origin, then we consider the subspace $L(X)$ of $\mathbb{R}^n$ where $L\colon \mathbb{R}^n \to \mathbb{R}^n$ is the translation map $y \to y - x_0$ which is clearly a homeomorphism preserving compactness and convexity of X. The subspace $L(X)$ now contains the origin of $\mathbb{R}^n$. If the existence of a lift for $L(X)$ is now assumed, then let $k : L(X) \to \mathbb{R}$ be a lift of $f \circ L^{-1}$ such that $k(0) = t_0$. Here, we are treating L as a map from X onto $L(X)$ only.

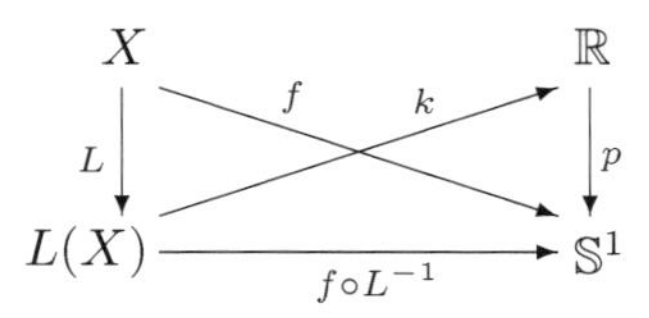

Then the composite map $k \circ L\colon X \to \mathbb{R}$ maps x_0 to t_0 and has the property that

$$p \circ (k \circ L) = (p \circ k) \circ L = (f \circ L^{-1}) \circ L = f.$$

Hence we now assume that X is a compact convex subset of $\mathbb{R}^n$ containing the origin. Since X is compact, the continuous map $f\colon X \to \mathbb{S}^1$ is uniformly continuous and so there is an $\epsilon > 0$ such that if $\|x - x'\| < \epsilon$ then $|f(x) - f(x')| < 2$. Here, $\|\cdot\|$ is the Euclidean norm of $\mathbb{R}^n$ and $|f(x)|$ denotes the modulus of the complex number $f(x)$. Since X is bounded, we can find an integer n such that $\|x\|/n < \epsilon$ for all $x \in X$. This means for each $j = 0, 1, 2, \ldots, n-1$, and all $x \in X$

$$\|\frac{(j+1)x}{n} - \frac{jx}{n}\| = \|x\|/n < \epsilon.$$

Therefore,

$$|f(\tfrac{(j+1)x}{n}) - f(\tfrac{jx}{n})| < 2.$$

It follows that the quotient $f\left(\frac{(j+1)x}{n}\right)/f(\frac{jx}{n})$ is a point in $\mathbb{S}^1 - \{e^{i\pi}\}$. Let $g_j\colon X \to \mathbb{S}^1 - \{e^{i\pi}\}$ for $0 \le j \le n-1$ be the maps defined by

$$g_j(x) = f(\tfrac{(j+1)x}{n})/f(\tfrac{jx}{n}).$$

Then for all $x \in X$, we can write $f(x)$ as the product

$$f(x) = f(0) \cdot g_0(x).g_1(x) \cdots g_{n-1}(x).$$

Let us define $f'\colon X \to \mathbb{R}$ by

$$f'(x) = t_0 + \ln g_0(x) + \ln g_1(x) + \cdots + \ln g_{n-1}(x).$$

Then f', being the sum of n continuous functions, is continuous and has the property that $f'(0) = t_0$, $p \circ f' = f$. ∎

Remark 2.6.8. When we consider X in the above theorem to be the unit interval $I = [0, 1] \subset \mathbb{R}$, we obtain the Path Lifting Property (Theorem 2.6.5) for the exponential map $p\colon \mathbb{R} \to \mathbb{S}^1$. On the other hand, if we take the unit square $I \times I \subset \mathbb{R}^2$ as X, we obtain the Homotopy Lifting Property for the exponential map recorded as Theorem 2.6.6.

Now, let $\alpha\colon I \to \mathbb{S}^1$ be a loop in $\mathbb{S}^1$ based at the point $p_0 = (1, 0) \in \mathbb{S}^1$. Let $\alpha'\colon I \to \mathbb{R}$ be the lift of α such that $\alpha'(0) = 0$. Such a lift is guaranteed by the path lifting property of the exponential map. Since $\alpha'(1) \in \mathbb{R}$ projects to the point (1,0) under the exponential map, it must be an integer. We call this integer $\alpha'(1)$ to be the **degree** of the loop α. Since there is only one lift of α starting at the origin of $\mathbb{R}$, $\alpha'(1)$ depends only on the loop α. If α, β are two equivalent loops in $\mathbb{S}^1$ based at p_0, then how are their degrees related? The next result answers this question.

Theorem 2.6.9. *Let α, β be two equivalent loops in $\mathbb{S}^1$ based at p_0. Then degree α = degree β.*

Proof. Since $\alpha \sim_{p_0} \beta$, there is a homotopy $F\colon I \times I \to \mathbb{S}^1$ such that $F(s, 0) = \alpha(s)$, $F(s, 1) = \beta(s)$, $F(0, t) = p_0 = F(1, t)$, for all $s, t \in I$. By the Homotopy Lifting Property for the exponential map p, there exists a homotopy $F'\colon I \times I \to \mathbb{R}$ such that $F'(0, 0) = 0$ and $p \circ F' = F$ (Fig. 2.19).

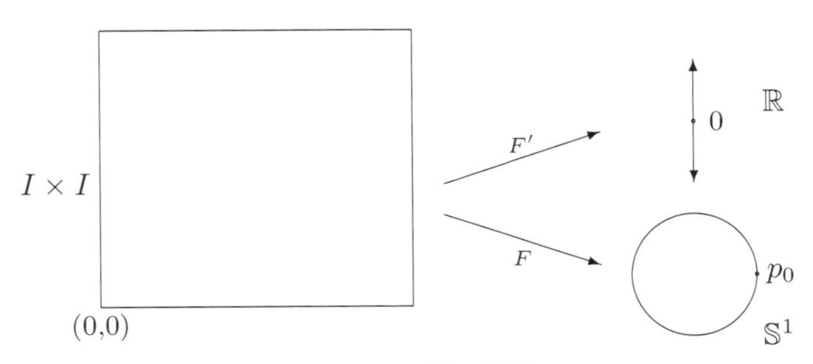

Fig. 2.19

The paths $\alpha', \beta'\colon I \to \mathbb{R}$ defined by $\alpha'(s) = F'(s, 0)$, $\beta'(s) = F'(s, 1)$ are evidently lifts of α and β respectively. Now, $\alpha'(0) = F'(0, 0) = 0$, and $p \circ F'(0, t) = F(0, t) = p_0$. It follows that $F'(0, t)$ is some integer for all $t \in I$. Since I is connected and $F'(0, 0) = 0$, we find that $F'(0, t) = 0$, $\forall t \in I$. In particular, $F'(0, 1) = 0 = \beta'(0)$, i.e., $\alpha'(0) = \beta'(0) = 0$. Hence, degree $\alpha = \alpha'(1)$, degree $\beta = \beta'(1)$. Since F maps the vertical line $\{(1, t) \mid t \in I\}$ of $I \times I$ into the point p_0, F' must map this line into a subset of $p^{-1}(p_0) = \mathbb{Z}$.

Moreover, since vertical line is connected, F' is continuous and $\mathbb{Z}$ is discrete, we find that $F'(1,t)$ is some integer, for all $t \in I$. This means degree $\alpha = \alpha'(1) = F'(1,0) = F'(1,1) = \beta'(1) =$ degree β. ■

The above theorem tells us that if we define a map $\deg\colon \pi_1(\mathbb{S}^1, p_0) \to \mathbb{Z}$ by $\deg([\alpha]) =$ degree α, then deg is a well-defined map. We call this map as the **degree map**. Now, we will prove our final step of the main theorem.

Theorem 2.6.10. *The degree map* $\deg\colon \pi_1(\mathbb{S}^1, p_0) \to \mathbb{Z}$ *is an isomorphism from the fundamental group of circle to the additive group* $\mathbb{Z}$ *of integers.*

Proof. The degree map is a homomorphism: Let $[\alpha], [\beta]$ be any two elements of $\pi_1(\mathbb{S}^1, p_0)$. Let $\alpha', \beta'\colon I \to \mathbb{R}$ be the unique paths starting from origin of $\mathbb{R}$ which are lifts of α and β, respectively. Define a new path $\omega\colon I \to \mathbb{R}$ by

$$\omega(t) = \begin{cases} \alpha'(2t) & 0 \le t \le 1/2 \\ \alpha'(1) + \beta'(2t-1) & 1/2 \le t \le 1. \end{cases}$$

Then obviously ω also begins at origin and

$$\begin{aligned} p \circ \omega(t) &= \begin{cases} p \circ \alpha'(2t) & 0 \le t \le 1/2 \\ p(\alpha'(1)) \cdot (p \circ \beta')(2t-1) & 1/2 \le t \le 1 \end{cases} \\ &= \begin{cases} \alpha(2t) & 0 \le t \le 1/2 \\ \beta(2t-1) & 1/2 \le t \le 1 \end{cases} \\ &= (\alpha * \beta)(t) \ \ \forall t \in I. \end{aligned}$$

Thus, ω is a lift of $\alpha * \beta$ starting from the origin. Consequently,

$$\begin{aligned} \deg([\alpha] * [\beta]) &= \deg([\alpha * \beta]) \\ &= \omega(1) \\ &= \alpha'(1) + \beta'(1) \\ &= \deg([\alpha]) + \deg([\beta]). \end{aligned}$$

The degree map is 1-1: Suppose $[\alpha], [\beta] \in \pi_1(\mathbb{S}^1, p_0)$ such that $\deg([\alpha]) = \deg([\beta])$. This means if $\alpha'\colon I \to \mathbb{R}$, $\beta' : I \to \mathbb{R}$ are lifts of α, β respectively starting from origin, then $\alpha'(1) = \beta'(1)$. Define a homotopy $H\colon I \times I \to \mathbb{R}$ by

$$H(s,t) = (1-t)\alpha'(s) + t\beta'(s), (s,t) \in I \times I.$$

Then H is a homotopy between α' and β' relative to {0,1}. It follows that $p \circ H\colon I \times I \to \mathbb{S}^1$ is a homotopy between α and β relative to {0,1}, i.e., $[\alpha] = [\beta]$.

The degree map is onto: Let $n \in \mathbb{Z}$. Define the loop $\gamma\colon I \to \mathbb{S}^1$ by

$$\gamma(t) = e^{2\pi i n t}$$

The path $\gamma' \colon I \to \mathbb{R}$ defined by $\gamma'(t) = nt$, starts at the origin of $\mathbb{R}$ and lifts the path γ. Therefore, $\deg([\gamma]) = \gamma'(1) = n$. Thus, we have proved that the degree map is an isomorphism. ■

The above theorem is the first nice result giving a nontrivial example of the fundamental group, viz., *the fundamental group of the circle is isomorphic to the additive group* $\mathbb{Z}$ *of integers*. As we shall see next, all other examples of fundamental groups that we discuss at this stage are based upon this example and the results proved earlier.

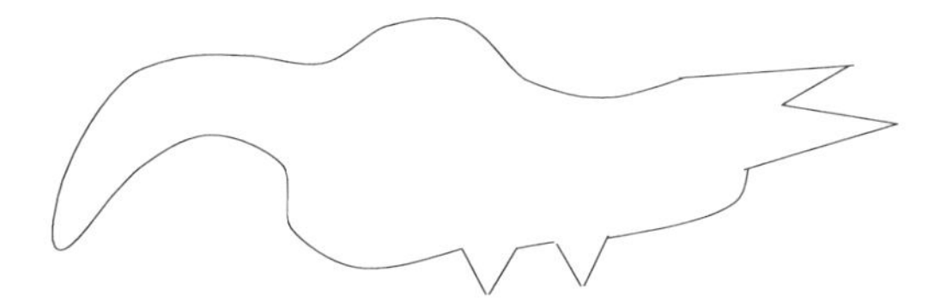

Fig. 2.20: Any closed curve without self-intersection

Remark 2.6.11. We have already proved (Theorem 2.4.21) that the fundamental groups of two homeomorphic spaces are isomorphic. Consequently, any space which is homeomorphic to the unit circle $\mathbb{S}^1$ will have its fundamental group as the additive group $\mathbb{Z}$ of integers. For example all closed curves which have the subspace topology induced from $\mathbb{R}^2$ and, as shown above (Fig. 2.20), have $\mathbb{Z}$ as their fundamental groups. In fact, take a pencil, put it at a point p in the plane $\mathbb{R}^2$, move your hand tracing some curve in any manner you like, but do not lift your hand and do not cross the curve already traced anywhere, finally come back to the point p. Then the curve so described is homeomorphic to the circle $\mathbb{S}^1$ and will have its fundamental group isomorphic to the additive group $\mathbb{Z}$ of integers.

One must take time and be convinced as to why, each of these spaces is homeomorphic to the unit circle $\mathbb{S}^1$. We also note that if we remove just one point from any of the closed curves, then the remaining space is homeomorphic to the open interval (0,1) and so is contractible. In other words, the fundamental group of the remaining space is trivial. Each of this fact is interesting, but far more interesting is the following: if two path-connected spaces are not homeomorphic but, more generally, are of the same homotopy type, even then they will have isomorphic fundamental groups (Theorem 2.4.31). We now consider results of this kind.

Fundamental Group of the Punctured Plane. By the **punctured plane** X, we mean a topological space which is homeomorphic to the plane $\mathbb{R}^2$ with some point $p \in \mathbb{R}^2$ deleted, i.e., $X = \mathbb{R}^2 - \{p\}$. It is, in fact, trivial to see that X is homeomorphic to the punctured plane $Y = \mathbb{R}^2 - (0, 0)$. Now, we point out that the unit circle $\mathbb{S}^1 \subset \mathbb{R}^2 - \{(0,0)\}$ is a strong deformation retract of Y. The homotopy $F\colon Y \times I \to Y$ defined by

$$F(x,t) = (1-t)x + t\frac{x}{\|x\|}$$

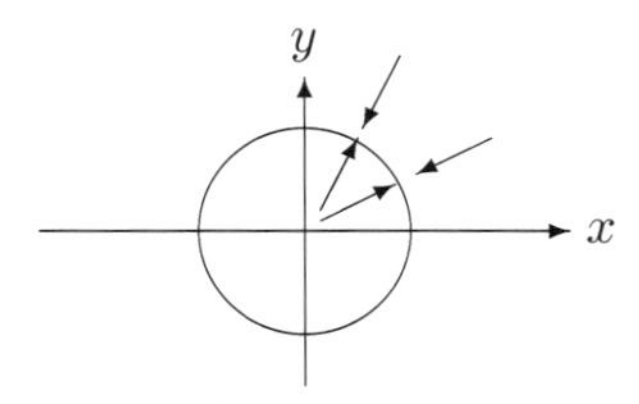

Fig. 2.21

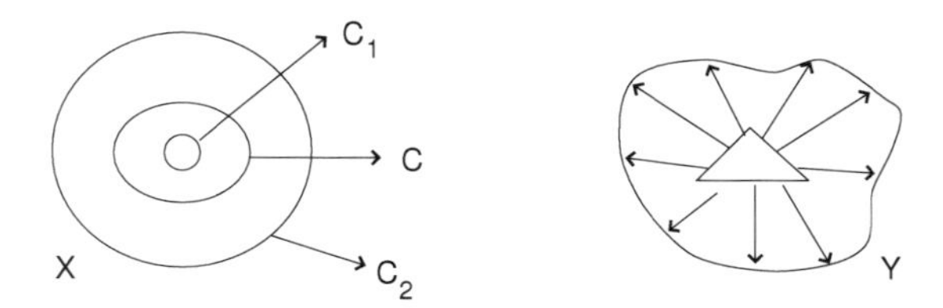

Fig. 2.22(a) and 2.22(b)

is a strong deformation retraction (Fig. 2.21). This means X is of the same homotopy type as the unit circle $\mathbb{S}^1$. Therefore, $\pi_1(X) = (\mathbb{Z}, +)$.

If we consider an annulus X (Fig. 2.22(a)) surrounded by two concentric circles C_1 and C_2, then any concentric circle C lying in the annulus is a strong deformation retract of X and hence, $\pi_1(X) = \mathbb{Z}$.

One can give a strong deformation retraction as follows: From the centre O draw a half line in each direction. Each such line intersects the circle C in a unique point, say, $r(x)$. Let $r\colon X \to C$ be the mapping which maps all points of the half line to the corresponding point $r(x)$. Now, define $H\colon X \times I \to X$ by $H(x,t) = (1-t)x + t \cdot r(x)$. Note that the space Y of Fig. 2.22(b) is not exactly an annulus, but has circle as its deformation retract. Hence, $\pi_1(Y) = \mathbb{Z}$.

Fundamental Group of Cylinder. Recall that a topological space X is said to be **cylinder** if it is homeomorphic to the subspace (see Fig. 2.23)

$$Y = \{(x,y,z) \in \mathbb{R}^3 : x^2 + y^2 = 1, -k \leq z \leq k, k \in \mathbb{R}\}.$$

Let $C = \{(x,y,0) \in \mathbb{R}^3 : x^2 + y^2 = 1\}$. Then C is a strong deformation retract of X under the homotopy $F((x,y,z),t) = (x,y,(1-t)z)$ and hence, $\pi_1(X) = \mathbb{Z}$.

Fundamental Group of Torus. We know that a torus T (a surface) is homeomorphic to the product $\mathbb{S}^1 \times \mathbb{S}^1$ of two circles $\mathbb{S}^1$, whereas the n-dimensional

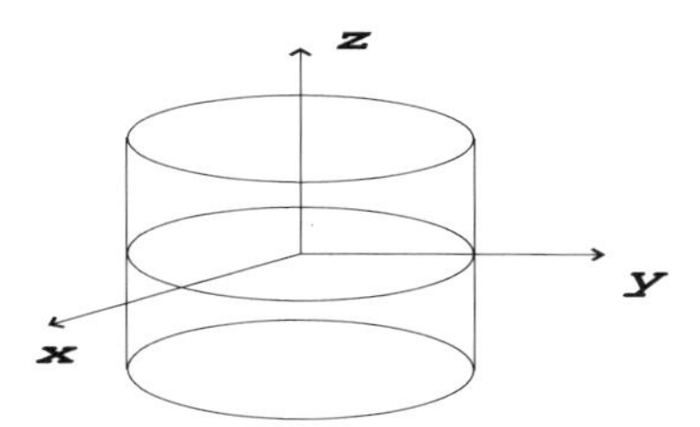

Fig. 2.23: Circle is a strong deformation retract

torus $T^n \cong \mathbb{S}^1 \times \mathbb{S}^1 \times \cdots \times \mathbb{S}^1$ (n copies). Each of them is a path-connected space. It follows, therefore, that $\pi_1(T) \cong \mathbb{Z} \times \mathbb{Z}$ and $\pi_1(T^n) \cong \mathbb{Z} \times \mathbb{Z} \times \cdots \times \mathbb{Z}$ (n copies). In the case of torus, two loops α, β based at a point (a, b) are the generators of $\pi_1(T)$ (see Fig. 4.7).

The next question is about the fundamental groups of other compact surfaces such as the Möbius band, the projective plane, the Klein bottle, etc., which we have not discussed so far. More generally, if we take an arbitrary compact surface X, then it is well-known that X is homeomorphic to a connected sum of tori or a connected sum of projective planes (see Section 1.1.7). Therefore, the general problem is to determine the fundamental groups of all compact surfaces, and this can be done. In order to answer these questions, however, we first need to determine the fundamental groups of connected graphs, which are 1-dimensional simplicial polyhedra. The solution is provided, among other things, by the following powerful result (see W.Massey [13] for a proof and other details), which gives the fundamental group of the union of two subspaces if we know the fundamental groups of individual subspaces. This result also extends Theorem 2.5.3.

Theorem 2.6.12 (Seifert-Van Kampen Theorem). *Let $X = U_1 \cup U_2$ with U_1, U_2 and $U_1 \cap U_2$ all nonempty path-connected open subsets. Then $\pi_1(X)$ is the free product of $\pi_1(U_1)$ and $\pi_2(U_2)$ amalgamated along $\pi_1(U_1 \cap U_2)$ via homomorphisms induced by the inclusion maps. In particular, if $U_1 \cap U_2$ is simply connected, then $\pi_1(X)$ is the free product of $\pi_1(U)$ and $\pi_2(U)$.*

Example 2.6.13. If $X = \mathbb{S}^1 \vee \mathbb{S}^1$ is the one-point union of two circles (figure of eight), then it follows from the above theorem and the fact that $\pi_1(\mathbb{S}^1) \cong \mathbb{Z}$ that $\pi_1(X)$ is the free group on two generators. This follows from the Van Kampen theorem and the fact that the figure of eight is the union of two circles with whiskers (open sets) having contractible intersection. Similarly, if X is a one-point union of three circles, then $\pi_1(X)$ is free group on 3-generators, etc.

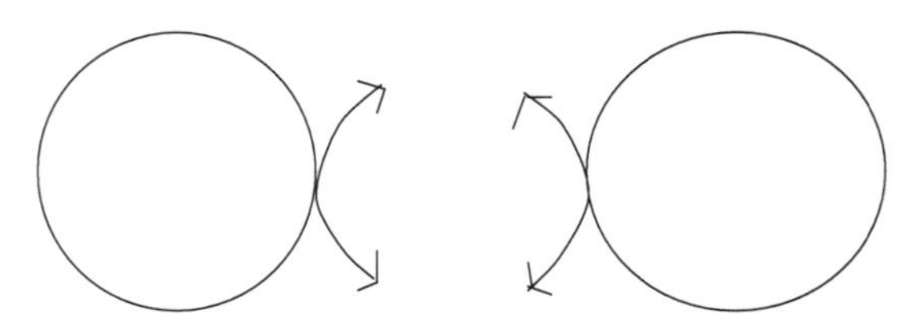

Fig. 2.24: Two circles with whiskers

The fundamental groups of projective planes, Klein bottle etc. are determined using Van Kampen theorem (see amongst the exercises given below).

Exercises

1. Show that if X is an indiscrete space, then $\pi_1(X) = 0$.
2. Suppose we remove the z-axis from $\mathbb{R}^3$ and denote the resulting space as X. Compute $\pi_1(X)$.
3. Let $f\colon X \to \mathbb{S}^n$ be a map which is not onto. Prove that f is null homotopic. (**Hint**: The image of f is contained in an Euclidean space which is contractible.)
4. Prove that any two maps $f, g\colon X \to C(Y)$ where $C(Y)$ is the cone over Y, are homotopic.
5. Prove that a map $f\colon X \to Y$ is null homotopic if and only if f can be extended to a map $\hat{f}\colon C(X) \to Y$.
6. Give an example of a space which is simply connected, but not contractible.
7. Let $p \in \mathbb{S}^1$ where $\mathbb{S}^1$ is the unit circle. Prove that $\mathbb{S}^1 \times \{p\}$ is a retract of $\mathbb{S}^1 \times \mathbb{S}^1$, but is not a deformation retract.
8. Let X be a 2-disk punctured at two interior points. Prove that a one-point union of two circles is embedded as a deformation retract of X and hence $\pi_1(X)$ is isomorphic to a free group on two generators.
9. Show that the figure of eight is a strong deformation retract of the unit disc punctured at two interior points.
10. Let p, q, r be three distinct points in the plane $\mathbb{R}^2$ and $X = \mathbb{R}^2 - \{p, q, r\}$. Use van-Kampen theorem to prove that $\pi_1(\mathbb{R}^2 - \{p, q, r\})$ is a free group on three generators.

11. Let $f\colon X \to Y$ be a homotopy equivalence. Show that if we remove a point $x_0 \in X$ then $X - \{x_0\}$ need not be homotopically equivalent to $Y - \{f(x_0)\}$.

12. Take the plane $\mathbb{R}^2$. Remove a point $p \in \mathbb{R}^2$. Then the remaining space $X = \mathbb{R}^2 - \{p\}$ is of the same homotopy type as circle $\mathbb{S}^1$. Now remove one point q from $\mathbb{S}^1$, then the remaining space $Y = \mathbb{S}^1 - \{q\}$ is contractible. However, if we remove two distinct points p and q from $\mathbb{R}^2$, then the remaining space $\mathbb{R}^2 - \{p, q\}$ is not contractible. Explain if something is wrong.

13. Prove that the fundamental group of the punctured n-space $\mathbb{R}^n - \{p\}$, $(n \geq 3)$, $p \in \mathbb{R}^n$, is trivial. Discuss the cases $n = 1, 2$.

14. Compute the fundamental group of the space $X = \mathbb{R}^3 - \{p, q\}$ where p, q are two distinct points of $\mathbb{R}^3$.

15. Prove that the cylinder C of height one and the Möbius band M are not homeomorphic, but they have isomorphic fundamental groups. (**Hint**: Möbius band has a circle as its deformation retract.)

16. Let X be a punctured torus. Prove that a one-point union of two circles is embedded in X as a deformation retract. Use this exercise and van-Kampen theorem to determine the fundamental group of torus.

17. Let X be the projective plane $\mathbb{RP}^2$ punctured at a point. Prove that circle is embedded in X as a deformation retract. Use this exercise and the van-Kampen theorem to determine the fundamental group of $\mathbb{RP}^2$.

18. Let X be the Klein bottle punctured at one point. Show that a one-point union of two circles is embedded in X as a strong deformation retract. Hence or otherwise determine the fundamental group of the Klein bottle.

19. Let G be a topological group with identity element e. Prove that $\pi_1(G, e)$ is abelian. (**Hint**: In the group $\pi_1(G, e)$, define another binary operation by $[\alpha] \cdot [\beta] = [\alpha \cdot \beta]$, where $(\alpha \cdot \beta)(t) = \alpha(t)\beta(t) \in G$, $\forall\, t \in I$. Then $\pi_1(G, e)$ is a group with respect to this operation also. Now verify directly that for arbitrary loops

$$([\alpha] \circ [\beta]) \cdot ([\gamma] \circ [\delta]) = ([\alpha] \cdot [\gamma]) \circ ([\beta] \cdot [\delta]).$$

Putting $\beta = \gamma =$ constant loop e, we find that $[\alpha] \circ [\delta] = [\alpha] \cdot [\delta]$. Now put $\alpha = \delta = e$.)

Chapter 3

Simplicial Complexes

3.1 Finite Simplicial Complexes

The word "simplicial" is derived from the word "simplex", and the notion of a simplex is fundamental not only to the subject of topology, but to several other branches of mathematics, like graph theory, approximation theory, real analysis, etc. We will, therefore, first proceed to recall (see Section 1.1.1) the meaning and definition of a simplex.

Let $\mathbb{R}^n$ denote the Euclidean space with its Euclidean topology, and recall its properties as an n-dimensional vector space over the field of real numbers. A subset $S = \{x_1, \ldots, x_k\}$ of $\mathbb{R}^n$ is said to be **linearly independent** if for arbitrary scalars $\alpha_1, \alpha_2, \ldots, \alpha_k$,

$$\sum_{i=1}^{k} \alpha_i x_i = 0 \Rightarrow \alpha_i = 0 \quad \forall\, i = 1, \ldots, k.$$

From this notion of linear independence, we now introduce the derived notion of geometric independence of a set of points $\{a_0, a_1, \ldots, a_k\}$ lying in $\mathbb{R}^n$ as follows:

Definition 3.1.1. *A set $A = \{a_0, a_1, \ldots, a_k\}$, $k \geq 1$, of points of $\mathbb{R}^n$ is said to be* **geometrically independent** *if and only if the set $S = \{a_1 - a_0, a_2 - a_0, \ldots, a_k - a_0\}$ of vectors of $\mathbb{R}^n$ is linearly independent. A set having only one point will be assumed to be geometrically independent.*

As an example, consider the set $\{a_0, a_1, a_2\}$ of three points of the plane $\mathbb{R}^2$. This will be a geometrically independent set if and only if $\{a_1 - a_0, a_2 - a_0\}$ is linearly independent. However, in $\mathbb{R}^2$, we know that a set having two vectors is linearly dependent if and only if one of the vectors is a scalar multiple of the other vector. Thus, the set $\{a_0, a_1, a_2\}$, as shown in Fig. 3.1(a), is geometrically independent whereas the one, as shown in Fig. 3.1(b), is not geometrically independent.

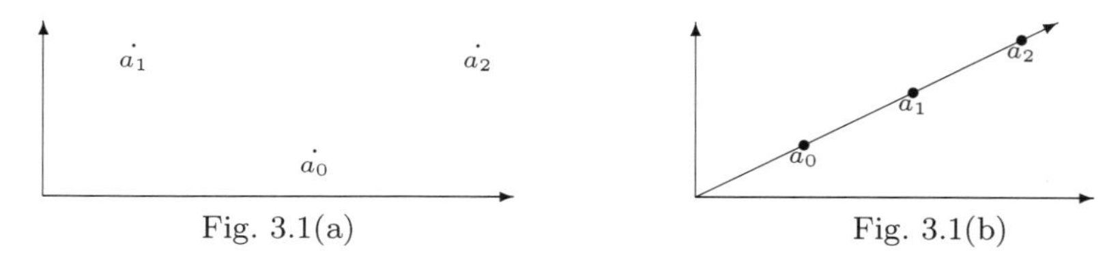

Fig. 3.1(a) Fig. 3.1(b)

Thus, in $\mathbb{R}^n$, for n sufficiently large, a set having two points will be geometrically independent if and only if the two points are distinct; a set having three points will be geometrically independent if and only if they do not lie on a line; a set having four points will be geometrically independent if and only if all the four points do not lie on a plane (2-dimensional). The four points a_0, a_1, a_2, a_3, as shown in Fig. 3.1(c), are geometrically independent, but the ones in Fig. 3.1(d) are geometrically dependent.

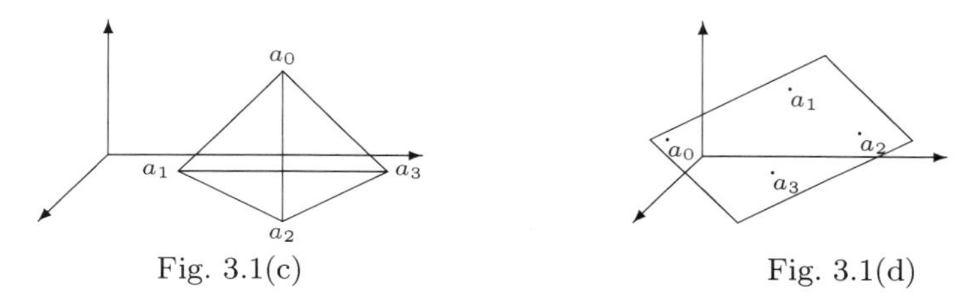

Fig. 3.1(c) Fig. 3.1(d)

Let us understand what is a "line" in $\mathbb{R}^3$, what is a "plane" in $\mathbb{R}^3$, etc.? In fact, by a *line* in $\mathbb{R}^n$, we mean a *translate* of some one-dimensional subspace of $\mathbb{R}^n$. In other words, a subset L of $\mathbb{R}^n$ is said to be a *line* (or, *hyperplane of dimension one*) if there exists a one-dimensional subspace V^1 of $\mathbb{R}^n$ and a vector $x \in \mathbb{R}^n$ such that

$$L = x + V^1.$$

Similarly, a subset P of $\mathbb{R}^n$ is said to be a *plane* (or, *hyperplane of dimension two*) if there exists a point $x \in \mathbb{R}^n$ and a two dimensional subspace V^2 of $\mathbb{R}^n$ such that

$$P = x + V^2.$$

More generally, we have

Definition 3.1.2. *A subset H of $\mathbb{R}^n$, n sufficiently large, is said to be a k-dimensional* **hyperplane** *of $\mathbb{R}^n$ if we can find a k-dimensional subspace V^k of $\mathbb{R}^n$ and a vector x of $\mathbb{R}^n$ such that*

$$H = x + V^k.$$

The next result, which generalizes the observations made earlier explains the reason as to why we use the term "geometrically independent".

Proposition 3.1.3. *A subset $A = \{a_0, a_1, \ldots, a_k\}$, $k \geq 1$, of $\mathbb{R}^n$ is geometrically independent if and only if all the points of A do not lie on a hyperplane of dimension $(k-1)$.*

Proof. Assume $A = \{a_0, a_1, \ldots, a_k\}$ is geometrically independent, i.e., the set $\{a_1 - a_0, a_2 - a_0, \ldots, a_k - a_0\}$ is linearly independent. If possible, let H^{k-1} be a hyperplane of $\mathbb{R}^n$ of dimension $(k-1)$ which contains $a_0, a_1, \ldots, a_k$. This means, there is a vector $x \in \mathbb{R}^n$ and a subspace V^{k-1} of dimension $(k-1)$ such that $a_0, a_1, \ldots, a_k \in x + V^{k-1}$. This is equivalent to saying that $a_0 - x, a_1 - x, \ldots, a_k - x \in V^{k-1}$. By subtracting the first vector from the others, we note that

$$a_1 - a_0, \ldots, a_k - a_0 \in V^{k-1}.$$

This is, however, a contradiction because a $(k-1)$-dimensional subspace cannot contain a linearly independent set having k vectors.

Conversely, suppose the set $\{a_1 - a_0, a_2 - a_0, \ldots, a_k - a_0\}$ is linearly dependent. Then the subspace of $\mathbb{R}^n$ generated by these vectors is of dimension less than or equal to $k-1$. Hence we can assume by adding some additional vectors, if necessary, that there is a subspace V^{k-1} of dimension $k-1$ which contains all the vectors $a_1 - a_0, \ldots, a_k - a_0$. Now, the hyperplane $a_0 + V^{k-1} = H^{k-1}$ obviously contains a_0. Moreover, since $a_i - a_0$ are in V^{k-1} for each $i = 1, 2, \ldots, k$, $a_i \in H^{k-1}$ for all $i = 0, 1, 2, \ldots, k$. Thus, all the vectors $a_0, a_1, \ldots, a_k$ lie on a hyperplane of dimension $k-1$. ■

Remark 3.1.4. It follows from the above proposition that if the set $\{a_0, a_1, \ldots, a_k\}$ of points of $\mathbb{R}^n$ is geometrically independent, then there is a hyperplane which contains all these points – it is just the hyperplane $a_0 + V^k$, where V^k is the vector subspace of $\mathbb{R}^n$ generated by the linearly independent set $\{a_1 - a_0, a_2 - a_0, \ldots, a_k - a_0\}$.

The geometric-independence condition on a set can be expressed algebraically and directly in terms of the elements of that set as follows:

Proposition 3.1.5. *A subset $S = \{a_0, a_1, \ldots, a_k\}$ of $\mathbb{R}^n$ is geometrically independent if and only if for arbitrary reals α_i,*

$$\text{(i) } \textstyle\sum_{i=0}^{k} \alpha_i a_i = 0 \text{ and (ii) } \sum_{i=0}^{k} \alpha_i = 0$$

imply $\alpha_i = 0$, for all $i = 0, 1, 2, \ldots, k$.

Proof. Suppose S is geometrically independent, i.e., the set $\{a_1 - a_0, a_2 - a_0, \ldots, a_k - a_0\}$ is linearly independent. Then, putting the value of α_0 from (ii) in (i), we see that

$$(-\alpha_1 - \alpha_2 - \ldots - \alpha_k)a_0 + \sum_{i=1}^{k} \alpha_i a_i = 0.$$

This means

$$\sum_{i=1}^{k} \alpha_i(a_i - a_0) = 0,$$

which implies $\alpha_i = 0$, for all $i = 1, 2, \ldots, k$. But, then it also follows from (ii) that $\alpha_0 = 0$.

Conversely, suppose the given condition is satisfied, and assume

$$\sum_{i=1}^{k} \beta_i(a_i - a_0) = 0,$$

where β_i's are some real numbers. Then we have

$$-\left(\sum_{i=1}^{k} \beta_i\right) a_0 + \sum_{i=1}^{k} \beta_i a_i = 0.$$

Putting $\beta_0 = -\sum_1^k \beta_i$, we conclude that $\sum_0^k \beta_i a_i = 0$ as well as $\sum_0^k \beta_i = 0$. Hence, by our hypothesis, $\beta_i = 0$, for all $i = 0, 1, \ldots, k$. This proves the desired result. ■

Any set of geometrically independent points of $\mathbb{R}^n$ determines a unique hyperplane of $\mathbb{R}^n$ passing through all the points of that set. This follows from the following:

Proposition 3.1.6. *Let $A = \{a_0, a_1, \ldots, a_k\}$ be a geometrically independent set in $\mathbb{R}^n$. Then there is a unique k-dimensional hyperplane which passes through all the points of A.*

Proof. Existence is immediate because, if V^k is the subspace of $\mathbb{R}^n$ spanned by the linearly independent set $\{a_1 - a_0, a_2 - a_0, \ldots, a_k - a_0\}$, then by the previous remark, the hyperplane $a_0 + V^k$ evidently passes through all the points $a_0, a_1, \ldots, a_k$.

For uniqueness, suppose there are two hyperplanes H^k and F^k of dimension k which contain the points $a_0, a_1, \ldots, a_k$. This means there are two subspaces V^k and W^k of dimension k each and two vectors $x, y \in \mathbb{R}^n$ such that

$$H^k = x + V^k$$
$$F^k = y + W^k.$$

Since $a_0, a_1, \ldots, a_k \in H^k$, it follows that $a_1 - a_0, a_2 - a_0, \ldots, a_k - a_0 \in V^k$. Similarly, $a_1 - a_0, a_2 - a_0, \ldots, a_k - a_0 \in W^k$. Since the set $\{a_1 - a_0, a_2 - a_0, \ldots, a_k - a_0\}$ is linearly independent and V^k, W^k are vector spaces each of dimension k containing that set, we must have $V^k = W^k$. Consequently, H^k and F^k are two cosets of the same subspace V^k. Since the intersection of two cosets is either identical or disjoint, and here the two cosets contain the points $a_0, a_1, \ldots, a_k$, we conclude that $H^k = F^k$. ■

The next proposition gives a nice representation of the points of a hyperplane as a linear combination of geometrically independent points. We have

Proposition 3.1.7. *Let $A = \{a_0, a_1, \ldots, a_k\}$ be a geometrically independent set in $\mathbb{R}^n$. Then each point of the hyperplane passing through all the points of A can be expressed uniquely as*

$$h = \sum_{i=0}^{k} \alpha_i a_i \text{ where } \sum_{i=0}^{k} \alpha_i = 1.$$

Proof. We know that the hyperplane passing through the set A is given by the set of all vectors h satisfying the condition

$$h = a_0 + \sum_{i=1}^{k} \alpha_i (a_i - a_0),$$

where α_i's are real numbers which are unique. This means h can be expressed uniquely as

$$\begin{aligned} h &= (1 - \sum_{i=1}^{k} \alpha_i) a_0 + \sum_{i=1}^{k} \alpha_i a_i \\ &= \alpha_0 a_0 + \sum_{i=1}^{k} \alpha_i a_i \\ &= \sum_{i=0}^{k} \alpha_i a_i, \end{aligned}$$

where $\sum_{i=0}^{k} \alpha_i = 1$. ∎

Remark 3.1.8. If $\{a_0, a_1, \ldots, a_k\}$ is a geometrically independent set in $\mathbb{R}^n$, $k \leq n$, then we can give a one-to-one correspondence between the vector space $\mathbb{R}^k$ and the hyperplane H^k passing through $a_0, a_1, \ldots, a_k$. The desired one-to-one correspondence will result as follows: Any element of $\mathbb{R}^k$ would be a set of k-tuples $(\alpha_1, \alpha_2, \ldots, \alpha_k)$ of real numbers. Let this k-tuple correspond to the point $(1 - \sum_{i=1}^{k} \alpha_i) a_0 + \sum_{i=1}^{k} \alpha_i a_i$ of H^k. It is also not difficult to see that this one-to-one correspondence, being a translation by some vector, is distance preserving and consequently is a homeomorphism.

The following concept of barycentric coordinates is important.

Definition 3.1.9. *Let $A = \{a_0, a_1, \ldots, a_k\}$ be a geometrically independent set of points of $\mathbb{R}^n$, and let h be any point in the unique hyperplane passing through $a_0, \ldots, a_k$. Write h as*

$$h = \sum_{i=0}^{k} \alpha_i a_i \text{ where } \sum_{i=0}^{k} \alpha_i = 1.$$

Then, the real numbers $\alpha_0, \alpha_1, \ldots, \alpha_k$, which are uniquely determined by A, are called the **barycentric coordinates** *of the point h with respect to the set A.*

Note that barycentric coordinates of a point are real numbers which could be positive, zero or negative. In the Euclidean space $\mathbb{R}^3$, for example, let $a_0 = (2, 0, 0), a_1 = (0, 3, 0)$. Then the one-dimensional hyperplane of $\mathbb{R}^3$ containing a_0, a_1 is the straight line passing through a_0 and a_1 (Fig. 3.2(a)). All the points h on the line, lying strictly between the two points a_0 and a_1, will have positive barycentric coordinates. The barycentric coordinates of a_0 are $(1, 0)$ whereas that of a_1 are $(0, 1)$. If $h = \alpha_0 a_0 + \alpha_1 a_1$, where $\alpha_0 + \alpha_1 = 1$, lies on the left of a_0, then the barycentric coordinates of h are α_0, α_1 where $\alpha_0 > 0$ but $\alpha_1 < 0$. But if h lies on the right of the point a_1, then $\alpha_0 < 0$ and $\alpha_1 > 0$.

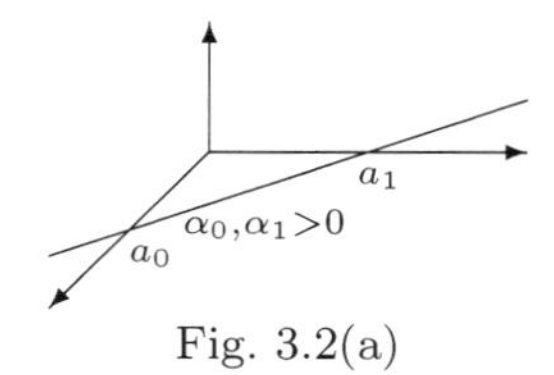

Fig. 3.2(a)

Consider a geometrically independent set $\{a_0, a_1, a_2\}$ in any $\mathbb{R}^n$, $n \geq 2$. Then the hyperplane of $\mathbb{R}^n$ containing a_0, a_1, a_2 is the unique plane passing through a_0, a_1 and a_2 (Fig. 3.2(b)).

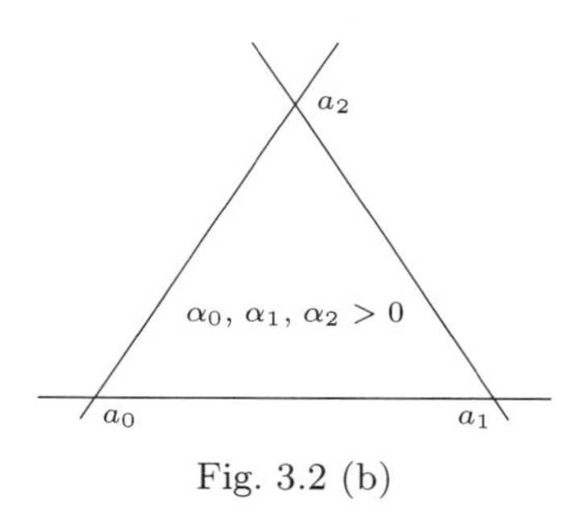

Fig. 3.2 (b)

The barycentric coordinates of a_0, a_1 and a_2 are (1,0,0), (0,1,0) and (0,0,1) respectively. The barycentric coordinates $\alpha_0, \alpha_1, \alpha_2$ of any point lying in the interior of the triangle are all positive; the barycentric coordinates of any point below the line joining a_0 and a_1 and opposite to a_2 are such that $\alpha_0 > 0$, $\alpha_1 > 0$, $\alpha_2 < 0$ etc.

As mentioned earlier, the notion of a simplex introduced below, is one of the most fundamental concepts.

Definition 3.1.10. *Let $A = \{a_0, a_1, \ldots, a_k\}$ be a geometrically independent set of points in $\mathbb{R}^n$, $n \geq k$. Then the k-***dimensional geometric simplex** *or k-***simplex**

spanned by the set A, and denoted by σ^k, is the set of all those points $x \in \mathbb{R}^n$ such that

$$x = \sum_{i=0}^{k} \alpha_i a_i \text{ where } \sum_{i=0}^{k} \alpha_i = 1 \text{ and } \alpha_i \geq 0$$

for each $i = 0, 1, 2, \ldots, k$.

In other words, the k-simplex σ^k consists of all those points of the hyperplane passing through $a_0, a_1, \ldots, a_k$ whose barycentric coordinates with respect to $a_0, a_1, \ldots, a_k$ are nonnegative. The points $a_0, a_1, \ldots, a_k$ are called the *vertices* of the simplex σ^k. We usually write $\sigma^k = \langle a_0, a_1, \ldots, a_k \rangle$ just to indicate that σ^k is the k-simplex with vertices $a_0, a_1, \ldots, a_k$.

Example 3.1.11. A 0-simplex in $\mathbb{R}^n$ is simply a singleton set or a point. Let a_0, a_1 be any two distinct points of $\mathbb{R}^n$. Then the 1-simplex determined by $\{a_0, a_1\}$ is just the closed straight line segment joining the points a_0 and a_1.

a_0 a_1

Let a_0, a_1, a_2 be any three distinct points in $\mathbb{R}^n$ not lying on a line. Then the 2-simplex $\langle a_0, a_1, a_2 \rangle$ determined by $\{a_0, a_1, a_2\}$ is simply the triangle (inside as well as the sides) joining a_0, a_1 and a_2.

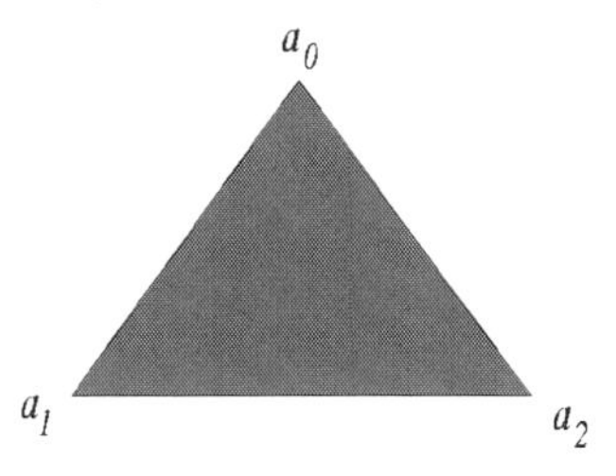

Similarly, if a_0, a_1, a_2, a_3 are four points which are non coplanar, then they determine a 3-simplex $\langle a_0, a_1, a_2, a_3 \rangle$, which is simply the solid tetrahedron with a_0, a_1, a_2 and a_3 as its vertices. It has four faces (triangles) and six edges.

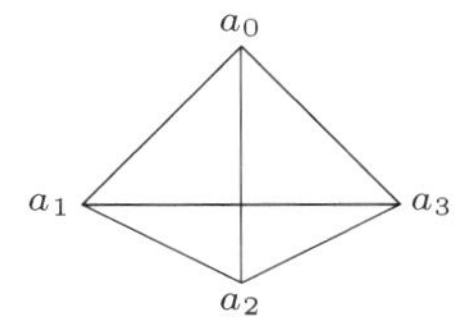

Before we look into further properties of a simplex, let us understand the relationship of a simplex with the so called convexity property. Recall that a subset S of $\mathbb{R}^n$ is said to be *convex* if $x_1, x_2 \in S$ implies that all the points on the line segment joining x_1 and x_2 also belong to S. Note that any point on the line segment joining x_1 and x_2 can be expressed uniquely as $tx_1 + (1-t)x_2$

where $0 \leq t \leq 1$. Thus, S is convex if and only if $x_1, x_2 \in S$ implies that $tx_1 + (1-t)x_2 \in S$ for all t, $0 \leq t \leq 1$. One can easily see that the intersection of any family of convex subsets of $\mathbb{R}^n$ is again a convex subset. If A is any subset of $\mathbb{R}^n$, then we can, therefore, consider the family of all convex subsets of $\mathbb{R}^n$ containing A. The set $\mathbb{R}^n$, being convex, belongs to this family, and so the family is nonempty. The intersection of the above family will be the smallest convex subset of $\mathbb{R}^n$ containing A. This intersection is called the **convex hull** of the set A. The next result describes the simplex $\langle a_0, a_1, \ldots, a_k \rangle$ as the convex hull of the geometrically independent set $\{a_0, a_1, \ldots, a_k\}$.

Proposition 3.1.12. *Let $A = \{a_0, a_1, \ldots, a_k\}$ be a geometrically independent set of points in $\mathbb{R}^n$. Then the simplex $\sigma^k = \langle a_0, a_1, \ldots, a_k \rangle$ is the convex hull of the set A.*

Proof. Recall that any point of the hyperplane H^k containing A can be uniquely written in terms of its barycentric coordinates. For $i = 0, 1, \ldots, k$, let us now consider the subset H_i of the hyperplane H^k consisting of those points whose i-th barycentric coordinate is nonnegative. In other words,

$$H_i = \{(\alpha_0, \alpha_1, \ldots, \alpha_k) \in H^k \mid \alpha_i \geq 0\}.$$

We observe that each one of these half-planes H_i is convex. For, let $x_1 = (\alpha_0, \alpha_1, \ldots, \alpha_k), x_2 = (\beta_0, \beta_1, \ldots, \beta_k)$ be any two points in H_i. Then α_i, β_i are nonnegative. Consequently, for each t, $0 \leq t \leq 1$, the i-th coordinate of $tx_1 + (1-t)x_2$, viz., $t\alpha_i + (1-t)\beta_i$, is also nonnegative, and so $tx_1 + (1-t)x_2 \in H_i$. Now we observe that $\sigma^k = H_0 \cap H_1 \cap \cdots \cap H_k$, which means σ^k, being the intersection of convex subsets of $\mathbb{R}^n$, is convex.

Next, we show that if B is any convex subset of $\mathbb{R}^n$ containing A, then B contains σ^k. We prove this by induction on k. For $k = 0$, the result is trivial. Assume $k \geq 1$ and the result to be true for $k - 1$, i.e., if B contains the set $\{a_1, a_2, \ldots, a_k\}$ then it also contains the simplex $\langle a_1, a_2, \ldots, a_k \rangle$. We assert that any point of σ^k can be written as $ta_0 + (1-t)x_1$, where $x_1 \in \langle a_1, \ldots, a_k \rangle$. This will prove the result because B is convex, contains a_0, and also contains x_1 by inductive hypothesis. We know that any point $x \in \sigma^k$ can be written as

$$x = \sum_{i=0}^{k} \alpha_i a_i, \quad \alpha_i \geq 0, \sum_{0}^{k} \alpha_i = 1.$$

If $\alpha_0 = 1$, then $x = a_0 \in B$ and if $\alpha_0 = 0$ then, obviously, $x \in \langle a_1, a_2, \ldots, a_k \rangle$ and so again $x \in B$. If both α_0, $\sum_1^k \alpha_i$ are nonzero, then

$$x_1 = \sum_{1}^{k} \left(\frac{\alpha_i}{\sum_1^k \alpha_j} \right) a_i$$

belongs to $\langle a_1, \ldots, a_k \rangle$. Because $\alpha_0 + \sum_1^k \alpha_j = 1$, $x = \alpha_0 a_0 + (1 - \alpha_0)x_1$. This proves the assertion and hence the result. ■

Definition 3.1.13. *Let $\sigma^k = \langle a_0, a_1, \ldots, a_k \rangle$ be a k-simplex. Then the set of those points of σ^k for which all barycentric coordinates are strictly positive, is called the* **open k-simplex** σ^k.

For example, the open interval joining the points a_0 and a_1 of $\mathbb{R}^n$ will be an open 1-simplex, the interior of the triangle joining a_0, a_1, a_2 will be an open 2-simplex, and so on. An important observation is that an open 0-simplex, determined by a 0-simplex $\langle a_0 \rangle$, will be the simplex $\langle a_0 \rangle$ itself.

Definition 3.1.14. *Let σ^p, σ^q, $p \leq q \leq n$ be two simplexes in $\mathbb{R}^n$. We say that σ^p is a* **p-dimensional face** *of σ^q (or merely a p-simplex of σ^q) if each vertex of σ^p is also a vertex of σ^q. If σ^p is a face of σ^q and $p < q$, then σ^p is a proper face of σ^q.*

Any 0-dimensional face of σ^q is simply a vertex of σ^q. A 1-dimensional face of a simplex is usually called an *edge* of that simplex. If we consider a 3-simplex $\sigma^3 = \langle a_0, a_1, a_2, a_3 \rangle$, then it has four 0-faces, six edges, four 2-faces which are proper faces and one 3-face which is σ^3 itself.

If $x \in \sigma^3$, then note that either x is a vertex of σ^3 or it lies in the interior of some edge or it lies in the interior of some 2-simplex or it lies in the interior of tetrahedron σ^3 itself. We also observe that any two open faces of σ^3 are either identical or disjoint. It follows that σ^3 can be seen as the disjoint union of all its open faces or simplexes. More generally, *any k-simplex σ^k can be written as the disjoint union of all its open simplexes.*

If x is an interior point of a simplex $\sigma^k = \langle v_0, v_1, \ldots, v_k \rangle$ and we write $x = \sum \alpha_i v_i$, $\alpha_i > 0$, $\sum \alpha_i = 1$, then $(\alpha_0, \alpha_1, \ldots, \alpha_k)$ are the barycentric coordinate of the point x. If one of the coordinates say α_i, is zero we get an interior point of the face of σ^k opposite the vertex v_i. If we put two coordinates α_i, α_j to be zero, we get the interior points of a face of σ^k. Finally, if we put all coordinates to be zero except one, say α_i, then we get the interior of the vertex $\langle v_i \rangle$ which is the point v_i itself. It follows that the points of the closed simplex $|\sigma^k|$ can be written in terms of barycentric coordinates as $x = (\alpha_0, \alpha_1, \ldots, \alpha_k)$ where $\alpha_i \geq 0$ and $\sum \alpha_i = 1$. This representation of $x \in |\sigma^k|$ is unique.

Now we will show that the map $p_i : |\sigma^k| \to [0, 1]$, defined by $p_i(\alpha_0, \alpha_1, \ldots, \alpha_k) = \alpha_i$, called the **projection on the i-th barycentric coordinate**, is continuous. We consider the unit vectors $e_0, e_1, \ldots, e_k$ of $\mathbb{R}^{k+1}$. Then there is a linear homeomorphism from the closed simplex $\langle v_0, v_1, \ldots, v_k \rangle$ to the closed simplex $\langle e_0, e_1, \ldots, e_k \rangle$ defined by $h(\alpha_0 v_0 + \alpha_1 v_1 + \ldots \alpha_k v_k) = \alpha_0 e_0 + \alpha_1 e_1 + \ldots + \alpha_k e_k$, $\alpha_i \geq 0, \sum \alpha_i = 1$. If $q_i : \mathbb{R}^{k+1} \to \mathbb{R}$ is the canonical projection to the i-th component, then clearly $q_i \circ h = p_i$. Hence p_i is continuous.

Simplexes are supposed to be the "building blocks" for a class of topological spaces considered a "nice" class of spaces. These spaces possess rich topological properties, but the problem of their classification, as usual, is difficult. The methods and concepts of algebraic topology are such that their power can be tested most effectively on these "nice" spaces. A variety of algebraic invariants have been defined for topological spaces, but often each of these is first defined for these nice spaces and then their generalizations are considered for arbitrary spaces. Let us now come to the main

Definition 3.1.15. *A* **simplicial complex** *or a* **geometric complex** *K is a finite collection of simplexes of $\mathbb{R}^m$, m sufficiently large, which satisfies the following conditions:*

(i) *If $\sigma \in K$, then all the faces of σ are also in K.*

(ii) *If σ and τ are in K, then either $\sigma \cap \tau = \emptyset$ or $\sigma \cap \tau$ is a common face of both σ and τ.*

We also define the **dimension** *of a simplicial complex K, denoted by* $\dim K$, *to be -1 if $K = \emptyset$ and to be $n \geq 0$ if n is the largest integer such that K has an n-simplex.*

We must mention that what we have defined above is actually a **finite** simplicial complex; we will define arbitrary simplicial complex later. Here, it is also necessary to point out that an n-dimensional simplicial complex K need not lie in $\mathbb{R}^n$. Since a simplicial complex K satisfies the two conditions of the foregoing definition, the complex could be forced to lie in $\mathbb{R}^m$ where m is far greater that n. We will see examples of this later.

Example 3.1.16. Let $\sigma^2 = \langle a_0, a_1, a_2 \rangle$ be a 2-simplex. Then the set K of all faces of σ^2, i.e.,

$$K = \{\langle a_0 \rangle, \langle a_1 \rangle, \langle a_2 \rangle, \langle a_0, a_1 \rangle, \langle a_0, a_2 \rangle, \langle a_1, a_2 \rangle, \langle a_0, a_1, a_2 \rangle\}$$

is a simplicial complex. This simplicial complex is denoted by $\mathrm{Cl}(\sigma^2)$ and is called the **closure** of σ^2. More generally, for any n, we can define the simplicial complex $\mathrm{Cl}(\sigma^n)$ to be the collection of all faces of σ^n.

Example 3.1.17. Consider the collection K of all simplexes shown in Fig. 3.3 (both pieces taken together) where shaded triangles indicate the 2-simplexes.

This is a finite collection of several 2-simplexes, 1-simplexes and 0-simplexes satisfying the two conditions and, therefore, is a geometric complex. On the other hand, the collection K' of simplexes shown in Fig. 3.4 (both pieces taken together) is not a simplicial complex.

Two simplexes σ^k and σ^l are said to be **properly joined** if their intersection is either empty or is a common face of both the simplexes. Condition (ii) of

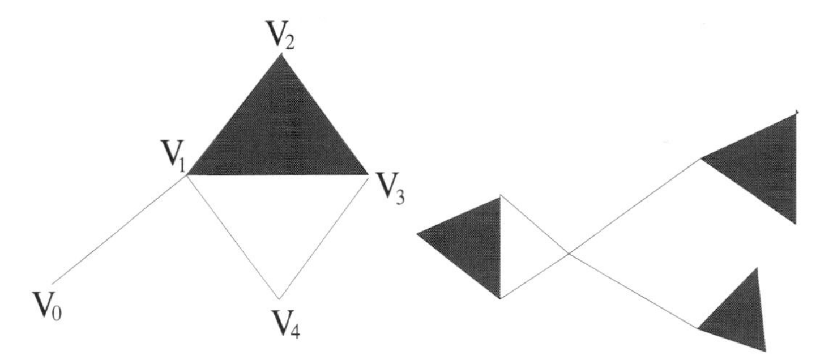

Fig. 3.3: The simplicial complex K

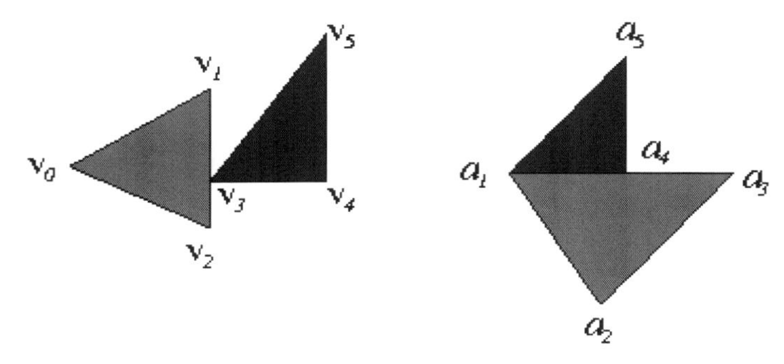

Fig. 3.4: The collection K'

the definition of a simplicial complex says that all the simplexes of K must be properly joined. In Fig. 3.4, the intersection of $\langle v_0, v_1, v_2 \rangle$ and $\langle v_3, v_4, v_5 \rangle$ is a point, which is a face of the latter simplex but not the former. Similarly, the 2-simplexes $\langle a_1, a_4, a_5 \rangle$ and $\langle a_1, a_2, a_3 \rangle$ of the same figure are not properly joined. The definition of a simplicial complex can be stated more intuitively as follows: *A finite collection K of properly joined simplexes is said to be a simplicial complex (or geometric complex or complex) if each face of a member of K is again a member of K.*

3.2 Polyhedra and Triangulations

Definition 3.2.1. *Let K be a simplicial complex. Let $|K| = \bigcup_{\sigma \in K} \sigma$ be the union of all simplexes of K. Then $|K|$, being a subset of $\mathbb{R}^n$ for some n, will be a topological space with the topology induced from $\mathbb{R}^n$. This space $|K|$ is called the* **geometric carrier** *of K. A subspace of $\mathbb{R}^n$, which is the geometric carrier of some simplicial complex, is called a* **rectilinear polyhedron**.

Definition 3.2.2. *A topological space X is said to be a* **polyhedron** *if there exists a simplicial complex K such that $|K|$ is homeomorphic to X. In this case, the space X is said to be* **triangulable** *and K is called a* **triangulation** *of X.*

We must point out the difference between a simplicial complex K and its geometric carrier $|K|$. The complex is a finite set of simplexes whereas $|K|$ is a subspace of some Euclidean space. Simplexes of K are simply elements of the finite set K and not subsets of K while each simplex is a subspace of the space $|K|$. Note that $|\mathrm{Cl}(\sigma^k)| = \sigma^k = |\sigma^k|$.

By the very definition, $|K|$ is a polyhedron for any K and the geometric shape of the rectilinear polyhedron $|K|$ consists of points, edges, plane triangles, tetrahedra, etc. Curved lines or surfaces will not be parts of $|K|$ unless these are contained in some higher dimensional simplexes of $|K|$. On the other hand, the geometric shape of an arbitrary polyhedron could be anything from nice tetrahedra to spheres, ellipsoids and much more curved surfaces. Let us consider some examples for further illustrations.

Example 3.2.3. Consider the curved portion X of the unit circle defined by (Fig.3.5)

$$X = \{(\cos\theta, \sin\theta) \in \mathbb{R}^2 \mid 0 \leq \theta \leq \pi/2\}.$$

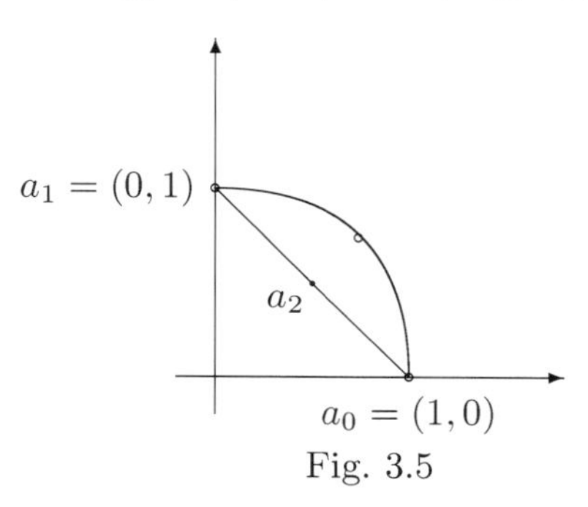

Fig. 3.5

Then X is a polyhedron, but not a rectilinear polyhedron. Let $\langle a_0, a_1\rangle$ be the simplex joining $(1,0)$ and $(0,1)$ and a_2 be any point on this simplex. Then the simplicial complex $K = \{\langle a_0\rangle, \langle a_1\rangle, \langle a_0, a_2\rangle\}$ is a triangulation of X since $|K|$ is homeomorphic to X. If we consider the simplicial complex $L = \{\langle a_0\rangle, \langle a_1\rangle, \langle a_2\rangle, \langle a_0, a_1\rangle, \langle a_1, a_2\rangle\}$, then also $|L|$ is homeomorphic to X and so L is another triangulation of X.

Example 3.2.4. (Cylinder) Consider the hollow prism (Fig.3.5(a)). We break each of the side rectangles into two triangles. Then

$$\begin{aligned} K \;=\; & \{\langle a_0\rangle, \langle a_1\rangle, \langle a_2\rangle, \langle b_0\rangle, \langle b_1\rangle, \langle b_2\rangle, \langle a_0, a_1\rangle, \langle a_1, a_2\rangle, \langle a_0, a_2\rangle, \\ & \langle b_0, b_1\rangle, \langle b_1, b_2\rangle, \langle b_0, b_2\rangle, \langle a_0, b_2\rangle, \langle a_1, b_2\rangle, \langle a_0, b_1\rangle, \langle a_0, b_0\rangle, \\ & \langle a_1, b_1\rangle, \langle a_2, b_2\rangle, \langle a_0, b_1, b_0\rangle, \langle a_0, a_1, b_1\rangle, \langle a_1, b_1, b_2\rangle, \\ & \langle a_1, a_2, b_2\rangle, \langle a_0, a_2, b_2\rangle, \langle a_0, b_0, b_2\rangle\} \end{aligned}$$

is a triangulation of the cylinder.

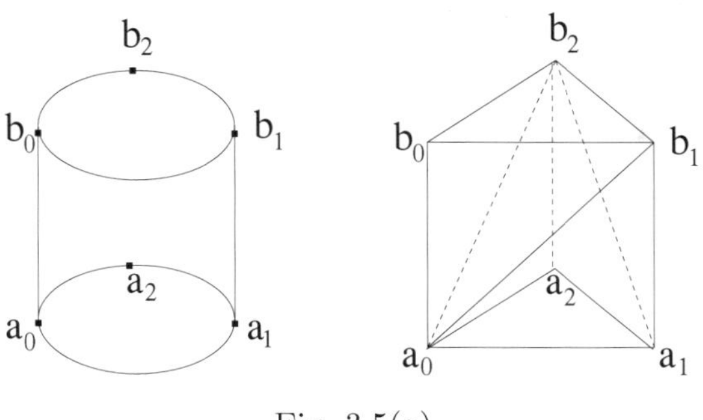

Fig. 3.5(a)

We need the following important topological result before giving further examples. This shows, in particular, that a square, a triangle, a pentagon etc., are all homeomorphic to the 2-disk $\mathbb{D}^2$; a cube, a tetrahedron, an octahedron, etc., are all homeomorphic to the 3-disk $\mathbb{D}^3$, and so on:

Proposition 3.2.5. *Let U be an open set of $\mathbb{R}^n$, $n \geq 1$ which is bounded and convex, and $w \in U$. Then*

(i) *Each half-ray L starting from w intersects the boundary of U at precisely one point.*

(ii) *There is a homeomorphism of $\overline{U}$ with the unit disk $\mathbb{D}^n$ which carries the boundary of U onto the unit sphere $\mathbb{S}^{n-1}$.*

Proof. (i) Let L be a fixed half-line starting from w and consider the intersection of L with open set U. Clearly, it is a convex, bounded and open subset of L and so it may be put in the form

$$\{w + t \cdot p : t \in [0, a)\},$$

where p is the unit vector along L. Then, obviously, L intersects the closure $\overline{U}$ of U in the point $x = w + a \cdot p$, say. To prove the uniqueness of the point x, suppose y is another such point. Then $y = w + b.p$ for some $b > a$ and x lies between w and y on the line L, i.e., $x = (1 - t)w + ty$, where $t = a/b$. Solving w for this t, we have $w = (x - ty)/(1 - t)$. Now, we can choose a sequence $\{y_n\}$ of points in U which converges to y. Then putting $w_n = (x - ty_n)/(1 - t)$, we find that w_n converges to w. Hence we have an n such that $w_n \in U$. But this is a contradiction because U being convex means $x = (1 - t)w_n + ty_n$ must be in U.

(ii) Without loss of generality, we can assume that w is the origin. The map $f \colon \mathbb{R}^n - \{0\} \to \mathbb{S}^{n-1}$ defined by $f(x) = x/\|x\|$ is onto and continuous. By (i) above, the restriction of f to the boundary $\text{Bd}(U)$ defines a bijection from $\text{Bd}(U)$ to $\mathbb{S}^{n-1}$, which must be a homeomorphism since $\text{Bd}(U)$ is compact. Let $g \colon \mathbb{S}^{n-1} \to \text{Bd}(U)$ be its inverse. Now, extend g to $G \colon \mathbb{D}^n \to \overline{U}$ by defining

$$G(x) = \begin{cases} \|g(x/\|x\|)\| \cdot x & x \neq 0 \\ 0 & x = 0. \end{cases}$$

In fact, G maps the line segment lying in $\mathbb{D}^n$ joining origin to a point u in $\mathbb{S}^{n-1}$ to the line segment in U joining origin to the point $g(u)$. Then G is a bijective continuous map, and so it is a homeomorphism whose restriction on $\mathbb{S}^{n-1}$ is the map g. ■

Example 3.2.6 (Discs). Consider the unit disc $\mathbb{D}^2$ in $\mathbb{R}^2$

$$\mathbb{D}^2 = \{(x, y) \in \mathbb{R}^2 \mid x^2 + y^2 \leq 1\}.$$

Let us consider the simplicial complex $\mathrm{Cl}(\sigma^2)$, where $\sigma^2 = \langle a_0, a_1, a_2 \rangle$ is a 2-simplex. Then $|\mathrm{Cl}(\sigma^2)|$ is the triangle Δ with vertices a_0, a_1, a_2. Since $\mathbb{D}^2$ is homeomorphic to the triangle Δ (see Example 1.4.4), we find that $\mathbb{D}^2$ is a polyhedron and $\mathrm{Cl}(\sigma^2)$ is a triangulation of $\mathbb{D}^2$. More generally, let $\mathbb{D}^n = \{(x_1, x_2, \ldots, x_n) \in \mathbb{R}^n \mid x_1^2 + x_2^2 + \ldots + x_n^2 \leq 1\}$ be the unit disc. Then $\mathbb{D}^n$ is a polyhedron – the simplicial complex K consisting of all faces of some n-simplex σ^n is a triangulation of $\mathbb{D}^n$. It now follows from Proposition 3.2.5 that every compact convex subset of Euclidean space is a polyhedron.

Definition 3.2.7. *Let K be a k-dimensional simplicial complex. For each r, $0 \leq r \leq k$, let K^r denote the set of all those simplexes of K which are of dimension $\leq r$. Then K^r is a simplicial complex which is known as r-dimensional* **skeleton** *of K. The space $|K^r|$ will be a rectilinear subpolyhedron of $|K|$.*

Example 3.2.8 (Sphere). Let σ^k be a k-simplex, $k \geq 1$, and $K = \mathrm{Cl}(\sigma^k)$ be the simplicial complex consisting of all faces of σ^k. Let K^{k-1} be the $(k-1)$-dimensional skeleton of K. Then K^{k-1} consists of all proper faces of σ^k. It can be shown (see Example 1.4.3) that $|K^{k-1}|$ is homeomorphic to

$$\mathbb{S}^{k-1} = \{(x_0, x_1, \ldots, x_{k-1}) \in \mathbb{R}^k \mid \Sigma_0^{k-1} x_i^2 = 1\},$$

which is a $(k-1)$-dimensional unit sphere. This proves that every k-sphere $\mathbb{S}^k$ is a polyhedron, and the collection of all proper faces of a $(k+1)$-simplex gives a triangulation of $\mathbb{S}^k$.

It is to be observed that if K is a simplicial complex, then $|K|$ is a compact metric space because it is finite union of simplexes which are clearly compact. This means any polyhedron (according to our definition in this book) is compact. If K is a triangulation of a polyhedron X, then $|K|$ is just a representation of the space X as a rectilinear polyhedron. Very often whenever we are required to prove some theorem about a polyhedron X, we prefer to prove that result for $|K|$ rather than for X because $|K|$ has more structure than just a topology on it. Another important point to be noted is that whenever we talk of the n-sphere

$$\mathbb{S}^n = \{(x_0, x_1, \ldots, x_n) \in \mathbb{R}^{n+1} \mid \sum_0^n x_i^2 = 1\},$$

this $\mathbb{S}^n$ is the standard unit sphere in $\mathbb{R}^{n+1}$. In topology, two spaces are not distinguished if they are homeomorphic. Now, if X is a topological space which is homeomorphic to $\mathbb{S}^n$, it is a common practice to say that "X is a topological n-sphere". This space X may appear geometrically very different from a genuine sphere allowing lot of flexibility but is still called an n-sphere. A similar practice is also prevalent regarding an n-disc $\mathbb{D}^n$. It will be very helpful if we look at a polyhedron X with a similar meaning, i.e., a polyhedron to a rectilinear polyhedron is the same thing as a topological sphere to a genuine sphere.

Giving examples of polyhedra as the union of a collection of simplexes satisfying the two conditions stated earlier, is not always a convenient method. One at once comes across complicated problems of analytic geometry for verifying "properly joined" condition of the definition. It is easier to specify polyhedra using abstract simplicial complexes, which we define as follows:

Definition 3.2.9. *An* **abstract simplicial complex** *is a collection K of finite nonempty sets such that if A is an element of K, then so is every nonempty subset of A.*

Using the above definition it is now easy to see that the following diagram defines a simplicial complex K where the vertices are $\langle a\rangle, \langle b\rangle, \langle c\rangle, \langle d\rangle, \langle e\rangle$ and $\langle f\rangle$ and the simplexes are $\langle a, b, d\rangle, \langle b, c, d\rangle, \langle c, d, e\rangle, \langle a, c, e\rangle, \langle a, e, f\rangle$ and $\langle a, f, d\rangle$ along with their nonempty subsets. We also notice that K is a triangulation of a cylinder, since $|K|$ is just a hollow prism. This triangulation of the cylinder is easier to describe as compared to Fig. 3.5(a).

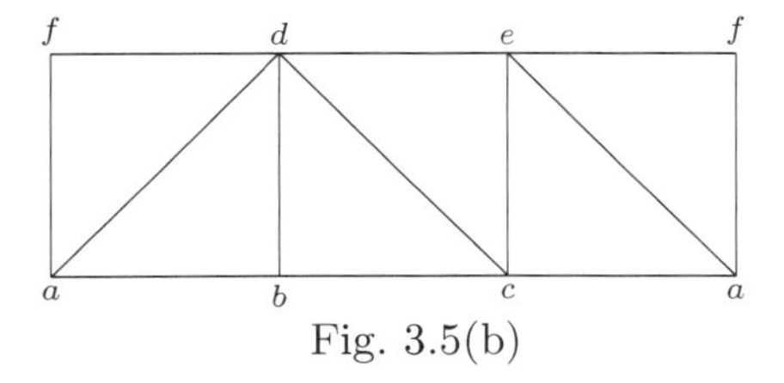

Fig. 3.5(b)

In the above definition of an abstract simplicial complex K the singletons are called **vertices** of K, and in this terminology, the set of vertices of an abstract simplicial complex K may be finite or infinite. We note that each n-simplex $\sigma = \langle v_0, v_1, \ldots, v_n\rangle$ can be identified with a simplex of $\mathbb{R}^n$. Since each $\mathbb{R}^n$ is embedded in $\mathbb{R}^\infty$, we find that $|K| = \bigcup\{\sigma \mid \sigma \in K\}$ is a subset of $\mathbb{R}^\infty$. The topology on $|K|$ is defined to be the weak topology induced by simplexes of K (see Section 3.5 for details).

Example 3.2.10 (Möbius band). We now give a triangulation of the Möbius band M. Recall (see Section 1.1.4) that M is a topological space obtained as follows: take a rectangle R and identify the two opposite sides of R by twisting one of the sides through 180°. In order to give a triangulation of M, we start with a triangulation L of the rectangle R itself (see Fig. 3.6) in which the two opposite sides are taken as 1-simplexes of L. Remember that when we identify the two opposite edges of R, two 1-simplexes of L are identified yielding a triangulation K of M. The labelling given in Fig. 3.6 indicates the identifications. The two conditions already satisfied by L continue to hold for K also except that the number of 1-simplexes of K is one less than the number of 1-simplexes of L. The number of 0-simplexes of K is also less than those of L. Thus, a triangulation of M is shown in Fig. 3.6.

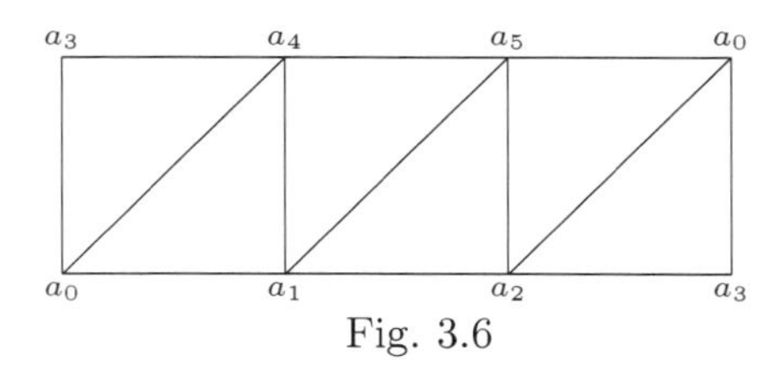

Fig. 3.6

Note that the edge $\langle a_0, a_3 \rangle$ on the left of the rectangle R is identified with the edge $\langle a_0, a_3 \rangle$ (order reversed) on the right, and the resulting quotient space M, the Möbius band, is homeomorphic to the geometric carrier of the simplicial complex K which consists of six triangles and all of their faces as in Fig. 3.6.

At this stage we go back to an important observation indicated earlier. Suppose we have a k-dimensional simplicial complex K. Then $|K|$ need not lie in some k-dimensional Euclidean space for the same k. For example, a circle is the geometric carrier of a 1-dimensional simplicial complex, but it cannot be embedded in the real line $\mathbb{R}$. In fact, $|K|$ need not be homeomorphic to a subspace of even a $(k+1)$-dimensional Euclidean space. For example, a 1-dimensional simplicial complex (a graph) may not be embedded in 2-dimensional Euclidean space, i.e., it need not be a planar graph (see the next example). A triangulation of the projective plane is a 2-dimensional simplicial complex K, but $|K|$ cannot be embedded even in 3-dimensional Euclidean space. This observation raises an interesting question: If K is a k-dimesnional simplicial complex, then what is the least n such that $|K|$ can be embedded in $\mathbb{R}^n$? The answer is known: a k-dimensional finite simplicial complex can always be embedded in $\mathbb{R}^{2k+1}$. Of course, if K is not finite, then it has to be assumed to be locally finite (see E.H.Spanier [18] p. 120). This is the best possible result because of the following example:

Example 3.2.11 (Kuratowski)**.** Consider the following simplicial complex:

$$\begin{aligned} K &= \{\langle v_0\rangle, \langle v_1\rangle, \langle v_2\rangle, \langle v_3\rangle, \langle v_4\rangle, \langle v_5\rangle, \langle v_0, v_3\rangle, \langle v_0, v_4\rangle, \langle v_0, v_5\rangle, \\ &\quad \langle v_1, v_3\rangle, \langle v_1, v_4\rangle, \langle v_1, v_5\rangle, \langle v_2, v_3\rangle, \langle v_2, v_4\rangle, \langle v_2, v_5\rangle\}. \end{aligned}$$

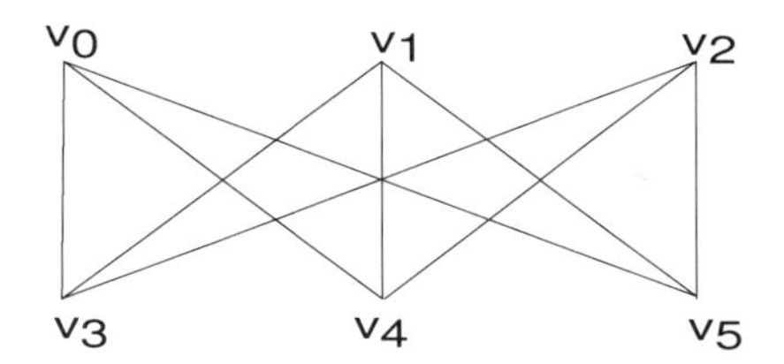

Fig. 3.6(a) : Kuratowski's non-planar graph

Clearly, K (Fig. 3.6(a)) is a connected 1-dimensional simplicial complex. It should be pointed out that any two edges of K intersect only at the vertices, not at any other point. The figure is not faithful since edges like $\langle v_0, v_4\rangle$ and $\langle v_1, v_3\rangle$ seem to intersect, but actually they don't. A combinatorial argument using Euler's theorem (see Chapter 4) shows that K cannot be a planar graph, i.e., $|K|$ cannot be embedded in the Euclidean plane $\mathbb{R}^2$.

Example 3.2.12 (Torus). A torus T is the quotient space obtained as follows (see Section 1.1.4): Take a rectangle $ABCD$ in the plane $\mathbb{R}^2$ and identify the opposite sides, i.e., identify AB with DC and AD with BC (Fig. 3.7).

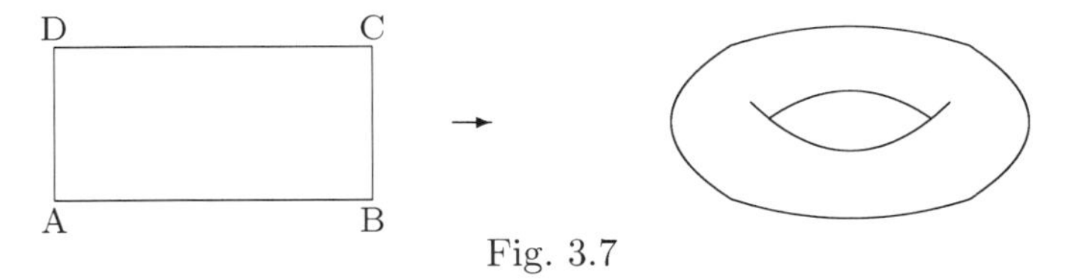

Fig. 3.7

Once again, we start with a triangulation L of the rectangle itself (see Fig. 3.8) in which each side of the rectangle is broken as the union of three 1-simplexes. The identification of opposite sides (identical labeling) collapses several pairs of 1-simplexes of L into 1-simplexes of a new complex K, and several 0-simplexes of L collapse to 0-simplexes of K yielding a triangulation K of the torus T. It is easily seen that conditions defining the simplicial complex L continue to hold for K also. Thus, the following diagram with indicated labeling gives a triangulation K of T:

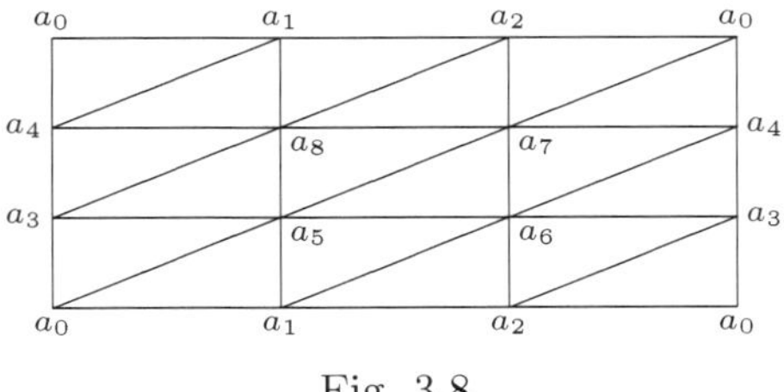

Fig. 3.8

It has 9 vertices, 27 edges and 18 triangles which are joined properly as shown in Fig. 3.8. Again note that $|K|$ lies in $\mathbb{R}^3$, not in $\mathbb{R}^2$(why?), though K is

2-dimensional. We also point out that besides the above, several other triangulations of T can be given and it is a good question to ask: what will be the minimum number of vertices, edges and triangles in a triangulation of T? This question will be answered in Chapter 4 (see Chapter 4, pseudomanifolds) for a large number of polyhedra.

The Klein bottle and the projective plane (see Section 1.1.4) are also good examples of polyhedra and the reader should give a triangulation of each of these too (see exercises at the end). These triangulations will be needed in understanding the simplicial homology discussed in the next chapter.

Definition 3.2.13. *Let K be a simplicial complex. A simplicial complex L is said to be a subcomplex of K if $L \subset K$. The* **boundary** *of a simplicial complex K, denoted by ∂K, is defined by*

$$\partial K = \{\tau \mid \tau \textit{ is a face of a simplex } \sigma^k \in K \textit{ which belongs to a unique } (k+1)\textit{-simplex of } K\}.$$

In other words, ∂K consists of all proper faces of maximal simplexes in K and all of their faces. Thus ∂K is a subcomplex of K of dimension one less than $dim(K)$.

For example, let $K = \mathrm{Cl}(\sigma^2)$ where $\sigma^2 = \langle a_0, a_1, a_2\rangle$. Then $\partial K = \{\langle a_0\rangle, \langle a_1\rangle, \langle a_2\rangle, \langle a_0, a_1\rangle, \langle a_1, a_2\rangle, \langle a_0, a_2\rangle\}$. Let K be the first piece of Example 3.1.17. Then $\partial K = \{\langle v_1, v_2\rangle, \langle v_1, v_3\rangle, \langle v_2, v_3\rangle, \langle v_1\rangle, \langle v_2\rangle, \langle v_3\rangle, \langle v_0\rangle, \langle v_4\rangle\}$.

Definition 3.2.14. *Let K and L be two simplicial complexes. By a* **simplicial map** *$f : K \to L$ we mean a map f from the vertices of K to the vertices of L such that if $\langle a_0, a_1, \ldots, a_k\rangle$ is a simplex of K, then $f(a_0), f(a_1), \ldots, f(a_k)$ are vertices of a simplex in L. In other words, canceling repetitions, $\langle f(a_0), f(a_1), \ldots, f(a_k)\rangle$ is a simplex in L whenever $\langle a_0, a_1, \ldots, a_k\rangle$ is a simplex in K.*

For any simplicial complex K, the identity map $I_K : K \to K$ is a simplicial map. We also note that if $f : K \to L$ and $g : L \to M$ are two simplicial maps, then their composite $g \circ f : K \to M$ is also a simplicial map.

Definition 3.2.15. *A simplicial map $f : K \to L$ is said to be an* **isomorphism** *if it is bijective on vertices and has the property that $\langle v_0, v_1, \ldots, v_k\rangle$ is a simplex in K if and only if $\langle f(v_0), f(v_1), \ldots, f(v_k)\rangle$ is a simplex in L. Two simplicial complexes K and L are said to be simplicially isomorphic (or just isomorphic) if there is a simplicial isomorphism $f : K \to L$.*

It is easy to see that "to be simplicially isomorphic" is an equivalence relation in the class of all simplicial complexes. As a result, the class of all simplicial complexes is decomposed into mutually disjoint isomorphism classes.

Let K be a finite simplicial complex and suppose $x \in |K|$. Then there is a simplex $\sigma^k = \langle v_0, v_1, \ldots, v_k \rangle$ such that x is in the open simplex σ^k. If $\alpha_0, \alpha_1, \ldots, \alpha_k$ are the barycentric coordinates of x w.r.t. $v_0, v_1, \ldots, v_k$, then we can write $x = \sum_0^k \alpha_i v_i$ where $\alpha_i > 0$. Now if w is any other vertex of K, not in σ^k, we put the w-th barycentric coordinate of x to be zero. Thus we find that each $x \in |K|$ has a unique representation as

$$x = \sum_i \alpha_i v_i, \ \alpha_i \geq 0, \ \sum \alpha_i = 1, \ v_i \in K.$$

Let v be a fixed vertex of K and define a function $f_v : |K| \to [0, 1]$ by $f_v(x) = v$-th barycentric coordinate of x. Note that f_v is well defined because if x is in two closed simplexes σ and τ, then the v-th barycentric coordinate of x depends only on the vertices of $\sigma \cap \tau$, not on other vertices. If $|\sigma^k|$ is a closed simplex of K and $x \in |\sigma^k|$, then $f_v(x)$ is simply the projection on the v-th barycentric coordinate of x, and is zero otherwise. Hence f_v restricted to each closed simplex σ of K is continuous. It follows that f_v itself is continuous. This proves the next

Proposition 3.2.16. *The vertex function $f_v : |K| \to [0, 1]$ is continuous.*

Next we have

Proposition 3.2.17. *Let $f : K \to L$ be a simplicial map. Then there is an induced continuous map $|f| : |K| \to |L|$ defined as follows: If $x \in |K|$, then x belongs to the interior of a unique simplex $\langle v_0, v_1, \ldots, v_k \rangle$. We write $x = \sum_{i=0}^k \alpha_i v_i$ and define $|f|(x) = \sum_{i=0}^k \alpha_i f(v_i)$.*

Proof. First we define $|f|$ on each closed simplex $|\sigma|$ by the formula: if $x = \alpha_0 v_0 + \alpha_1 v_1 + \ldots + \alpha_k v_k$, $\alpha_i \geq 0, \sum \alpha_i = 1$, then $|f|(x) = \alpha_0 f(v_0) + \alpha_1 f(v_1) + \ldots + \alpha_k f(v_k)$. This shows that $|f|$ is well-defined. If x lies in the intersection $|\sigma| \cap |\tau|$ of two closed simplexes σ and τ, then the barycentric coordinates of x depend only on the common vertices $v_0, v_1, \ldots, v_l$ of $\sigma \cap \tau$. Hence $|f|(x) = \alpha_0 f(v_0) + \alpha_1 f(v_1) + \ldots + \alpha_l f(v_l)$ is unique whether x is treated as a point of $|\sigma|$ or a point of $|\tau|$. Therefore, now it suffices to prove that $|f||_{|\sigma|}$ is continuous for each simplex $\sigma \in K$. To see this note that the barycentric coordinates of $|f|(x)$ are $\alpha_i, i = 0, 1, \ldots k$ or the sum of these coordinates depending on whether $f(v_i)$ are all distinct or not. In any case, when we consider the composite $q_i \circ |f| : |\sigma^k| \to \mathbb{R}$ where $q_i : \mathbb{R}^{k+1} \to \mathbb{R}$ are the canonical projection, then this is simply the projection map $(\alpha_0, \alpha_1, \ldots, \alpha_k) \to \alpha_i$ or is a sum of these projection maps. Hence $q_i \circ |f|$ is continuous for each $i = 0, 1, 2, \ldots, k$. This mean $|f| : |K| \to |L|$ is continuous. ∎

Remark 3.2.18. If we look at $|K|$ and $|L|$ as topological spaces, then a continuous map $f : |K| \to |L|$ may or may not be induced by some simplicial map $g : K \to L$. In case the map f is induced by some simplicial map, then sometimes we abuse the language and refer f to be a "simplicial map". In this terminology, the set of all continuous maps from $|K|$ to $|L|$ is divided into two classes, viz., the "simplicial" maps and the "non-simplicial" maps. Since we are dealing with only finite simplicial complexes, the number of simplicial maps from K to L is always finite, whereas the number of non-simplicial maps is always infinite, unless, of course, $|L|$ consists of only 0-simplexes (Why?)

Let $f : K \to L$ and $g : L \to M$ be two simplicial maps. Then it follows from the definitions that the induced map $|g \circ f| : |K| \to |M|$ is simply the composite of the induced maps $|f| : |K| \to |L|$ and $|g| : |L| \to |M|$, i.e., $|g \circ f| = |g| \circ |f|$. Also, if I_K is the identity map, then the induced map $|I_K| : |K| \to |K|$ is also the identity map $I_{|K|}$. These observations say that $K \to |K|$ is a covariant functor from the category of simplicial complexes and simplicial maps to the category of topological spaces and continuous maps.

Definition 3.2.19. *A simplicial complex K is said to be* **connected** *if for each pair of vertices $a, b \in K$, there exists a sequence of 1-simplexes $\langle a_i, a_{i+1} \rangle$ for $i = 0, 1, 2, \ldots, p-1$ in K such that $a = a_0, b = a_p$.*

Let $n \geq 1$ and $K = \mathrm{Cl}(\sigma^n)$ for some n-simplex σ^n. Then the r-dimensional skeleton $K^r, r \geq 1$ of K is connected whereas K^0 is not connected. It is easily seen that any triangulation of sphere, Möbius band or torus is connected. Maximal connected subcomplexes of a simplicial complex are called the **combinatorial** components of that complex.

3.3 Simplicial Approximation

We have already noted that if $f : K \to L$ is a simplicial map, then the induced map $|f| : |K| \to |L|$ in their geometric carriers (associated polyhedra) is continuous. However, given a continuous map $h : |K| \to |L|$, there may not be any simplicial map from K to L which induces h. In fact, when $|K|$ and $|L|$ are genuine geometric complexes, any induced continuous map $|f| : |K| \to |L|$ will always take a straight line segment contained in a simplex of $|K|$ into a vertex or straight line segment of $|L|$, and in a linear fashion. A continuous map $h : |K| \to |L|$, on the other hand, may take a straight line segment to a curved line segment and, therefore, need not be simplicial. Consider the unit interval $|K| = [0, 1]$, where $K = \{\langle 0 \rangle, \langle 1 \rangle, \langle 0, 1 \rangle\}$ and the unit square $|L| = I \times I$ in the plane with the four vertices, viz., $(0, 0)$, $(1, 0)$, $(1, 1)$, $(0, 1)$, four edges and two triangles with one common edge, as shown in Fig. 3.9.

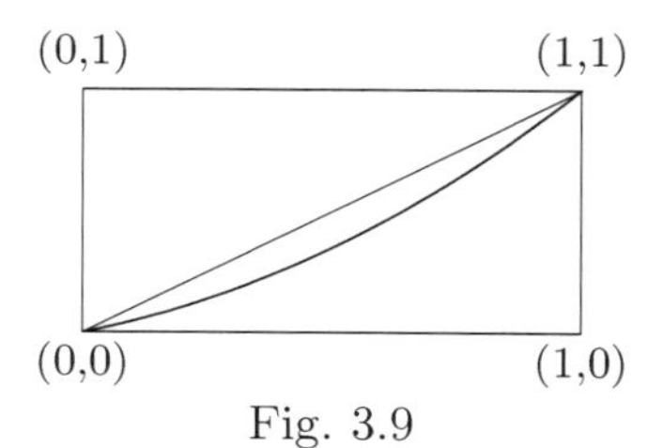

Fig. 3.9

For each $n \geq 1$, we define continuous maps $f_n : |K| \to |L|$ by putting $f_n(t) = (t, t^n)$. Then f_1 is simplicial, but $f_n(n \neq 1)$ is never simplicial. The unit interval can also be considered as the geometric carrier $|K'|$ of some other simplicial complex, say, $K' = \{\langle 0\rangle, \langle 1/3\rangle, \langle 1\rangle, \langle 0, 1/3\rangle, \langle 1/3, 1\rangle\}$. Even then there is no simplicial map from K' to L which induces f_2. In fact, there is no representation of $[0,1]$ as the geometric carrier of any simplicial complex for which f_2 or $f_n(n \geq 3)$ is simplicial because each of these maps takes any 1-simplex, however small, to a curved arc which is not straight.

We have the following

Definition 3.3.1. *Let K be a simplicial complex and v be a vertex of K. By the* **star** *of v, denoted by* $\mathrm{st}(v)$, *we mean the following subset of K:*

$$\mathrm{st}(v) = \{\sigma : \sigma \text{ is a simplex of } K \text{ and } v \text{ is a vertex of } \sigma\}.$$

By the **open star** *of v, denoted by* $\mathrm{ost}(v)$, *we mean the following subset of $|K|$:*

$$\mathrm{ost}(v) = \bigcup\{\mathrm{int}(\sigma) : \sigma \text{ is a simplex of } K \text{ with } v \text{ as a vertex of } \sigma\}.$$

Example 3.3.2. As an example, consider the simplicial complex K, where $|K|$ is shown in Fig.3.10. Then

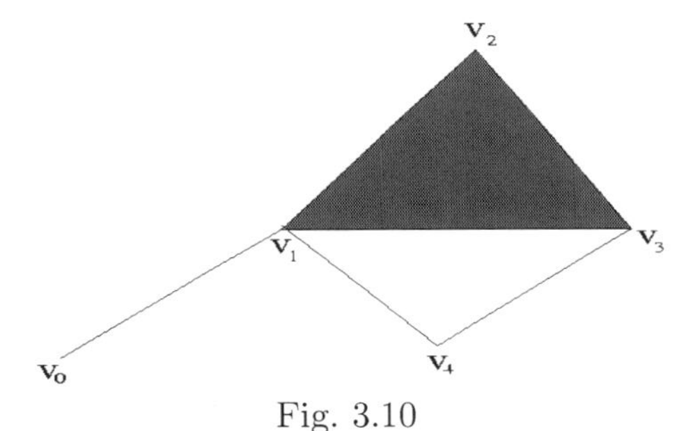

Fig. 3.10

$$\begin{aligned} \mathrm{st}(v_0) &= \{\langle v_0\rangle, \langle v_0, v_1\rangle\}, \\ \mathrm{st}(v_1) &= K - \{\langle v_0\rangle, \langle v_2\rangle, \langle v_3\rangle, \langle v_4\rangle, \langle v_2, v_3\rangle, \langle v_3, v_4\rangle\}. \end{aligned}$$

On the other hand, $\mathrm{ost}(v_0) = \langle v_0\rangle \cup \mathrm{int}\langle v_0, v_1\rangle$, and $\mathrm{ost}(v_1) = \langle v_1\rangle \cup \mathrm{int}\langle v_0, v_1\rangle \cup \mathrm{int}\langle v_1, v_2\rangle \cup \mathrm{int}\langle v_1, v_3\rangle \cup \mathrm{int}\langle v_1, v_4\rangle \cup \mathrm{int}\langle v_1, v_2, v_3\rangle$.

It is very important to point out that the open star $\mathrm{ost}(v)$ of a vertex v of a simplicial complex K is not only a subset of $|K|$, but also is an open set in $|K|$. This follows from the fact that for any vertex v of K, the function $f_v : |K| \to [0,1]$ given by $f_v(x) =$ the vth barycentric coordinate of x, is continuous, and $f_v^{-1}(0,1] = \mathrm{ost}(v)$.

In this section, we are going to show that if $h : |K| \to |L|$ is any given continuous map, then there exists a representation $|K'|$ of $|K|$, (i.e., there is a simplicial complex K' such that $|K'| = |K|$) and a simplicial map $g : K' \to L$ such that for each vertex v of K', $h(\text{ost}(v)) \subset \text{ost}(g(v))$, in particular $|g| : |K'| = |K| \to |L|$ is homotopic to h. Such a result is known as a simplicial approximation theorem. We have the following

Definition 3.3.3. *Let $h : |K| \to |L|$ be a continuous map. Then a simplicial map $g : K \to L$ is said to be a* **simplicial approximation** *of h if for each vertex v of K, $h(\text{ost}(v)) \subset \text{ost}(g(v))$.*

We will show later that if a simplicial map $g : K \to L$ is a simplicial approximation of h, then $|g|$ is homotopic to h.

Example 3.3.4. Let K be any simplicial complex and $L = \text{Cl}(\sigma^p)$, where $\sigma^p = \langle a_0, \ldots, a_p \rangle$. Suppose $h : |K| \to |L|$ is any continuous map such that $h(|K|)$ does not intersect the face $\langle a_1, a_2, \ldots, a_p \rangle$. Let $g : K \to L$ be the simplicial map which maps all vertices of K to the vertex a_0 of L. Then g is a simplicial approximation of h because for each vertex v of K, $h(\text{ost}(v)) \subset \text{ost}(g(v))$.

The following result is easily verified:

Proposition 3.3.5. *Suppose $g_1 : K \to L$ is a simplicial approximation of $f_1 : |K| \to |L|$ and $g_2 : L \to M$ is a simplicial approximation of $f_2 : |L| \to |M|$. The $g_2 \circ g_1 : K \to M$ is a simplicial approximation to $f_2 \circ f_1$.*

Proof. The composite map $g_2 \circ g_1$ is clearly simplicial. We know that for each vertex v of K, $f_1(\text{ost}(v)) \subset \text{ost}(g_1(v))$, and for each vertex w of L, $f_2(\text{ost}(w)) \subset \text{ost}(g_2(w))$. Since $g_1(v)$ is a vertex of L, one has $f_2(\text{ost}(g_1(v))) \subset \text{ost}(g_2(g_1(v)))$. This means $f_2 \circ f_1(\text{ost}(v)) \subset \text{ost}(g_2 \circ g_1(v))$, which proves the result. ■

We now proceed to develop necessary definitions and concepts for giving a proof of the simplicial approximation theorem mentioned earlier.

Definition 3.3.6. *Let K and L be two simplicial complexes and $f : |K| \to |L|$ be a continuous map. Then we say that K is* **star-related** *to L with respect to f, if for each vertex v of K, there is a vertex v' of L such that*

$$f(\text{ost}(v)) \subset \text{ost}(v').$$

The above definition will help us in constructing a simplicial map. We will also require the following result;

Proposition 3.3.7. *A set $\{v_0, v_1, \ldots, v_n\}$ of vertices of a simplicial complex K forms a simplex of K if and only if $\bigcap_{i=0}^{n} \text{ost}(v_i)$ is nonempty.*

Proof. If $\sigma^n = \langle v_0, v_1, \ldots, v_n \rangle$ is a simplex of K, then we know that $\text{int}(\sigma^n) \subset \text{ost}(v_i)$ for each $i = 0, 1, \ldots n$ and so $\phi \neq \text{int}(\sigma^n) \subset \bigcap_{i=0}^n \text{ost}(v_i)$. Conversely, suppose $\bigcap_{i=0}^n \text{ost}(v_i) \neq \emptyset$ and let x be a point of this intersection. For each $i = 0, 1, \ldots, n$, there is a simplex σ_i in K such that $x \in \text{int}(\sigma_i)$ and v_i is a vertex of σ_i. Recall that the set of interiors of all simplexes of a simplicial complex K forms a partition of $|K|$ and hence, there is a unique simplex of K whose interior contains x, i.e., $\sigma_0 = \sigma_1 = \sigma_2 = \ldots = \sigma_n$. This implies that $v_0, v_1, \ldots, v_n$ are vertices of the simplex σ_0. ∎

We now start by first proving a basic special case of the Simplicial Approximation Theorem as follows:

Theorem 3.3.8. *Let $f : |K| \to |L|$ be a continuous map. Suppose K is star-related to L relative to f. Then there is a simplicial map $g : K \to L$ such that g is a simplicial approximation of f. In particular, $|g| : |K| \to |L|$ is homotopic to f.*

Proof. Since K is star-related to L relative to f, we can define a map g from the vertices of K to the vertices of L as follows: for each vertex v of K, let $g(v)$ be a vertex of L such that

$$f(\text{ost}(v)) \subset \text{ost}(g(v)).$$

We now prove that the map g actually takes a simplex of K to a simplex of L. For this, let $v_0, v_1, \ldots, v_n$ be the vertices of a simplex of K. By Proposition 3.3.7, this means

$$\bigcap_{i=0}^{n} \text{ost}(v_i) \neq \phi.$$

Hence,

$$\emptyset \neq f(\textstyle\bigcap_{i=0}^n \text{ost}(v_i)) \subset \bigcap_{i=0}^n f(\text{ost}(v_i)) \subset \bigcap_{i=0}^n \text{ost}(g(v_i)),$$

i.e., $\bigcap_{i=0}^n \text{ost}(g(v_i))$ is nonempty, and therefore $g(v_0), \ldots, g(v_n)$ are vertices of some simplex of L. This proves that $g : K \to L$ is a simplicial map, and so g is a simplicial approximation of f.

Next, we want to show that $|g| : |K| \to |L|$ is homotopic to f. Let $x \in |K|$ and let $\sigma = \langle v_0, v_1, \ldots, v_k \rangle$ be the simplex of smallest dimension which contains x. Let v_i be any vertex of σ. Note that $x \in \text{ost}(v_i)$ which means $f(x) \in f(\text{ost}(v_i)) \subset \text{ost}(g(v_i))$. Thus $f(x) \in \text{int}\langle g(v_0), \ldots, g(v_k) \rangle$. On the other hand if we write

$$x = \alpha_0 v_0 + \ldots + \alpha_k v_k, \alpha_i > 0, \ \ \forall i,$$

then $|g|(x) = \alpha_0 g(v_0) + \ldots + \alpha_k g(v_k)$, in which some of the $g(v_i)'s$ may be equal. In any case the coefficient of $g(v_i)$ will be strictly positive, i.e., $|g|(x) \in \text{int}\langle g(v_0), \ldots, g(v_k) \rangle$. This means for each $x \in |K|$, both $f(x)$ and $|g|(x)$ are in the same simplex of L. Since a simplex is a convex set, we can define a map $H : |K| \times I \to |L|$ by $H(x,t) = (1-t)f(x) + t|g|(x)$. H is continuous because both f and $|g|$ are continuous and $|L|$ is a subset of some Euclidean space. Therefore H is a homotopy from f to $|g|$. ∎

It is easy to give an example of a continuous map $f : |K| \to |L|$ and a simplicial map $g : K \to L$ such that $|g| : |K| \to |L|$ is homotopic to f, but g is not a simplicial approximation of f.

As remarked earlier, the above theorem is a special case of the *Simplicial Approximation Theorem* (to be proved later) because for any given continuous map $f : |K| \to |L|$, it may be the case that K is not star-related to L relative to f. What are we to do in that case? The answer to this question lies in creating the concept of subdivision of a simplicial complex. Geometrically speaking, what we will do is as follows: We will subdivide each simplex σ of K in a set of smaller simplices thereby getting a new simplicial complex K' which will have the same geometric carrier as K, i.e, $|K'| = |K|$, and the simplexes of K' will be so small that K' will be star-related to L relative to f.

3.4 Barycentric Subdivision – Simplicial Approximation Theorem

The concept of subdividing a simplicial complex K is an extremely important and useful method of changing the simplicial structure of K without changing the underlying set of $|K|$ or its topology. As an example, consider the 2-simplex $\sigma^2 = \langle v_0, v_1, v_2 \rangle$. We know that the simplicial complex

$$\mathrm{Cl}(\sigma^2) = \{\langle v_0 \rangle, \langle v_1 \rangle, \langle v_2 \rangle, \langle v_0 v_1 \rangle, \langle v_1 v_2 \rangle, \langle v_0 v_2 \rangle, \langle v_0 v_1 v_2 \rangle\}$$

has the simplex σ^2 as its geometric carrier, i.e., $|\mathrm{Cl}(\sigma^2)| = \sigma^2$. Now, if we introduce a vertex v_3 in the interior of the triangle σ^2, and consider the set K' of all triangles, edges and vertices so created, then this new set K' is again a simplicial complex (Fig.3.11) which has the same geometric carrier as the triangle σ^2, i.e., $|K'| = \sigma^2$.

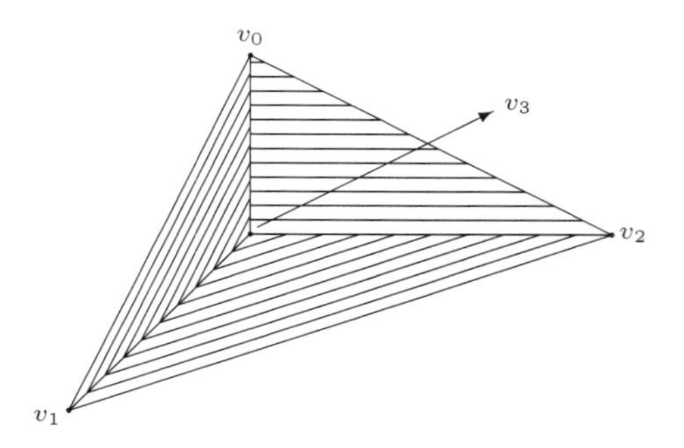

Fig. 3.11

In fact, K' has four vertices, six edges and three triangles and is entirely different from K because K had only three vertices, three edges and one triangle. Imagine what will be K' when the newly introduced vertex v_3 lies on one of the edges of the simplex σ^2. If K is a simplicial complex and we subdivide each simplex in the same manner as indicated above, then we get a new simplicial complex K' which is called a **subdivision** of K. Among infinitely many possibilities of subdividing a simplicial complex, there is one type of subdivision, which is iterative, and is called the **barycentric subdivision**. This subdivision stands out as the most useful, highly regular and conveniently describable subdivision.

Definition 3.4.1. *Let $\sigma = \langle v_0, v_1, \ldots, v_k \rangle$ be a k-simplex in $\mathbb{R}^n$. Then the* **barycentre** *of σ, denoted by $\dot{\sigma}$, is defined as the point of σ given by*

$$\dot{\sigma} = \sum_{i=0}^{k} \frac{1}{k+1} v_i.$$

In other words, the barycentre $\dot{\sigma}^k$ of the k-simplex σ^k is that point of σ^k whose barycentric coordinates, with respect to each of the vertices of σ^k, are equal.

For example, the barycentre of a 0-simplex $\langle v_0 \rangle$ will be the point v_0 itself. The barycentre of a 1-simplex $\langle v_0, v_1 \rangle$ is simply the midpoint of the line segment joining v_0 and v_1. Similarly, the barycentre of the 2-simplex $\sigma^2 = \langle v_0, v_1, v_2 \rangle$ is the centroid of the triangle with vertices v_0, v_1 and v_2, and so on.

Let $\sigma^k = \langle v_0, v_1, \ldots, v_k \rangle$ and $K = \text{Cl}(\sigma^k)$. Consider the set of all barycentres of all faces of σ^k. Note that this new set contains all the vertices of σ^k and corresponding to each edge, one new vertex; corresponding to each 2-face, one new vertex, and so on. We now consider a new simplicial complex K' whose vertices are barycentres of all faces of σ^k and whose simplexes are as follows: $\langle \dot{\sigma_0}, \dot{\sigma_1}, \ldots, \dot{\sigma_q} \rangle$ will be a q-simplex if σ_i's are all distinct and σ_i is a face of σ_{i+1} for $i = 0, 1, \ldots, q-1$. For example, in the case of 2-simplex $\sigma^2 = \langle v_0, v_1, v_2 \rangle$, the new simplicial complex K' has vertices, edges and 2-simplexes as shown in Fig. 3.12.

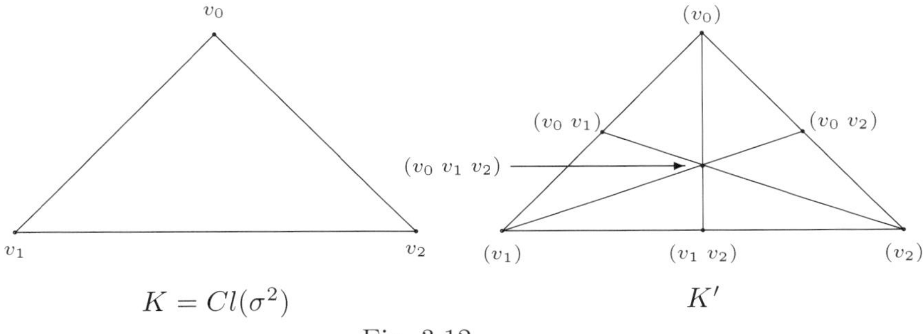

Fig. 3.12

In fact, the starting simplicial complex is

$$K = \{\langle v_0\rangle, \langle v_1\rangle, \langle v_2\rangle, \langle v_0, v_1\rangle, \langle v_1, v_2\rangle, \langle v_0, v_2\rangle, \langle v_0, v_1, v_2\rangle\},$$

and the new simplicial complex is K', whose vertices are barycentres of each simplex of K. If we use the notation (v_0v_1) to denote the barycentre of $\langle v_0v_1\rangle$, etc., then K' has the seven vertices: $(v_0), (v_1), (v_2), (v_0v_1), (v_1v_2), (v_0v_2), (v_0v_1v_2)$. It has 1-simplexes such as $\langle(v_0), (v_0v_2)\rangle, \langle(v_1), (v_0v_1)\rangle, \ldots, \langle(v_0v_1), (v_0v_1v_2)\rangle, \ldots$ etc., and 2-simplexes such as $\langle(v_0), (v_0v_1), (v_0v_1v_2)\rangle, \langle(v_1), (v_1v_2), (v_0v_1v_2)\rangle, \ldots$, and so on. The Fig. 3.12 gives a nice description of the new simplicial complex K'. It is obvious that $|K| = |K'|$. More generally, we have

Definition 3.4.2. *Let K be a simplicial complex. Let $K^{(1)}$ be a simplicial complex whose vertices are barycentres of all simplexes of K, and for distinct simplexes $\sigma_0, \sigma_1, \ldots, \sigma_n$ of K, $\langle\dot{\sigma_0}, \dot{\sigma_1}, \ldots, \dot{\sigma_n}\rangle$ is a simplex of $K^{(1)}$ if and only if σ_i is a face of σ_{i+1} for each $i = 0, 1, 2, \ldots, n-1$. Then $K^{(1)}$ is a simplicial complex and is called the* **first barycentric subdivision** *of K. By induction, we define* n**th barycentric subdivision** *$K^{(n)}$ of K to be the first barycentric subdivision of $K^{(n-1)}$ for each $n > 1$. We also put $K^{(0)} = K$ for convenience.*

Note that $|K^{(n)}| = |K|$ for each $n \geq 0$. We also observe that when we take the first barycentric subdivision $K^{(1)}$ of a positive-dimensional complex, the simplexes of $K^{(1)}$ become strictly smaller in size than those in K. In fact, their sizes approach towards zero when we go on making further and further subdivisions. We now make these observations precise.

Recall that if A is a subset of $\mathbb{R}^n$, then

$$\text{diam}(A) = \sup\{d(x, y) : x, y \in A\}$$

is called the *diameter* of A; here $d(x, y) = \|x - y\| = (\sum_{i=1}^n (x_i - y_i)^2)^{1/2}$, where $x = (x_1, \ldots, x_n)$ and $y = (y_1, \ldots, y_n)$.

Definition 3.4.3. *Let K be a simplicial complex. We define the* **mesh** *of K, denoted by* mesh(K), *as follows:*

$$\text{mesh}(K) = \ max\{\text{diam}(\sigma) : \sigma \ is\ a\ simplex\ of\ K\}.$$

Note that if K is a 0-dimensional simplicial complex, then $\text{mesh}(K) = 0 = \text{mesh}(K^{(1)}) = \text{mesh}(K^{(2)}) = \ldots$. First, we have

Lemma 3.4.4. *Let σ be any positive-dimensional simplex. Then* $\text{diam}(\sigma) = \|v - w\|$ *for some pair of vertices v and w of σ. In other words, the diameter of a simplex σ is the length of its largest edge.*

Proof. Let $\sigma = \langle v_0, v_1, \ldots, v_q\rangle$ and $x, y \in \sigma$. Let $y = \sum_{i=0}^{q} b_i v_i$, where b_i are barycentric coordinates of y. Then

$$\begin{aligned}
\|x - y\| &= \|(\textstyle\sum_{i=0}^{q} b_i)x - \sum_{i=0}^{q} b_i v_i\| \\
&= \|\textstyle\sum_{i=0}^{q} b_i(x - v_i)\| \\
&\le \textstyle\sum_{i=0}^{q} b_i\|x - v_i\| \\
&\le \textstyle\sum b_i \cdot \max\{\|x - v_i\| : 0 \le i \le q\} \\
&\le \max(\|x - v_i\| : 0 \le i \le q).
\end{aligned}$$

Replacing x by v_i and y by x, we find similarly

$$\|x - v_i\| \le \max\{\|v_j - v_i\| : 0 \le j \le q\}.$$

Hence, by combining the two, we find

$$\|x - y\| \le \max\{\|v_j - v_i\| : 0 \le j,\ i \le q\}.$$

This evidently implies that diam$(\sigma) = \|v_r - v_s\|$ for some vertices v_r and v_s of the simplex σ. ■

Now, we can prove the fact which was intuitively mentioned earlier.

Theorem 3.4.5. *Let K be any positive-dimensional simplicial complex. Then* $\lim_{n\to\infty} \text{mesh}(K^{(n)}) = 0$.

Proof. Let us assume that $\dim(K) = m$. First, we compare the mesh $(K^{(1)})$ with the mesh of K. Since mesh $(K^{(1)})$ is determined by the length of its largest 1-simplex, we consider any 1-simplex $\langle\dot{\sigma}, \dot{\tau}\rangle$ of $K^{(1)}$ where σ, τ are simplexes of K and σ is a proper face of $\tau = \langle v_0, v_1, \ldots, v_p\rangle$. We know that

$$\dot{\tau} = \frac{1}{(p+1)} \sum_{i=0}^{p} v_i,$$

and so, because $\dot{\sigma}, \dot{\tau}$ both are points of τ, by the earlier lemma, there is a vertex v of τ such that

$$\|\dot{\tau} - \dot{\sigma}\| \le \|\dot{\tau} - v\|.$$

Then

$$\begin{aligned}
\|\dot{\tau} - \dot{\sigma}\| &\leq \|\dot{\tau} - v\| \\
&= \|\frac{1}{p+1}(\sum_{i=0}^{p} v_i) - v\| \\
&= \|\frac{1}{p+1}\sum_{i=0}^{p}(v_i - v)\| \\
&\leq \frac{1}{p+1}\sum_{i=0}^{p}\|v_i - v\| \\
&\leq \frac{1}{p+1}(p \cdot \text{mesh}(K)) \\
&\leq \frac{m}{m+1} \cdot \text{mesh}(K),
\end{aligned}$$

since $p \leq m$. This implies that mesh $(K^{(1)}) \leq \frac{m}{m+1} \cdot$ mesh (K). By induction, therefore, we find that mesh$(K^{(n)}) \leq (\frac{m}{m+1})^n$mesh (K). Since

$$\lim_{n\to\infty}\left(\frac{m}{m+1}\right)^n = 0,$$

we have the desired result. ■

Now, we have all the necessary ingredients to prove our general

Theorem 3.4.6. (Simplicial Approximation Theorem). *Let $f : |K| \to |L|$ be any continuous map. Then there is a barycentric subdivision $K^{(k)}$ of K and a simplicial map $g : K^{(k)} \to L$ such that g is a simplicial approximation of f. In particular, $|g| : |K^{(k)}| = |K| \to |L|$ is homotopic to f.*

Proof. Since our simplicial complexes are finite, $|K|$, $|L|$ etc., are all compact subsets of some euclidean space $\mathbb{R}^n$. We have already proved in Theorem 3.3.8 that if K is star-related to L relative to $f : |K| \to |L|$ then there is a simplicial map $g : K \to L$ such that $|g|$ is homotopic to f. If K is not starr-related to L relative to f, then we proceed as follows: since $|L|$ is a compact metric space, we apply Lebesgue's Covering Lemma (Lemma 1.7.2) to the open covering $\{\text{ost}(v) : v \text{ is a vertex of } L\}$ of $|L|$ to get a Lebesgue number $\eta > 0$ so that any open ball in $|L|$ of radius η about any point of $|L|$ is contained in ost(v) for some vertex v of L. Since any continuous map from a compact metric space to any space is uniformly continuous (see Proposition 1.7.3), the map $f : |K| \to |L|$ is also uniformly continuous. This means that we can find a positive number δ such that $\|x - y\| < \delta$ in $|K|$ implies that $\|f(x) - f(y)\| < \eta$ in $|L|$. Now, we take k so large that the mesh of the barycentric subdivision $K^{(k)}$ of K is less than $\delta/2$. If v is a vertex of $K^{(k)}$, then ost(v) lies in a ball of diameter δ, which means $f(\text{ost}(v))$ lies in a ball of diameter η, i.e.,

$f(\text{ost}(v)) \subset \text{ost}(v')$ for some vertex v' of L. Then $K^{(k)}$ is evidently star-related to L relative to $f : |K^{(k)}| = |K| \to |L|$ and the proof follows from Theorem 3.3.8. ■

In the proof of the above Simplicial Approximation Theorem, the necessity of going to the barycentric subdivision $K^{(k)}$ of K is an important and indispensable feature. Let us explain this by an example: consider the unit circle $\mathbb{S}^1$ and let K be the 1-skeleton of $\text{Cl}(\sigma^2)$ where $\sigma^2 = \langle a_0, a_1, a_2 \rangle$ is a 2-simplex. Then we know that $\mathbb{S}^1$ is homeomorphic to $|K|$ by some homeomorphism, say, $h : \mathbb{S}^1 \to |K|$. Now, let $f, g : \mathbb{S}^1 \to \mathbb{S}^1$ be any two continuous maps. Then since composite of homotopic maps is homotopic, we find that f is homotopic to g if and only if hfh^{-1} is homotopic to hgh^{-1}. In other words, the set of all homotopy classes of maps from $\mathbb{S}^1$ to itself is in 1-1 correspondence with the set of all homotopy classes of maps from $|K|$ to itself.

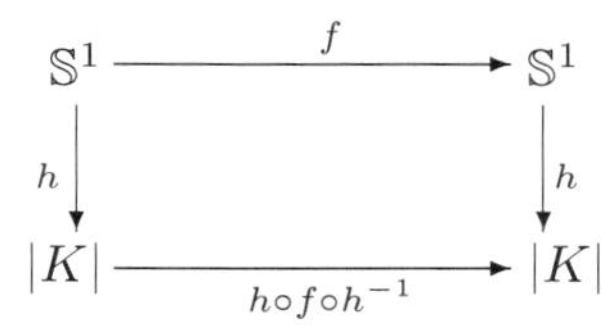

We also note that there is a bijection between $\pi_1(\mathbb{S}^1)$ and the set $[\mathbb{S}^1, \mathbb{S}^1]$ of all homotopy classes of maps from $\mathbb{S}^1$ to $\mathbb{S}^1$. Since $\pi_1(\mathbb{S}^1) \cong \mathbb{Z}$, we see that **there are infinite number of mutually non-homotopic maps from $|K|$ to itself, while for any k there are only finite number of simplicial maps from $|K^{(k)}|$ to itself**. The Simplicial Approximation Theorem says that for any given map $f : |K| \to |K|$, there is a k (depending on f) and a simplicial map $g : K^{(k)} \to K$ which will approximate f; it does not say that there is a fixed k such that every $f : |K| \to |K|$ can be approximated by some simplicial map $g : K^{(k)} \to K$. One may have to go to a higher order barycentric subdivision of K for finding out a simplicial approximation to a given continuous map $f : |K| \to |K|$. This fact can also be expressed by asserting that for any given k, there exists a continuous map $f : |K| \to |K|$ which does not have a simplicial approximation from $K^{(k)}$ to K.

We now present an interesting application of the Simplicial Approximation Theorem. Recall that a space X is said to be **simply connected** if X is path-connected and $\pi_1(X)$ is trivial. This is the same thing as saying that any map from $\mathbb{S}^0$ to X and any map from $\mathbb{S}^1$ to X are null-homotopic. More generally, we have

Definition 3.4.7. *A space X is said to be n-**connected** $(n \geq 0)$ if for each $k \leq n$, any map from $\mathbb{S}^k$ to X is null-homotopic.*

Thus, 0-connectedness is the path-connectedness of X and 1-connectedness is the same thing as simply connectedness. Now, we have

Theorem 3.4.8. *For each $n \geq 1$, $\mathbb{S}^n$ is $(n-1)$-connected.*

Proof. Let $0 \leq k < n$. We will prove that any map $f : \mathbb{S}^k \to \mathbb{S}^n$ is null-homotopic. Let σ^{k+1} be a $(k+1)$-simplex and σ^{n+1} be a $(n+1)$-simplex. If K represents a k-skeleton of $\mathrm{Cl}(\sigma^{k+1})$ and L represents the n-skeleton of $\mathrm{Cl}(\sigma^{n+1})$, then we know that $|K|$ is homeomorphic to $\mathbb{S}^k$ and $|L|$ is homeomorphic to $\mathbb{S}^n$. Let $h : |K| \to \mathbb{S}^k$ and $h' : |L| \to \mathbb{S}^n$ be homeomorphisms.

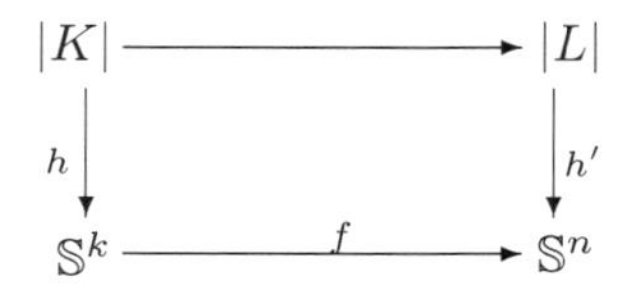

Let $g : K^{(m)} \longrightarrow L$ be a simplicial approximation to the map $(h')^{-1} \circ f \circ h : |K| \to |L|$. Now observe that $\dim(K^{(m)}) = k < \dim(L) = n$ and any simplicial map $g : K^{(m)} \to L$ cannot raise the dimension of the simplicial complex. This means g cannot be onto and hence $|g| : |K^{(m)}| \to |L|$ cannot be onto. Consequently, there is a point $a \in |L|$ such that $|g|$ maps $|K|$ into $|L| - \{a\}$. Because $|L| - \{a\}$ is homeomorphic to $\mathbb{R}^n$ and the latter is contractible, $|g|$ is null-homotopic. This means $(h')^{-1} \circ f \circ h : |K| \to |L|$ is null-homotopic and, consequently, f is null-homotopic. ■

We have already proved (see Example 2.5.1) the next result, but it follows from the above theorem also.

Corollary 3.4.9. *The n-sphere $\mathbb{S}^n$, $n \geq 2$, is simply connected.*

Concluding Remarks: The basic topological spaces dealt with in this chapter are the compact polyhedra (finite simplicial complexes) which form the most accessible class of topological spaces. They include all graphs, compact surfaces and compact 3-manifolds. The real projective space $\mathbb{RP}^n$, and the complex projective space $\mathbb{CP}^n$, $n \geq 2$ are classical examples of polyhedra. This restricted class of topological spaces (polyhedra) has two beautiful generalizations: the first one is to allow a simplicial complex to have infinite number of simplexes, in which case the topology of its geometric carrier (still called a polyhedra) is no longer compact (see Section 3.5). However, these are paracompact and most of the results proved in this chapter are true for them also. The second generalization, resulting in a more powerful and appropriate class of topological spaces from the viewpoint of algebraic topology, is known as ***CW*-complexes**. They carry the *CW*-structure which is a far-reaching generalization of the simplicial structure discussed here, and they encompass a remarkably large class of nice spaces. The simplicial approximation theorem remains valid for them also. These CW-complexes are also paracompact and become compact if the *CW*-structure is finite. Chapter 2 of James Munkres [16] (for simplicial complexes) and the book of A.T. Lundell and S. Weingram [12](for CW-complexes) are excellent references for a study of these spaces and their extensive applications.

Exercises

1. Let $\sigma^n = \langle v_0, v_1, \ldots, v_n \rangle$ be n-simplex. Find a formula giving the number of faces of σ^n.

2. Prove that every compact convex subset of the Euclidean space $\mathbb{R}^n$ is triangulable.

3. Prove that the following rectangular strip in which the left side ada is identified with the right side ada, and the top side $abca$ is identified with the bottom side $abca$

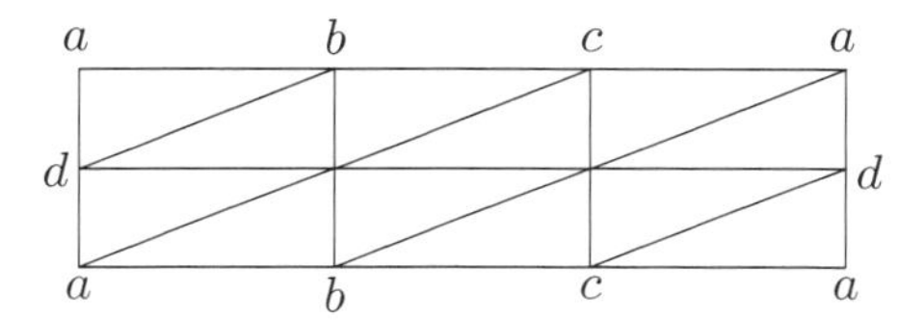

is not a triangulation of torus.

4. Show that the following strip in which the left side ba is identified with the right side ab

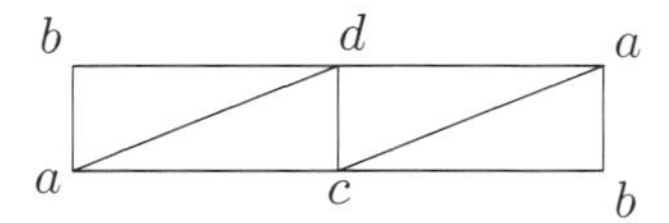

is not a triangulation of Möbius band.

5. If we take a rectangle $ABCD = I \times I$ in plane and identify all points of

the side AD with the corresponding points of the side BC, and also identify all the points of the side AB of the type $(x, 0)$ with the points $(1-x, 1)$ of the side DC, then the resulting space is called the **Klein bottle** K (see Section 1.1.4). Give a triangulation of the Klein bottle. (**Hint**: Modify the triangulation of torus).

6. Take a disc $\mathbb{D}^2 = \{(x, y) \in \mathbb{R}^2 : x^2 + y^2 \leq 1\}$. Identify a point $(\cos\theta, \sin\theta)$ of the boundary of $\mathbb{D}^2$ with the point $(-\cos\theta, -\sin\theta)$. Then the resulting

quotient space (Section 1.1.4), denoted by $\mathbb{RP}^2$, is called the **projective plane**. Prove that $\mathbb{RP}^2$ is a polyhedron. (**Hint**: start with a suitable triangulation of the disc itself.)

7. Let $K = \mathrm{Cl}(\sigma^2)$, where $\sigma^2 = \langle v_0, v_1, v_2 \rangle$ is a 2-simplex. Show that the open simplex $\langle v_0, v_1 \rangle$, which is a subset of $|K|$, is not an open set of $|K|$. However, the open simplex $\langle v_0, v_1, v_2 \rangle$ is indeed an open set in $|K|$.

8. Show that any r-skeleton $K^{(r)}$, $r \geq 0$, of a simplicial complex K is also a simplicial complex.

9. Prove that Kuratowski graph (Fig. 3.6(a)) is not a planar graph, i.e., it cannot be embedded in the plane $\mathbb{R}^2$.

10. Prove that a simplicial complex K is connected if and only if $|K|$ is path-connected.

11. Let K be a simplicial complex. Prove that for each vertex v of K, $\mathrm{ost}(v)$ is an open set of $|K|$. Also prove that $\{\mathrm{ost}(v) : v \text{ is a vertex of } K\}$ is an open covering of $|K|$.

12. Let $f : |K| \to |L|$ be a continuous map. If a simplicial map $\phi : K \to L$ is a simplicial approximation of f, then show that $|\phi|$ is homotopic to f. Give an example to show that the converse need not be true.

13. Let $S = \{a_0, a_1, \ldots, a_n\}$ be a set with $n+1$ distinct points. Consider the set
$$F(S) = \{f : S \to [0,1] : \textstyle\sum_{i=0}^{n} f(a_i) = 1\}.$$
Prove that the set $F(S)$ is in one-to-one correspondence with the points of an n-simplex $\sigma^n = \langle v_0, v_1, \ldots, v_n \rangle$ in $\mathbb{R}^{n+1}$. (This exercise tells that an abstract simplex can be given a topology from a simplex of $\mathbb{R}^n$ via the above one-to-one correspondence).

14. Prove that the following compact subspace X (Topologist's Sine Curve) of $\mathbb{R}^2$ cannot be a polyhedron:
$$X = \{(x,y) \in \mathbb{R}^2 : y = \sin 1/x, x > 0\} \cup \{(0,y) : |y| \leq 1\}.$$
(**Hint**: a polyhedron is locally contractible.)

15. A set $S = \{a_0, \ldots, a_k\}, k > n$ of points of $\mathbb{R}^n$ is said to be **in general position** if any $n+1$ points of S are geometrically independent. Show that for any n, we can find an infinite set of points which are in general position in $\mathbb{R}^n$.

16. Let $a_1, a_2, \ldots, a_{n+1}$ be points of $\mathbb{R}^n$ and let the coordinates of a_i be written as $a_i = (a_i^1, a_i^2, \ldots, a_i^n)$ for each i. Prove that $S = \{a_1, a_2, \ldots, a_{n+1}\}$ is geometrically independent if and only if the matrix

$$\begin{pmatrix} a_1^1 & a_1^2 & . & a_1^n & 1 \\ a_2^1 & a_2^2 & . & a_2^n & 1 \\ . & . & . & . & . \\ . & . & . & . & . \\ a_{n+1}^1 & a_{n+1}^2 & . & a_{n+1}^n & 1 \end{pmatrix}$$

is nonsingular.

3.5 General Simplicial Complexes

So far we assumed that a simplicial complex K is just a **finite set of simplexes** satisfying the condition that $\sigma \in K$ implies all faces of σ are in K, and σ, $\tau \in K$ imply that either $\sigma \cap \tau = \phi$ or $\sigma \cap \tau$ is a face of both σ and τ. A **general simplicial complex** K will be defined to be an **arbitrary set of simplexes** satisfying the same two conditions as above. In this case, the number of vertices of K can be of arbitrary cardinality and the dimension of K need not be finite. Hence, we have to understand clearly where such a simplicial complex will lie, and how to define its topology on $|K|$. This can be explained by the idea of an **abstract simplicial complex** defined as follows:

Definition 3.5.1. *An* **abstract simplicial complex** $\mathcal{K}$ *consists of a set* $\{v\}$ *of vertices and a set* $\{s\}$ *of nonempty finite subsets of vertices, called simplexes, such that*

(a) *any set consisting of exactly one vertex is a simplex, and*

(b) *every nonempty subset of a simplex is a simplex.*

Now, let $\{v_\alpha \mid \alpha \in J\}$ be the set of vertices of $\mathcal{K}$ indexed by an arbitrary set J. Consider the set

$$\mathbb{R}^J = \{f : J \to \mathbb{R} \mid f(v_\alpha) = 0 \text{ for all } \alpha \text{ except finitely many}\}.$$

Then, clearly $\mathbb{R}^J$ is a vector space over $\mathbb{R}$, and it is the direct sum of as many copies of $\mathbb{R}$ as the cardinality of J. There is a natural basis $\{\epsilon_\alpha \mid \alpha \in J\}$ of this vector space defined by

$$\epsilon_\alpha(\beta) = \begin{cases} 1 & \text{if } \beta = \alpha \\ 0 & \text{if } \beta \neq \alpha. \end{cases}$$

We can define a metric on $\mathbb{R}^J$ such that

$$d(f, g) = \sqrt{\sum_{\alpha \in J} (f(v_\alpha) - g(v_\alpha))^2},$$

which makes $\mathbb{R}^J$ a metric space. Note that corresponding to each finite subset $\{\epsilon_{\alpha_1}, \ldots, \epsilon_{\alpha_k}\}$ of basis elements, $\mathbb{R}^J$ has $\mathbb{R}^k$ as a subspace with Euclidean topology. Observe that if $(v_{\alpha_0}, v_{\alpha_1}, \ldots, v_{\alpha_k})$ is a k-simplex of $\mathcal{K}$, then we may consider the geometric k-simplex $(\epsilon_{\alpha_0}, \epsilon_{\alpha_1}, \ldots, \epsilon_{\alpha_k})$ spanned by the unit vectors $\epsilon_{\alpha_0}, \epsilon_{\alpha_1}, \ldots, \epsilon_{\alpha_k}$, which really consists of all points f of $\mathbb{R}^J$ such that

$$\sum_0^k f(v_{\alpha_i}) = 1, \quad 0 \leq f(v_{\alpha_i}) \leq 1 \;\; \forall i.$$

This is indeed the closed simplex of the Euclidean space generated by the vertices $(\epsilon_{\alpha_0}, \epsilon_{\alpha_1}, \ldots, \epsilon_{\alpha_k})$. Now, let K be the set of all geometric simplexes in $\mathbb{R}^J$ defined by all simplexes of $\mathcal{K}$, and let $|K|$ be their union in $\mathbb{R}^J$. Then, $|K|$ is called the **geometric realization** of the abstract simplicial complex $\mathcal{K}$. Note that K has arbitrary number of vertices and simplexes of any finite dimension. Also, K is now a general simplicial complex satisfying both the conditions of a simplicial complex. Since $|K| \subset \mathbb{R}^J$, it has the subspace topology, which is a metric topology. We denote the resulting space by $|K|_d$.

Topology of a Polyhedron

There is yet another and more important topology on $|K|$. To see this, note that the geometric realization $(\epsilon_{\alpha_0}, \epsilon_{\alpha_1}, \ldots, \epsilon_{\alpha_k})$ of each simplex s of $\mathcal{K}$ is a compact subspace of the Euclidean subspace $\mathbb{R}^k$ and, therefore, it has a unique topology of its own. Hence, we can consider the weak topology on $|K|$ defined by these closed simplexes. We call this topolgy on $|K|$ as the **coherent topology** on $|K|$. The resulting space, denoted by just $|K|$, is called a **general simplicial complex** or a **polyhderon**. Thus, the topology of a polyhedron $|K|$ is the largest topology on $|K|$ which makes all inclusions $i : (\epsilon_{\alpha_0}, \epsilon_{\alpha_1}, \ldots, \epsilon_{\alpha_k}) \to |K|$ continuous. In other words, a subset F of $|K|$ is closed iff $F \cap \sigma$ is closed in every geometric simplex σ of $|K|$.

Definition 3.5.2. *A topological space X is said to be* **triangulable** *(or a polyhedron) if there exists a simplicial complex K and a homeomorphism $f : |K| \to X$, where $|K|$ has the coherent topology defined above.*

Example 3.5.3. Let $K = \{\langle n \rangle, \langle n, n+1 \rangle \mid n \in \mathbb{Z}\}$. The vertices of K are all integers and the closed intervals $[n, n+1]$ are all its 1-simplexes. Then, clearly, K is an infinite simplicial complex and $|K|$ is homeomorphic to the real line $\mathbb{R}$ with usual topology under the identity map $i : |K| \to \mathbb{R}$. The two topologies on $|K|$ are indeed identical. If F is closed in $\mathbb{R}$, then, clearly, $F \cap [n, n+1]$ is closed in $[n, n+1]$ for each n, and so is closed in the coherent topology on $|K|$. Conversely, if $F \subset |K|$ such that $F \cap [n, n+1]$ is closed in $[n, n+1]$ for all n, then, since $F = \bigcup_{n \in \mathbb{Z}} F \cap [n, n+1]$ is a locally finite family of closed sets of $\mathbb{R}$, F itself is closed in $\mathbb{R}$. This shows that the real line $\mathbb{R}$ is triangulable, and K is a triangulation of $\mathbb{R}$.

Example 3.5.4. By a similar argument, one can easily see that the following figure gives a triangulation of $\mathbb{R}^2$:

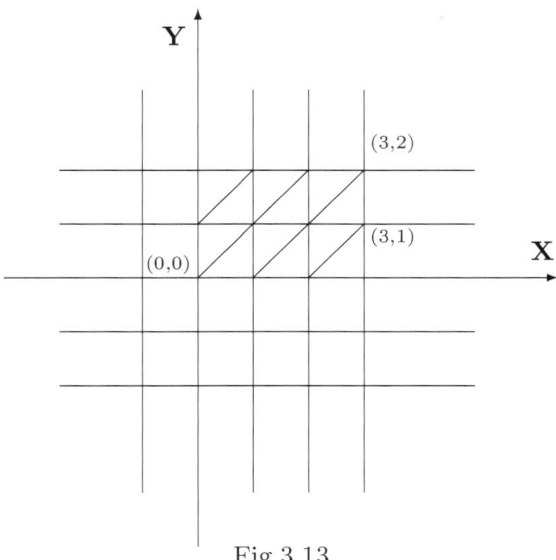

Fig.3.13

More generally, we can show that the Euclidean space $\mathbb{R}^n$, $n \geq 1$, is also triangulable. To see this, consider the points of $\mathbb{R}^n$ having integral coordinates, i.e., the set $\mathbb{Z}^n$. Define $x \leq y$, $x = (x_1, x_2, \ldots, x_n)$, $y = (y_1, y_2, \ldots, y_n) \in \mathbb{Z}^n$ if $x_i \leq y_i \;\; \forall\, i = 1, 2, \ldots, n$. Then, clearly, $\mathbb{Z}^n$ is a partially ordered set. Let us take points of $\mathbb{Z}^n$ as vertices, and let a set $\{x^0, x^1, \ldots, x^q\}$ of vertices be a simplex if $x^0 \leq x^1 \leq \ldots \leq x^q$ is a totally ordered chain, i.e., $x_i^q - x_i^0 = 0$ or 1 for each i. Since a subset of totally ordered set is totally ordered, this construction gives a simplicial complex K. Moreover, since the set of simplicies of K is a locally finite family of closed sets of $\mathbb{R}^n$, K gives a triangulation of $\mathbb{R}^n$. Thus, $\mathbb{R}^n$ is a polyhedron $\forall\, n \geq 1$.

From this latter example, one can observe that any partially ordered set $(P, \leq)$ always defines a simplicial complex in a natural way as described above.

Remark 3.5.5. Suppose $|K| \subset \mathbb{R}^n$ for some n. Then, $|K|$ has two topologies: one, the subspace topology induced from $\mathbb{R}^n$ which is a metric topology, and the other is the coherent topology defined by all simplices of K. We observe that in this case the coherent topology on $|K|$ is always finer than the subspace topology since F is closed in $\mathbb{R}^n$ means $F \cap \sigma$ will be closed in σ for all simplexes σ of K, i.e., $F \cap |K|$ will be closed in the coherent topology of $|K|$. The following example shows that the coherent topology can be strictly finer than the subspace topology.

Example 3.5.6. Let K be a simplicial complex whose vertices are $\{(0,0),(1,1),(1,1/2),\ldots,(1,\frac{1}{n}),\ldots)\} \cup \{(1,0)\}$ in the plane $\mathbb{R}^2$, and the edges are the line segments joining $(0,0)$ to $(1,\frac{1}{n})$ for all $n \in \mathbb{N}$.

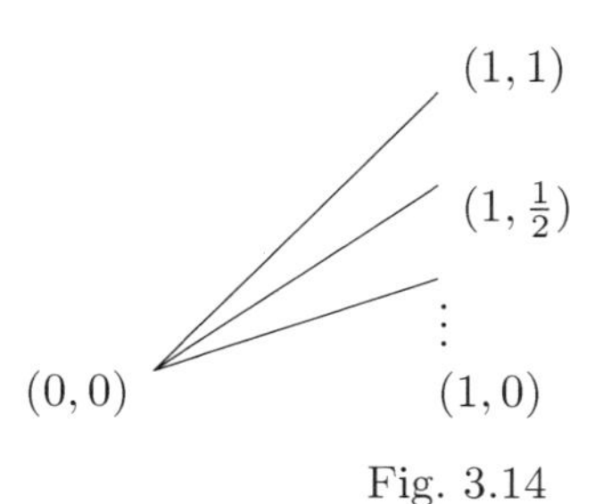

Fig. 3.14

Then, note that the set $F = \{(1,1),(1,\frac{1}{2}),\ldots,(1,\frac{1}{n}),\ldots\}$ is closed in $|K|$ since its intersection with each simplex is either empty or just a point and so is closed in that simplex. However, F is not closed in the subspace topology on $|K|$ induced from the plane $\mathbb{R}^2$ because $(1,0)$ is the limit point of F which is not in F.

The above example also shows that a polyhedron $|K|$ need not be metrizable. Suppose, on the contrary, the above $|K|$ is metrizable. Then, $|K|$ must be first countable at the vertex $v = (0,0)$. Let $\{U_1, U_2, \ldots, U_n, \ldots\}$ be a countable base for $|K|$ at the point v. We can clearly assume that

$$U_1 \supseteq U_2 \supseteq \ldots \supseteq U_n \supseteq \ldots$$

Note that v is a limit point of the open simplex $((0,0),(1,\frac{1}{n}))$ for each $n \geq 1$. Hence, we can choose a point $a_n \in U_n \cap ((0,0),(1,\frac{1}{n}))\ \forall\ n \in \mathbb{N}$. Then, we get a sequence $\{a_1, a_2, \ldots, a_n, \ldots\}$ in $|K|$ which converges to v since in any neighbourhood V of v there is a $U_n \subseteq V$ such that $a_i \in U_n\ \forall\ i \geq n$. But, the sequence $B = \{a_1, a_2, \ldots, a_n, \ldots\}$ is a closed set in $|K|$ since its intersection with each simplex of K is just a point and so is closed in that simplex. Thus, B contains all of its limit points, a contradiction to the fact that B converges to v.

A very useful result about the coherent topology on $|K|$ is the following:

Theorem 3.5.7. *Let X be a topological space and $f : |K| \to X$ be a map. Then f is continuous iff $f|_\sigma$ is continuous for each simplex σ of K.*

Proof. Since restriction of a continuous map is continuous, it is enough to prove the converse part. Hence, suppose $f|_\sigma$ is continuous for every $\sigma \in K$. Let F be a closed set of X. Then, for each σ,

$$(f|_\sigma)^{-1}(F) = f^{-1}(F) \cap \sigma$$

is closed in σ. Hence, by the definition of coherent topology on $|K|$, $f^{-1}(F)$ is closed in $|K|$. This means f is continuous. ■

Proposition 3.5.8. *Let L be a subcomplex of K. Then $|L|$ is a closed subspace of $|K|$.*

Proof. We should prove two facts. First, if F is a closed subset of $|K|$, then $F \cap |L|$ is closed in $|L|$; second, if A is a closed subset of $|L|$, then $A = F \cap |L|$ for some closed set F of $|K|$. First is obvious by definition of coherent topologies of $|K|$ and $|L|$. To prove the second, we will prove a stronger result, viz., $A \subset |L|$ is closed in $|L|$ implies A is closed in $|K|$. Let $\sigma \in K$ be any simplex. Then, for a face τ of σ, which lies in L, $A \cap \tau$, is closed in τ and hence in σ. Thus $A \cap \sigma$ is the finite union of such sets $A \cap \tau$, where τ is a face of σ. Being a finite union of closed sets of $\sigma, A \cap \sigma$ is closed in σ. It follows that A is closed in $|K|$. ■

Theorem 3.5.9. *A polyhedron $|K|$ is Hausdorff.*

Proof. Let us recall the barycentric coordinates of a point $x \in |K|$. Since x belongs to the interior a unique simplex $\sigma = \langle v_0, \ldots, v_k \rangle$, its barycentric coordinates with each vertex of σ are positive, whereas its barycentric coordinates with any other vertex of K is zero. Now, let v be a fixed vertex of K. Then, we can define a function $f_v : |K| \to [0, 1]$ such that $f_v(x)$ is equal to the barycentric coordinates of x with respect to the vertex v. Note that $f_v|_\sigma(x)$ is the vth barycentric coordinate of x if x belongs to σ and is zero if x is not in σ. Hence, $f_v|_\sigma$ is continuous for each $\sigma \in K$. Therefore, f_v is a continuous function. Now, take two distinct points $x, y \in |K|$. Then, there exists a vertex v of K such that $f_v(x) \neq f_v(y)$. Now, choose a real number r such that $f_v(x) < r < f_v(y)$. Then, clearly, the open sets $f_v^{-1}([0, r))$ and $f_v^{-1}((r, 1])$ of $|K|$ are disjoint and separate x and y. Hence, $|K|$ is Hausdorff. ■

Proposition 3.5.10. *A polyhderon $|K|$ is compact iff K is finite. More generally, if a subset $A \subset |K|$ is compact then $A \subseteq |L|$ for some finite subcomplex L of K.*

Proof. If K is finite, then since $|K|$ is the union of finite number of simplexes and each of these simplexes is compact, $|K|$ is compact. Conversely, suppose a compact subset A is not contained in a finite subcomplex of K. This means there will be an infinite number of simplexes $\{\sigma_1, \sigma_2, \ldots, \sigma_n, \ldots\}$ such that $A \cap \sigma_i \neq \phi \ \forall \ i$. We choose a point $x_i \in A \cap \sigma_i \ \forall \ i$ and consider the set $B = \{x_i \mid i \in \mathbb{N}\}$. Then, since B intersects each simplex only in finite number of points, $B \cap \sigma_i$ is closed in σ_i . Hence, B is closed in A. In fact, every subset of B is closed in A, which means B is an infinite discrete closed subset of A having no limit point, a contradiction to the compactness of A. ■

Theorem 3.5.11. *A polyhedron $|K|$ is normal.*

Proof. We will use the Tietze extension theorem to prove the normality of $|K|$. Let $A \subset |K|$ be a closed set of $|K|$ and $f : A \to [0, 1]$ be a continuous function. We will extend f continuously to the whole of $|K|$. For this, it is enough to define a family $\{f_\sigma \mid \sigma \in K\}$ of continuous maps $f_\sigma : \sigma \to [0, 1]$ such that

(a) $f_\sigma|_\tau = f_\tau$ whenever τ is a face of σ, and

(b) $f_\sigma|_{A\cap\sigma} = f|_{A\cap\sigma}$ for each σ.

Because then, we can define a map $g : |K| \to [0,1]$ by the condition that $g|_\sigma = f_\sigma \; \forall \, \sigma \in |K|$. This map g will be a continuous map by Proposition 3.5.7 which will extend the given map.

We define f_σ by induction on dimension of σ. If σ is a vertex, say $\{v\}$, then we define $f_\sigma(v) = f(v)$ if $v \in A$, and $f_\sigma(v)$ to be any point in $[0,1]$ if $v \notin A$. Thus, f_σ is defined for all 0-simplexes. Now, suppose $n > 1$ and f_{σ^q} is defined for all $q < n$ which satisfies both the conditions (a) and (b). Let σ be an n-simplex and consider the set $\sigma \cap A \cup \dot{\sigma}$. This is a closed set of σ, and f_τ is defined for all faces τ of $\dot{\sigma}$ (boundary of σ) satisfying (a) and (b). So we have a continuous map $f' : \sigma \cap A \cup \dot{\sigma} \to [0,1]$ such that $f'|_\tau = f_\tau$ for all faces τ of $\dot{\sigma}$ and $f'|_{\sigma\cap A} = f$. Now, using Tietze extension theorem, we can extend f' to a continuous map $f_\sigma : \sigma \to [0,1]$ satisfying conditions (a) and (b). This proves the theorem. ■

The following result yields an example of a compact topological space which can not be triangulated.

Theorem 3.5.12. *Any polyhedron $|K|$ is locally contractible.*

Proof. Let us take a point $x \in |K|$. Assume that U is a neighbourhood of x in K. We must show that there is an open neighbourhood V of x in U such that V can be deformed to the point x in U. Let us assume that $x = v$ is a vertex of K. Then, we know that $\text{ost}(v) = A$ is an open set in $|K|$, and we have a homotopy $H : A \times I \to |K|$ defined by

$$H(a,t) = (1-t)a + tv.$$

This homotopy deforms A into the point v such that $H(v \times I) = \{v\}$. Since U is a neighbourhood of v, H is continuous and I is compact, there exists a neighbourhood V of v in A such that $H(V \times I) \subset U$. Since V is open in A and A is open in $|K|$, V is an open neighbourhood of v which is deformable to the point v in U. This proves that $|K|$ is locally contractible at each vertex of $|K|$.

Next, suppose that x is not a vertex. Let σ be a simplex of smallest dimension such that $x \in \text{int}\, \sigma$. Now, we take any subdivision K' of $\dot{\sigma}$. Then $K' * x$ is a subdivision of $\bar{\sigma}$ and hence, we get a subdivision K'' of K such that x is a vertex of K''. Since $|K''| = |K|$, and $|K''|$ is locally contractible at its vertex x, we conclude that $|K|$ is locally contractible at x also. ■

Corollary 3.5.13. *The topologist sine curve*

$$T = \{(0,y) \mid |y| \leq 1\} \cup \{(x, \sin \tfrac{1}{x}),\; x > 0\}$$

in the plane $\mathbb{R}^2$ is not triangulable since it is not locally path connected (a locally contractible space must be locally path connected).

Theorem 3.5.14. *Any polyhedron $|K|$ is paracompact.*

Proof. The proof will be based on the famous theorem of A.H. Stone which says that every metric space is paracompact. We already know that the identity map $i : |K| \to |K|_d$, where $|K|$ has the coherent topology and $|K|_d$ has the metric topology, is continuous. Let $\mathcal{U}$ be an open cover of $|K|$. We will prove that $\mathcal{U}$ has a nbd-finite open refinement. Note that open stars of vertices of K are always open in $|K|$.

First, we show that K has a subdivision $\mathrm{Sd}(K)$ such that the closed stars of vertices of $\mathrm{Sd}(K)$ in $|\mathrm{Sd}(K)| = |K|$ are contained in some member of $\mathcal{U}$. We will do this by a clever induction on the nth skeleton $K^{(n)}$ of K for each $n \geq 0$. For each vertex $v \in K^{(0)}$, we choose a member $U(v) \in \mathcal{U}$ such that $v \in U(v)$. Let $\langle v_0, v_1 \rangle$ be a 1-simplex of K. By the Lebesgue covering lemma, we can find an $\epsilon > 0$ such that any closed interval of length ϵ is contained in some member of $\mathcal{U}$. By subdividing the interval into uniform mesh of size $\epsilon/2$, we get a subdivision K' of $\langle v_0, v_1 \rangle$ such that closed stars of vertices of K' lying in the open interval (v_0, v_1) are contained in some member of $\mathcal{U}$, while closed star of vertices in $K^{(0)}$ are contained in the already chosen members of $\mathcal{U}$. Subdividing each 1-simplex as above, we find a subdivision K' of $K^{(1)}$ such that closed stars of vertices of K' are contained in some member of $\mathcal{U}$. This starts the induction at $n = 1$.

We suppose that $n > 1$, and $K^{(n-1)}$ has a subdivision K' such that closed star of any vertex of K' is contained in a member of $\mathcal{U}$. Let σ^n be an n-simplex, and consider the boundary $\dot{\sigma}^n$ of σ^n. This boundary, being contained in $K^{(n-1)}$, has already been subdivided such that closed vertex stars of vertices of $K' \cap \dot{\sigma}^n$ are contained in some member of $\mathcal{U}$. For each vertex p of $K' \cap \dot{\sigma}^n$, let us choose a member $U(p)$ of $\mathcal{U}$ such that closed star of p is contained in $U(p)$. Let b be the barycentre of σ, and consider the cone with vertex b over the boundary of the new subdivision of $\dot{\sigma}^n$ induced by K'. If p is such a vertex of $\dot{\sigma}^n$, then note that closed vertex star of p is contained in some member of $\mathcal{U}$. Hence, there is an $\epsilon > 0$ such that the ϵ-nbd of that closed star of p is also contained in the same member of $\mathcal{U}$. Now, we take the barycentric subdivision of all open simplexes contained in the above cone. Note that this subdivision does not affect the subdivision of $\dot{\sigma}^n$ already obtained. Repeating this process of taking barycentric subdivisions of open simplexes in σ^n, we can find an $\epsilon > 0$ such that simplexes of this new subdivision of diameter $< \epsilon/2$ are contained in some member of $\mathcal{U}$. Combining both subdivisions given above, we can find an $\epsilon > 0$ such that closed vertex stars of new vertices are contained in some member of $\mathcal{U}$ while the closed vertex stars of vertices p lying on the boundary of σ^n are contained in members $U(p)$ of $\mathcal{U}$ already chosen. Thus, the new subdivision of σ^n does not disturb the subdivision of $\dot{\sigma}^n$ already obtained. We do this for each σ^n and then conclude that the nth skeleton $K^{(n)}$ has a subdivision K'' such that the closed stars of all vertices of K'' are contained in some member of $\mathcal{U}$. Hence, by induction, it follows that we can find a subdivision $\mathrm{Sd}(K)$ of K such that closed stars of vertices of $\mathrm{Sd}(K)$ form a refinement of the open cover $\mathcal{U}$.

Now, consider the open stars $\{\text{ost}(v')\}$ of all the vertices v' of $\text{Sd}(K)$ in $|K|_d$. These open stars are also open in $|\text{Sd}(K)|_d$ since the coordinate functions $f_{v'} : |\text{Sd}(K)|_d = |K|_d \to [0,1]$ are continuous functions. Since $|\text{Sd}(K)|_d$ is paracompact, this covering has a locally finite open refinement, say $\mathcal{V} = \{V\}$ in $|K|_d$. Then, clearly, $\{i^{-1}(V)\}$ is a locally finite open refinement of $\mathcal{U}$ in $|K|$.
■

Exercises

1. Prove that the subspace $(0,1]$ of the real line $\mathbb{R}$ is a triangulable space.
2. Let v be a vertex of any simplicial complex K. Show that $\text{ost}(v)$ and $\overline{\text{ost}(v)}$ are path-connected in $|K|$.
3. Let s be a simplex. Prove that the closed simplex $|s|$ is homeomorphic to the cone with base $|\text{Bd}(s)|$, i.e., $|s|$ is homeomorphic to $|\text{Bd}(s)| * w$.
4. Let U be a bounded open subset of $\mathbb{R}^n$ and suppose U is star-convex with respect to the origin. Show that
   (a) a ray from origin may intersect $\text{Bd}(U)$ in more than one point.
   (b) there is an example of a U such that $\overline{(U)}$ need not be homeomorphic to the disc $\mathbb{D}^n$.
5. A simplicial complex K is said to be **locally finite** if each vertex of K is the vertex of only finitely many simplexes of K. If K is locally finite, prove that every point of $|K|$ has a nbd of the form $|L|_d$ where L is a finite subcomplex of K.
6. Let K be a simplicial complex. Show that the following conditions are equivalent:
   (a) K is locally finite.
   (b) $|K|$ is locally compact.
   (c) $|K| \to |K|_d$ is a homeomorphism.
   (d) $|K|$ is metrizable.
   (e) $|K|$ satisfies the first axiom of countability.
7. Prove that every polyhedron $|K|$ is a perfectly normal space.
8. Suppose K is a countable simplicial complex which is locally finite and $\dim(\text{K}) \leq n$. Then prove that $|K|$ can be linearly embedded in $\mathbb{R}^{2n+1}$.
9. Show that the assignment $K \to |K|$ is a covariant functor from the category of simplicial complexes and simplicial maps to the category of topological spaces and continuous maps. More generally, show that $(K, L) \to (|K|, |L|)$ is a covariant functor from the category of simplicial pairs and simplicial maps to the category of topological pairs and continuous maps of pairs.
10. Let (X, A) be a polyhedral pair. Prove that A is a strong deformation retract of some nbd of A in X.

Chapter 4

Simplicial Homology

4.1 Introduction

"Homology groups" associated to a given simplicial complex K, constitute the first comprehensive topic of the subject of algebraic topology. In this chapter, we will explain how we can associate a sequence of abelian groups $\{H_n(K) : n \geq 0\}$ to a given simplicial complex K. These groups, called **homology groups** of the simplicial complex K, will have interesting functorial properties, viz., for each simplicial map $f\colon K \to L$, there will be an induced group homomorphism $(f_*)_n\colon H_n(K) \to H_n(L)$ for each $n \geq 0$ satisfying the following two properties:

(i) If $f\colon K \to L$ and $g\colon L \to M$ are two simplicial maps, then for each $n \geq 0$,

$$((g \circ f)_*)_n = (g_*)_n \circ (f_*)_n\colon H_n(K) \to H_n(M).$$

(ii) If $I_K\colon K \to K$ is the identity map, then for each $n \geq 0$, the induced map $((I_K)_*)_n$ is the identity map on $H_n(K)$.

These two functorial properties, which are usual for any kind of homology groups, yield the result that any simplicial isomorphism from K to L induces isomorphisms between their homology groups $H_n(K)$ and $H_n(L)$ for each $n \geq 0$. Therefore, the homology groups $H_n(K)$ of the simplicial complex K are invariants of the simplicial structure of the complex K. This is a typical phenomenon with most of the algebraic invariants which an algebraic topologist tends to define and study. The simplicial homology groups happen to be one of the first classical algebraic objects to be invented and studied. The most important and interesting feature of these groups is that they lend themselves to easy computations.

Let X be a polyhedron and K_1, K_2 be two triangulations of X. We will finally prove that $H_n(K_1)$ is isomorphic to $H_n(K_2)$ for each $n \geq 0$. This allows us to define the homology groups $\{H_n(X) : n \geq 0\}$ of a polyhedron X simply by declaring $H_n(X)$ to be $H_n(K)$, where K is some triangulation of X. This means that the simplicial homology groups of a simplicial complex K are not only the invariants of simplicial structure of K, but more importantly, these are topological invariants of the space $|K|$. In particular, we will conclude that the Euler-characteristic of a polyhedron X, viz., (# of vertices − # of edges + # of triangles − ...) in a triangulation of X is a topological invariant. By a somewhat lengthy technical procedure, we will also show that any continuous map $f: X \to Y$ between two polyhedra $X = |K|$, and $Y = |L|$ induces group homomorphisms $(f_*)_n: H_n(K) \to H_n(L)$ for each $n \geq 0$, which satisfy the analogues of the two functorial properties mentioned earlier. One of the first uses of these homology groups will be made in solving some "classification problems" of point set topology. For instance, it will be shown that two spheres $\mathbb{S}^m$ and $\mathbb{S}^n$ are homeomorphic if and only if $m = n$, with a similar result for Euclidean spaces $\mathbb{R}^m$ and $\mathbb{R}^n$. We will present several other interesting applications of these homology groups and induced homomorphisms. Detailed proofs of the famous theorems like Brouwer's Fixed Point Theorem, Lefschetz Fixed Point Theorem, Borsuk-Ulam Theorem, etc. will be given. The question of existence of tangent vector fields on spheres will also be answered.

4.2 Orientation of Simplicial Complexes

Let us consider an n-simplex $\sigma^n = \langle v_0, v_1, \ldots, v_n \rangle$. Suppose the vertices of σ^n are ordered by declaring $v_0 < v_1 < v_2 < \ldots < v_n$. Then this ordering determines a certain direction among the vertices of σ^n. Any other arrangement of these vertices gives rise to a permutation of vertices, which is either an even permutation or an odd permutation of the given ordering. When we consider the simplex σ^n together with the equivalence class of all even permutations of its vertices, we say that σ^n is **positively oriented** and write the pair as $+\sigma^n$. On the other hand, when σ^n is considered with the equivalence class of all odd permutations, we say that σ^n is **negatively oriented** and we write this as $-\sigma^n$. Thus, by an oriented simplex σ^n, we mean the simplex σ^n together with a certain order of its vertices determining the positive orientation. The same simplex σ^n, when we change the ordering of its vertices by a transposition, will have positive orientation with respect to the new ordering whereas it has negative orientation with respect to the earlier ordering.

As an example, consider the 1-simplex $\sigma^1 = \langle v_0, v_1 \rangle$. If we order $v_0 < v_1$, then $+\sigma^1$ is $\langle v_0, v_1 \rangle$ and $-\sigma^1$ is $\langle v_1, v_0 \rangle$. The two simplexes $+\sigma^1$ and $-\sigma^1$ are now different 1-simplexes, one having the direction from v_0 to v_1 and the other from v_1 to v_0.

$+\sigma^1$ v_0 → v_1 $-\sigma^1$ v_0 ← v_1

As another example, consider the 2-simplex $\sigma^2 = \langle v_0, v_1, v_2 \rangle$ (Fig. 4.1).

If we order the vertices as $v_0 < v_1 < v_2$, then $+\sigma^2 = \langle v_0, v_1, v_2 \rangle = \langle v_1, v_2, v_0 \rangle = \langle v_2, v_0, v_1 \rangle$ and $-\sigma^2 = \langle v_1, v_0, v_2 \rangle = \langle v_0, v_2, v_1 \rangle = \langle v_2, v_1, v_0 \rangle$. In fact, orienting a simplex means fixing the positive direction of its vertices, and then the negative direction is automatically fixed. Moreover, there are only two directions which can be assigned to its vertices.

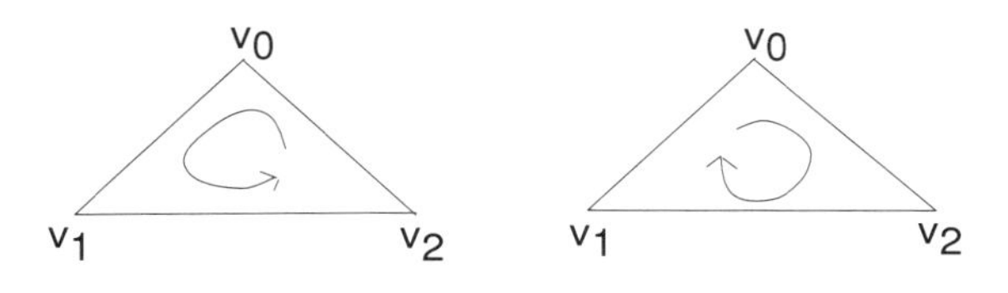

Fig. 4.1

A 0-simplex $\langle v_0 \rangle$ is always taken to be positively oriented, by assumption. Note that if an n-simplex $\sigma^n = \langle v_0, v_1, \ldots, v_n \rangle$ is oriented by the ordering $v_0 < v_1 < \ldots < v_n$, then all of its faces are automatically oriented by the ordering induced by the above ordering. We have

Definition 4.2.1. *A simplicial complex K is said to be* **oriented** *if each of its simplexes is assigned an orientation.*

There are several ways of orienting a simplicial complex K. As remarked a while ago, if we assign a certain ordering to the set of all vertices of K, then that ordering will induce an ordering in the vertices of every simplex of K. Consequently, each simplex is automatically assigned an orientation and the complex K becomes an **oriented complex**. Another way of orienting a complex K is to assign an orientation to each of its simplexes arbitrarily. When we orient by the latter method and a simplex τ of K is a face of a simplex σ of K, it might happen that the orientation on τ induced by σ may coincide or may be opposite to the original orientation of τ.

It may be mentioned that assigning an orientation to a simplicial complex K is a preparation towards the main objective of associating some algebraic structure to the complex K. We also note that the following definition depends on the orientation of K.

Definition 4.2.2. *Let K be an oriented simplicial complex and σ^p, σ^{p+1} be two simplexes whose dimensions differ by 1. With each such pair (σ^{p+1}, σ^p), we associate its* **incidence number***, denoted by $[\sigma^{p+1}, \sigma^p]$, as follows: If σ^p is not a face of σ^{p+1}, we put $[\sigma^{p+1}, \sigma^p] = 0$. If σ^p is a face of σ^{p+1}, we label the vertices of σ^p so that $+\sigma^p = \langle v_0, v_1, \ldots, v_p \rangle$. Let v be the additional vertex of σ^{p+1}.*

Then $\langle v, v_0, \ldots, v_p\rangle$ is either $+\sigma^{p+1}$ or $-\sigma^{p+1}$. We define

$$[\sigma^{p+1}, \sigma^p] = \begin{cases} 1 & \text{if } \langle v, v_0, \ldots, v_p\rangle = +\sigma^{p+1} \\ -1 & \text{if } \langle v, v_0, \ldots, v_p\rangle = -\sigma^{p+1}. \end{cases}$$

Example 4.2.3. Let the closure of σ^2, where $\sigma^2 = \langle v_0, v_1, v_2\rangle$, be oriented by the ordering $v_0 < v_1 < v_2$. Then

$$\begin{array}{ll} [\sigma^2, \langle v_0, v_1\rangle] = +1 & [\langle v_0, v_1\rangle, \langle v_1\rangle] = +1 \\ [\sigma^2, \langle v_1, v_2\rangle] = +1 & [\langle v_0, v_1\rangle, \langle v_0\rangle] = -1 \\ [\sigma^2, \langle v_0, v_2\rangle] = -1 & [\langle v_0, v_1\rangle, \langle v_2\rangle] = 0. \end{array}$$

The following result on these incidence numbers is of crucial importance:

Theorem 4.2.4. *Let K be an oriented simplicial complex. If σ^{p-2} is a $(p-2)$-face of a simplex σ^p of K, then*

$$\sum[\sigma^p, \sigma^{p-1}][\sigma^{p-1}, \sigma^{p-2}] = 0,$$

where the summation is over all $(p-1)$-simplexes σ^{p-1} of K.

Proof. Let $v_0, v_1, \ldots, v_{p-2}$ be the vertices of σ^{p-2} so that $+\sigma^{p-2} = \langle v_0, \ldots, v_{p-2}\rangle$ represents the positive orientation of σ^{p-2}. Let the two additional vertices of σ^p be a and b. Without any loss of generality, we can assume that $+\sigma^p = \langle a, b, v_0, \ldots, v_{p-2}\rangle$. There are only two $(p-1)$-simplexes of K which yield the nonzero terms in the above summation, viz.,

$$\sigma_1^{p-1} = \langle a, v_0, \ldots, v_{p-2}\rangle, \qquad \sigma_2^{p-1} = \langle b, v_0, \ldots, v_{p-2}\rangle.$$

The indicated ordering of the vertices of σ_1^{p-1} (respectively σ_2^{p-1}) may give either the positive orientation or negative orientation of σ_1^{p-1} (respectively σ_2^{p-1}). Thus, there are four distinct possibilities which we must examine:

Case I. Suppose $+\sigma_1^{p-1} = \langle a, v_0, \ldots, v_{p-2}\rangle$ and $+\sigma_2^{p-1} = \langle b, v_0, \ldots, v_{p-2}\rangle$. Then it is easily seen that

$$\begin{array}{ll} [\sigma^p, \sigma_1^{p-1}] = -1, & [\sigma_1^{p-1}, \sigma^{p-2}] = +1 \\ [\sigma^p, \sigma_2^{p-1}] = +1, & [\sigma_2^{p-1}, \sigma^{p-2}] = +1. \end{array}$$

Hence the two nonzero terms in the summation are -1 and $+1$, respectively, i.e., the summation is zero.

Case II. Suppose $+\sigma_1^{p-1} = \langle a, v_0, \ldots, v_{p-2}\rangle$ and $-\sigma_2^{p-1} = \langle b, v_0, \ldots, v_{p-2}\rangle$. Then

$$\begin{array}{ll} [\sigma^p, \sigma_1^{p-1}] = -1, & [\sigma_1^{p-1}, \sigma^{p-2}] = +1 \\ [\sigma^p, \sigma_2^{p-1}] = -1, & [\sigma_2^{p-1}, \sigma^{p-2}] = -1 \end{array}$$

and the desired summation is again zero.

Case III. Suppose $-\sigma_1^{p-1} = \langle a, v_0, \ldots, v_{p-2}\rangle$ and $+\sigma_2^{p-1} = \langle b, v_0, \ldots, v_{p-2}\rangle$. Then

$$[\sigma^p, \sigma_1^{p-1}] = +1, \qquad [\sigma_1^{p-1}, \sigma^{p-2}] = -1$$
$$[\sigma^p, \sigma_2^{p-1}] = +1, \qquad [\sigma_2^{p-1}, \sigma^{p-2}] = +1$$

and the summation is zero in this case also.

Case IV. Finally, suppose $-\sigma_1^{p-1} = \langle a, v_0, \ldots, v_{p-2}\rangle$ and $-\sigma_2^{p-1} = \langle b, v_0, \ldots, v_{p-2}\rangle$. Then

$$[\sigma^p, \sigma_1^{p-1}] = +1, \qquad [\sigma_1^{p-1}, \sigma^{p-2}] = -1$$
$$[\sigma^p, \sigma_2^{p-1}] = -1, \qquad [\sigma_2^{p-1}, \sigma^{p-2}] = -1$$

and the summation is again zero. ∎

4.3 Simplicial Chain Complex and Homology

Let K be a simplicial complex with a fixed orientation. Let $\tilde{S}_q$ denote the set of all oriented q-simplexes of K. Since each q-simplex, $q \geq 1$, can be oriented in exactly two distinct ways, the number of elements in the set $\tilde{S}_q$ is twice the number of q-simplexes in K, whereas the number of oriented 0-simplexes is the same as the number of vertices of K because 0-simplexes are assumed to be always positively oriented.

Definition 4.3.1. *Let $0 \leq q \leq \dim K$ and $\mathbb{Z}$ be the additive group of integers. Any map $f\colon \tilde{S}_q \to \mathbb{Z}$ with the property that if $q \geq 1$, then $f(-\sigma^q) = -f(\sigma^q)$ for each $\sigma^q \in \tilde{S}_q$, is called a q-**chain** of K. For $q = 0$, a 0-chain is just a mapping from the set of all 0-simplexes of K to $\mathbb{Z}$. The set of all q-chains of K is denoted by $C_q(K)$. If $q < 0$ or $q > \dim K$, we define $C_q(K) = 0$.*

It is easy to see that the set $C_q(K)$ of all q-chains is an abelian group with point-wise operations, viz.,

$$(f + g)(\sigma^q) = f(\sigma^q) + g(\sigma^q)$$

for all $\sigma^q \in \tilde{S}_q$. The zero map $0\colon \tilde{S}_q \to \mathbb{Z}$ defined by $0(\sigma^q) = 0$ for all q-simplexes σ^q is the additive identity, and the inverse $-f$ of f is defined by

$$(-f)(\sigma^q) = -f(\sigma^q).$$

The group $C_q(K)$ is called the q-dimensional **chain group** of K. If we imagine the collection of all chain groups $C_q(K)$ arranged in descending order, we have an infinite sequence of abelian groups

$$C(K): \qquad \ldots, 0, C_n(K), C_{n-1}(K), \ldots, C_q(K), \ldots, C_0(K), 0, \ldots,$$

where all $C_q(K)$ are zero except for $q = 0, 1, 2, \ldots, n = \dim(K)$.

We can give a nice description of the elements of $C_q(K)$ which is easy to work with. For each positively oriented q-simplex σ^q, define a q-chain $\bar{\sigma}^q$ (this notation is only temporary) as follows:

$$\bar{\sigma}^q(\tau^q) = \begin{cases} +1, & \text{if } \tau^q = \sigma^q \\ -1, & \text{if } \tau^q = -\sigma^q \\ 0, & \text{otherwise.} \end{cases}$$

Then $\bar{\sigma}^q$ is indeed a q-chain, and is called an **elementary q-chain** of K. The next result shows that $C_q(K)$ is a free abelian group generated by all elementary q-chains. Let S_q denote the set of all positively oriented q-simplexes of K.

Proposition 4.3.2. *For each $q \geq 0$, $C_q(K) = \bigoplus \mathbb{Z} \cdot \bar{\sigma}^q$, $\sigma^q \in S_q$.*

Proof. It suffices to show that each $f \in C_q(K)$ can be written uniquely as an integral linear combination of the elementary q-chains $\bar{\sigma}^q$. Suppose $f(\sigma^q) = n_q$, where $\sigma^q \in S_q$. Then, by definition, $f(-\sigma^q) = -n_q$. Consider the element $\sum_{\sigma^q \in S_q} n_q \bar{\sigma}^q$. We claim that $f = \sum n_q \bar{\sigma}^q$. To see this, we must show that f and $\sum n_q \bar{\sigma}^q$ both have the same value at each oriented q-simplex of K. If $\sigma^q \in S_q$, then

$$(\textstyle\sum n_q \bar{\sigma}^q)(\sigma^q) = n_q \bar{\sigma}^q(\sigma^q) = n_q = f(\sigma^q).$$

If $\sigma^q \notin S_q$, then $-\sigma^q \in S_q$ and

$$\begin{aligned} f(\sigma^q) &= f(-(-\sigma^q)) \\ &= -f(-\sigma^q), \quad \text{by definition} \\ &= -(\textstyle\sum n_q \bar{\sigma}^q)(-\sigma^q), \quad \text{by what we proved earlier} \\ &= (\textstyle\sum n_q \bar{\sigma}^q)(\sigma^q). \end{aligned}$$

This proves our claim. To prove the uniqueness, suppose $f = \sum n_q \bar{\sigma}^q = \sum m_q \bar{\sigma}^q$. Then $\sum (n_q - m_q)\bar{\sigma}^q = 0$ in $C_q(K)$ and so $(\sum (n_q - m_q)\bar{\sigma}^q)(\sigma^q) = 0$ in $\mathbb{Z}$. However, this means $n_q - m_q = 0$. Since σ^q is arbitrary, we find that the corresponding coefficients of $\bar{\sigma}^q$ in the two expressions of f are identical. ∎

If we had a different orientation of the complex K, we would have ended up with a different set of generators for $C_q(K)$, viz., the new generators would be the same $\bar{\sigma}^q$ if the orientation of the simplex σ^q remains the same and would be $-(\bar{\sigma}^q)$ if σ^q gets the opposite orientation. In either case, however, the group $C_q(K)$ would be isomorphic to the direct sum of as many copies of $\mathbb{Z}$ as there are q-simplexes in K. This also says that the group stucture of $C_q(K)$ depends on the simplicial complex K, not on the particular orientation of K.

The elementary q-chains $\bar{\sigma}^q$ are functions from the set of all oriented q-simplexes of K to the additive group $\mathbb{Z}$ of integers. We could as well consider q-chains as functions from the set of all oriented q-simplexes to any abelian group G. In that case, the chain groups, denoted by $C_q(K; G)$, would be isomorphic to the direct sum of as many copies of G as there are q-simplexes in K. The group G is called the **coefficient group** for the chain groups. In this terminology, $C_q(K)$ is really $C_q(K;\mathbb{Z})$. We will have occasion to use coefficients other than the group $\mathbb{Z}$ of integers, and sometimes their use will be preferable. It is natural to ask how the groups $C_q(K) = C_q(K;\mathbb{Z})$ and $C_q(K;G)$ are related. The answer is provided in terms of the so called **tensor product**, viz., $C(K,G) = C(K) \otimes G$, an important well-known algebraic concept.

The elementary q-chain $\bar{\sigma}^q$ depends on the q-simplex σ^q and its orientation. If we fix an orientation for each q-simplex σ^q, then the group $C_q(K) = \bigoplus_{\sigma^q \in S_q} \mathbb{Z} \cdot \bar{\sigma}^q$ is isomorphic to the group $\bigoplus \mathbb{Z} \cdot \sigma^q$, where $\mathbb{Z} \cdot \sigma^q$ is the infinite cyclic group generated by the simplex σ^q. Since the elements of the latter group are simply the formal integral linear combinations of q-simplexes σ^q, rather than elementary chains $\bar{\sigma}^q$, there is no loss of generality if we identify $\bar{\sigma}^q$ with $\sigma^q = 1 \cdot \sigma^q$ and write $C_q(K) = \oplus_{\sigma^q \in S_q} \mathbb{Z} \cdot \sigma^q$. This is what we shall do from now onwards and drop the temporary notation $\bar{\sigma}^q$ used earlier.

One might ask at this stage the following question: The q-dimensional chain group $C_q(K)$ associated to the oriented complex K is indeed isomorphic to the direct sum of as many copies of $\mathbb{Z}$ as there are q-simplexes in K regardless of what orientation we give to K. Then, why to introduce an orientation in K and make the definition of $C_q(K)$ so lengthy and complicated? The answer to this question will become clear as we proceed to define the **boundary map** and realize the geometrical interpretation of the same. In what follows, whenever we write a q-simplex as σ^q, we mean σ^q is positively oriented.

Definition 4.3.3. *For each q, $0 < q \leq \dim K$, we now define a homomorphism $\partial_q \colon C_q(K) \to C_{q-1}(K)$, called the* **boundary homomorphism**, *as follows: on generators σ^q of $C_q(K)$, we define*

$$(*) \qquad \partial_q(\sigma^q) = \sum_{i=0}^{q} [\sigma^q, \sigma_i^{q-1}] \sigma_i^{q-1},$$

where σ_i^{q-1} runs over all positively oriented $(q-1)$-faces of σ^q. Then we extend it over $C_q(K)$ by linearity, i.e., we set

$$\partial_q\left(\sum n_q \sigma^q\right) = \sum n_q \partial_q(\sigma^q),$$

where $\partial_q(\sigma^q)$ is defined by $(*)$. For $q \leq 0$ or $q > \dim K$, we define ∂_q to be the zero homomorphism (which is incidently the only possible map).

We note that in the summation $(*)$, σ_i^{q-1} runs over all $(q-1)$-faces of σ^q. If we allow σ_i^{q-1} to run over all $(q-1)$-simplexes of K, the summation will remain unchanged since $[\sigma^q, \sigma_i^{q-1}] = 0$ if σ_i^{q-1} is not a face of σ^q. Thus, we can write $(*)$ as

$$\partial_q(\sigma^q) = \sum_{\sigma_i^{q-1} \in K} [\sigma^q, \sigma_i^{q-1}] \sigma_i^{q-1}.$$

Let us observe that if $\sigma^q = \langle v_0, v_1, \ldots, v_q\rangle$ is oriented by the ordering $v_0 < v_1 < \ldots < v_q$, then σ^q has $(q+1)$-faces $\langle v_0, v_1, \ldots, \hat{v}_i, \ldots, v_q\rangle$, $i = 0, 1, \ldots, q$, where $\hat{v}_i$ means the vertex v_i has been omitted from $\langle v_0, v_1, \ldots, v_i, \ldots, v_q\rangle$ and each of these faces is oriented by the induced ordering. Evidently the incidence number $[\sigma^q, \sigma_i^{q-1}]$ is $(-1)^i$ and, therefore, $(*)$ reads as:

$$(**) \qquad \partial_q\langle v_0, v_1, \ldots, v_q\rangle = \sum_{i=0}^{q}(-1)^i\langle v_0, v_1, \ldots, \hat{v}_i, \ldots, v_q\rangle.$$

Many authors use this representation of the boundary map rather than $(*)$ to avoid the concept of incidence number.

There are, however, two points which we must answer before ∂_q becomes a well-defined map by $(**)$. One, a generator $\langle v_0, v_1, \ldots, v_q\rangle$ of $C_q(K)$ can be written in several ways as, for example,

$$\langle v_0, v_1, \ldots, v_q\rangle = \langle v_{\pi(0)}, v_{\pi(1)}, \ldots, v_{\pi(q)}\rangle,$$

where π is any even permutation on the symbols $(0, 1, 2, \ldots, q)$. Therefore, we must ensure that $\partial_q\langle v_0, v_1, \ldots, v_q\rangle$ is independent of these representations. Second, if π is an odd permutation, then $\langle v_{\pi(0)}, v_{\pi(1)}, \ldots, v_{\pi(q)}\rangle$ is also a generator of $C_q(K)$, and the relation $\langle v_{\pi(0)}, v_{\pi(1)}, \ldots, v_{\pi(q)}\rangle = -\langle v_0, v_1, \ldots, v_q\rangle$ holds among such generators. Hence we must show that for an odd permutation π,

$$\partial_q\langle v_{\pi(0)}, v_{\pi(1)}, \ldots, v_{\pi(q)}\rangle = -\partial_q\langle v_0, v_1, \ldots, v_q\rangle.$$

Since any permutation can be written as a product of transpositions, both the points raised earlier follow from the next result.

Proposition 4.3.4. *If π is a transposition on the symbols $(0, 1, 2, \cdots, q)$, then $\partial_q\langle v_{\pi(0)}, v_{\pi(1)}, \ldots, v_{\pi(q)}\rangle = -\partial_q\langle v_0, v_1, \ldots, v_q\rangle$.*

Proof. Let $\sigma^q = \langle v_0, v_1, \ldots, v_s, \ldots, v_t, \ldots, v_q\rangle$, $0 \le s < t \le q$ and π be the transposition which interchanges v_s and v_t. Suppose $\tau^q = \langle v_0, v_1, \ldots, v_t, \ldots, v_s, \ldots, v_q\rangle$ is the new oriented simplex by the indicated ordering. We show that $\partial_q(\sigma^q) = \sum_{i=0}^q[\sigma^q, \sigma_i^{q-1}]\sigma_i^{q-1}$ is the negative of $\partial_q(\tau^q) = \sum_{i=0}^q[\tau^q, \tau_i^{q-1}]\tau_i^{q-1}$. If $s \neq i \neq t$, then observe that $[\sigma^q, \sigma_i^{q-1}]\sigma_i^{q-1} = (-1)^i\sigma_i^{q-1}$ and $[\tau^q, \tau_i^{q-1}]\tau_i^{q-1} = (-1)^i\tau_i^{q-1}$. Because for each such i, σ_i^{q-1} can be obtained from τ_i^{q-1} by just one transposition, the two terms differ only in sign. If $i = s$, then $[\sigma^q, \sigma_s^{q-1}]\sigma_s^{q-1} = (-1)^s\langle v_0, \ldots, \hat{v}_s, \ldots, v_t, \ldots, v_q\rangle$ corresponds to the term

$$[\tau^q, \tau_s^{q-1}]\tau_s^{q-1} = (-1)^s\langle v_0, \ldots, v_t, \ldots, \hat{v}_s, \ldots, v_q\rangle.$$

Note that we need $(t-s-1)$ transpositions to change $\langle v_0, \ldots, v_t, \ldots, \hat{v}_s, \ldots, v_q\rangle$ into $\langle v_0, \ldots, \hat{v}_s, \ldots, v_t, \ldots, v_q\rangle$, i.e.,

$$\begin{aligned}
[\sigma^q, \sigma_s^{q-1}]\sigma_s^{q-1} &= (-1)^{(t+s)+(t-s-1)}[\tau^q, \tau_t^{q-1}]\tau_t^{q-1} \\
&= (-1)^{2t-1}[\tau^q, \tau_t^{q-1}]\tau_t^{q-1} \\
&= -[\tau^q, \tau_t^{q-1}]\tau_t^{q-1}.
\end{aligned}$$

Finally, if $i = t$, then a similar argument would show that

$$[\sigma^q, \sigma_t^{q-1}]\sigma_t^{q-1} = -[\tau^q, \tau_s^{q-1}]\tau_s^{q-1}.$$

Summing all the terms yields the desired result. ■

The next result is basic in defining the simplicial homology groups of an oriented simplicial complex K.

Lemma 4.3.5. *For each q, the composite homomorphism $\partial_{q-1} \circ \partial_q \colon C_q(K) \to C_{q-2}(K)$ is the zero map.*

Proof. Let σ^q be a generator of chain group $C_q(K)$. It suffices to show that $\partial_{q-1} \circ \partial_q(\sigma^q) = 0$. Now

$$\begin{aligned} \partial_{q-1}(\partial_q(\sigma^q)) &= \partial_{q-1}\Big(\sum_{\sigma_i^{q-1} \in K} [\sigma^q, \sigma_i^{q-1}]\sigma_i^{q-1} \Big) \\ &= \sum_{\sigma_i^{q-1} \in K} [\sigma^q, \sigma_i^{q-1}]\partial_{q-1}(\sigma_i^{q-1}) \\ &= \sum_{\sigma_i^{q-1} \in K} [\sigma^q, \sigma_i^{q-1}]\Big(\sum_{\sigma_j^{q-1} \in K} [\sigma_i^{q-1}, \sigma_j^{q-2}]\sigma_j^{q-2} \Big). \end{aligned}$$

Changing the order of summation and collecting the coefficients of each $(q-2)$-simplex σ_j^{q-2} shows that the above term is equal to

$$\sum_{\sigma_j^{q-2} \in K} \Big(\sum_{\sigma_i^{q-1} \in K} [\sigma^q, \sigma_i^{q-1}][\sigma_i^{q-1}, \sigma_j^{q-2}] \Big) \sigma_j^{q-2}.$$

By Theorem 4.2.4,

$$\sum_{\sigma_i^{q-1} \in K} [\sigma^q, \sigma_i^{q-1}][\sigma_i^{q-1}, \sigma_j^{q-2}] = 0,$$

for each σ_j^{q-2} and so $\partial_{q-1} \circ \partial_q(\sigma^q) = 0$. ■

The notations and terminology introduced below will occur frequently in the sequel.

Definition 4.3.6. *Let K be an oriented complex. A q-chain $z_q \in C_q(K)$ is called a* **q-dimensional cycle** *of K (or just a* **q-cycle***) if $\partial_q(z_q) = 0$. The set of all q-cycles of K, denoted by $Z_q(K)$, is the kernel of the boundary homomorphism $\partial_q \colon C_q(K) \to C_{q-1}(K)$ and, therefore, is a subgroup of $C_q(K)$. This $Z_q(K)$ is referred to as* **group of q-cycles** *of K. An element $b_q \in C_q(K)$ is said to be a* **q-dimensional boundary** *(or just a* **q-boundary***) of K if there exists a $c' \in C_{q+1}(K)$ such that $\partial_{q+1}(c') = b_q$. The set of all q-boundaries, being the homomorphic image $\partial_{q+1}(C_{q+1}(K))$, is also a subgroup of $C_q(K)$. It is denoted by $B_q(K)$ and is referred to as* **group of q-boundaries of** *K.*

If $\dim K = n$, then note that there is a sequence

$$\begin{aligned} C(K) \quad : \quad & \ldots 0 \to C_n(K) \stackrel{\partial_n}{\to} \ldots \to C_{q+1}(K) \stackrel{\partial_{q+1}}{\to} C_q(K) \stackrel{\partial_q}{\to} C_{q-1}(K) \\ & \to \ldots \to C_0(K) \to 0 \to \ldots \end{aligned}$$

of free abelian groups and group homomorphisms in which the composite of any two consecutive homomorphisms is zero (Lemma 4.3.5). This long sequence is called the **oriented simplicial chain complex of** K. Since $\partial_q \circ \partial_{q+1} = 0$ for each q, $\mathrm{Im}\, \partial_{q+1} \subseteq \ker \partial_q$, i.e., $B_q(K) \subseteq Z_q(K)$. Consequently, one can talk of the factor group $Z_q(K)/B_q(K)$.

The factor group mentioned above leads us to the following basic

Definition 4.3.7. *Let K be an oriented simplicial complex. Then the q-dimensional* **homology group** *of K, denoted by $H_q(K)$, is defined to be the quotient group*

$$H_q(K) = Z_q(K)/B_q(K).$$

If we consider the chain complex $C_*(K; G)$ where G is any coefficient group, then we can similarly define

$$H_q(K; G) = Z_q(K; G)/B_q(K; G)$$

which will be called the **homology groups of K with G as coefficients.**

Observe that for $q < 0$ or $q > \dim K$, all q-dimensional chain groups $C_q(K)$ are evidently zero groups and, therefore, $H_q(K) = 0$ for all such q. Possible nontrivial homology group $H_q(K)$ of K can occur only when $0 \leq q \leq \dim K$. Moreover, note that $B_n(K) = 0$ and $Z_0(K) = C_0(K)$. Let us also point out that for an oriented simplicial complex K, we have now defined four kinds of groups, viz., $C_q(K), Z_q(K), B_q(K)$ and $H_q(K)$ for each q, $0 \leq q \leq \dim K$. It turns out, however, that it is the homology groups $H_q(K)$ which are of fundamental importance rather than other three groups. Later on we will see that the homology groups are indeed invariants of the topology of $|K|$, and they really detect some kind of "holes" in the space $|K|$ whereas other groups mentioned above do not represent anything of $|K|$ like this – they are only the means to define the homology groups. An element of $H_q(K)$ is a coset of $B_q(K)$ in $Z_q(K)$ and is called a **homology class**. If this homology class is $z_q + B_q(K)$, we normally denote it by $\{z_q\}$. We note that a q-cycle z_q will represent a nontrivial element of $H_q(K)$ if and only if it is not a boundary. Two q-cycles z_q and z'_q are said to be **homologous** if they yield the same homology class, i.e., $z_q - z'_q$ is a boundary. A q-cycle is homologous to zero if and only if it is a boundary.

An important question regarding the simplicial homology groups $H_n(K)$ of a simplicial complex K, which must be answered right here, is: do these homology groups $H_n(K)$, which have been defined with respect to a given orientation of K, depend on the orientation of K? We prove below that the answer to this question is "no". Note that a q-simplex σ^q has exactly two orientations, and that a simplicial complex K is oriented by assigning an arbitrary orientation to each of the simplexes of K. Therefore, a simplicial complex can be oriented in several different ways. We have

Theorem 4.3.8. *Let K_1 and K_2 denote the same simplicial complex K equipped with different orientations. Then $H_q(K_1) \cong H_q(K_2), \forall q \geq 0$.*

Proof. If σ^q is an arbitrary simplex of K, then it has one orientation as a simplex of K_1 and has same or different orientation with respect to K_2. If ${}^i\sigma^q$ denotes the simplex σ^q having positive orientation from $K_i, i = 1, 2$, then, clearly, ${}^1\sigma^q = \phi(\sigma) \cdot {}^2\sigma^q$, where $\phi(\sigma) = +1$ or -1 depending on whether σ^q has the same or different orientation received from K_1 and K_2. Thus, we get a map $\phi\colon K \to \{1, -1\}$. Now, we define a sequence $f_q\colon C_q(K_1) \to C_q(K_2)$ of homomorphisms by

$$f_q(\sum g_i \cdot {}^1\sigma_i^q) = \sum g_i \cdot \phi(\sigma^q) \cdot {}^2\sigma_i^q,$$

where $c = \sum g_i \cdot {}^1\sigma_i^q$ is a chain in $C_q(K_1)$. Next, we claim that the above sequence of homomorphisms is a chain map. We will prove that $f\partial = \partial f$ by showing that both of these maps agree on elementary chains ${}^1\sigma^q$. We have

$$\begin{aligned}
f_{q-1}\partial({}^1\sigma^q) &= f_{q-1}\Big(\sum_{\sigma^{q-1}\in K} [{}^1\sigma^q, {}^1\sigma^{q-1}] \cdot {}^1\sigma^{q-1}\Big) \\
&= \sum \phi(\sigma^{q-1})[{}^1\sigma^q, {}^1\sigma^{q-1}] \cdot {}^2\sigma^{q-1}) \\
&= \sum \phi(\sigma^{q-1})[\phi(\sigma^q) \cdot {}^2\sigma^q, \phi(\sigma^{q-1}) \cdot {}^2\sigma^{q-1}] \cdot {}^2\sigma^{q-1}) \\
&= \sum_{\sigma^{q-1}\in K} \phi(\sigma^{q-1})\phi(\sigma^q)\phi(\sigma^{q-1})[{}^2\sigma^q, {}^2\sigma^{q-1}] \cdot {}^2\sigma^{q-1}) \\
&= \phi(\sigma^q)\Big(\sum_{\sigma^{q-1}\in K} [{}^2\sigma^q, {}^2\sigma^{q-1}] \cdot {}^2\sigma^{q-1}\Big) \\
&= \partial(\phi(\sigma^q) \cdot {}^2\sigma^q) \\
&= \partial f_q({}^1\sigma^q).
\end{aligned}$$

We can also define a chain map $g_q\colon C_q(K_2) \to C_q(K_1)$ by putting

$$g_q({}^2\sigma^q) = \phi(\sigma^q) \cdot {}^1\sigma^q$$

on elementary q-chains, and then extending linearly to $C_q(K_2)$. Then, since $\phi(\sigma^q) \cdot \phi(\sigma^q) = 1$, the two chain maps $\{f_q\}$ and $\{g_q\}$ are inverse of each other. Hence the induced homomorphism $f_*\colon H_q(K_1) \to H_q(K_2)$ is an isomorphism for all $q \geq 0$, proving the theorem. ■

4.4 Some Examples

Let us now have some examples to illustrate how these homology groups can be computed in some simple cases with $\mathbb{Z}$ as the coefficient group. These simple cases are not just typical illustrations of computations, but also present interesting hints of some general results and concepts to be identified and studied later on. The reader is advised to carry out every detail of these examples himself.

Example 4.4.1. Consider the simplicial complex K having only one vertex v, i.e., $K = \{v\}$. There is only one possible orientation on K and with that orientation, we have $C_q(K) = 0$ for all $q \neq 0$ and $C_0(K) = \mathbb{Z} \cdot v$ is the free abelian group on the single generator v. Hence we see that for $q \neq 0$, $H_q(K) = 0$ and

$$H_0(K) = \ker \partial_0 / \operatorname{Im} \partial_1 \cong \mathbb{Z}.$$

In fact, in this case, $C_0(K) = Z_0(K) = H_0(K) \cong \mathbb{Z}$.

Example 4.4.2. Consider a two simplex $\sigma^2 = \langle v_0, v_1, v_2 \rangle$ and let $K = \mathrm{Cl}(\sigma^2)$ (Fig. 4.2).

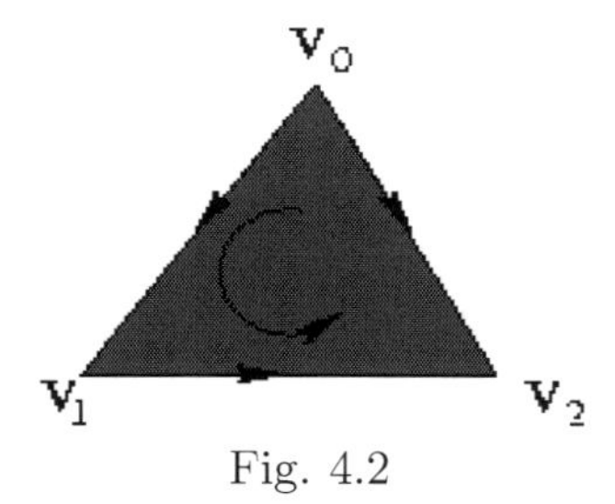

Fig. 4.2

Let us orient K by the ordering $v_0 < v_1 < v_2$ (it has two distinct orientations). We will first compute the non trivial chain groups, viz., $C_0(K), C_1(K), C_2(K)$ and see how the boundary homomorphisms act on these groups. Since there are three vertices in K, $C_0(K)$ is the free abelian group on the three vertices, i.e.,

$$C_0(K) = \mathbb{Z} \cdot v_0 \oplus \mathbb{Z} \cdot v_1 \oplus \mathbb{Z} \cdot v_2.$$

Similarly,

$$C_1(K) = \mathbb{Z} \cdot \langle v_0, v_1 \rangle \oplus \mathbb{Z} \cdot \langle v_1, v_2 \rangle \oplus \mathbb{Z} \cdot \langle v_0, v_2 \rangle$$

and

$$C_2(K) = \mathbb{Z} \cdot \langle v_0, v_1, v_2 \rangle.$$

The nontrivial part of our simplicial chain complex now looks like

$$0 \to C_2(K) \xrightarrow{\partial_2} C_1(K) \xrightarrow{\partial_1} C_0(K) \to 0.$$

Note that any 2-chain c is of the type $c = n\langle v_0, v_1, v_2\rangle$, where $n \in \mathbb{Z}$, and so $\partial_2(c) = n\partial\langle v_0, v_1, v_2\rangle = n[\langle v_1, v_2\rangle - \langle v_0, v_2\rangle + \langle v_0, v_1\rangle]$, which is zero if and only if $n = 0$. Hence, $Z_2(K) = 0$ and it follows at once that $H_2(K) = \ker\partial_2/\text{Im }\partial_3 = 0$. Next, let us see what $B_1(K)$ and $Z_1(K)$ are. If $c_1 = m_0\langle v_0, v_1\rangle + m_1\langle v_1, v_2\rangle + m_2\langle v_0, v_2\rangle$ is an arbitrary 1-chain then note that

(i) $\partial(c_1) = m_0[\langle v_1\rangle - \langle v_0\rangle] + m_1[\langle v_2\rangle - \langle v_1\rangle] + m_2[\langle v_2\rangle - \langle v_0\rangle] = -(m_0 + m_2)\langle v_0\rangle + (m_0 - m_1)\langle v_1\rangle + (m_1 + m_2)\langle v_2\rangle$ which is zero if and only if $m_0 + m_2 = 0, m_0 - m_1 = 0$ and $m_1 + m_2 = 0$, i.e., if and only if $m_0 = m_1 = -m_2$. Hence a 1-cycle is of the form $m\langle v_0, v_1\rangle + m\langle v_1, v_2\rangle - m\langle v_0, v_2\rangle, m \in \mathbb{Z}$. It follows from this that $Z_1(K)$ is isomorphic to the additive group $\mathbb{Z}$ of integers with 1-cycle $\langle v_0, v_1\rangle + \langle v_1, v_2\rangle + \langle v_2, v_0\rangle$ as a generator. On the other hand, to compute $B_1(K)$, note that any 2-chain is of the form $c_2 = n\langle v_0, v_1, v_2\rangle, n \in \mathbb{Z}$. Consequently, $\partial_2(c_2) = n[\langle v_1, v_2\rangle - \langle v_0, v_2\rangle + \langle v_0, v_1\rangle]$. In other words, $B_1(K)$ is also isomorphic to the additive group $\mathbb{Z}$ of integers, and with the same generator as that of $Z_1(K)$. Therefore, $B_1(K) = Z_1(K)$, and hence $H_1(K) = 0$. Finally, we compute $H_0(K)$ which is slightly tricky. We already know that $Z_0(K) = C_0(K)$ is isomorphic to the direct sum of three copies of $\mathbb{Z}$. Let us compute $B_0(K)$. If $c_0 = h_0\langle v_0\rangle + h_1\langle v_1\rangle + h_2\langle v_2\rangle$, where h_i's are integers, is an element of $B_0(K)$, then there must be a 1-chain, say, $c_1 = m_0\langle v_0, v_1\rangle + m_1\langle v_1, v_2\rangle + m_2\langle v_0, v_2\rangle$, where m_0, m_1, m_2 are integers, such that $\partial_1(c_1) = c_0$. Using (i) and comparing the coefficients, we find

(ii) $h_0 = -(m_0 + m_1), h_1 = (m_0 - m_1), h_2 = (m_1 + m_2)$, i.e., $h_0 + h_1 = -h_2$. Thus, if $c_0 \in B_0(K)$, then the third coefficient of c_0 is completely determined by the first two coefficients. This tells us that $B_0(K)$ will be isomorphic to the direct sum of two copies of $\mathbb{Z}$. Now, $Z_0(K)$ and $B_0(K)$ are determined, the first is isomorphic to the direct sum of three copies of $\mathbb{Z}$ and the second is isomorphic to two copies of $\mathbb{Z}$. One might now conjecture that the quotient group $Z_0(K)/B_0(K)$ will be isomorphic to one copy of $\mathbb{Z}$, but this requires a proof. For this, notice that any 0-cycle $c_0 = n_0\langle v_0\rangle + n_1\langle v_1\rangle + n_2\langle v_2\rangle$ can be written as $c_0 = \partial_1[n_1\langle v_0, v_1\rangle + n_2\langle v_0, v_2\rangle] + (n_0 + n_1 + n_2)\langle v_0\rangle$. This means that c_0 is homologous to the 0-chain $(n_0 + n_1 + n_2)\langle v_0\rangle$. Thus, any 0-cycle is homologous to an integral multiple of $\langle v_0\rangle$. On the other hand, by the criterion (ii), any two distinct integral multiples of $\langle v_0\rangle$ cannot be homologous. Thus, $H_0(K)$ is isomorphic to the additive group $\mathbb{Z}$ of integers. We summarize

$$H_q(K) = \begin{cases} \mathbb{Z} & \text{if } q = 0 \\ 0 & \text{if } q \neq 0. \end{cases}$$

Note. The reader should compute the homology groups of the above simplicial complex K, if it is given some other orientation and see whether or not the final result is different.

Example 4.4.3. Let K' be the 1-skeleton of the complex $K = \text{Cl}(\sigma^2)$. Then K' has three vertices v_0, v_1, v_2 and three edges $\langle v_0, v_1\rangle, \langle v_1, v_2\rangle$ and $\langle v_0, v_2\rangle$ (Fig.

4.3) and K' is a triangulation of the space $|K'|$ which is homeomorphic to the circle $\mathbb{S}^1$. Once again, we orient K' by the ordering $v_0 < v_1 < v_2$. Here, the

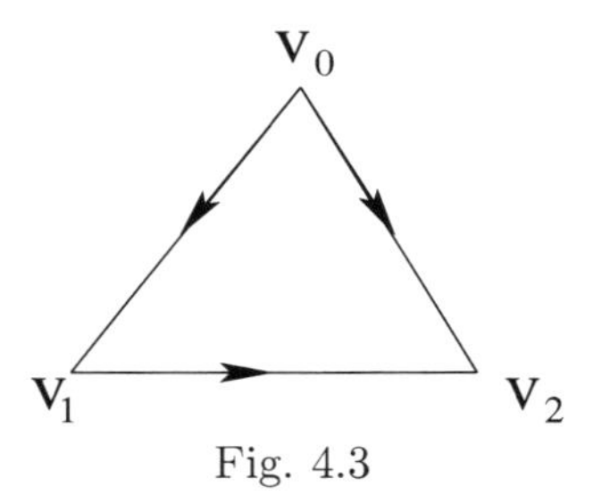

Fig. 4.3

nontrivial part of the simplicial chain complex $C_*(K')$ looks like

$$0 \to C_1(K') \xrightarrow{\partial_1} C_0(K') \to 0,$$

where

$$C_1(K') = \mathbb{Z} \cdot \langle v_0, v_1 \rangle \oplus \mathbb{Z} \cdot \langle v_1, v_2 \rangle \oplus \mathbb{Z} \cdot \langle v_0, v_2 \rangle$$

and

$$C_0(K') = \mathbb{Z} \cdot \langle v_0 \rangle \oplus \mathbb{Z} \cdot \langle v_1 \rangle \oplus \mathbb{Z} \cdot \langle v_2 \rangle.$$

It follows, exactly as in the previous example, that $H_0(K') \cong \mathbb{Z}$. Moreover, a 1-chain $c_1 = n_0\langle v_0, v_1 \rangle + n_1\langle v_1, v_2 \rangle + n_2\langle v_0, v_2 \rangle$ is a 1-cycle if and only if $n_0 = n_1 = -n_2$ (as in the previous example). This means elements of the form $n[\langle v_0, v_1 \rangle + \langle v_1, v_2 \rangle - \langle v_0, v_2 \rangle], n \in \mathbb{Z}$, generate $Z_1(K')$, i.e., $Z_1(K') \cong \mathbb{Z}$. Since $B_1(K') = 0$, $H_1(K') \cong \mathbb{Z}$. Therefore, we conclude that

$$H_q(K') = \begin{cases} \mathbb{Z}, & \text{if } q = 0 \\ \mathbb{Z}, & \text{if } q = 1 \\ 0, & \text{otherwise.} \end{cases}$$

Remark. We must point out that with the orientation given by the ordering $v_0 < v_1 < v_2$, the 1-cycle $\langle v_0, v_1 \rangle + \langle v_1, v_2 \rangle - \langle v_0, v_2 \rangle = \langle v_0, v_1 \rangle + \langle v_1, v_2 \rangle + \langle v_2, v_0 \rangle$ is indeed a "closed circuit" or cycle of K' which is not a boundary, i.e., this cycle represents a "hole" in K'. The same 1-cycle in the earlier example was not a "hole" in that complex $K = \mathrm{Cl}(\sigma^2)$ because there it was the boundary of the 2-simplex $\langle v_0, v_1, v_2 \rangle$ of that complex.

Example 4.4.4. Once again, let $\sigma^3 = \langle v_0, v_1, v_2, v_3 \rangle$ be a 3-simplex and L be the 2-dimensional skeleton of K (Fig. 4.4). Then $|L|$ is homeomorphic to a 2-sphere $\mathbb{S}^2$, and L is a triangulation of $\mathbb{S}^2$. We orient L by ordering $v_0 < v_1 < v_2 < v_3$. We are going to show that

$$H_q(L) = \begin{cases} \mathbb{Z}, & \text{if } q = 0 \text{ or } 2 \\ 0, & \text{otherwise.} \end{cases}$$

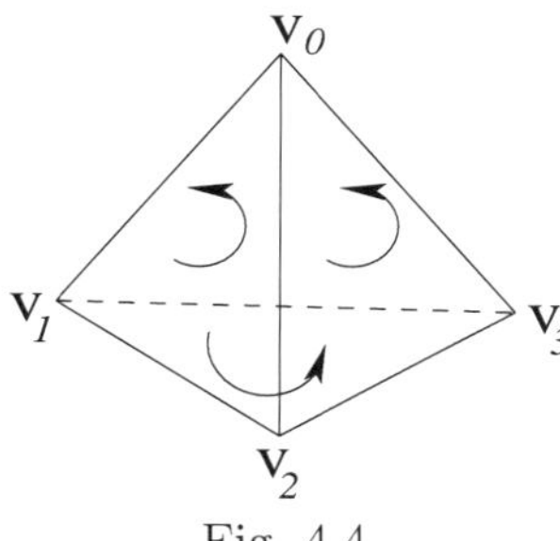

Fig. 4.4

In this case, observe that the nontrivial part of our oriented simplicial chain complex is

$$0 \to C_2(L) \overset{\partial_2}{\to} C_1(L) \overset{\partial_1}{\to} C_0(L) \to 0,$$

where

$$\begin{aligned}
C_0(L) &= \mathbb{Z}\cdot\langle v_0\rangle \oplus \mathbb{Z}\cdot\langle v_1\rangle \oplus \mathbb{Z}\cdot\langle v_2\rangle \oplus \mathbb{Z}\cdot\langle v_3\rangle \\
C_1(L) &= \mathbb{Z}\cdot\langle v_0v_1\rangle \oplus \mathbb{Z}\cdot\langle v_0v_2\rangle \oplus \mathbb{Z}\cdot\langle v_0v_3\rangle \oplus \mathbb{Z}\cdot\langle v_1v_2\rangle \\
&\quad \oplus\mathbb{Z}\cdot\langle v_1v_3\rangle \oplus \mathbb{Z}\cdot\langle v_2v_3\rangle \\
C_2(L) &= \mathbb{Z}\cdot\langle v_0v_1v_2\rangle \oplus \mathbb{Z}\cdot\langle v_1v_2v_3\rangle \oplus \mathbb{Z}\cdot\langle v_0v_1v_3\rangle \oplus \mathbb{Z}\cdot\langle v_0v_2v_3\rangle.
\end{aligned}$$

To compute $H_0(L)$, note that every 0-chain is a zero cycle. For each $i = 1, 2, 3$, since $\partial(\langle v_i, v_0\rangle) = \langle v_0\rangle - \langle v_i\rangle$, we find that the difference $\langle v_0\rangle - \langle v_i\rangle$ is a boundary, i.e., $\langle v_i\rangle$ is homologous to $\langle v_0\rangle$ for each i. Thus, any 0-cycle $c_0 = n_0\langle v_0\rangle + n_1\langle v_1\rangle + n_2\langle v_2\rangle + n_3\langle v_3\rangle$ is homologous to $(n_0+n_1+n_2+n_3)\langle v_0\rangle$. In other words, each 0-cycle is homologous to an integral multiple of $\langle v_0\rangle$. Also, it can be seen, as in Example 4.4.2, that a 0-cycle $n_0\langle v_0\rangle + n_1\langle v_1\rangle + n_2\langle v_2\rangle + n_3\langle v_3\rangle$ is a boundary if and only if $n_0 + n_1 + n_2 + n_3 = 0$. This means that the two integral multiples $m\langle v_0\rangle$ and $n\langle v_0\rangle$ of $\langle v_0\rangle$ are not homologous unless $m = n$. From these observations it follows at once that the map $H_0(L) \to \mathbb{Z}$ defined by $\{n\langle v_0\rangle\} \to n$ is an isomorphism.

To compute $H_2(L)$, observe that $H_2(L) \cong Z_2(L)$. Now, if $z = a\langle v_0, v_1, v_2\rangle + b\langle v_0, v_1, v_3\rangle + c\langle v_1, v_2, v_3\rangle + d\langle v_0, v_2, v_3\rangle$ is a 2-cycle, then $\partial_2(z) = 0$ i.e., $a[\langle v_1, v_2\rangle - \langle v_0, v_2\rangle + \langle v_0, v_1\rangle] + b[\langle v_1, v_3\rangle - \langle v_0, v_3\rangle + \langle v_0, v_1\rangle] + c[\langle v_2, v_3\rangle - \langle v_1, v_3\rangle + \langle v_1, v_2\rangle] + d[\langle v_2, v_3\rangle - \langle v_0, v_3\rangle + \langle v_0, v_2\rangle] = 0$. Equating the coefficients of various 1-simplexes to zero, we get $a = -b = -c = d$. This means $z = a[\langle v_0, v_1, v_2\rangle + \langle v_1, v_3, v_2\rangle + \langle v_0, v_3, v_1\rangle + \langle v_0, v_2, v_3\rangle]$. Since $a \in \mathbb{Z}$ is arbitrary, we conclude that $H_2(L) \cong Z_2(L) \cong \mathbb{Z}$.

Finally, to prove that $H_1(L) = 0$, we shall show that every 1-cycle is a boundary. Let $z = a_0\langle v_0, v_1\rangle + a_1\langle v_0, v_2\rangle + a_2\langle v_0, v_3\rangle + a_3\langle v_1, v_2\rangle + a_4\langle v_1, v_3\rangle + a_5\langle v_2, v_3\rangle$ be a 1-cycle. Note that $\partial_2\langle v_1, v_2, v_3\rangle = \langle v_2, v_3\rangle - \langle v_1, v_3\rangle + \langle v_1, v_2\rangle$, which means $\langle v_2, v_3\rangle$ is homologous to $\langle v_1, v_3\rangle - \langle v_1, v_2\rangle$, and so the cycle z is homologous to a cycle in which the term $\langle v_2, v_3\rangle$ is replaced by $\langle v_1, v_3\rangle - \langle v_1, v_2\rangle$.

Similarly, the term $\langle v_1, v_2\rangle$ can also be replaced by $\langle v_1, v_0\rangle + \langle v_0, v_2\rangle$. Thus, z is homologous to z' which does not contain any terms of $\langle v_2, v_3\rangle$ or $\langle v_1, v_2\rangle$. But, then z' cannot contain any term like $\langle v_0, v_2\rangle$, otherwise $\partial_1(z') = 0$ will mean that the coefficient of $\langle v_2\rangle$ in $\partial_1(z')$ must be zero, which would be impossible. Thus, z' must be of the form $z' = b_0\langle v_0, v_1\rangle + b_1\langle v_1, v_3\rangle + b_2\langle v_3, v_0\rangle$. The fact that $\partial_1(z') = 0$ will further imply, as in Example 4.4.2, that $b_0 = b_1 = b_2$. Consequently, z' is a multiple of $x = \langle v_0, v_1\rangle + \langle v_1, v_3\rangle + \langle v_3, v_0\rangle = \partial_2(\langle v_0, v_1, v_3\rangle)$ which, evidently, shows that z' is a boundary and therefore z is a boundary.

Note. It may be observed that since $H_2(L) \cong \mathbb{Z}$, there is 2-dimensional "hole" in the simplicial complex L which is a triangulation of the 2-spheres $\mathbb{S}^2$; it has no 1-dimensional "holes", no 3-dimensional or higher dimensional "holes" either.

Example 4.4.5. We now consider the torus $T = \mathbb{S}^1 \times \mathbb{S}^1$. It has been already observed that a triangulation K of T is given by Fig. 4.5.

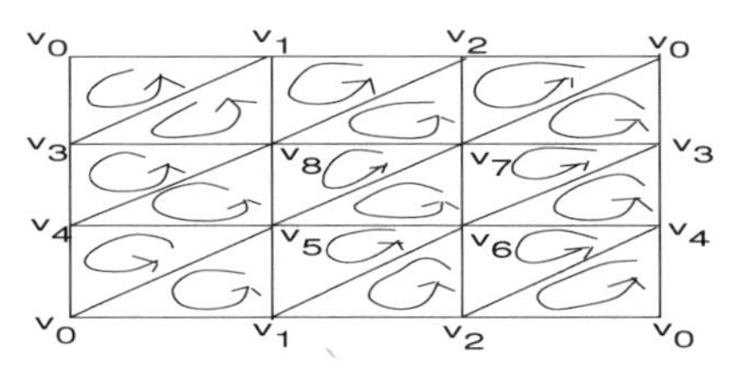

Fig. 4.5

We orient triangles of K as indicated by the arrows in Fig. 4.5, not by the ordering of its vertices, simply for computational convenience. We may orient all edges by the linear ordering of its vertices. Our simplicial chain complex is

$$0 \to C_2(K) \xrightarrow{\partial_2} C_1(K) \xrightarrow{\partial_1} C_0(K) \to 0,$$

where $C_i(K)$ is a free abelian group of rank equal to the number of i-simplexes in K. Evidently, $H_q(K) = 0$ if $q > 2$.

Note that each 1-simplex of K is an edge of exactly two 2-simplexes of K and the orientation of the edge induced by one of the 2-simplexes is opposite to that of the other 2-simplex. Now, we consider the 2-chain P which is the sum of all positively oriented 2-simplexes of K. We claim that it is a 2-cycle. For, the boundary $\partial_2(P)$ would be the sum of various 1-simplexes σ^1 of K. Any 1-simplex σ^1 will occur in the boundary of exactly two 2-simplexes, say, σ_1^2 and σ_2^2 of the sum P. Since the incidence numbers $[\sigma_1^2, \sigma^1]$ and $[\sigma_2^2, \sigma^1]$ are opposite to each other, the coefficient of σ^1 in $\partial_2(P)$ is zero. Consequently, $\partial_2(P) = 0$. On the other hand, if $z \in C_2(K)$ is an arbitrary 2-cycle, then write z as a linear combination of all 2-simplexes of K and consider the equation $\partial_2(z) = 0$. Since $\partial_2(z)$ is a linear combination of various 1-simplexes of K and each 1-simplex occurs in the boundary of exactly two 2-simplexes, the coefficient of each 1-simplex will be zero only when the coefficients of various 2-simplexes in z are identical, i.e, z is an integral multiple of P. Thus, $H_2(K) \cong Z_2(K) \cong \mathbb{Z}$.

To compute $H_1(K)$, note that any 1-cycle z of K is an integral linear combination of all of its 1-simplexes. Since $\partial_2(\langle v_0, v_1, v_3\rangle) = \langle v_1, v_3\rangle - \langle v_0, v_3\rangle + \langle v_0, v_1\rangle$ is a boundary, $\langle v_1, v_3\rangle$ can be replaced in z by $\langle v_0, v_3\rangle - \langle v_0, v_1\rangle$ without changing the homology class. Similarly, $\langle v_3, v_8\rangle$ can be replaced by $\langle v_3, v_0\rangle + \langle v_0, v_1\rangle + \langle v_1, v_8\rangle$, etc. Making such replacements, one can easily see that any 1-cycle is homologous to a cycle of the type $z' = a_1\langle v_0, v_3\rangle + a_2\langle v_3, v_4\rangle + a_3\langle v_4, v_0\rangle + a_4\langle v_0, v_1\rangle + a_5\langle v_1, v_2\rangle + a_6\langle v_2, v_0\rangle + a_7\langle v_1, v_8\rangle + a_8\langle v_8, v_5\rangle + a_9\langle v_6, v_7\rangle + a_{10}\langle v_7, v_2\rangle$. Equating various coefficients of vertices in the equation $\partial_1(z') = 0$ to zero, we see that $a_7 = 0 = a_8 = a_9 = a_{10}$, $a_1 = a_2 = a_3$ and $a_4 = a_5 = a_6$. Let us put $u = \langle v_0, v_1\rangle + \langle v_1, v_2\rangle + \langle v_2, v_0\rangle$ and $v = \langle v_0, v_3\rangle + \langle v_3, v_4\rangle + \langle v_4, v_0\rangle$ and observe that each one of these is a closed circuit (hence 1-cycle) which is not a boundary. Hence, z is homologous to $au + bv \in Z_1(K)$ for some $a, b \in \mathbb{Z}$, i.e., $Z_1(K) = \mathbb{Z}\cdot u + \mathbb{Z}\cdot v$. To see that the sum is direct, we note that if $au+bv$ is boundary for some $a, b \in \mathbb{Z}$, then $au + bv = \partial_2(x)$ for some $x \in C_2(K)$. Now, if σ is an oriented 2-simplex appearing in x, then at least one edge of σ, say e, is different from those in u and v. This means e will also be an edge of some other 2-simplex, say, τ. Since σ, τ have opposite orientations and the coefficient of e in $\partial_2(x)$ is zero, the coefficients of σ and τ must be identical, i.e., $x = kP$ for some $k \in \mathbb{Z}$, where P is the 2-cycle used earlier. But then $\partial_2(x) = k\partial_2(P) = 0$. This proves that $H_1(K) \cong \mathbb{Z} \oplus \mathbb{Z}$.

By an argument parallel to Examples 4.4.2, 4.4.3 and 4.4.4, one can easily show that $H_0(K) \cong \mathbb{Z}$. Combining all these we have

$$H_q(K) = \begin{cases} \mathbb{Z}, & q = 0 \\ \mathbb{Z} \oplus \mathbb{Z}, & q = 1 \\ \mathbb{Z}, & q = 2 \\ 0, & \text{otherwise.} \end{cases}$$

Example 4.4.6. Recall that the projective plane $\mathbb{P}^2$ is obtained by identifying each pair of diametrically opposite points of the disc $\mathbb{D}^2$. We have seen (exercises of Chapter 3) that $\mathbb{P}^2$ is triangulable and a triangulation K of $\mathbb{P}^2$ is given in Fig. 4.6. Once again, we orient 2-simplexes of K as indicated by the arrows and 1-simplexes by the linear ordering of their vertices.
We will show that

$$H_q(K) = \begin{cases} \mathbb{Z}, & q = 0 \\ \mathbb{Z}/2\mathbb{Z}, & q = 1 \\ 0, & \text{otherwise.} \end{cases}$$

We observe, once again, that every 1-simplex is a face of exactly two 2-simplexes of K. Moreover, each 1-simplex $\langle v_3, v_4\rangle$ and $\langle v_4, v_5\rangle$ or $\langle v_5, v_3\rangle$ receives the same

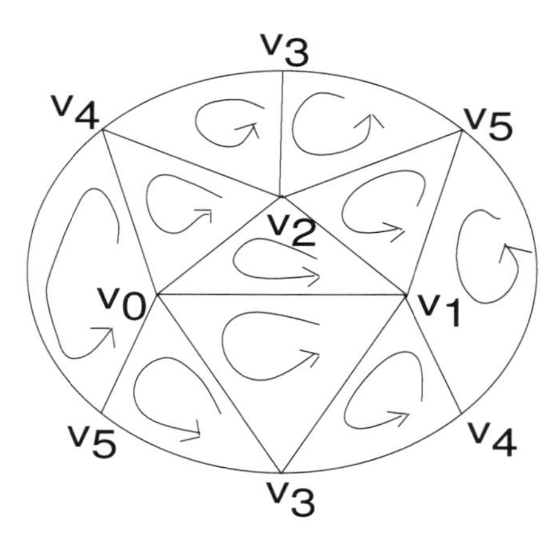

Fig. 4.6

orientation from both the 2-simplexes of which it is a common face whereas all other 1-simplexes receive opposite orientations from their corresponding pairs of 2-simplexes. Here, the simplicial chain complex is

$$0 \to C_2(K) \xrightarrow{\partial_2} C_1(K) \xrightarrow{\partial_1} C_0(K) \to 0,$$

each $C_i(K)$ being the free abelian group on as many generators as there are i-simplexes of K.

Let z be a 2-cycle of K. Write z as an integral linear combination of all 2-simplexes of K and put $\partial_2(z) = 0$. Then we claim that the coefficients of various 2-simplexes in z must be equal. This follows from the fact that any 2-simplex has at least a 1-face lying in the interior of the disk, which receives opposite orientations from the other 2-simplexes having that 1-simplex as a common face. This yields $z = n(\sum_{i=0}^{10} \sigma_i^2), n \in \mathbb{Z}$ and so $\partial_2(z) = 2n[\langle v_3, v_4 \rangle + \langle v_4, v_5 \rangle + \langle v_5, v_3 \rangle]$ because the coefficients of all other 1-simplexes receiving opposite orientations from the 2-simplexes must be zero. Now, $\partial_2(z) = 0$ means $n = 0$ and so $z = 0$. Therefore, $H_2(K) = 0$.

Note that $z = \langle v_3, v_4 \rangle + \langle v_4, v_5 \rangle + \langle v_5, v_3 \rangle$, making a complete circuit, is a 1-cycle of K lying on the boundary of the disk. Moreover, earlier considerations show that $2z$, and so, any even multiple of z is a boundary. We use intuition and geometry of K to assert that any 1-cycle is homologous to an integral multiple of z. Let us see how. As an example, note that $z' = \langle v_3, v_2 \rangle + \langle v_2, v_0 \rangle + \langle v_0, v_3 \rangle$ is also a 1-cycle. However, we can replace $\langle v_3, v_2 \rangle$ by $\langle v_3, v_4 \rangle + \langle v_4, v_2 \rangle$, $\langle v_2, v_0 \rangle$ by $\langle v_2, v_4 \rangle + \langle v_4, v_0 \rangle$, $\langle v_4, v_0 \rangle$ by $\langle v_4, v_5 \rangle + \langle v_5, v_0 \rangle$, $\langle v_0, v_3 \rangle$ by $\langle v_0, v_5 \rangle + \langle v_5, v_3 \rangle$ and see that z' is homologous to z. This tells us that $H_1(K) = Z_1(K)/B_1(K) \cong \mathbb{Z}/2\mathbb{Z}$. Finally, as in the earlier examples, one can easily see that $H_0(K) \cong \mathbb{Z}$. This proves our claim.

Remark. In all of the above examples, the coefficient group for the simplicial homology $H_*(K)$ has been the additive group $\mathbb{Z}$ of integers. Let G be any abelian group. Then by defining $C_q(K, G)$ to be the direct sum $\bigoplus G$ of as many copies of G as there are q-simplexes in K, we could have used the abelian group G instead of $\mathbb{Z}$ to compute the above groups. In that case the answer would have been different. The abelian group $\mathbb{Z}_2$ is frequently used as the coefficient group and the resulting homology groups of K are known as (mod 2) homology groups of K. We will also have occasion to use the additive group of rationals $(\mathbb{Q}, +)$ as the coefficient group.

Zero-dimensional Homology Groups

From the foregoing examples, one might observe that, in each case, the simplicial complex is connected and the zero-dimensional homology group is isomorphic to $\mathbb{Z}$. In fact, this is a special case of the general result to be proved below. Recall that, by definition, our simplicial complexes have finite number of simplexes.

Theorem 4.4.7. *Let K be a simplicial complex. Then $H_0(K)$ is always a free abelian group. More precisely, if K has r combinatorial components $K_i, i = 1, \ldots, r$ and we choose one vertex v_i from each K_i, then $H_0(K)$ is isomorphic to the direct sum $\bigoplus_1^r \mathbb{Z}$ of r-copies of $\mathbb{Z} \cdot \{v_i\}$.*

Proof. We note that all 0-chains v_i are 0-cycles. Let $w \in K_i$ be any other vertex. Since K_i is a combinatorial component of K, there is a sequence $a_0, a_1, \ldots, a_n$ such that $w = a_0, v_i = a_n$ and $\langle a_i, a_{i+1} \rangle$ is a 1-simplex for each $i = 0, 1, \ldots, n-1$. Then the boundary is $\partial(\sum \langle a_i, a_{i+1} \rangle) = v_i - w$, which means w is homologous to v_i. This says that any 0-cycle $\sum n_j w_j$ lying in the component K_i is homologous to $(\sum n_j) v_i$. More generally, this implies that any 0-chain (which is a 0-cycle) in K is homologous to a linear combination of the elementary 0-chains v_i.

Next, we prove that the cycles $\{v_i\}_{i=1}^r$ are linearly independent, i.e., no nontrivial linear combination $c = \sum n_i v_i$ can be a boundary. Suppose $c = \partial(d)$ for some 1-chain d. Since each 1-simplex of K lies in a unique component K_i, we can write $d = \sum d_i$, where $d_i \in K_i$. Hence $\partial(d) = \sum \partial(d_i)$, which means $\partial(d_i) \in K_i$, and therefore, $\partial(d_i) = n_i v_i$, for all $i = 1, 2, \ldots, r$. Now, we define a homomorphism from $C_0(K)$ to $\mathbb{Z}$ by putting $\epsilon(v) = 1$ for each vertex of K. Then it is easily seen that for each elementary 1-chain $\langle v, w \rangle$, $\epsilon\partial(\langle v, w \rangle) = \epsilon(w - v) = 1 - 1 = 0$. In particular, $0 = \epsilon(\partial d_i) = \epsilon(n_i v_i) = n_i$, for all $i = 1, 2, \ldots, r$.

The first para above shows that the homomorphism $f \colon \bigoplus \mathbb{Z} \to H_0(K)$ defined by

$$f(m_1, \ldots, m_r) = \sum m_i \{v_i\}$$

is onto while the second para shows that it is 1-1. It follows that $H_0(K) \cong \bigoplus_1^r \mathbb{Z} \cdot \{v_i\}$. ∎

Remark. We have already seen that the combinatorial components of a simplicial complex K are in 1-1 correspondence with the path-components of $|K|$. Hence it follows from the above theorem that if we take two simplicial complexes K and K' such that $|K|$ is homeomorphic to $|K'|$, the number of path components of $|K|$ is equal to the number of path components of $|K'|$ and so $H_0(K) \cong H_0(K')$. This says that the zero-dimensional homology group of a simplicial complex K depends on the topology of $|K|$, not on a particular triangulation of $|K|$. In other words, **the group $H_0(K)$ is a topological invariant of** $|K|$. It is worthwhile to indicate here that, though the higher-dimensional homology groups are also topological invariants of $|K|$, but the proofs, as we shall see later, are not that easy. One should also know a better result, viz., $H_0(K)$ is not only a topological invariant of $|K|$, but also a homotopy invariant, i.e., if $|K|$ and $|K'|$ are of the same homotopy type, then $H_0(K) \cong H_0(K')$. This will become clear in due course.

The following version of homology groups will be quite useful in the sequel:

Definition 4.4.8. *Let K be a simplicial complex. Define a homomorphism ϵ : $C_0(K) \to \mathbb{Z}$, called* **augmentation map** *by putting $\epsilon(v) = 1$ for each vertex v of K and extending linearly to $C_0(K)$.*

Note that this augmentation homomorphism is onto. If $\partial_1 \colon C_1(K) \to C_0(K)$ is the boundary homomorphism, then evidently $\epsilon\partial = 0$.The quotient group $\ker \epsilon / \text{Im } \partial_1$, denoted by $\tilde{H}_0(K)$, is called the zero-dimensional **reduced homology group** of K. If we put $\tilde{H}_p(K) = H_p(K)$ for each $p > 1$, then the groups $\{\tilde{H}_i(K) : i = 0, 1, \ldots\}$ are called the **reduced homology groups** of K. Since $\text{Im } \partial_1 \subset \ker \epsilon \subset C_0(K)$, there is an induced inclusion map $\tilde{H}_0(K) \to H_0(K)$. In fact, we have

Theorem 4.4.9. *For a simplicial complex K, zero-dimensional reduced homology group is also free abelian, and we have the following formulae:*

$$H_0(K) \cong \tilde{H}_0(K) \oplus \mathbb{Z}, \qquad H_p(K) = \tilde{H}_p(K), \ p \geq 1.$$

Proof. The sequence

$$0 \to \ker \epsilon \to C_0(K) \xrightarrow{\epsilon} \mathbb{Z} \to 0$$

is exact. This implies that

$$0 \to \frac{\ker \epsilon}{\text{Im } \partial_1} \to \frac{C_0(K)}{\text{Im } \partial_1} \xrightarrow{\epsilon} \mathbb{Z} \to 0$$

is exact. Since $\mathbb{Z}$ is free, the above sequence splits and we find that $H_0(K) \cong \tilde{H}_0(K) \oplus \mathbb{Z}$. Moreover, if $\{v_i\}_0^r$ is a basis of $H_0(K)$, then $\{v_i - v_0\}_1^r$ is a basis of $\tilde{H}_0(K)$. ∎

Exercises

1. Consider X as the union of two line segments AB and AC in $\mathbb{R}^n$, $n \geq 2$ having the point A common. Write down a triangulation K for X and compute the homology groups $H_q(K)$ for all $q \geq 0$.

2. Consider the following simplicial complexes (vertices and edges are shown below, and with no triangles) K_1 and K_2. Compute their homology groups.

3. Give a triangulation K of the Möbius band M and compute all the homology groups $H_q(K)$, $q \geq 0$.

4. Give a triangulation of cylinder and compute all of its homology groups.

5. Give a triangulation L of the Klein bottle K and compute all of its homology groups.

6. Work out Examples 4.4.4, 4.4.5, and 4.4.6 using the coefficient group $G = \mathbb{Z}_2, G = \mathbb{Q}$ instead of $\mathbb{Z}$.

7. Construct a simplicial complex K such that $H_0(K) \cong \mathbb{Z}, H_1(K) \cong \mathbb{Z}/3\mathbb{Z}$ and $H_q(K) = 0 \ \forall \ q \geq 2$. Generalize this result, i.e., for any n find a simplicial complex K_n such that

$$H_0(K_n) \cong \mathbb{Z}, \ \ H_1(K_n) \cong \mathbb{Z}/n\mathbb{Z}, \ \ H_q(K_n) = 0 \ \text{ for all } q \geq 2.$$

(**Hint:** Consider a triangle ABC, identify AB with BC, BC with CA.)

8. Let K^r be the r-dimensional skeleton of an n-dimensional simplicial complex K, $0 \leq r < n$. Determine, how the homology groups $H_q(K^r)$ are related to $H_q(K^n)$.

9. Let K be a triangulation of the connected sum $T\#T$ of two tori. Compute the homology groups of K.

10. Let K be a triangulation of the connected sum $\mathbb{RP}^2\#\mathbb{RP}^2$ of two projective planes. Compute the homology groups of K.

4.5 Properties of Integral Homology Groups

Looking at the various examples that we have discussed so far, we observe that direct sums of copies of $\mathbb{Z}$ and finite cyclic groups have occurred as homology groups of simplicial complexes considered so far. It is, therefore, natural to ask: what kind of groups can occur as the homology groups of an arbitrary finite simplicial complex when we use $\mathbb{Z}$ as the coefficient group?

This question can be readily answered: note that, a finite simplicial complex K has only finite number of simplexes. If $\dim K = n$, then there are no q-simplexes for $q > n$ and so $H_q(K) = 0$ for all $q > n$. For each $q \leq n$, there are only finite number of q-simplexes in K and, consequently, the qth chain group $C_q(K)$ is a finitely generated free abelian group of rank equal to the number of q-simplexes in K. Because $Z_q(K)$ and $B_q(K)$ are subgroups of $C_q(K)$ and a subgroup of a free group is always free, the group of cycles and boundaries are also finitely generated free abelian groups. The factor group $H_q(K) = Z_q(K)/B_q(K)$, therefore, must be a finitely generated abelian group which may or may not be free. In any case we conclude that *the qth homology group $H_q(K)$ of a simplicial complex K is a finitely generated abelian group for all $q \geq 0$.*

In view of the Structure Theorem for Finitely Generated Abelian Groups, the above information about $H_q(K)$ alone determines the nature of $H_q(K)$ completely. The theorem asserts that $H_q(K)$ will be isomorphic to the direct sum of a finite number of copies of $\mathbb{Z}$ and a finite number of cyclic groups of descending orders. In other words,

$$(*) \qquad H_q(K) \cong \overbrace{\mathbb{Z} \oplus \ldots \oplus \mathbb{Z}}^{k \text{ copies}} \oplus \mathbb{Z}/m_1\mathbb{Z} \oplus \ldots \oplus \mathbb{Z}/m_n\mathbb{Z},$$

for each $q \geq 0$, where m_i is divisible by $m_{i+1}, i = 1, \ldots, n-1$.

Definition 4.5.1. *If we write $H_q(K) \cong F_q(K) \oplus T_q(K)$, where $F_q(K)$ is the direct sum of copies of $\mathbb{Z}$ (called the* **free part** *of $H_q(K)$) and $T_q(K)$ is the direct sum of finite cyclic groups (called* **torsion subgroup** *of $H_q(K)$) as in $(*)$ above, then the rank of $F_q(K)$ is called the qth* **Betti number** *(in honour of E. Betti* (1823–1892)*) of the simplicial complex K and the numbers m_i occurring in the torsion subgroup $T_q(K)$ are called the qth* **torsion coefficients** *of K.*

It is interesting to emphasize that, as a consequence of the structure theorem for finitely generated abelian groups, knowing both the qth Betti number as well as the qth torsion coefficients of K is equivalent to knowing the qth homology group of K itself. Hence it is one and the same thing to determine all the homology groups $H_q(K)$ of the simplicial complex K or to compute all of its Betti numbers and torsion coefficients of K. As a matter of fact, historically, it is the Betti numbers and torsion coefficients of a complex K which were emphasized and given importance first. It is due to the famous lady mathematician Emmy Noether that, later on, the viewpoint changed completely and the emphasis shifted to the **group structure** of $H_q(K)$ rather than those numbers. The main reason for this transition of emphasis was that a more useful algebra in the form of "homological algebra" was naturally developed and introduced in the study of topological problems. Concepts like "homology with coefficients", "cohomology product" and "higher order cohomology operations", etc., to name a few, thus emerged and gave a lot more information about arbitrary topological spaces than what could be deduced from the knowledge of Betti numbers and torsion coefficients alone.

As pointed out in the introduction, we will finally prove (see Section 4.9) that homology groups $H_q(K)$ of a simplicial complex K depend on the topology of the polyhedron $|K|$, not on the simplicial structure of K or its orientation. This is one of the most beautiful and exciting results of algebraic topology. This fundamental fact at once increases our (topologist's) respect for homology groups enormously. First of all, the problem of computing homology groups $H_q(K)$ of a simplicial complex K becomes easier because we can choose any suitable triangulation K' of $|K|$ which might be easier to work with and then compute $H_q(K')$ rather than $H_q(K)$. Secondly, any information about the homology groups $H_q(K)$ should be interpreted as reflecting some interesting property of the topological space $|K|$. For instance, one can say that the nontrivial homology groups of a simplicial complex K detect a kind of "holes" in the underlying space $|K|$. For instance, all positive-dimensional homology groups of a triangulation K of the 2-disk $D^2 \approx |K|$ are zero (Example 4.4.2) means the disk D^2 has no "holes" in any dimension; the circle $\mathbb{S}^1$ is homeomorphic to $|K|$, where K is the 1-skeleton of the geometric complex of a 2-simplex and we have seen (Example 4.4.3) that $H_1(K) \cong \mathbb{Z}$, which means $\mathbb{S}^1$ has a 1-dimensional "hole" and a generator of $\mathbb{Z}$ corresponds to that hole. Moreover, $\mathbb{S}^1$ has no holes of higher dimension. Similarly, the second homology group $H_2(K)$ of a triangulation K of the sphere $\mathbb{S}^2$ (Example 4.4.4) is isomorphic to $\mathbb{Z}$, which means a generator of $H_2(K)(\cong \mathbb{Z})$ corresponds to a 2-dimensional hole of $\mathbb{S}^2$ and no holes in other dimensions ≥ 2. Finally, for a triangulation K of the torus T, we saw (Example 4.4.5) that $H_1(K) = \mathbb{Z} \cdot u \oplus \mathbb{Z} \cdot v$ which means T has two "1-dimensional holes". The generator u corresponds to one hole and v corresponds to the other hole (Fig. 4.7). Since $H_2(K) \cong \mathbb{Z}$, the torus has a "two-dimensional hole" also.

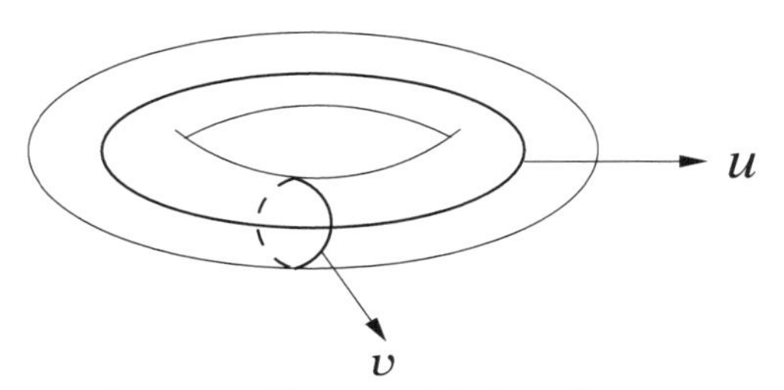

Fig. 4.7: The two 1-holes of torus

In case of the triangulation K (Example 4.4.6) of the projective plane $\mathbb{P}^2$, $H_1(K) = \mathbb{Z}_2$ and that means $\mathbb{P}^2$ has a "twisting" along its 1-dimensional hole. In fact, for any complex K, the free part of $H_q(K)$ indicates that there are as many q-dimensional holes in $|K|$ as there are copies of $\mathbb{Z}$ in $H_q(K)$

and the torsion part indicates that there are "twistings" around other holes in $|K|$. The 0-dimensional homology group $H_0(K)$ of K, as we have already seen, reflects a nice topological property of $|K|$, viz., the rank of $H_0(K)$ is the number of combinatorial components of K, which is equal to the number of connected components of $|K|$.

Next, we give some interesting applications of homology groups by answering a few topological questions mentioned earlier. However, before coming to these applications, we need to compute the homology groups of an n-disk and an n-sphere for any $n \geq 1$.

Simplicial Homology of Discs and Spheres

Let $\mathbb{D}^n = \{(x_1, \dots, x_n) \in \mathbb{R}^n \mid \sum |x_i|^2 \leq 1\}$ and let $\mathbb{S}^{n-1} = \{(x_1, \dots, x_n) \in \mathbb{R}^n \mid \sum |x_i|^2 = 1\}$ be its boundary. We know that if σ^n denotes the n-simplex $\langle v_0, \dots, v_n \rangle$, then $K = \mathrm{Cl}(\sigma^n)$ is the simplicial complex consisting of all faces of σ^n. Also, the collection K' of all proper faces of σ^n is a simplicial complex. We have already seen that $|K| \approx \mathbb{D}^n$ and $|K'| \approx \mathbb{S}^{n-1}$. It follows that K is a triangulation of $\mathbb{D}^n$ and K' is a triangulation of $\mathbb{S}^{n-1}$. As mentioned in Section 4.1, later on we will prove that if K_1 and K_2 are any two triangulations of a compact polyhedron X, then $H_q(K_1) \cong H_q(K_2)\ \forall\ q \geq 0$. Therefore, we can define the simplicial homology of any polyhedron X by putting $H_q(X) = H_q(K)$, where K is some triangulation of X. In view of this observation we find that $H_q(\mathbb{D}^n) \cong H_q(K)$ and $H_q(\mathbb{S}^{n-1}) \cong H_q(K')$ in all dimensions and we now compute $H_q(K)$ and $H_q(K')$.

Let us have

Definition 4.5.2. *Suppose K is a complex in $\mathbb{R}^n$ for n large, and w is a point in $\mathbb{R}^n$ such that each ray emanating from w intersects $|K|$ in at most one point. Let $w * K$ denote the collection of all simplexes of the form $\langle w, a_0, \dots, a_p \rangle$, where $\langle a_0, \dots, a_p \rangle$ is a simplex of K, along with all faces of such simplexes. It can be seen that $w * K$ is a simplicial complex. This complex is called a* **cone** *over K with vertex w.*

Note that if $\sigma^n = \langle v_0, \dots, v_n \rangle$ is a n-simplex then the complex $K = \mathrm{Cl}(\sigma^n)$ consisting of all faces of σ^n is the cone $v_0 * \mathrm{Cl}\langle v_1, \dots, v_n \rangle$ where $\langle v_1, \dots, v_n \rangle$ is the face opposite to the vertex v_0. In fact, for any vertex v_i, $\mathrm{Cl}(\sigma^n) = v_i * \mathrm{Cl}(\tau)$, where τ is the face opposite to v_i. The following result determines the homology groups of a n-disk $\mathbb{D}^n$ completely.

Theorem 4.5.3. *Let K be any complex. Then the reduced homology groups of the cone $w * K$ are trivial in all dimensions, and hence*

$$H_q(w * K) = \begin{cases} \mathbb{Z}, & q = 0 \\ 0, & \textit{otherwise.} \end{cases}$$

Proof. It suffices to show that the reduced homology groups $\tilde{H}_q(w*K)$ vanish in all dimensions. Since the complex $w*K$ is connected, the 0-dimensional reduced homology group is evidently trivial. Now, let $q > 0$. Let z_q be a q-cycle of $w*K$. We will show that z_q is a boundary. For any oriented simplex $\sigma = \langle v_0, \dots, v_q \rangle$ of K, let $[w, \sigma]$ denote the oriented simplex $\langle w, v_0, \dots, v_q \rangle$ of $w*K$. Note that this operation, called **bracket operation**, is well defined and gives a homomorphism from $C_p(K) \to C_{p+1}(w*K)$ by putting $[w, \sum n_i \sigma_i] = \sum n_i [w, \sigma_i]$. Also, we note that for a simplex σ,

$$\partial[w, \sigma] = \begin{cases} \sigma - w & \text{if } \dim \sigma = 0 \\ \sigma - [w, \partial\sigma] & \text{if } \dim \sigma > 0 \end{cases}$$

and so, more generally, for a chain c_q,

$$\partial[w, c_q] = \begin{cases} c_q - \epsilon(c_0)w, & q = 0 \\ c_q - [w, \partial c_q], & q > 0. \end{cases}$$

Let us write $z_q = c_q + [w, d_{q-1}]$ where c_q consists of those terms of z_q which are in the complex K and d_{q-1} is a chain of K. By direct calculation using the above formulae, we get

$$\begin{aligned} z_q - \partial[w, c_q] &= c_q + [w, d_{q-1}] - c_q + [w, \partial c_q] \\ &= [w, c_{q-1}] \end{aligned}$$

where $c_{q-1} = d_{q-1} + \partial c_q$ is a chain in K. Applying the boundary operator ∂ on both sides above, and noting that z_q is a cycle, we have

$$0 = \begin{cases} c_{q-1} - \epsilon(c_{q-1})w & q = 1 \\ c_{q-1} - [w, \partial c_{q-1}] & q > 1. \end{cases}$$

Note that c_{q-1} is a chain with all its simplices in K. However, $\epsilon(c_{q-1})w$ and $[w, \partial_{q-1}]$ are chains whose all simplices, having w as a vertex, are not in K. This is possible only when c_{q-1} and the latter chains are themselves 0-chains, i.e., $c_{q-1} = 0$. But that implies

$$z_q - \partial[w, c_q] = [w, c_{q-1}] = 0$$

i.e., z_q is a boundary. This completes the proof. ■

Corollary 4.5.4. *Let $\mathbb{D}^n$ be the Euclidean n-disk. Then*

$$H_q(\mathbb{D}^n) = \begin{cases} \mathbb{Z} & q = 0 \\ 0 & \textit{otherwise.} \end{cases}$$

Proof. The proof follows from the above theorem and the fact that $K = \mathrm{Cl}(\sigma^n)$ gives a triangulation of $\mathbb{D}^n$. Furthermore, the complex K itself is a cone. ■

The next result determines the homology groups of the n-sphere $\mathbb{S}^n$ completely.

Theorem 4.5.5. *Let $K(n), n \geq 1$, denote the n-skeleton of the complex $K = \mathrm{Cl}(\sigma^{n+1})$. Then*

$$H_q(K(n)) = \begin{cases} \mathbb{Z} & \text{if } q = 0 \text{ or } n \\ 0 & \text{otherwise.} \end{cases}$$

Proof. It suffices to prove that $\tilde{H}_q(K(n)) \cong \mathbb{Z}$ if $q = n$ and is zero otherwise. Note that $K(n)$ is a subcomplex of $K = \mathrm{Cl}(\sigma^{n+1})$. We consider the following commutative diagram of chain complexes defining the homology groups of K and $K(n)$. The vertical maps are induced by inclusions.

$$\begin{array}{ccccccccc} C_{n+1}(K) & \xrightarrow{\partial_{n+1}} & C_n(K) & \xrightarrow{\partial_n} & \cdots & \longrightarrow & C_0(K) & \xrightarrow{\epsilon} & \mathbb{Z} \\ & & \uparrow = & & & & \uparrow = & & \\ 0 & \longrightarrow & C_n(K(n)) & \xrightarrow[\partial_n]{} & \cdots & \longrightarrow & C_0(K(n)) & \xrightarrow[\epsilon]{} & \mathbb{Z} \end{array}$$

Note that both chain complexes are identical except in dimension $n + 1$. It follows, therefore, that

$$\tilde{H}_i(K(n)) \cong \tilde{H}_i(K) = 0$$

for all $i \neq n$. In dimension n, we have

$$\begin{aligned} H_n(K(n)) &\cong Z_n(K(n)) \\ &= \ker \partial_n \\ &= \mathrm{Im}\ \partial_{n+1}, \end{aligned}$$

since $H_n(K) = 0$. Now, $C_{n+1}(K)$ is infinite cyclic generated by σ^{n+1} and ∂_{n+1} is a nonzero homomorphism. Hence Im ∂_{n+1} is infinite cyclic because $C_n(K)$ has no torsion, i.e., $H_n(K(n)) \cong \mathbb{Z}$. ∎

Corollary 4.5.6. *Let $\mathbb{S}^n$, $n \geq 1$, denote the n-sphere. Then*

$$H_q(\mathbb{S}^n) = \begin{cases} \mathbb{Z} & \text{if } n = 0 \text{ or } q \\ 0 & \text{otherwise.} \end{cases}$$

Proof. This follows at once from the above theorem since $K(n)$ in the above theorem is a triangulation of $\mathbb{S}^n$. ∎

The Euler-Poincaré Theorem

We know that boundary of a tetrahedron is homeomorphic to a 2-sphere $\mathbb{S}^2$ via the radial map which takes vertices to points of $\mathbb{S}^2$, edges to spherical edges, and the triangles to spherical triangles lying on $\mathbb{S}^2$. Note that for this spherical triangulation of $\mathbb{S}^2$, we have the following obvious relation:

$$(\#\text{ of vertices} - \#\text{ of edges} + \#\text{ of faces}) = 2.$$

L. Euler (1707–1783) proved in 1752 that if one takes an arbitrary decomposition K' of $\mathbb{S}^2$ in terms of properly joined polyhedral faces, which need not necessarily be triangles but could be rectangles or any convex polygons, then the same above relation holds for K' also – this is the celebrated Euler's theorem, which we will prove shortly. Now, we observe that the number 2 appearing in the right-hand side of the above relation is, in fact, the alternate sum of the Betti numbers of $\mathbb{S}^2$, and these Betti numbers, as we have pointed out earlier, are independent of the triangulation of the 2-sphere. One can now ask whether or not a similar relation holds for a triangulation of torus, a triangulation of Klein bottle or a triangulation of any polyhedron. H. Poincaré made the first use of the homology groups in proving that the answer to the above question is in affirmative. Let us have

Definition 4.5.7. *Let K be an oriented complex of dimension n, and let $R_q(K)$ be the (Betti numbers) rank of the abelian group $H_q(K)$, $q = 0, 1, \ldots, n$. Then the alternate sum $\sum_{q=0}^{n}(-1)^q R_q(K)$, denoted by $\chi(K)$, is called the* **Euler characteristic** *of K.*

We observe that, in the definition of Euler characteristic $\chi(K)$, we have used the homology groups $H_i(K)$ with integer coefficients. If we consider these groups with rational coefficients $\mathbb{Q}$, then each $H_i(K;\mathbb{Q})$ will be a vector space over $\mathbb{Q}$, and in that case the rank $H_i(K;\mathbb{Q})$ is the dimension of the vector space $H_i(K;\mathbb{Q})$. It can be seen (Exercise 1 of Section 6.10) that rank $H_i(K;\mathbb{Z}) = \dim_{\mathbb{Q}} H_i(K;\mathbb{Q}) =$ ith Betti number of the complex K. Hence the Euler characteristic can be defined using rational coefficients also. The most important point to be kept in mind is that if K and K' are triangulations of a space X, then by the topological invariance of simplicial homology, $H_i(K) \cong H_i(K'), \forall\, i \geq 0$, and hence the Euler characteristic really depends on the topology of the space $X = |K|$, not on any triangulation K of X. In other words, *the Euler characteristic of a compact polyhedron is a topological invariant.*

The next theorem generalizes the Euler's Theorem mentioned above to arbitrary polyhedra.

Theorem 4.5.8. (Euler-Poincaré Theorem). *Let K be an oriented complex of dimension n. Suppose for each $q = 0, 1, \ldots, n$, α_q denotes the number of q-simplexes of K. Then*

$$\sum_{0}^{n}(-1)^q \alpha_q = \sum_{0}^{n}(-1)^q R_q(K).$$

Proof. Note that the right hand side of the relation to be proved is in terms of rank of homology groups $H_q(K)$ of K, $q \geq 0$. As pointed out earlier, rank $H_q(K;\mathbb{Z}) = \dim_{\mathbb{Q}} H_q(K;\mathbb{Q})$ and hence, we can assume that all the homology groups are with coefficients in $\mathbb{Q}$. It follows that all the groups $C_q(K), Z_q(K), B_q(K)$ and $H_q(K)$ associated to the complex K are vector spaces over $\mathbb{Q}$. In what follows, we will suppress the complex K as well as the coefficient $\mathbb{Q}$ while using above groups. Now, consider the following chain complex

$$0 \to C_n(K) \to \ldots \to C_{q+1}(K) \xrightarrow{\partial_{q+1}} C_q(K) \to \ldots \to C_0(K).$$

Note that $\alpha_q = \dim C_q(K)$, $q \geq 0$. Also, observe that

$$B_q = \text{Im } (\partial_{q+1}) \cong C_{q+1}/\ker \partial_{q+1} = C_{q+1}/Z_{q+1}.$$

Therefore, for all $q \geq 0$, we have

$$\dim B_q = \dim C_{q+1} - \dim Z_{q+1}.$$

On the other hand, $H_q \cong Z_q/B_q$ and so

$$\dim H_q = \dim Z_q - \dim B_q.$$

Substituting the value of $\dim B_q$ from the latter equation in the former one, we find that for all $q \geq -1$,

$$\dim C_{q+1} = \dim Z_q - \dim H_q + \dim Z_{q+1}.$$

Putting $q = -1, 0, \ldots, n$ and taking the alternate sum, we find that

$$\sum_0^n (-1)^q \dim C_q = \sum_0^n (-1)^q \dim H_q.$$

Since $\dim C_q = \alpha_q$ and $\dim H_q = R_q(K)$, we get the desired result. ∎

Now, we are going to derive the classical theorem due to L. Euler indicated earlier from the Euler-Poincaré Theorem. We need the following:

Definition 4.5.9. *By a* **rectilinear polyhedron** *P in the Euclidean 3-space, we mean a solid in $\mathbb{R}^3$ which is bounded by properly joined 2-dimensional convex polygons. The bounding polygons which may be triangles, rectangles or any n-gon, are called* **faces** *of P. The intersection of any two faces is called an* **edge** *of P and the intersection of any two edges is called a* **vertex**. *The polyhedron is said to be* **simple** *if its boundary is homeomorphic to the 2-sphere $\mathbb{S}^2$. A rectilinear polyhedron is said to be* **regular** *if its faces are regular congruent polygons and all polyhedral angles are equal.*

In the Euler-Poincaré Theorem, if we take K to be the boundary complex of $\text{Cl}(\sigma^3)$, where σ^3 is a 3-simplex, then we have already proved that $|K| \approx \mathbb{S}^2$ and $H_0(K) \cong \mathbb{Z}, H_1(K) = 0, H_2(K) \cong \mathbb{Z}$ and $H_i(K) = 0, \forall\ i \geq 3$. Hence the Euler characteristic $\chi(K) = 2$. Moreover, since the homology groups are independent of the triangulation, the Euler characteristic $\chi(K)$ is independent of K, i.e., $\chi(K) = \chi(K')$ for any other triangulation K' of $\mathbb{S}^2$.

Theorem 4.5.10. (Euler's Theorem). *If S is any simple polyhedron with V vertices, E edges and F faces, then $V - E + F = 2$.*

Proof. Note that the boundary of S is a polyhedral complex K, i.e., it consists of properly joined 2-dimensional convex polygons. If all these bounding polygons are triangles, then K is a simplicial complex such that $|K|$ is homeomorphic to $\mathbb{S}^2$ and the theorem follows directly from the Euler-Poincaré Theorem because $V = \alpha_0, E = \alpha_1$ and $F = \alpha_2$. Hence we should consider the case when K consists of polygonal faces which are not triangular. Suppose τ is one such face having n_0 vertices and n_1 edges. When we compute (Vertices − Edges + Faces) for the face τ, we find that the number is $n_0 - n_1 + 1$. Note that τ can always be subdivided into triangles by taking a vertex v in the interior of τ and joining v with each vertex of τ. See Fig. 4.7(a).

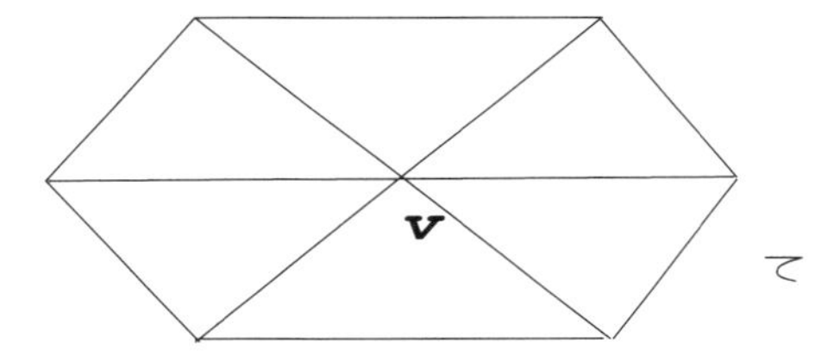

Fig. 4.7(a): Contribution of τ is invariant

In the above triangulation of τ, one new vertex and n_0 new edges have been added. Also, the face τ has been now replaced by n_0 triangles. Hence the contribution of τ in the computation of (vertices − edges + faces) = $(n_0 + 1) - (n_1 + n_0) + n_0 = n_0 - n_1 + 1$. This shows that the sum $V - E + F$ is not changed when we triangulate every face of K. In other words, we can assume that K is a simplicial complex. Now, the theorem follows directly from the Euler-Poincaré Theorem because $V - E + F = \alpha_0 - \alpha_1 + \alpha_2$ which is equal to the Euler characteristic of $\mathbb{S}^2$, viz., 2. Therefore, $V - E + F = 2$ for any simple polyhedron. ■

Euler's Theorem has a very interesting consequence. Note that a solid tetrahedron and a solid cube both are simple polyhedron which are regular. In the first case the faces are equilateral triangles and in the other case the faces are squares. Now the question is: Can we have simple regular polyhedra whose faces are regular pentagons, regular hexagons, or any n-gons, $n \geq 3$? We can also ask whether there are other simple regular polyhedra, besides tetrahedron, whose faces are triangular? It is known since the time of Greeks that besides tetrahedron and cube, there are also octahedron, dodecahedron and icosahedron which are simple regular polyhedra. It is a consequence of Euler's Theorem that these are the only five simple regular polyhedra – these are known as **Platonic Solids**. We have

Theorem 4.5.11. *There are only five simple, regular polyhedra.*

Proof. Suppose S is a simple regular polyhedron with V vertices, E edges and F faces. Let m be the common number of edges which meet every vertex, and let n be the number of edges in each face of S. Also, note that $n \geq 3$. By Euler's theorem,

$$V - E + F = 2$$

and by our assumptions,

$$mV = 2E = nF.$$

This means $(nF)/m - (nF)/2 + F = 2$, i.e., $F(2n - mn + 2m) = 4m$. Clearly, we must have $2n - mn + 2m > 0$. Since $n \geq 3$,

$$2m > n(m-2) \geq 3(m-2) = 3m - 6$$

which at once implies that $m < 6$. Geometrically, this says that in S not more than 5 edges can meet at a vertex. Next, the relation

$$F(2n - mn + 2m) = 4m, \quad n \geq 3, \ m < 6,$$

permits the following possibilities for the values of (m, n, F), viz., $(m, n, F) = (3,3,4), (3,4,6), (4,3,8), (3,5,12)$ and $(5,3,20)$. To examine all these cases, let $n = 3$. Then we should have $F(6 - m) = 4m$. Now, this last equation says that m cannot be 1 or 2. The case $m = 3$ shows that $F = 4$, and $m = 4$ shows that $F = 8$. In other words, when the faces are triangles, we have only two possibilities, viz., tetrahedron and octahedron. Proceeding likewise, we can similarly prove that there are only the five possibilities mentioned above, i.e., the only five simple regular polyhedra are tetrahedron, cube, octahedron, dodecahedron and icosahedron (see Fig. 4.8). ∎

Orientability and Homology

We now prove a result which is at the heart of a class of important theorems in algebraic topology called "duality theorems". We will not prove here the theorems themselves, but will explain the basic concept of **orientability**, which is needed in their hypotheses and show that this is indeed a topological concept. We have already seen what is meant by an oriented complex, viz., a complex K in which every simplex is given a fixed but arbitrary orientation. Here, we will discuss the idea of orientability for a class of complexes, motivated by manifolds.

Definition 4.5.12. *A simplicial complex K is said to be an n-**pseudomanifold** if it satisfies the following conditions:*

1. *Each simplex of K is a face of some n-simplex of K.*
2. *Each $(n-1)$-simplex is a face of exactly two n-simplexes of K.*

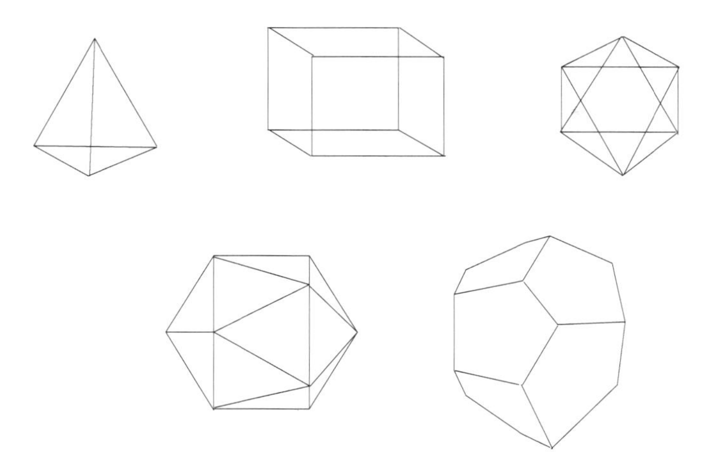

Fig. 4.8: The five Platonic solids

3. *Given a pair* σ_1^n *and* σ_2^n *of* n*-simplexes of* K*, there is a sequence of* n*-simplexes* $\tau_1, \tau_2, \ldots, \tau_k$ *such that* $\tau_1 = \sigma_1^n, \tau_k = \sigma_2^n$ *and for each* i*,* $\tau_i \cap \tau_{i+1}$ *is a* $(n-1)$*-face of both* τ_i *and* τ_{i+1}*,* $i = 1, 2, \ldots, k-1$.

The last condition says that K is connected. One can easily verify that if K is the 2-skeleton of $\text{Cl}(\sigma^3)$, where $\sigma^3 = \langle v_0, v_1, v_2, v_3 \rangle$, then K is a 2-pseudomanifold. Similarly, the triangulation K of the projective plane $\mathbb{P}^2$ (see Fig. 4.6) is a 2-pseudomanifold. Likewise, one can easily verify that the triangulations of torus T and Klein Bottle given by Fig. 4.5 and Exercise 4 are also 2-pseudomanifolds. An example of an n-dimensional pseudomanifold K is obtained by letting K be the n-skeleton of $\text{Cl}(\sigma^{n+1})$, where $\sigma^{n+1} = \langle v_0, \ldots, v_{n+1} \rangle$ is an $(n+1)$-simplex.

Let us recall (see Chapter 1) that a connected Hausdorff topological space X is said to be a **topological n-manifold** if each point $x \in X$ has an open neighbourhood U such that U is homeomorphic to the Euclidean space $\mathbb{R}^n$. Some people also require that X be a paracompact space or second countable or metrizable; in fact, all these three conditions are equivalent for a manifold X (see Theorem 1.8.2). The manifolds, which appeared in the work of G.F.B. Riemann (1826-1866), form a very important class of topological spaces. There is a nice relationship between compact n-manifolds and n-pseudomanifolds, viz., *if* K *is an* n*-pseudomanifold, then* $|K|$ *is a compact, connected polyhedron, which is not necessarily a topological* n*-manifold.* (If we take two hollow tetrahedra with a common base, remove interior of the triangles constituting the common base and identify the two remaining vertices, then the resulting complex K is a 2-pseudomanifold, but $|K|$ is not a 2- manifold.) *However, if* X *is a compact, connected, topological* n*-manifold, which is also triangulable by a complex* K*, then* K *is a* n*-pseudomanifold.* In fact, any triangulation of X would be an n-

pseudomanifold. This fact is some times very useful in computing the homology groups of a triangulable n-manifold. We are going to illustrate this somewhat in detail. First, let us have the following result concerning compact **surfaces**, every triangulation of which is a 2-pseudomanifold.

Theorem 4.5.13. *Let K be a 2-pseudomanifold with α_0 vertices, α_1 edges and α_2 triangles. Then*

(i) $3\alpha_2 = 2\alpha_1$

(ii) $\alpha_1 = 3(\alpha_0 - \chi(K))$

(iii) $\alpha_0 \geq (1/2)(7 + \sqrt{49 - 24\chi(K)})$.

Proof. A triangle has 3 edges and so there are a total of $3\alpha_2$ number of edges in the complex K. But we note that every edge has been counted twice since each edge is a face of exactly two triangles. Therefore, $\alpha_1 = (3\alpha_2)/2$. We already know by Euler-Poincaré Theorem that

$$\alpha_0 - \alpha_1 + \alpha_2 = \chi(K).$$

This means $\alpha_0 - \alpha_1 + (2/3)\alpha_1 = \chi(K)$, i.e., $\alpha_1 = 3(\alpha_0 - \chi(K))$, which proves (ii).

To prove the part (iii), note that $\alpha_0 \geq 4$. Obviously, the maximal number of possible edges in K is $\alpha_0(\alpha_0 - 1)/2$. Hence, $\alpha_1 \leq \alpha_0(\alpha_0 - 1)/2$. Now, $6\alpha_2 = 4\alpha_1$, i.e., $2\alpha_1 = 6\alpha_1 - 6\alpha_2$ or $\alpha_0(\alpha_0 - 1) \geq 6\alpha_1 - 6\alpha_2$. This means

$$\begin{aligned} \alpha_0^2 - \alpha_0 - 6\alpha_0 &\geq 6\alpha_1 - 6\alpha_2 - 6\alpha_0 \\ &= -6\chi(K) \\ \alpha_0^2 - 7\alpha_0 &\geq -6\chi(K). \end{aligned}$$

Therefore,

$$(2\alpha_0 - 7)^2 \geq 49 - 24\chi(K)$$

and so

$$\alpha_0 \geq (1/2)(7 + \sqrt{49 - 24\chi(K)}).$$

■

The above theorem, which is a consequence of the Euler-Poincaré Theorem, is quite useful in determining a triangulation of a given 2-pseudomanifold with minimum number of vertices, edges, and triangles. Since the computation of homology groups of a complex K could be quite complicated, we can minimize the complexity of calculation by choosing a suitable triangulation having minimum number of simplexes. It is evidently true though that finding a minimum triangulation itself could very well be a problem.

Example 4.5.14. Consider the 2-sphere $\mathbb{S}^2$ and note that $\chi(\mathbb{S}^2) = 2$. Hence, by the above theorem, we see that in any triangulation of $\mathbb{S}^2$ we must have

$$\alpha_0 \geq 1/2(7 + \sqrt{49 - 24\chi(K)}) = 4$$

$$\alpha_1 = 3(\alpha_0 - \chi(K)) \geq 3(4 - 2) = 6$$

$$\alpha_2 = (2/3)\alpha_1 \geq 4.$$

Thus, in any triangulation of $\mathbb{S}^2$, there must be at least 4 vertices, 6 edges and 4 triangles. This shows that a tetrahedron gives a minimal triangulation of $\mathbb{S}^2$.

Example 4.5.15. Now, consider the projective plane $\mathbb{P}^2$ which is also a 2-manifold. We have already seen that $H_0(\mathbb{P}^2) \cong \mathbb{Z}$, $H_1(\mathbb{P}^2) \cong \mathbb{Z}_2$ and $H_2(\mathbb{P}^2) = 0$. Hence, $\chi(\mathbb{P}^2) = 1$. Therefore, by the above theorem,

$$\alpha_0 \geq 6, \ \alpha_1 \geq 15, \ \alpha_2 \geq 10.$$

This shows that in any triangulation of $\mathbb{P}^2$, there must be at least 6 vertices, 15 edges and 10 triangles. Hence the triangulation given in Fig.4.6 is minimal.

We can similarly try and find out possible triangulations of torus, Klein bottle which are also minimal. Note that the lower bound for the number of vertices α_0 in the case of $T\#T$ is an irrational number. Thus, lower bounds α_0, α_1, α_2 may not actually be always attained.

Now let us consider the following concept of "orientability" of a complex.

Definition 4.5.16. *Let K be an n-pseudomanifold. If σ^{n-1} is a $(n-1)$-simplex, then it must be the common face of two n-simplexes, say, σ_1^n, σ_2^n. An orientation of K, which has the property that*

$$[\sigma_1^n, \sigma^{n-1}] = -[\sigma_2^n, \sigma^{n-1}]$$

for every $(n-1)$-simplex σ^{n-1}, is called a **coherent orientation** *on K. An n-pseudomanifold K is said to be* **orientable** *if it admits a coherent orientation.*

The above definition can be extended to topological manifolds which are triangulable. We will say that a n-manifold X is **orientable** if it has a triangulation K having a coherent orientation. In fact, it can be proved (see Theorem 4.5.18) that orientability of a manifold is a topological property, i.e., if a particular triangulation of an n-manifold X has a coherent orientation, then any other triangulation of X also has a coherent orientation. An n-manifold M with boundary ∂M is said to be orientable if its double DM is orientable; here DM is the n-manifold obtained from two copies of M by identifying their boundaries. It is easily seen that 2-spheres, torus, etc., are orientable whereas projective plane, Klein bottle and Möbius band (a manifold with boundary) are nonorientable 2-manifolds. More generally, we have

Proposition 4.5.17. *The n-sphere $\mathbb{S}^n$, $n \geq 1$, is orientable.*

Proof. Let us consider the $(n+1)$-simplex $\sigma^{n+1} = \langle v_0, \ldots, v_{n+1}\rangle$. We will prove that its boundary complex K is orientable. For each integer j, $0 \leq j \leq n+1$, let

$$\sigma_j = \langle v_0, \ldots, \hat{v}_j, \ldots, v_{n+1}\rangle$$

be the n-simplex where v_j is omitted. Let us give σ_j the positive orientation defined by the above ordering when j is even and give it opposite orientation if j is odd. Then $(n-1)$-simplex, positively oriented by

$$+\sigma_{ij} = +\langle v_0, \ldots, \hat{v}_i, \ldots, \hat{v}_j, \ldots, v_{n+1}\rangle$$

is a face of two simplexes σ_i and σ_j. We have, therefore, given an orientation to all n-simplexes as well as to all $(n-1)$-simplexes of the boundary complex K of σ^{n+1}. Now, give arbitrary orientation to the remaining simplexes. One can easily see that with this orientation on K,

$$[\sigma_i, \sigma_{ij}] = -[\sigma_j, \sigma_{ij}]$$

for each pair of n-simplexes σ_i, σ_j and their common face σ_{ij}. Hence K has a coherent orientation. ■

The above result, which should be visually verified in the case of $\mathbb{S}^2$, also follows from the next theorem which determines the orientability of a compact n-manifold in terms of its n-th integral homology and shows that orientability of manifolds is a topological concept.

Theorem 4.5.18. *A triangulable connected n-manifold X is orientable if and only if $H_n(K)$ is nontrivial for some triangulation (hence, every triangulation) K of X.*

Proof. Suppose X is triangulable. This means there is a triangulation K of X which is an n-pseudomanifold. Now, suppose K has a coherent orientation. Thus, if σ^{n-1} is a $(n-1)$-face of two n-simplexes σ_1^n and σ_2^n of K, then

$$[\sigma_1^n, \sigma^{n-1}] = -[\sigma_2^n, \sigma^{n-1}].$$

As a result, note that for any fixed integer m, the n-chain

$$c = \sum_{\sigma^n \in K} m \cdot \sigma^n$$

is easily seen to be a n-cycle. This means $Z_n(K) \neq 0$. Since $B_n(K) = 0$, we find that $H_n(K) \neq 0$. This proves the direct part.

Conversely, suppose for some given orientation of K, $H_n(K) \neq 0$ and let $z = \sum_{\sigma_i^n \in K} m_i \sigma_i^n$ be an n-cycle which is nonzero. Now, using the fact that $\partial(z) = 0$, and that each $(n-1)$-simplex is a face of exactly two n-simplexes and also the fact that $|K|$ is connected, it follows that $m_i = \pm m_0, \forall\ i \neq 0$ where m_0 is a fixed coefficient. By changing the orientation of σ_i^n for which $m_i = -m_0$, we find that

$$\sum_{\sigma_i^n \in K} m_i \sigma_i^n = m_0 \big(\sum_{\sigma_i^n \in K} \sigma_i^n \big)$$

is an n-cycle. However, this implies that any $(n-1)$-face τ of K must have positive incidence number w.r.t. one n-simplex and negative incidence number w.r.t. the other n-simplex. This means the changed orientation of K is coherent, and so X is orientable. ■

The above theorem tells us that the projective plane $\mathbb{P}^2$ and Klein bottle, which have trivial homology in dimension 2 with coefficients in $\mathbb{Z}$, are nonorientable. Here, the integer coefficients are important because even though $\mathbb{P}^2$ is nonorientable, $H_2(\mathbb{P}^2; \mathbb{Z}_2) \cong \mathbb{Z}_2$, which is nonzero. We should also mention here that none of the even-dimensional projective spaces $\mathbb{P}^{2n}$, $n \geq 1$, is orientable, but all odd-dimensional ones $\mathbb{P}^{2n+1}$ are orientable. This follows from their homology groups and the above theorem (Exercises Section 4.7).

Exercises

1. Determine which of the following surfaces is orientable:
   (a) $\{(x, y, z) \in \mathbb{R}^3 : x^2/a^2 + y^2/b^2 + z^2/c^2 = 1,\ a > b > c\}$
   (b) Torus (c) Projective plane (d) Klein bottle (e) Möbius band M
   (f) Double torus (Connected sum of two tori.)

2. Use Euler-Poincaré theorem to compute the Euler characterstic of the spaces given in Exercise 1.

3. Show that for a simplicial complex K the Euler characterstic of K computed using homology with integer coefficients is the same as that computed using homology with coefficients in $\mathbb{Q}$.

4. Show that torus and Klein bottle are 2-pseudomanifolds. Draw minimal triangulations of torus and Klein bottle, and justify your answer.

5. If K is a 2-pseudomanifold then show that $\chi(K) \leq 2$.

6. Give an example of a 3-pseudomanifold K so that $|K|$ is not a 3-manifold.

7. Triangulate the Möbius band in a simple way so that its centre circle is a subcomplex. Orient the boundary circle and the central circle. Let the resulting cycles be denoted by z_1 and z. Prove that z_1 is homologous to $2z$ or $-2z$.

8. Compute the homology groups of the following simplicial complexes:
   (a) Three copies of the boundary of a triangle all joined at a vertex.
   (b) Two hollow tetrahedra glued together along an edge.

9. Let S_1, S_2 be 2 compact surfaces with triangulations K_1, K_2 respectively. Choose triangles ABC and $A'B'C'$ from each of $|K_1|, |K_2|$ respectively, remove their interiors and identify the boundaries of the two triangles by the obvious linear homeomorphism. Then the resulting quotient space, denoted by $|K_1|\#|K_2|$, is called the **connected sum** of $|K_1|$ and $|K_2|$. Prove that this connected sum gives a triangulation of $S_1\#S_2$, and hence deduce that

$$\chi(S_1\#S_2) = \chi(S_1) + \chi(S_2) - 2.$$

10. Using the classification theorem for compact surfaces compute the Euler characterstic of all compact 2-manifolds.

11. Let us define a rectilinear polyhedron to be **topologically regular** if the number of edges meeting at each vertex is same, e.g., a tetrahedron whose faces are triangles, not necessarily equilateral triangles; a solid lamina whose faces are rectangles, not necessarily squares, are examples of topological regular polyhedra. Prove that there are exactly five classes of simple topologically regular polyhedra.

4.6 Induced Homomorphisms

Recall that a simplicial map $f\colon K \to L$ is a map from the set of vertices of K to the set of vertices of L which has the property that if $\sigma^q = \langle v_0, \ldots, v_q\rangle$ is a q-simplex of K, then $\langle f(v_0), \ldots, f(v_q)\rangle$ is a simplex of L of dimension $r \leq q$: If all the vertices $f(v_0), \ldots, f(v_q)$ are not distinct then by deleting repetitions, we find that $r < q$, otherwise $r = q$.

Now, we have

Definition 4.6.1. *Let $f\colon K \to L$ be a simplicial map. Then for each $q \geq 0$, we define a homomorphism $f_\#\colon C_q(K) \to C_q(L)$ by putting*

$$f_\#(\langle v_0, \ldots, v_q\rangle) = \begin{cases} \langle f(v_0), \ldots, f(v_q)\rangle, & \textit{if } f(v_i)\textit{'s are distinct} \\ 0, & \textit{otherwise,} \end{cases}$$

for each simplex $\langle v_0, \ldots, v_q\rangle$ of K and then extending it linearly on $C_q(K)$.

It is seen at once that $f_\#$ is a well-defined map because exchanging any pair of vertices in the left hand side of the above definition results in a corresponding change in the right hand side. The collection $\{f_\#\colon C_q(K) \to C_q(L)\}$ of homomorphisms (we have suppressed the index q deliberately from $f_\#$ and also from ∂ because suffixing them would make the notations cumbersome) is called the **chain map** induced by the simplicial map f. This is proved in the next

Proposition 4.6.2. *If $f\colon K \to L$ is a simplicial map, then the induced sequence $\{f_\#\colon C_q(K) \to C_q(L)\}$ of homomorphisms commutes with the boundary homomorphism ∂ of the chain complex and hence, $f_\#$ induces a homomorphism $f_*\colon H_q(K) \to H_q(L)$ in each dimension q.*

Proof. Let us consider the following diagram $q \geq 1$:

$$\begin{array}{ccccccc}
\longrightarrow & C_q(K) & \xrightarrow{\partial} & C_{q-1}(K) & \longrightarrow \\
& {\scriptstyle f_\#}\downarrow & & \downarrow{\scriptstyle f_\#} & \\
\longrightarrow & C_q(L) & \xrightarrow[\partial]{} & C_{q-1}(L) & \longrightarrow
\end{array}$$

We must show that $f_\#\partial = \partial f_\#$, i.e.,

$$(*) \qquad f_\#(\partial\langle v_0, \ldots, v_q\rangle) = \partial(\langle f(v_0), \ldots, f(v_q)\rangle)$$

for each oriented q-simplex $\langle v_0, \ldots, v_q\rangle$ of K. Since there may be repetitions in the vertices $f(v_0), \ldots, f(v_q)$, we let τ be the simplex of L spanned by these vertices, and note that $\dim \tau \leq q$. We consider the following three cases:

Case I: $\dim \tau = q$. In this case all the vertices $f(v_0), \ldots, f(v_q)$ are distinct and the result follows directly from definitions of $f_\#$ and ∂.

Case II: $\dim \tau \leq q - 2$. In this case both sides of $(*)$ are zero because in the set of vertices $f(v_0), \ldots, f(v_q)$, at least three are identical.

Case III: $\dim \tau = q - 1$. In this case two vertices from $f(v_0), \ldots, f(v_q)$ are identical and we can assume that $f(v_0) = f(v_1)$, and $f(v_2), \ldots f(v_q)$ are distinct. Then, by definition, the right side of $(*)$ vanishes. The left side has only two nonzero terms, viz.,

$$\langle f(v_1), f(v_2), \ldots, f(v_q)\rangle - \langle f(v_0), f(v_2), \ldots, f(v_q)\rangle.$$

Since $f(v_0) = f(v_1)$ these terms also cancel each other and the left side is also zero.

Finally, to prove that for each $q \geq 0$, $f_\#$ induces a homomorphism $f_*\colon H_q(K) \to H_q(L)$, we note that $H_q(K) = Z_q(K)/B_q(K)$ and $H_q(L) = Z_q(L)/B_q(L)$. It is easily verified that $f_\#(Z_q(K)) \subseteq Z_q(L)$ and $f_\#(B_q(K)) \subseteq B_q(L)$, i.e., $f_\#$ maps cycles into cycles and boundaries into boundaries. It follows, therefore, that $f_\#$ induces homomorphism $f_*\colon H_q(K) \to H_q(L)$ defined by $f_*(\{z_q\}) = \{f_\#(z_q)\}$. ∎

The next result now follows from the definitions of induced homomorphism.

Theorem 4.6.3. (a) *If $I_K\colon K \to K$ be the identity simplicial map, then the induced homomorphism $(I_K)_*\colon H_q(K) \to H_q(K)$ is identity for each $q \geq 0$.*

(b) *If $f\colon K \to L$ and $g\colon L \to M$ are simplicial maps, then for all $q \geq 0$,*

$$(g \circ f)_* = g_* \circ f_* \colon H_q(K) \to H_q(M).$$

Topological Invariance of Homology Groups

Now, suppose K, L are two complexes and $f\colon K \to L$ is a simplicial map. Then we know that f induces a continuous map $|f|\colon |K| \to |L|$. Conversely, given any continuous map $h\colon |K| \to |L|$, we want to know whether there is a induced homomorphism $h_*\colon H_q(K) \to H_q(L)$ in homology. If h is induced by a simplicial map, as indicated above, then the answer to the above question is trivially, "yes". If, however, h is not induced by a simplicial map, as is mostly the case, then what can we do? Take a simplicial approximation $g\colon K^{(k)} \to L$ of h and note that the induced map $|g|\colon |K| = |K^{(k)}| \to |L|$ is homotopic to h. Now, by the previous result there is the induced homomorphism $g_*\colon H_q(K^{(k)}) \to H_q(L)$. If we could prove (i) *there is an isomorphism $\mu\colon H_q(K) \to H_q(K^{(k)})$ whatever be the integer k, and* (ii) *$g\colon K^{(k)} \to L$, $g'\colon K^{(m)} \to L$ are any two simplicial approximations of h implies $g_*\mu = g'_*\mu'$*, then we could unambiguously define a homomorphism $h_*\colon H_q(K) \to H_q(L)$ by the commutativity of the following diagram:

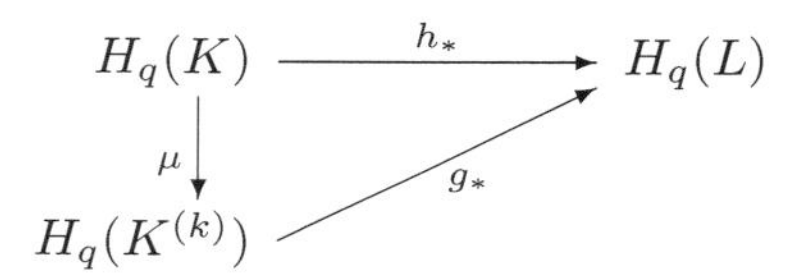

i.e., we define $h_* = g_* \circ \mu$.

Later, in Sections 4.9.1 and 4.9.2, we will show that (i) and (ii) above can actually be proved. Hence presently we assume that any continuous map $h\colon |K| \to |L|$ induces a homomorphism $h_*\colon H_q(K) \to H_q(L)$ for all $q \geq 0$ as defined above. We will also assume the following in which part (i) is, of course, obvious.

Theorem 4.6.4. (i) *The identity map $I_{|K|}\colon |K| \to |K|$ induces the identity homomorphism $I_{H_q(K)}\colon H_q(K) \to H_q(K)$ in each dimension q.*

(ii) *If $f\colon |K| \to |L|$ and $g\colon |L| \to |M|$ are any two continuous maps, then the induced homomorphism $(g \circ f)_* = g_* \circ f_*\colon H_q(K) \to H_q(M)$ in each dimension $q \geq 0$.*

We have only stated the above theorem. The proof is postponed to Section 4.9 for the simple reason that besides being lengthy, it involves several more concepts and algebraic results, which we have not discussed so far. The statement is given because, without interrupting continuity, we want to illustrate some of the classical applications of homology groups right now.

The most important result which we should mention here and which follows from the preceding theorem is that the **homology groups of a polyhedron are its topological invariants**. We have

Theorem 4.6.5. *Let X be a polyhedron and K, L be any two triangulations of X. Then for each $q \geq 0$, $H_q(K) \cong H_q(L)$.*

Proof. Since K and L are triangulations of the space X, there are homeomorphisms $f\colon |K| \to X$, $g\colon |L| \to X$. This means $h = g^{-1} \circ f\colon |K| \to |L|$ is a homeomorphism. Let $h^{-1}\colon |L| \to |K|$ be the inverse of h. Consider the induced homomorphism $h_*\colon H_q(K) \to H_q(L)$ and $h_*^{-1}\colon H_q(L) \to H_q(K)$. We know that $h^{-1} \circ h = I_{|K|}$ and $h \circ h^{-1} = I_{|L|}$. Hence, by (i) and (ii) of the preceding theorem $h_*^{-1} \circ h_* = I_{H_q(K)}$ and $h_* \circ h_*^{-1} = I_{H_q(L)}$. This says that $h_*\colon H_q(K) \to H_q(L)$ is an isomorphism. ■

Homology of a polyhedron

The above theorem says that if X is a compact polyhedron, then we can define simplicial homology groups $H_q(X), q \geq 0$, of X simply by setting $H_q(X) = H_q(K)$, where K is some triangulation of X. By the above result, the simplicial homology groups of X are well-defined up to isomorphism only.

It may be noticed that if K is a triangulation of X, then $|K|$ is a rectilinear polyhedron whereas X is a polyhedron which may not be rectilinear, i.e., the simplexes of X may be curved. The two spaces X and $|K|$ are not identical, but they are homeomorphic. The standard n-sphere $\mathbb{S}^n$ and the standard n-disk $\mathbb{D}^n$ are all compact polyhedra, and we have already computed homology groups of a triangulation (see Section 4.5) of $\mathbb{S}^n$ and a triangulation of $\mathbb{D}^n$. Thus, we can now unambiguously assert that the simplicial homology groups (up to isomorphism) of these spaces are given by:

$$H_q(\mathbb{S}^n) \cong \begin{cases} \mathbb{Z}, & q = 0, n \\ 0, & \text{otherwise} \end{cases}$$

$$H_q(\mathbb{D}^n) \cong \begin{cases} \mathbb{Z} & q = 0 \\ 0, & \text{otherwise.} \end{cases}$$

Likewise we can also say (see Section 4.4 and exercises therein) that the simplicial homology of the torus T, the projective plane $\mathbb{P}^2$, the Möbius band M and the Klein bottle K are given as follows:

$$H_q(T) \cong \begin{cases} \mathbb{Z}, & q = 0, 2 \\ \mathbb{Z} \oplus \mathbb{Z}, & q = 1 \\ 0, & \text{otherwise,} \end{cases}$$

$$H_q(\mathbb{P}^2) \cong \begin{cases} \mathbb{Z}, & q = 0 \\ \mathbb{Z}/2\mathbb{Z}, & q = 1 \\ 0, & \text{otherwise}, \end{cases}$$

$$H_q(M) \cong \begin{cases} \mathbb{Z}, & q = 0, 1 \\ 0, & \text{otherwise}, \end{cases}$$

$$H_q(K) \cong \begin{cases} \mathbb{Z}, & q = 0 \\ \mathbb{Z} \oplus \mathbb{Z}/2\mathbb{Z}, & q = 1 \\ 0, & \text{otherwise}. \end{cases}$$

Induced homomorphism

We should also observe that if $f\colon X \to Y$ is a continuous map from a polyhedron X to another polyhedron Y, then choosing triangulations $h\colon |K| \to X$ and $k\colon |L| \to Y$, the continuous map $k^{-1}fh\colon |K| \to |L|$ induces a homomorphism $(k^{-1}fh)_*\colon H_q(K) \to H_q(L)$ and we say that $f\colon X \to Y$ induces the homomorphism $f_*\colon H_q(X) \to H_q(Y)$ by putting $f_* = (k^{-1}fh)_*$. This is all right in so far as we are studying homology groups up to isomorphism only; if we want to distinguish between two isomorphisms, then this identification is not acceptable. To see it further, let us now choose a different triangulation, say, $h_1\colon |K_1| \to X, k_1\colon |L_1| \to Y$. Then the induced homomorphism would be $(k_1^{-1}fh_1)_*\colon H_q(K_1) \to H_q(L_1)$. We know that there are homeomorphisms $h_1^{-1}h\colon |K| \to |K_1|, k_1^{-1}k\colon |L| \to |L_1|$ which induce isomorphisms in homology and so we have a commutative diagram

$$\begin{array}{ccc} |K| & \xrightarrow{(k^{-1}fh)_*} & |L| \\ {\scriptstyle (h_1^{-1}h)_*}\downarrow & & \downarrow{\scriptstyle (k_1^{-1}k)_*} \\ |K_1| & \xrightarrow[(k_1^{-1}fh_1)_*]{} & |L_1| \end{array}$$

because $(k_1^{-1}k)^{-1}(k_1^{-1}fh_1)(h_1^{-1}h) = k^{-1}fh$. Therefore, if we identify the groups $H_q(K)$ and $H_q(K_1)$, etc., then we can as well identify the induced homomorphisms $(k^{-1}fh)_*$ and $(k_1^{-1}fh_1)_*$. Thus the induced homomorphisn f_* is defined up to this identification of isomorphisms.

4.7 Some Applications

As stated earlier, having determined the homology groups of disks $\mathbb{D}^n$ and spheres $\mathbb{S}^n$, now we come to a few classical applications of homology groups and induced homomorphisms. We call attention to Theorem 4.5.5 for these applications.

Theorem 4.7.1. (Invariance of dimension). *If $m \neq n$, then*

(i) $\mathbb{S}^m$ *is not homeomorphic to* $\mathbb{S}^n$, *and*

(ii) $\mathbb{R}^m$ *is not homeomorphic to* $\mathbb{R}^n$.

Proof. (i) Suppose, on the contrary, there is a homeomorphism $f\colon \mathbb{S}^m \to \mathbb{S}^n$. Let $h\colon |K| \to \mathbb{S}^m$ and $k\colon |L| \to \mathbb{S}^n$ be triangulations of $\mathbb{S}^m$ and $\mathbb{S}^n$, respectively. We have already computed the homology groups of K and L. Since f is a homeomorphism, the map $g = k^{-1}fh\colon |K| \to |L|$ is also a homeomorphism. Let $g^{-1}\colon |L| \to |K|$ be its inverse. We consider the induced homomorphisms $g_*\colon H_m(K) \to H_m(L)$ and $g_*^{-1}\colon H_m(L) \to H_m(K)$ in simplicial homology. Since $g^{-1}g = I_{|K|}$, we find by theorem 4.6.4, that

$$g_*^{-1}g_* = (g^{-1}g)_* = (I_{|K|})_* = I_{H_m(K)}.$$

Similarly, we find that $g_*g_*^{-1}$ is also identity on $H_m(L)$. Therefore, $g_*\colon H_m(K) \to H_m(L)$ is an isomorphism. But this is a contradiction, because $H_m(K) = \mathbb{Z}$ whereas $H_m(L) = 0$, since $m \neq n$. This proves the theorem.

(ii) Again, suppose $\mathbb{R}^m$ is homeomorphic to $\mathbb{R}^n$. Since these are locally compact Hausdorff spaces, their one-point compactifications, viz., $\mathbb{S}^m$ and $\mathbb{S}^n$ must also be homeomorphic, which is a contradiction to (i) proved above. Hence, if $m \neq n$, $\mathbb{R}^m$ cannot be homeomorphic to $\mathbb{R}^n$. ■

If $\mathbb{D}^n$ denotes the n-dimensional disk (closed), then its boundary is homeomorphic to the $(n-1)$-dimensional sphere $\mathbb{S}^{n-1}$. Thus, $\mathbb{S}^{n-1}$ is a compact subset of $\mathbb{D}^n$. Recall that a subspace A of a space X is said to be **retract** of X if there is a continuous map $r\colon X \to A$ such that $r(a) = a$ for $a \in A$. Now, one can ask the question: Is $\mathbb{S}^{n-1}$ is a retract of $\mathbb{D}^n$? If $n = 1$, this is clearly impossible because $D^1 = [-1, 1]$ is connected whereas $\mathbb{S}^0 = \{-1, 1\}$ is disconnected. If $n \geq 2$, then also the answer to the above question is "no". We have

Theorem 4.7.2. (No-retraction Theorem). *The sphere* $\mathbb{S}^{n-1}$ *cannot be a retract of* $\mathbb{D}^n$, *for any* $n \geq 1$.

Proof. If possible, suppose there is a retraction $r\colon \mathbb{D}^n \to \mathbb{S}^{n-1}$. Let $i\colon \mathbb{S}^{n-1} \to \mathbb{D}^n$ be the inclusion map. Then, clearly, $ri = I_{\mathbb{S}^{n-1}}$. The case when $n = 1$ is clear because $\mathbb{D}^1$ is connected whereas $\mathbb{S}^0$ is not. Hence we assume that $n > 1$. Let $h\colon |K| \to \mathbb{D}^n$ and $k\colon |L| \to \mathbb{S}^{n-1}$ be triangulations of the disc and sphere. Notice that we have continuous maps $k^{-1}rh\colon |K| \to |L|$ and $h^{-1}ik\colon |L| \to |K|$ such that their composite is identity, i.e., $(k^{-1}rh)(h^{-1}ik) = I_{|L|}$. This means the composite

$$\mathbb{Z} = H_{n-1}(L) \to H_{n-1}(K) \to H_{n-1}(L) = \mathbb{Z}$$

is identity on $\mathbb{Z}$ by theorem 4.6.4. But this last map factors through the zero group $H_{n-1}(K) = 0$, which is a contradiction. ■

Recall that a space X is said to have fixed-point property if for every continuous map $f\colon X \to X$, there exists a point $x_0 \in X$ such that $f(x_0) = x_0$. From the first course in real analysis, we know that the unit interval [0,1] has the fixed-point property. The following important theorem is a far reaching generalization of this result.

Theorem 4.7.3. (Brouwer's Fixed-Point Theorem). *Let $f\colon \mathbb{D}^n \to \mathbb{D}^n$, $n \geq 1$ be any continuous map. Then f has at least one fixed point.*

Proof. Suppose f has no fixed points. This means for every x, $f(x) \neq x$. Now, we define a map $g\colon \mathbb{D}^n \to \mathbb{S}^{n-1}$ as follows: For any $x \in \mathbb{D}^n$, consider the line segment (vector) joining $f(x)$ to x and extend it. Then, this vector when produced will meet $\mathbb{S}^{n-1}$ exactly at one point, which we call $g(x)$ (see Fig. 4.9). First, we prove that $g\colon \mathbb{D}^n \to \mathbb{S}^{n-1}$ is continuous. Note that for each x, the nonzero vector $x - f(x)$ can be multiplied by a unique positive scalar, say λ, such that $g(x) = f(x) + \lambda(x - f(x))$, and this λ depends on x. Since $g(x)$ lies on $\mathbb{S}^{n-1}$, its norm is 1 and so $||f(x) + \lambda(x - f(x))|| = 1$. This implies that

$$||f(x)||^2 + \lambda^2||(x - f(x))||^2 + 2\lambda f(x) \cdot (x - f(x)) = 1.$$

This is a quadratic equation in λ having only one positive real root λ. Therefore, by the formula for the roots of a quadratic equation, we get $\lambda = \frac{-f(x)\cdot(x-f(x))}{||x-f(x)||^2}$. This proves that λ is a continuous function of x and therefore g is a continuous function. But g is clearly a retraction of $\mathbb{D}^n$ onto $\mathbb{S}^{n-1}$, a contradiction to the no-retraction theorem. ■

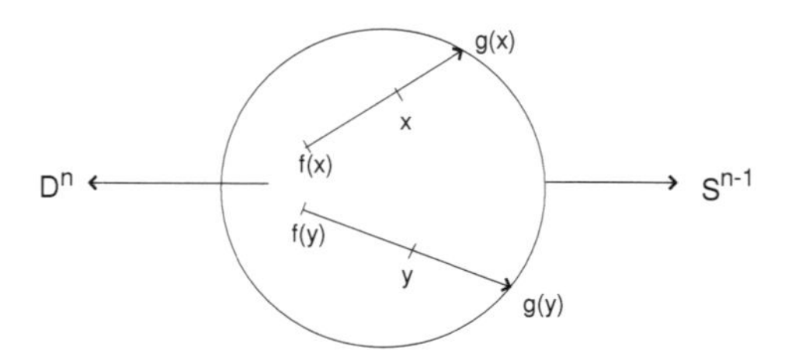

Fig. 4.9: The map g becomes a retraction

The following result gives an interesting application of the Brouwer fixed-point theorem to a result of linear algebra. We have

Proposition 4.7.4. *Let A be an $n \times n$ real matrix with positive entries. Then A has a positive eigen value.*

Proof. Consider the Euclidean space $\mathbb{R}^n$, and observe that A determines a linear transformation from $\mathbb{R}^n$ to itself. Also, note that if $x = (x_1, \ldots, x_n) \in \mathbb{R}^n$, where $x_i \geq 0$ for all $i = 1, \ldots, n$, then, since all entries of A are positive, Ax also has the same property, i.e., A maps the positive octant P including its boundary to itself. Furthermore, if one entry of x is positive, then all entries of Ax are positive. Let us also point out that if $\mathbb{S}^{n-1}$ denotes the unit sphere in $\mathbb{R}^n$, then $\mathbb{S}^{n-1} \cap P$ is homeomorphic to the $(n-1)$-disk D^{n-1}. Now, we can define a map $f\colon \mathbb{S}^{n-1} \cap P \to \mathbb{S}^{n-1} \cap P$ by putting $f(x) = Ax/\|Ax\|$. Then evidently f is continuous. Hence, by the Brouwer's fixed point theorem, there exists a non-zero vector $x_0 \in \mathbb{S}^{n-1} \cap P$ such that $f(x_0) = x_0$, i.e., $Ax_0/\|Ax_0\| = x_0$. This says that $Ax_0 = \|Ax_0\|x_0$ which means $\|Ax_0\|(\neq 0)$ is an eigen value of A. ■

Knowing that every continuous map $f\colon X \to X$ has a fixed point is an important property of the space X, and it has interesting applications. Let us illustrate this by another example. Suppose we have a set of continuous functions $f_i\colon \mathbb{R}^n \to \mathbb{R}, i = 1, 2, \ldots, n$, each of which has a nonempty zero set, i.e., there exists some points of $\mathbb{R}^n$ where f_i is zero. Now, the question is: Is there a common zero of all these functions f_i? In other words, do the following system of simultaneous equations has a solution in $\mathbb{R}^n$:

$$\begin{aligned} f_1(x_1, x_2, \ldots, x_n) &= 0 \\ f_2(x_1, x_2, \ldots, x_n) &= 0 \\ &\vdots \\ f_n(x_1, x_2, \ldots, x_n) &= 0 \end{aligned}$$

This question is really a question whether the space $\mathbb{R}^n$ has the fixed-point property. To see why this is so, let us consider the following continuous maps $g_i\colon \mathbb{R}^n \to \mathbb{R}$ defined by

$$g_i(x_1, x_2, \ldots, x_n) = f_i(x_1, \ldots, x_n) + x_i$$

for $i = 1, 2, \ldots, n$. We consider the map $h\colon \mathbb{R}^n \to \mathbb{R}^n$ defined by

$$h(x_1, \ldots, x_n) = (g_1(x_1, \ldots, x_n), \ldots, g_n(x_1, \ldots, x_n)).$$

Notice that h is continuous because each g_i is continuous. Now, observe that h has a fixed point $(a_1, \ldots, a_n) \in \mathbb{R}^n$ if and only if for each $i = 1, \ldots, n$, we have

$$g_i(a_1, \ldots, a_n) = a_i,$$

and that will happen if and only if

$$f_1(a_1, \ldots, a_n) = 0 = f_2(a_1, \ldots, a_n) = \ldots = f_n(a_1, \ldots, a_n),$$

i.e., all the $f_i's$ have a common zero. It may be remarked that the Euclidean space $\mathbb{R}^n$ used in this example does not have the fixed-point property because translations by nonzero vectors do not have fixed points. On the other hand, by Brouwer's theorem, the cube I^n, where $I = [-1, 1]$, has the fixed-point property and so can very well be used for the space X to conclude that the simultaneous equations have a solution.

Exercises

1. If a space X has the fixed-point property, then show that any space Y which is homeomorphic to X also has the fixed-point property.

2. If A is a retract of X and if X has the fixed-point property, then show that A also has the fixed point property.

3. Let X be a compact metric space and $f\colon X \to X$ be a fixed-point free map. Prove that there is an $\epsilon > 0$ such that $d(x, f(x)) > \epsilon$ for all $x \in X$.

4. By giving concrete examples prove that the 2-sphere, the torus and the Klein bottle do not have the fixed-point property.

5. Prove that the projective plane $\mathbb{P}^2$ has the fixed-point property (this may be bit difficult at this stage!).

6. If a polyhedron $A \subset X$ is a retract of a polyhedron X, then show that for $q \geq 0$, $H_q(A)$ is a direct summand of $H_q(X)$.

7. Prove that an injective map between two polyhedra does not necessarily induce an injective map between their homology groups.

8. Show that the following conditions are equivalent:
   (a) $\mathbb{S}^{n-1}$ is not a retract of $\mathbb{D}^n$.
   (b) $\mathbb{D}^n$ has the fixed-point property.
   (c) The n-simplex δ_n has the fixed-point property.

9. Show that the map $g\colon \mathbb{RP}^3 \to \mathbb{RP}^3$ defined by the linear transformation $T\colon \mathbb{R}^4 \to \mathbb{R}^4$ given by $T(x_1, x_2, x_3, x_4) = (-x_2, x_1, -x_4, x_3)$ does not have the fixed-point property. More generally, show that $\mathbb{RP}^{2k+1}$ does not have the fixed-point property.

10. Prove that the homology groups of the real projective spaces $\mathbb{RP}^n$, $n \geq 2$, and that of the complex projective spaces $\mathbb{CP}^n$, $n \geq 1$ are given by the following (See Examples 1.1.4 (a) and 1.1.4 (b). The cases $n \geq 3$ need concepts not covered so far.)

    (i) n is odd:

$$H_q(\mathbb{R}P^n) \cong \begin{cases} \mathbb{Z}, & q = 0, n \\ \mathbb{Z}_2, & 0 < q < n, q \text{ odd} \\ 0, & \text{otherwise}, \end{cases}$$

    (ii) n is even:

$$H_q(\mathbb{R}P^n) \cong \begin{cases} \mathbb{Z}, & q = 0 \\ \mathbb{Z}_2, & 0 < q < n, q \text{ odd} \\ 0, & \text{otherwise}, \end{cases}$$

(iii) The complex case

$$H_q(\mathbb{C}P^n) \cong \begin{cases} \mathbb{Z}, & q \text{ is even} \\ 0, & \text{otherwise,} \end{cases}$$

11. Show that any nonsingular linear transformation $T\colon \mathbb{R}^3 \to \mathbb{R}^3$ defines a continuous map $g\colon \mathbb{RP}^2 \to \mathbb{RP}^2$ which has a fixed point. More generally, prove that $\mathbb{RP}^{2k}$ has the fixed-point property for any $k \geq 0$.

12. Prove that any compact locally contractible space has the fixed-point property.

4.8 Degree of a Map and its Applications

Recall that a homomorphism $f\colon \mathbb{Z} \to \mathbb{Z}$ of the infinite cyclic group is completely determined by the image of its generator $1 \in \mathbb{Z}$ under the map f, i.e., f is simply multiplication by the integer $f(1) = n$. This fact is used in the following:

Definition 4.8.1. *Let $f\colon \mathbb{S}^n \to \mathbb{S}^n, (n \geq 1)$ be a continuous map and $h\colon |K| \to \mathbb{S}^n$ be any triangulation of $\mathbb{S}^n$. Then we know that f induces a homomorphism $(h^{-1}fh)_*\colon H_n(K) \to H_n(K)$. Since $H_n(K) \cong \mathbb{Z}$, there is a unique integer d such that for every element $\alpha \in H_n(K)$, $(h^{-1}fh)_*(\alpha) = d\alpha$. This unique integer d is called the* **degree** *of f and we denote it by* $\deg f$.

We must prove that the integer d does not depend on the chosen triangulation $h\colon |K| \to \mathbb{S}^n$. For this, let $k\colon |L| \to \mathbb{S}^n$ be another triangulation of $\mathbb{S}^n$. Note that $\phi = k^{-1}h\colon |K| \to |L|$ and $\phi^{-1} = h^{-1}k\colon |L| \to |K|$ are homeomorphisms. Therefore, $(k^{-1}fk)_*(\alpha) = (k^{-1}h)_*(h^{-1}fh)_*(h^{-1}k)_*(\alpha) = \phi_*(h^{-1}fh)_*\phi_*^{-1}(\alpha) = \phi_*(d \cdot (\phi^{-1})_*(\alpha)) = \phi_*(\phi^{-1})_*(d \cdot \alpha) = d \cdot \alpha$, which says that $(k^{-1}fk)_*$ is again multiplication by d. This proves the claim.

The following result is a consequence of the definitions.

Proposition 4.8.2. (a) *The identity map $I_{\mathbb{S}^n}\colon \mathbb{S}^n \to \mathbb{S}^n$ has degree $+1$.*

(b) *If $f\colon \mathbb{S}^n \to \mathbb{S}^n$, $g\colon \mathbb{S}^n \to \mathbb{S}^n$ are continuous maps, then* $\deg(g \circ f) = \deg g \cdot \deg f$.

(c) *The degree of any homeomorphism is ± 1.*

Proof. (a) This follows from the fact that the identity map $I_{\mathbb{S}^n}$ induces the identity map in homology.

(b) Let $k\colon |K| \to \mathbb{S}^n$ be a triangulation of $\mathbb{S}^n$. Suppose $\deg f = n_1$, $\deg g = n_2$. Then for any $\alpha \in H_n(K)$, we have

$$\begin{aligned}(k^{-1}g \circ fk)_*(\alpha) &= (k^{-1}gk)_*(k^{-1}fk)_*(\alpha)) \\ &= (k^{-1}gk)_*(n_1(\alpha)) \\ &= n_2(n_1\alpha) \\ &= (n_2n_1)\alpha.\end{aligned}$$

Hence, $\deg(g \circ f) = \deg g \cdot \deg f$.

(c) Let $h\colon \mathbb{S}^n \to \mathbb{S}^n$ be a homeomorphism. Then $h^{-1} \circ h = I_{\mathbb{S}^n}$ and hence, by (a), we have

$$\deg(h^{-1} \circ h) = 1 = \deg h^{-1} \cdot \deg h.$$

Since $\deg h, \deg h^{-1}$ both are integers, we must have $\deg h = \deg h^{-1} = +1$ or -1. ∎

Next, we are going to prove that if $f, g\colon \mathbb{S}^n \to \mathbb{S}^n$ are two homotopic maps then $\deg f = \deg g$. First, observe that if $h\colon |K| \to \mathbb{S}^n$ is any triangulation of $\mathbb{S}^n$, then $h^{-1}fh,\ h^{-1}gh\colon |K| \to |K|$ are also homotopic. We will prove later (see Theorem 4.8.4) that two homotopic maps induce identical homomorphism in homology. Hence the result follows at once. However, proving that homotopic maps induce identical homomorphisms in homology is much more involved than proving the above result on degree directly. We will, therefore, use the classical definition of the degree of a map due to L.E.J. Brouwer, which is more intuitive than the "homology definition" given earlier, to prove the above result on degree. Indeed, Brouwer's definition of degree of a map $f\colon \mathbb{S}^n \to \mathbb{S}^n$ really means the number of times the domain sphere "wraps" around the range sphere. We have

Definition 4.8.3. *Let $f\colon \mathbb{S}^n \to \mathbb{S}^n$ be a continuous map and let $h\colon |K| \to \mathbb{S}^n$ be a triangulation of $\mathbb{S}^n$. Let $\phi\colon K^{(k)} \to K$ be a simplicial approximation of f, where $K^{(k)}$ is the kth barycentric subdivision of K. For any positively oriented n-simplex τ of K, let p be the number of positively oriented simplexes σ of $K^{(k)}$ such that $\phi(\sigma) = \tau$ and let q be the number of negatively oriented simplexes σ of $K^{(k)}$ such that $\phi(\sigma) = \tau$. Then the integer $p - q$ is independent of the choice of τ, K, $K^{(k)}$ and ϕ. This integer is called the* **degree** *of the map f.*

It is true that the two definitions of the degree of a map are equivalent, but we will not prove this fact here. The curious reader may like to see (Hocking and Young [10]) for a detailed discussion and proofs of all the "independent statements" made above.

It follows from Brouwer's definition that the map $f\colon \mathbb{S}^1 \to \mathbb{S}^1$ defined by $f(z) = z^n$ has degree n; the degree of a constant map $f\colon \mathbb{S}^n \to \mathbb{S}^n, n \geq 1$, is zero; the degree of identity map $I_{\mathbb{S}^n} : \mathbb{S}^n \to \mathbb{S}^n$ is one, etc. Now, we can prove

Theorem 4.8.4. *If two continuous maps $f,\ g\colon \mathbb{S}^n \to \mathbb{S}^n$ are homotopic, then* $\deg f = \deg g$.

Proof. Let $H\colon \mathbb{S}^n \times I \to \mathbb{S}^n$ be a homotopy starting from f and terminating into g. If we put $H(x,t) = h_t(x)$, $x \in \mathbb{S}^n, t \in I$, then each $h_t\colon \mathbb{S}^n \to \mathbb{S}^n$ is a continuous map and $h_0 = f, h_1 = g$. Now, it suffices to show that the map $I \to \mathbb{Z}$ defined by $t \rightsquigarrow \deg h_t$ is constant. The last fact will follow if we can prove that the map from $I \to \mathbb{Z}$ is continuous because I is connected and $\mathbb{Z}$ has the discrete topology. This is what we now proceed to do: Let K be a triangulation of $\mathbb{S}^n$, and consider the open cover $\{\text{ost}(w_i) : w_i \text{ is a vertex of } K\}$. Suppose ϵ is a Lebesgue number for the open cover. Since H is uniformly continuous, we can find a positive real δ such that $A \subset \mathbb{S}^n, B \subset I$ with $\text{diam}(A) < \delta$, $\text{diam}(B) < \delta$ imply that $\text{diam}(H(A \times B)) < \epsilon$. Let $K^{(k)}$ be a barycentric subdivision of K with mesh less than $\delta/2$ so that for each vertex v of $K^{(k)}$, $\text{diam}(\text{ost}(v)) < \delta$. Select a partition

$$0 = t_0 < t_1 < \ldots < t_q = 1$$

of I for which $|t_j - t_{j-1}| < \delta, \forall j = 0, 1, \ldots, q$. Then for each vertex v_i of K, the set $h(\text{ost}(v_i) \times [t_{j-1}, t_j])$ has diameter less than ϵ, for each j. Therefore, there is a vertex w_{ij} of K such that $h(\text{ost}(v_i) \times [t_{j-1}, t_j]) \subseteq \text{ost}(w_{ij})$. Hence, for each $t \in [t_{j-1}, t_j]$, we can define a simplicial map ϕ_t by putting $\phi_t(v_i) = w_{ij}$, which is evidently a simplicial approximation to h_t, for all $t \in [t_{j-1}, t_j]$. Since ϕ_t is same for all $t \in [t_{j-1}, t_j]$, it follows from the Brouwer's definition of the degree of a map that h_t has the same degree for all $t \in [t_{j-1}, t_j]$, showing that $t \rightsquigarrow \deg h_t$ is a continuous map from I to $\mathbb{Z}$. ∎

We must point out here that the converse of the above theorem was also studied by Brouwer. In fact, he had proved that if $f,\ g\colon \mathbb{S}^2 \to \mathbb{S}^2$ are two continuous maps such that $\deg f = \deg g$, then f and g are homotopic. It was H. Hopf (1894-1971) who proved the converse in full generality in the year 1927. We are not including the proof of the converse here, but give only the statement of this beautiful theorem:

Theorem 4.8.5. (Hopf's Classification Theorem). *Two continuous maps $f,\ g\colon \mathbb{S}^n \to \mathbb{S}^n$ are homotopic if and only if they have the same degree.*

We will remain contented only by stating the above theorem, though its several generalizations are also known. The most important aspect of the above theorem is to observe that the homotopy classes of maps from $\mathbb{S}^n$ to itself are completely classified by the set of integers! Since we are leaving the above theorem without giving its proof, we can as well state the following theorem without proof which is regarded as one of the most satisfying theorems of topology because it accomplishes the complete classification of compact 2-manifolds without boundary (called **closed 2-manifolds or closed surfaces**) up to homeomorphisms.

Theorem 4.8.6. *Two closed surfaces are homeomorphic if and only if they have the same Betti numbers in all dimensions.*

The above theorem says that any two closed 2-manifolds X and Y are homeomorphic if and only if $H_q(X) \cong H_q(Y), \forall q \geq 0$. This is one instance where the homology groups provide complete classification of closed 2-manifolds. Note that we have defined homology groups only for those compact topological spaces which are triangulable, i.e., are polyhedra. Hence it is natural to ask as to which manifolds are triangulable? Poincaré had asked this question for any n-manifold, $n \geq 2$, not necessarily compact – there is a suitable definition of simplicial complexes which are not necessarily finite. It is now well known through the works of several mathematicians that every 2-manifold, as well as every 3-manifold, is triangulable. Hence a surface is always triangulable whether it is compact or not. This is the reason that in the above theorem triangulability of the surfaces is not a part of the hypothesis. We must also mention here that the above theorem really shows the power of homology groups. In fact, this theorem was known as early as 1890 through the works of C. Jordan (1858–1922) and A.F. Möbius (1790–1860). As we know, C. Jordan is known for his "Jordan Curve" Theorem, and A.F. Möbius is known for his "Möbius band" through which he introduced the idea of "orientability". The modern definition of orientability in terms of homology groups was introduced by J.W. Alexander (1888–1971). A complete proof of the above theorem in a more precise form can be found in W.H. Massey [13].

Definition 4.8.7. *Let $\mathbb{S}^n$ be the unit n-sphere embedded in $\mathbb{R}^{n+1}$. Then the map $A\colon \mathbb{S}^n \to \mathbb{S}^n$ defined by $A(x) = -x$, $x \in \mathbb{S}^n$ is called the* **antipodal map**.

Note that the antipodal map is induced by a linear transformation $T\colon \mathbb{R}^{n+1} \to \mathbb{R}^{n+1}$ whose determinant is $(-1)^{n+1}$. The following result is, therefore, quite interesting.

Theorem 4.8.8. *The degree of the antipodal map $A\colon \mathbb{S}^n \to \mathbb{S}^n$ is $(-1)^{n+1}$, $n \geq 1$.*

Proof. We know that $\mathbb{S}^n = \{(x_1, \ldots, x_{n+1}) \in \mathbb{R}^{n+1} \mid \sum_1^{n+1} x_i^2 = 1\}$. A map $r_i\colon \mathbb{S}^n \to \mathbb{S}^n$ defined by $r_i(x_1, \ldots, x_i, \ldots, x_{n+1}) = (x_1, \ldots, -x_i, \ldots, x_{n+1})$ is called a **reflection map**. We will show that the map $r\colon \mathbb{S}^n \to \mathbb{S}^n$ defined by $r(x_1, \ldots, x_{n+1}) = (x_1, \ldots, -x_{n+1})$ has degree -1. This will mean that each reflection r_i has degree -1. To see this, note that if h is the homeomorphism of $\mathbb{S}^n$ which interchanges the coordinates x_i and x_{n+1} for a fixed i, then $h^{-1}rh = r_i$ and so $\deg r_i = \deg h^{-1} \cdot \deg r \cdot \deg h = \deg r$, proving that any reflection has degree (-1). Now, since the antipodal map $A = r_1 \circ r_2 \circ \ldots r_{n+1}$, it follows that $\deg A = (-1)^{n+1}$. Next, we are going to determine a triangulation $S(K)$ of $\mathbb{S}^n$ and a simplicial map $g\colon S(K) \to S(K)$ which will induce the map r. To do this, let K be a triangulation of $\mathbb{S}^{n-1}$ embedded in $\mathbb{S}^n$ in the standard way. Let w_0 and w_1 be two distinct vertices of the cones $w_0 * K$ and $w_1 * K$. Put $S(K) = (w_0 * K) \cup (w_1 * K)$. Let g be the simplicial map from $S(K)$ to itself which interchanges the vertices w_0 and w_1 and keeps all other vertices fixed. Let $h\colon |K| \to \mathbb{S}^{n-1}$ be a triangulating homeomorphism. We define $k\colon |S(K)| \to \mathbb{S}^n$ by the following formula: If $y = (1-t)x + tw_0$ for some $x \in |K|$, then define

$$k(y) = (\sqrt{1-t^2}\, h(x), t),$$

and if $y = (1-t)x + tw_1$, then put

$$k(y) = (\sqrt{1-t^2}\, h(x), -t).$$

Then k is a homeomorphism making the diagram

$$\begin{array}{ccc} |S(K)| & \xrightarrow{\quad k \quad} & \mathbb{S}^n \\ {\scriptstyle g}\downarrow & & \downarrow{\scriptstyle r} \\ |S(K)| & \xrightarrow[\quad k \quad]{} & \mathbb{S}^n \end{array}$$

commutative. Hence it suffices to show that $\deg g = -1$. Let z be an n-cycle of $S(K)$. Then z is a chain of the form

$$z = [w_0, c_m] + [w_1, d_m],$$

where c_m and d_m are chains in K and $m = n-1$. Assume $n > 1$. Since z is a cycle, we have

$$0 = \partial(z) = c_m - [w_0, \partial c_m] + d_m - [w_1, \partial d_m].$$

Restricting this chain to K, we get

$$0 = c_m + d_m.$$

Therefore,

$$z = [w_0, c_m] - [w_1, c_m].$$

Since g simply exchanges w_0, w_1, we find that

$$g_*(z) = [w_1, c_m] - [w_0, c_m] = -z.$$

Hence, $\deg g = -1$. The case $n = 1$ is proved similarly. ■

Tangent Vector Fields on Spheres

Now, we present an interesting application of the above theorem on the existence of tangent vector fields on Euclidean spheres. We have

Definition 4.8.9. *A continuous mapping $f\colon \mathbb{S}^n \to \mathbb{S}^n$ is said to be a* **tangent vector field** *on $\mathbb{S}^n$ if for each vector $x \in \mathbb{S}^n$, the two vectors x and $f(x)$ are perpendicular to each other.*

If one thinks of the point $x \in \mathbb{S}^n$ as a vector in $\mathbb{R}^{n+1}$, then a vector field f on $\mathbb{S}^n$ is simply a continuous map $f\colon \mathbb{S}^n \to \mathbb{R}^{n+1}$ such that $f(x)$ is another vector in $\mathbb{R}^{n+1}$. If $f(x)$ is transported in a parallel direction so that it starts from the point x of $\mathbb{S}^n$, then x and $f(x)$ are perpendicular really means that $f(x)$ is tangent to $\mathbb{S}^n$ at the point x. For the case $n = 1$, it is illustrated in Fig. 4.10

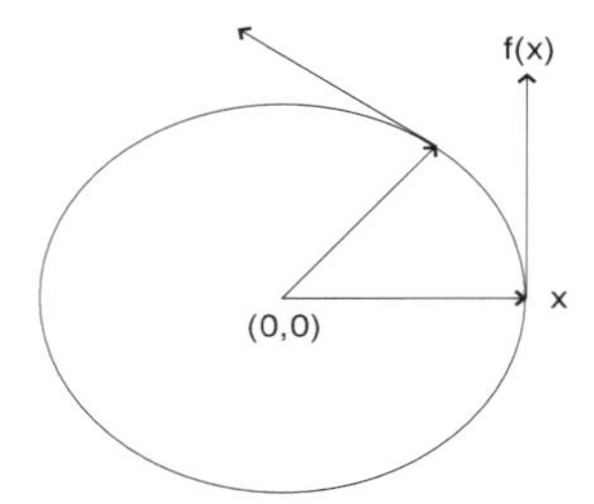

Fig. 4.10: Tangent vector field

Now, we have the following question: Given any sphere $\mathbb{S}^n$, does there always exist a nonzero tangent vector field on $\mathbb{S}^n$? The following result gives a complete answer to this question.

Theorem 4.8.10. (Vector fields on spheres). *The n-sphere $\mathbb{S}^n$ admits a nonzero tangent vector field if and only if n is odd.*

Proof. Suppose $n = 2m+1$ is odd. Then $\mathbb{S}^n$ is given by the set

$$\{(x_1, x_2, \ldots, x_{2m+1}, x_{2m+2}) \in \mathbb{R}^{2m+2} \mid \sum_{1}^{2m+2} x_i^2 = 1\}.$$

We can now define a map $f\colon \mathbb{S}^n \to \mathbb{S}^n$ by the formula

$$f(x_1, x_2, \ldots, x_{2m+1}, x_{2m+2}) = (x_2, -x_1, \ldots, x_{2m+2}, -x_{2m+1}).$$

Then, evidently, f is a continuous map and has the property that for each $x \in \mathbb{S}^n$ the scalar product $x \cdot f(x) = 0$. Thus, $\mathbb{S}^n$ has a nonzero tangent vector field.

Conversely, suppose n is even and let $g\colon \mathbb{S}^n \to \mathbb{S}^n$ be a nonzero tangent vector field. Let us define a homotopy $F\colon \mathbb{S}^n \times I \to \mathbb{S}^n$ by

$$F(x,t) = x\cos(\pi t) + g(x)\sin(\pi t),$$

where $x \in \mathbb{S}^n$ and $t \in I$. Using the fact that $x \cdot g(x) = 0$, it is easily verified that $\|F(x,t)\| = 1$, i.e., F is a well-defined homotopy. Moreover, $F(x,0) = x$ and $F(x,1) = -x$, for all $x \in \mathbb{S}^n$. This says that the identity map $I_{\mathbb{S}^n}$ is homotopic to the antipodal map $A\colon \mathbb{S}^n \to \mathbb{S}^n$. Since homotopic maps have the same degree, it follows that $1 = (-1)^{n+1}$, which is a contradiction because n is even. Hence no tangent vector field on $\mathbb{S}^n$ can exist if n is even. ■

The above theorem has an amusing description for the case $n = 2$. Suppose one takes a ball in which there is a nonzero vector emanating from each point of the ball. Now, think of each vector as a hair on the ball. The word "nonzero" means there are hairs everywhere. Finding a tangent vector field means one can "comb the hairs" on the ball in such a way that each hair is tangent to the 2-sphere, which is the boundary of the ball, and their directions vary continuously. The above theorem asserts that such a "hairstyle" is impossible, unless there is a discontinuity.

The Fundamental Theorem of Algebra

We now use the notion of "degree" to prove the fundamental theorem of algebra which is a well-known basic result of complex analysis. We have

Theorem 4.8.11. (Fundamental Theorem of Algebra). *Every polynomial of positive degree with complex coefficients has a zero in* $\mathbb{C}$.

Proof. It is clearly enough to prove the theorem for a polynomial of the type $p(z) = z^n + a_1 z^{n-1} + \ldots + a_n$, where $n \geq 1$, and $a_1, a_2, \ldots, a_n$ are all complex numbers, i.e., $p(z)$ is a nonconstant monic polynomial. Suppose the assertion is not true. Then $z \mapsto p(z)$ is a mapping from the complex plane $\mathbb{C}$ to $\mathbb{C} - \{0\}$. The restriction of this mapping to various circles $|z| = r$, $r \geq 0$ are loops in $\mathbb{C} - \{0\}$ and, therefore, the restriction of map $z \mapsto \frac{p(z)}{\|p(z)\|}$ to circles $|z| = r$ are loops in the unit circle $\mathbb{S}^1 \subseteq \mathbb{C}$. This means the map $f_r(z) = \frac{p(z)/p(r)}{\|p(z)/p(r)\|}$ defined on the circle $|z| = r$ is a loop in $\mathbb{S}^1$ based at $1 \in \mathbb{S}^1$. Note that f_0 is clearly the constant loop and $r \mapsto f_r$ defines a homotopy from f_0 to f_r for any $r \geq 0$. This says that, for all $r \geq 0$, the loop f_r is null-homotopic, i.e., $\deg f_r = 0$.

Next, note that if we take $|z| = r > \max\{|a_1| + |a_2| + \ldots + |a_n|, 1\}$, then we observe that

$$|z|^n = rr^{n-1} \geq (|a_1| + \ldots + |a_n|)|z|^{n-1} \geq |a_1 z^{n-1} + \ldots + a_n|,$$

which means for $0 \leq t \leq 1$, the map

$$z \mapsto z^n + t(a_1 z^{n-1} + \ldots + a_n)$$

is never zero on such a circle $|z| = r$. Since, with $p_t(z) = z^n + t(a_1 z^{n-1} + \ldots + a_n)$,

$$F(z,t) = \frac{p_t(z)/p_t(r)}{\| p_t(z)/p_t(r) \|},$$

is a homotopy between the loops $z \mapsto z^n / \| z^n \|$, i.e., $s \mapsto e^{2\pi i n s}$ and $z \to \frac{p(z)/p(r)}{\|p(z)/p(r)\|}$, i.e., $s \to f_r(s)$, we find that f_r is homotopic to the loop $s \mapsto e^{2\pi i n s}$ in $\mathbb{S}^1$. Since the degree of this last loop is n, which is positive, we have a contradiction to the fact proved earlier that $\deg f_r = 0$ for all $r \geq 0$. ■

Exercises

1. Prove that the union of two nonintersecting circles in the torus $\mathbb{T}$ cannot be a retract of $\mathbb{T}$. (**Hint.** Use set topology.)
2. Prove that the central circle in the Möbius band $\mathbb{M}$ is a retract of $\mathbb{M}$.
3. Show that a 2-sphere $\mathbb{S}^2$ cannot be homeomorphic to the torus $\mathbb{T}$ or projective plane $\mathbb{P}^2$.
4. Prove that an orientable surface cannot be homeomorphic to a nonorientable surface.
5. Prove that an orientable surface of genus m cannot be homeomorphic to an orientable surface of genus n, $m \neq n$.
6. Prove that a nonorintable surface of genus m cannot be homeomorphic to a nonorintable surface of genus n, $m \neq n$.
7. Let A be a nonsingular $n \times n$ matrix. Then A defines a linear homeomorphism from $\mathbb{R}^n \to \mathbb{R}^n$, which can be extended to a map $f\colon \mathbb{S}^n \to \mathbb{S}^n$. Prove that deg(f) = sign Det(A) (This generalizes Theorem 4.8.4).
8. Show that the long line (Example 1.7.7) has the fixed-point property. (This is difficult, see [18] p. 56.)

4.9 Invariance of Homology Groups

In this section, which was postponed earlier, we give detailed proofs of two results. First, we show that the barycentric subdivision of a complex K does not change its simplicial homology. Secondly, we prove that any continuous map $f\colon |K| \to |L|$ between two compact polyhedra induces a well-defined homomorphism $f_*\colon H_q(K) \to H_q(L)$, $\forall\, q \geq 0$ from the simplicial homology groups of K to those of L. More importantly, we go on to show that these induced homomorphisms in homology satisfy the desired functorial properties. The last result is then utilized to prove that if X is a compact polyhedron and K, K' are two triangulations of X, then $H_q(K) \cong H_q(K')$, $\forall\, q \geq 0$. This establishes the fact that the simplicial homology of a polyhedra $|K|$ does not depend on the triangulation of $|K|$, it rather depends on the topology of $|K|$. As a matter of fact we prove much more: the simplicial homology groups of $|K|$ are homotopy invariants of $|K|$ – a stronger result than saying that these are topological invariants. In sections 4.9.4 and 4.9.5 we prove the Lefschetz fixed-point theorem and the Borsuk-Ulam theorem respectively.

4.9.1 Subdivision Chain Map

In order to show that simplicial homology groups of a complex K do not change when we pass on from K to its barycentric subdivision, we define two chain maps. The first chain map $\mu_q \colon C_q(K) \to C_q(K^{(1)})$, called the **subdivision chain map**, will be defined using the definition of barycentric subdivision $K^{(1)}$ of K. It will not arise from any simplicial map from K to $K^{(1)}$ which is normally the case. The other chain map $\theta_q \colon C_q(K^{(1)}) \to C_q(K)$, in the opposite direction, will be defined using a simplicial map $\theta \colon K^{(1)} \to K$ called **standard simplicial map**. Then we show that these two chain maps induce homomorphisms in the homology which are inverse of each other, and thus we find that homology groups of K and $K^{(1)}$ are isomorphic. We have

Theorem 4.9.1. *Let $K^{(1)}$ denote the first barycentric subdivision of a simplicial complex K. Then $H_q(K) \cong H_q(K^{(1)})$ for all $q \geq 0$.*

Proof. Recall that the vertices of $K^{(1)}$ are just the barycentres $\dot{\sigma^q}$ of all simplexes σ^q of K, $q \geq 0$, and $'\sigma^q = \langle \dot{\sigma_0}, \dot{\sigma_1}, \ldots, \dot{\sigma_q} \rangle$ is a q-simplex of $K^{(1)}$ if there is a permutation t of $\{0, 1, \ldots, q\}$ so that $\sigma_{t(0)}$ is a face of $\sigma_{t(1)}, \ldots, \sigma_{t(q-1)}$ is a face of $\sigma_{t(q)}$. It would be helpful to have the case of a 2-simplex $\sigma^2 = \langle v_0, v_1, v_2 \rangle$ of K in mind (Fig. 4.11).

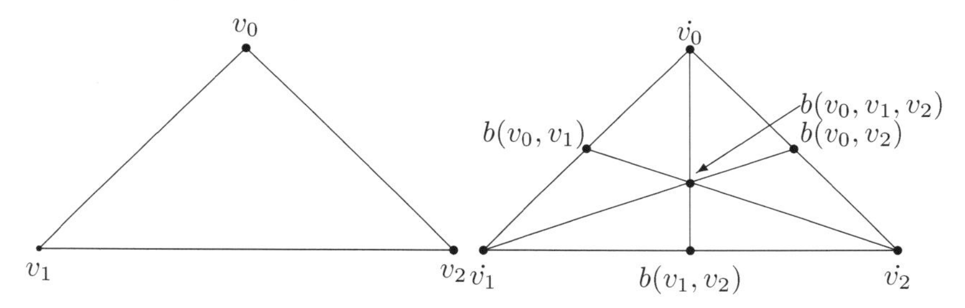

Fig. 4.11

We have denoted, just for convenience, the barycentre of a vertex v_i by $\dot{v_i}$, and shown the barycentre of $\langle v_i, v_j \rangle$ by $b(v_i, v_j)$, etc., in the figure.

A vertex v_i of K is a 0-simplex σ^0 of K. The barycentre $\dot{\sigma^0}$ of σ^0 is evidently σ^0 itself. Hence all the 0-simplexes of K are also the 0-simplexes of $K^{(1)}$. For any q-simplex σ^q of K, the elementary 1-chain $1 \cdot \sigma^q$ will be denoted simply by σ^q. Finally, if $\sigma^q = \langle v_0, \ldots, v_q \rangle$ and v is an additional vertex such that $\{v, v_0, \ldots, v_q\}$ is a geometrically independent set, then the $(q+1)$-simplex $\langle v, v_0, \ldots, v_q \rangle$ will be denoted by $v \cdot \sigma^q$. More generally, for a q-chain $c = \sum n_i \sigma_i^q$, we will denote by $v \cdot c$ the $(q+1)$-chain $\sum n_i (v \cdot \sigma_i^q)$. One can see that for any chain c in $C_q(K)$, $\partial(v \cdot c) = c - v.\partial(c)$. Now, we define the subdivision chain map $\mu_q \colon C_q(K) \to C_q(K^{(1)})$ for $q \geq 0$ by induction on q. For a vertex v_i of K, we define $\mu_0(v_i) = \dot{v_i}$, and then extend it linearly to $C_0(K)$. Let $\sigma^1 = \langle v_0, v_1 \rangle$ be a positively-oriented 1-simplex of K. We define $\mu_1(\sigma^1) = \langle \dot{\sigma}^1, \dot{v_1} > - < \dot{\sigma}^1, \dot{v_0} \rangle$

and extend it linearly to $C_1(K)$. Note that for each 1-simplex σ^1 of K,

$$\begin{aligned}\partial\mu_1(\sigma^1) &= \partial(\langle\dot{\sigma}^1, \dot{v_1}\rangle) - \partial(\langle\dot{\sigma}^1, \dot{v_0}\rangle)\\ &= \dot{v_1} - \dot{\sigma}^1 - \dot{v_0} + \dot{\sigma}^1\\ &= \mu_0(\langle v_1\rangle) - \mu_0(\langle v_0\rangle)\\ &= \mu_0\partial(\sigma^1).\end{aligned}$$

This means $\partial\mu_1 = \mu_0\partial$ and so μ_0, μ_1 both satisfy the condition of a chain map. Now we assume that μ_q has been defined for all $q < p, (p > 1)$, satisfying the condition $\partial\mu_q = \mu_{q-1}\partial$ for all q-chains of $C_q(K)$. Now, let σ^p be a p-simplex of K and define $\mu(\sigma^p) = \dot{\sigma}^p\mu_{p-1}\partial(\sigma^p)$, where the right hand side, as explained before, means the sum of all p-simplexes of $K^{(1)}$ which are faces of the cone over the complex $\mu_{p-1}\partial(\sigma^p)$. Then we find that

$$\begin{aligned}\partial\mu_p(\sigma^p) &= \partial(\dot{\sigma}^p\mu_{p-1}\partial(\sigma^p))\\ &= \mu_{p-1}(\partial(\sigma^p)) - \dot{\sigma}^p\partial\mu_{p-1}\partial(\sigma^p)\\ &= \mu_{p-1}\partial(\sigma^p),\end{aligned}$$

since, by induction hypothesis, $\partial\mu_{p-1}\partial = \mu_{p-2}\partial\partial = 0$. Hence μ_p, satisfying the condition $\partial\mu_p = \mu_{p-1}\partial$, is also defined. Thus, by induction, $\mu_q\colon C_q(K) \to C_q(K^{(1)})$ is defined for all $q \geq 0$ and is a chain map. It is helpful to actually verify the case of a 2-simplex σ^2 also to see what is happening.

Next, we define a chain map $\theta\colon C(K^{(1)}) \to C(K)$ which, in fact, arises from a simplicial map $\theta\colon K^{(1)} \to K$. We are using the same symbol θ for the induced chain map also just for convenience and this should cause no confusion. Let $\dot{\sigma}^q$ be any vertex of $K^{(1)}$. We choose a vertex v_0 of σ^q and define $\theta(\dot{\sigma}^q) = v_0$. This induces a simplicial map $\theta\colon K^{(1)} \to K$. To see this, note that if $\sigma^1 = \langle v_0, v_1\rangle$ is a 1-simplex,

then, under the map θ, $\dot{\sigma}^1$ goes to a vertex, say, v_0, of σ^1, and $\dot{v_0}$ goes to v_0, $\dot{v_1}$ goes to v_1. In other words, the simplex $\langle\dot{\sigma}^1, v_1\rangle$ goes to the 1-simplex $\langle v_0, v_1\rangle$, but the 1-simplex $\langle v_0, \dot{\sigma}^1\rangle$ goes to 0-simplex $\langle v_0\rangle$, i.e., $\langle v_0, \dot{\sigma}^1\rangle$ collapses to a vertex v_0. Next, if $\sigma^2 = \langle v_0, v_1, v_2\rangle$ is a 2-simplex, and we put $\theta(\dot{\sigma}^2) = v_0$, then assuming that $b(v_1, v_2)$ goes to v_2 we find that the 2-simplex $\langle v_1, b(v_1, v_2), \dot{\sigma}^2\rangle$ goes to $\langle v_1, v_2, v_0\rangle$ whereas the 2-simplex $\langle v_2, b(v_1, v_2), \dot{\sigma}^2\rangle$ goes to a 1-simplex $\langle v_2, v_0\rangle$, i.e., it collapses to a lower-dimensional simplex. Now, we can check that rest of 2-simplices in σ^2 also collapse to either a 1-simplex or to a 0-simplex of σ^2. Now, it is clear what would be $\theta(1\sigma^q)$, where $1\sigma^q$ is a q-simplex of $K^{(1)}$. Thus, we have defined a simplicial map $\theta\colon K^{(1)} \to K$. It is, of course, evident that θ is not uniquely defined because there are choices involved. But once we fix our choice for each q-simplex of $K^{(1)}$, $q \geq 0$, then θ is a well defined simplicial map. Hence, θ induces a chain map $\theta\colon C_q(K^{(1)}) \to C_q(K)$, $q \geq 0$.

In fact, the simplicial map θ is a simplicial approximation of the identity map $|K'| \to |K|$.

We observe that the composite chain map $\theta \circ \mu\colon C(K) \to C(K)$ is the identity map since for any q-simplex σ^q of K, out of several simplexes occurring in $\mu(\sigma^q)$, there is only one q-simplex in $K^{(1)}$ which is mapped back to σ, and the rest of the q-simplexes are collapsed to lower-dimensional simplexes, contributing nothing to $\theta\mu(\sigma^q)$, i.e., $\theta\mu(\sigma^q) = \sigma^q$, proving our statement.

Finally, we show that the chain map $\mu\theta\colon C(K^{(1)}) \to C(K^{(1)})$ is chain homotopic to the identity map on $C(K^{(1)})$. Once again we will define a chain homotopy $D_q\colon C_q(K^{(1)}) \to C_{q+1}(K^{(1)})$, $q \geq 0$, by induction on q. Given a vertex $\dot{\sigma}$ of $K^{(1)}$, we define $D_0\colon C_0(K^{(1)}) \to C_1(K^{(1)})$ such that $\partial D_0(\dot{\sigma}) = \dot{\sigma} - \mu\theta(\dot{\sigma})$. To see this, note that $\theta(\dot{\sigma})$ is a vertex v_0 of σ. Hence, $\mu\theta(\dot{\sigma}) = \langle \dot{v_0} \rangle = \langle v_0 \rangle$. Thus, $\langle v_0, \dot{\sigma} \rangle$ is a 1-simplex in $K^{(1)}$ and we put $D_0(\dot{\sigma}) = \langle v_0, \dot{\sigma} \rangle$ which clearly has the desired property. Then we extend D_0 to all the zero chains in $C_0(K^{(1)})$ linearly. Having defined D_0 we assume that D_q is defined for all $q < p$ $(0 < p)$ satisfying the condition that

$$\partial D_q(\sigma') + D_{q-1}\partial(\sigma') = \sigma' - \mu\theta(\sigma')$$

for all q-simplexes σ' of $K^{(1)}$. Now, note that $1\sigma^p - \mu\theta(1\sigma^p) - D_{p-1}\partial(1\sigma^p)$ is a chain on the cone $\mu(\sigma^k) = \dot{\sigma}^k \cdot \mu(\partial\sigma^k)$ for some $\sigma^k \in K$ which is well known to be acyclic. Hence any cycle of this cone will be a boundary. We compute

$$\begin{aligned}
&\partial[1\sigma^p - \mu\theta(1\sigma^p) - D_{p-1}\partial(1\sigma^p)] \\
&\quad = \partial(1\sigma^p) - \partial\mu\theta(1\sigma^p) - \partial D_{p-1}\partial(1\sigma^p) \\
&\quad = \partial(1\sigma^p) - \partial\mu\theta(1\sigma^p) - [\partial(1\sigma^p) - \mu\theta\partial(1\sigma^p) - D_{p-2}\partial\partial(1\sigma^p)],
\end{aligned}$$

by inductive hypothesis. Since $\partial\partial = 0$, and since $\mu\theta$ is a chain map, right hand side of the above equation is zero, i.e., the quantity $1\sigma^p - \mu\theta(1\sigma^p) - D_{p-1}\partial(1\sigma^p)$ is a cycle. However, this is a cycle on the cone $\mu(\sigma^k) = \dot{\sigma}^k\mu\partial(\sigma^k)$ and, therefore, the same must be a boundary in that cone, i.e., we can find a chain c on $\mu(\sigma^k)$ so that $\partial(c)$ is the above cycle. Then we define $D_p(\sigma^p) = c$, which yields us

$$\partial D_p(1\sigma^p) = 1\sigma^p - \mu\theta(1\sigma^p) - D_{p-1}\partial(1\sigma^p).$$

This completes the inductive definition of the chain homotopy D_q, $q \geq 0$, satisfying the desired property. It is, therefore, clear that the homomorphisms $\mu_*\colon H_q(K) \to H_q(K^{(1)})$ and $\theta_*\colon H_q(K^{(1)}) \to H_q(K)$ induced by the chain maps μ and θ are inverse of each other. Hence, $H_q(K) \cong H_q(K^{(1)})\ \forall\, q \geq 0$. ■

Note that if $K = K^{(0)}, K^{(1)}, K^{(2)}, \ldots, K^{(m)}$ are the successive barycentric subdivisions of K and $\mu_i\colon C(K^{(i)}) \to C(K^{(i+1)})$ are the subdivision chain maps, then the composite $\mu_{m-1} \circ \ldots \circ \mu_1 \circ \mu_0$ is again a chain map which also is denoted by the same symbol $\mu\colon C(K) \to C(K^{(m)})$ and will be called subdivision chain map. Each of these chain maps μ_i has a chain homotopy inverse $\theta_i\colon C(K^{(i+1)}) \to C(K^{(i)})$ induced by a simplicial map and so their composite $\theta = \theta_0 \circ \theta_1 \circ \ldots \circ \theta_{m-1}\colon C(K^{(m)}) \to C(K)$ will be a chain homotopy inverse of μ. The composite simplicial map $\theta\colon K^{(m)} \to K$ will be called the **standard simplicial map**. The same notation for the simplicial map θ and the induced chain map θ should cause no confusion. Combining these results, the following is now immediate:

Corollary 4.9.2. *Let $K^{(m)}$ denote the mth order barycentric subdivision of the complex K. Then $H_q(K) \cong H_q(K^{(m)})$, $\forall\, q \geq 0$.*

Contiguity Classes

Let K_1, K_2 be two simplicial complexes. Then the two simplicial maps $\phi_1,\ \phi_2\colon K_1 \to K_2$ are said to be **contiguous** (or close to each other) if for each simplex $s \in K_1$, $\phi_1(s) \cup \phi_2(s)$ is a simplex of K_2. This relation in the set of all simplicial maps from K_1 to K_2 is reflexive and symmetric, but not necessarily transitive. The smallest equivalence relation generated by this relation is **contiguity relation** and the corresponding equivalence classes are called **contiguity classes**.

From now onwards, $\mathrm{st}(v)$ will stand for $\mathrm{ost}(v)$ for convenience. We recall that a simplicial map $\phi\colon K_1 \to K_2$ is a simplicial approximation of $f\colon |K_1| \to |K_2|$ if and only if for every vertex v of K_1, $f(\mathrm{st}\ v) \subset \mathrm{st}\ \phi(v)$.

Proposition 4.9.3. *Any two simplicial approximations from $K_1 \to K_2$ of the same continuous map $f\colon |K_1| \to |K_2|$ are contiguous.*

Proof. Let $\phi,\ \phi'\colon K_1 \to K_2$ be simplicial approximations to $f\colon |K_1| \to |K_2|$, and let $s = \langle v_0, \ldots, v_n\rangle$ be a simplex of K_1. Then $\bigcap_0^n \mathrm{st}\ v_i \neq \emptyset$ and so

$$\emptyset \neq f(\bigcap \mathrm{st}\ v_i) \subset \bigcap f(\mathrm{st}\ v_i) \subset \bigcap(\mathrm{st}\ \phi(v_i) \cap \mathrm{st}\ \phi'(v_i)),$$

which shows that $\phi(s) \cup \phi'(s)$ is a simplex of K_2. ■

Proposition 4.9.4. *Let $f,\ f'\colon |K| \to |L|$ be homotopic maps. Then there exists an integer n such that $f,\ f'$ have simplicial approximations $\phi,\ \phi'\colon K^{(n)} \to L$ and there are finite number of simplicial maps $\phi_1, \phi_2, \ldots, \phi_k$ such that $\phi = \phi_1$ is contiguous to ϕ_2, ϕ_2 is contiguous to ϕ_3, $\ldots$, ϕ_{k-1} is contiguous to $\phi_k = \phi'$.*

Proof. Let $F\colon |K| \times I \to |L|$ be a homotopy from f to f'. Since $|K|$ is compact, there exists a sequence $0 = t_0 < t_1 < \ldots < t_k = 1$ of points of I such that for $\alpha \in |K|$, and $i = 1, 2, \ldots, k$ there is a vertex $v \in L$ such that $F(\alpha, t_{i-1})$ and $F(\alpha, t_i)$ both belong to $st\ v$. Let $f_i\colon |K| \to |L|$ be defined by $f_i(\alpha) = F(\alpha, t_i)$. Then $f = f_0$, $f' = f_k$ and for $i = 1, 2, \ldots, k$, the set

$$\mathcal{U}_i = \{f_i^{-1}(\mathrm{st}\ v) \cap f_{i-1}^{-1}(\mathrm{st}\ v) \mid v \in L\}$$

is an open cover of $|K|$. Let N be chosen sufficiently large so that $K^{(N)}$ is finer than $\mathcal{U}_1, \mathcal{U}_2, \ldots, \mathcal{U}_k$. For $i = 1, 2, \ldots k$ let ϕ_i be a vertex map from $K^{(N)} \to L$ such that

$$f_i(\text{st } v) \cup f_{i-1}(\text{st } v) \subset \text{st } \phi_i(v)$$

for each vertex v of $K^{(N)}$. Such a vertex map exists since $\text{Sd}^N(K)$ is finer than $\mathcal{U}_i$. Then $\phi_i \colon K^{(N)} \to L$ is a simplicial approximation to both f_i and f_{i-1}. Since ϕ_i, ϕ_{i+1} are simplicial approximations to f_i, these are contiguous by previous proposition, and this completes the proof. ■

Proposition 4.9.5. *Let ϕ, $\phi' \colon K \to L$ be contiguous simplicial maps. Then the induced chain maps $\phi_\#$, $\phi'_\# \colon C(K) \to C(L)$ in the simplicial chain complexes are chain homotopic.*

Proof. We define maps $D_q \colon C_q(K) \to C_{q+1}(L)$, $q \geq 0$, by the formula

$$D_q(\langle v_0, v_1, \ldots, v_q \rangle) = \sum (-1)^i \langle \phi'(v_0), \ldots, \phi'(v_i), \phi(v_i), \ldots, \phi(v_q) \rangle,$$

where $\langle v_0, v_1, \ldots, v_q \rangle$ is a q-simplex of K, and then extend it to $C_q(K)$ by linearity. Then it is easily seen that on $\sigma^q = \langle v_0, \ldots, v_q \rangle$ we have

$$\begin{aligned} &\partial_{q+1} D_q(\sigma^q) + D_{q-1}\partial_q(\sigma^q) \\ &\quad = \langle \phi(v_0), \ldots, \phi(v_q) \rangle - \langle \phi'(v_0), \ldots, \phi'(v_q) \rangle \\ &\quad = (\phi_\# - \phi'_\#)(\sigma^q). \end{aligned}$$

Hence, by linearity again, $\{D_q\}$ is a chain homotopy from $\phi_\#$ to $\phi'_\#$. ■

4.9.2 Homomorphism Induced by a Continuous Map

We have already seen how a simplicial map $|K| \to |L|$ induces a homomorphism in simplicial homology groups $H_q(K) \to H_q(L)$ $\forall$ $q \geq 0$. Now, let us take a continuous map $f \colon |K| \to |L|$ which is not necessarily a simplicial map. Let us choose a simplicial approximation $s \colon |K^{(m)}| \to |L|$ of f, and let $\mu \colon C(K) \to C(K^{(m)})$ be the subdivision chain map. Then, we define the induced homomorphism $f_* \colon H_q(K) \to H_q(L)$ as the composite map $s_*\mu_* \colon H_q(K) \to H_q(K^{(m)}) \to H_q(L)$.

In the above definition of induced homomorphism $f_* \colon H_q(K) \to H_q(L)$, there is a choice involved, viz., the choice of the simplicial approximation $s \colon |K^{(m)}| \to |L|$. Therefore, in order that f_* is well defined, we must show that f_* is independent of this choice. For this, let $t \colon |K^{(n)}| \to |L|$ be another simplicial approximation to f. We may assume that $n \geq m$. Let $\mu' \colon C(K^{(m)}) \to C(K^{(n)})$ be the subdivision chain map and let $\theta \colon |K^{(n)}| \to |K^{(m)}|$ be the standard simplicial map. Now, observe that $s\theta \colon |K^{(n)}| \to |L|$ is also a simplicial approximation to f. Therefore, by Proposition 4.9.3 both $s\theta$ and t are contiguous and so by Proposition 4.9.5, the induced homomorphisms $(s\theta)_*, t_* \colon H_q(K^{(n)}) \to H_q(L)$ must be equal. This evidently means

$(s\theta)_* \circ (\mu'\mu)_* = t_* \circ (\mu'\mu)_*$, i.e., $s_*\mu_* = t_*\mu'_*\mu_*$ since $\theta_*\mu'_* = I_d$. The last equation says that $f_*\colon H_q(K) \to H_q(L)$ is indeed well-defined.

4.9.3 Homotopy Invariance

We can now prove the desired functorial properties of the induced homomorphisms and much more. We have

Theorem 4.9.6. (i) *If $f\colon |K| \to |K|$ is the identity map, then the induced homomorphism $f_*\colon H_q(K) \to H_q(K)$ is also identity for $q \geq 0$.*

(ii) *If $f\colon |K| \to |L|, g\colon |L| \to |M|$ are two continuous maps, then for all $q \geq 0$,*

$$(g \circ f)_* = g_* \circ f_*\colon H_q(K) \to H_q(M).$$

Proof. (i) This follows at once from the definition of induced homomorphism.

(ii) Choose a simplicial approximation $t\colon |L^{(n)}| \to |M|$ to $g\colon |L| \to |M|$. Then find a simplicial approximation $s\colon |K^{(m)}| \to |L^{(n)}|$ to the map $f\colon |K| \to |L| = |L^{(n)}|$. Let $\mu_1\colon C(K) \to C(K^{(m)}), \mu_2\colon C(L) \to C(L^{(n)})$ be the subdivision chain maps and let $\theta\colon |L^{(n)}| \to |L|$ be the standard simplicial map. Consider the following diagram

$$\begin{array}{ccccc} H_q(K^{(m)}) & \xrightarrow{s_*} & H_q(L^{(n)}) & & \\ \uparrow{\scriptstyle \mu_{1*}} & & {\scriptstyle \mu_{2*}}\uparrow\downarrow{\scriptstyle \theta_*} & \searrow^{t_*} & \\ H_q(K) & \xrightarrow[f_*]{} & H_q(L) & \xrightarrow[g_*]{} & H_q(M). \end{array}$$

Note that θs is a simplicial approximation to $f\colon |K| \to |L|$ and ts is a simplicial approximation to $gf\colon |K| \to |M|$. Hence,

$$\begin{aligned} g_* \circ f_* &= t_*\mu_{2*}\theta_*s_*\mu_{1*} \\ &= t_*s_*\mu_{1*} = (ts)_*\mu_{1*} \\ &= (gf)_*. \end{aligned}$$ ■

Theorem 4.9.7. *If $f,\ g\colon |K| \to |L|$ are homotopic maps, then $f_* = g_*\colon H_q(K) \to H_q(L) \forall\ q \geq 0$.*

Proof. Since f, g are homotopic we know, by Proposition 4.9.4, that there is a subdivision $K^{(m)}$ of K and simplicial maps $s_1, s_2, \ldots, s_n\colon |K^{(m)}| \to |L|$ such that s_1 is a simplicial approximation of f, s_n is a simplicial approximation of g, and s_i is contiguous to s_{i+1}, $i = 1, 2, \ldots, n-1$. Let $\mu\colon C(K) \to C(K^{(m)})$ be the subdivision chain map. Then, it follows from the definition of induced homomorphism, that $f_* = s_{1*}\mu_* = s_{2*}\mu_* = \ldots = s_{n*}\mu_* = g_*$. ■

The above theorem has remarkable consequences. The first and foremost says that if two polyhedra $|K|$ and $|L|$ are homotopically equivalent (in particular, if K and L are triangulations of the same polyhedron X), then they have isomorphic simplicial homology, i.e., $H_q(K) \cong H_q(L)$, $\forall\, q \geq 0$. To see this, let $f\colon |K| \to |L|$ and $g\colon |L| \to |K|$ be two maps such that $g \circ f$ is homotopic to $I_{|K|}$ and $f \circ g$ is homotopic to the identity map $I_{|L|}$. By the functorial properties of induced homomorphisms, we find that the composites

$$H_q(K) \xrightarrow{f_*} H_q(L) \xrightarrow{g_*} H_q(K),$$
$$H_q(L) \xrightarrow{g_*} H_q(K) \xrightarrow{f_*} H_q(L)$$

are identity maps on $H_q(K)$ and $H_q(L)$, respectively, and, therefore, $H_q(K) \cong H_q(L)$ for all $q \geq 0$. In particular, this says that simplicial homology groups of a compact polyhedron X are not only topological invariants, but also they are homotopy invariants. For instance, it follows from this result that the simplicial homology groups of an n-disc $\mathbb{D}^n$ are the same as those of a point-space.

Exercises

1. Let $f\colon X \to Y$ be a constant map between polyhedra X and Y. Show that the induced homomorphism $f_*\colon H_q(X) \to H_q(Y)$ is zero, for all $q > 0$. What happens when $q = 0$?

2. Prove that the Euler characteristic of a compact contractible polyhedron X is 1.

3. Let K be a simplicial complex, and $w * K$ denote the cone complex over K. Define a simplicial map $\phi\colon w * K \to \{w\}$ by mapping every simplex of $w * K$ to the 0-simplex w. If $i\colon \{w\} \to w * K$ denotes the inclusion map, then prove that induced chain map $\phi_{\#}\colon C(w * K) \to C(\{w\})$ is a chain equivalence. Hence deduce that $H_q(w * K) = 0$, $\forall\, q > 0$. (This result is stronger than Theorem 4.5.1.)

4. For any simplicial complex K, let $C(K)$ denote the simplicial chain complex of K. Then prove that $C(K, G) = C(K) \otimes G$ is a chain complex for any coefficient group G and $H_q(X, G) = H_q(C(K) \otimes G)$, $q \geq 0$. Using this, show that for any short exact sequence $0 \to G' \to G \to G'' \to 0$ of coefficient modules, there is a long exact sequence

   $$\ldots \to H_q(K; G') \to H_q(K; G) \to H_q(K; G'') \to H_{q-1}(K; G') \to \ldots$$

   of simplicial homology groups of K with coefficients in G', G and G'' respectively.

4.9.4 Lefschetz Fixed-Point Theorem

We are now going to present one of the most important fixed-point theorems of algebraic topology, called Lefschetz Fixed-Point Theorem, named after the American mathematician Solomon Lefschetz (1884–1972), who discovered it in 1926. Brouwer's Fixed Point Theorem, proved earlier, will be a simple corollary of this theorem. Let X be a compact polyhedron and $f\colon X \to X$ be a continuous map. Associated with such a map, Lefschetz defined a number $\lambda(f)$ and then proved the result that if $\lambda(f) \neq 0$, then f must have a fixed point. It must be pointed out that the number $\lambda(f)$, called the **Lefschetz number** of f, will be an integer whose non-vanishing will imply that f will have a fixed point – it will not give us the number of fixed points.

To define the Lefschetz number $\lambda(f)$, let $h\colon |K| \to X$ be a triangulation of X, where dim $X = n$, and let us consider the simplicial homology $H_q(K;\mathbb{Q})$ of K with coefficients in the additive group of rational numbers. By result of section 4.9.2 proved earlier, $h^{-1}fh\colon |K| \to |K|$ induces a homomorphism $(h^{-1}fh)_*\colon H_q(K;\mathbb{Q}) \to H_q(K;\mathbb{Q})$, for all $q \geq 0$. Since the homology groups $H_q(K;\mathbb{Q})$ are vector spaces, $(h^{-1}fh)_*$ is a linear transformation for each q. We consider the alternate sum of traces of all these linear transformations for each $q \geq 0$ and define,

$$\lambda(f) = \sum_{q=0}^{n} (-1)^q \operatorname{Tr}\,(h^{-1}fh)_*.$$

Here, $(h^{-1}fh)_*$ really means $(h^{-1}fh)_{*,q}\colon H_q(K;\mathbb{Q}) \to H_q(K;\mathbb{Q})$ and n denotes the dimension of the complex K, which means the sum is finite as $H_q(K) = 0 \;\forall\, q > n$.

In order that $\lambda(f)$ is well defined, we must show that $\lambda(f)$ is independent of any triangulation $h\colon |K| \to X$. For this, let $k\colon |L| \to X$ be yet another triangulation of X. It suffices to show that the two linear maps $(h^{-1}fh)_*\colon H_q(K;\mathbb{Q}) \to H_q(K;\mathbb{Q})$ and $(k^{-1}fk)_*\colon H_q(L;\mathbb{Q}) \to H_q(L;\mathbb{Q})$ have identical traces for all $q \geq 0$. Note that

$$k^{-1}fk = (k^{-1}h)(h^{-1}fh)(h^{-1}k)$$

and hence, at the homology level,

$$(k^{-1}fk)_* = (k^{-1}h)_*(h^{-1}fh)_*(k^{-1}h)_*^{-1},$$

where $(k^{-1}h)_*\colon H_q(K;\mathbb{Q}) \to H_q(L;\mathbb{Q})$ is an isomorphism. However, the above equation shows that the matrices of linear maps $(h^{-1}fh)_*$ and $(k^{-1}fk)_*$ are indeed similar and, therefore, they have the same traces. Now, we have

Theorem 4.9.8. (Lefschetz Fixed-Point Theorem). *If X is a compact polyhedron and $f\colon X \to X$ is a continuous map such that $\lambda(f) \neq 0$, then f has a fixed point.*

In order to prove the above theorem, we require a result known as **Hopf's Trace Formula** and let us explain this formula first. Let

$$C: \quad 0 \to C_n \to C_{n-1} \to \ldots \to C_1 \to C_0 \to 0$$

be a finite chain complex of finite-dimensional vector spaces over $\mathbb{Q}$ and let $\phi_q : C_q \to C_q, 0 \leq q \leq n$ be a chain map. Then, here also, we can consider the well defined number $\sum_{i=0}^{n}(-1)^q \operatorname{Tr}(\phi_q)$. Since this chain map induces homomorphism $\phi_{q*}\colon H_q(C) \to H_q(C)$ for all q, $0 \leq q \leq n$, we also have the number $\sum(-1)^q \operatorname{Tr}(\phi_{q*})$. The Hopf's formula is given by:

Proposition 4.9.9. (H. Hopf). *Let C be a chain complex of finite length n in which each C_q is a finite dimensional vector space over $\mathbb{Q}$. Then for any chain map $\phi\colon C \to C$,*

$$\sum_{q=0}^{n}(-1)^q \operatorname{Tr}(\phi_q) = \sum_{q=0}^{n}(-1)^q \operatorname{Tr}(\phi_{q*}).$$

Proof. We choose a suitable basis for the vector space C_q. We have subspace $B_q \subset Z_q \subset C_q$ of q-boundaries, q-cycles and q-chains. Start with a basis $\partial c_1^{q+1}, \ldots, \partial c_{r_{q+1}}^{q+1}$ of B_q, then extend it to a basis of Z_q by adding cycles $z_1^q, \ldots, z_{\beta_q}^q$, and finally extend this to a basis of C_q by adding q-chains $c_1^q, \ldots, c_{r_q}^q$. Thus, for each q, we work with basis of C_q:

$$B = \{\partial c_1^{q+1}, \ldots, \partial c_{r_{q+1}}^{q+1}, z_1^q, \ldots, z_{\beta_q}^q, c_1^q, \ldots, c_{r_q}^q\}.$$

The diagonal elements of the matrix of ϕ_q with respect to above basis are obtained by taking any element w of the above basis and expressing $\phi_q(w)$ as a linear combination of elements of B, and finding the coefficients of w in $\phi_q(w)$. Let us denote this coefficient by $\lambda(w)$. Then, clearly, the trace of ϕ_q is

$$\sum \lambda(\partial c_j^{q+1}) + \sum \lambda(z_j^q) + \sum \lambda(c_j^q).$$

Since ϕ is a chain map, we find that $\lambda(\partial c_j^{q+1}) = \lambda(c_j^{q+1})$. Therefore,

$$\sum_{q=0}^{n}(-1)^q \operatorname{Tr}(\phi_q) = \sum_{q=o}^{n}(-1)^q \sum_{j=1}^{\beta_q} \lambda(z_j^q),$$

since the other terms cancel away in pairs. Because $\{z_1^q\}, \ldots, \{z_{\beta_q}^q\}$ is a basis of $H_q(C)$, we have

$$\sum_{j=1}^{\beta_q} \lambda(z_j^q) = \text{ trace of } \phi_{q*}.$$

This establishes the desired formula. ■

Proof of the Lefschetz Fixed-Point Theorem. Let $h\colon |K| \to X$ be a triangulation of X and $f\colon X \to X$ the given continuous map. Then $g = h^{-1} \circ f \circ h\colon |K| \to |K|$ has a fixed point if and only if f has a fixed point. We prove the theorem by contradiction. Let us assume that $g\colon |K| \to |K|$ has no fixed points. Let d be the metric on $|K|$ regarded as a subspace of some Euclidean space. Since $|K|$ is a compact space and the continuous function defined by $x \to d(x, g(x))$ is never zero, it has a positive minimum, say, $\delta > 0$. By going to sufficiently high order barycentric subdivision, we may as well assume that mesh K is less than $\delta/3$. Let $s\colon |K^{(m)}| \to |K|$ be a simplicial approximation of $g\colon |K| \to |K|$ and $\mu\colon C(K; \mathbb{Q}) \to C(K^{(m)}; \mathbb{Q})$ be the subdivision chain map. Since, for all $q \geq 0$, $g_{q*}\colon H_q(K) \to H_q(K)$ is the composite $g_{q*} = s_{q*}\mu_{q*}$, it suffices, by Hopf's Trace Formula, to show that trace $s_q \circ \mu_q\colon C_q(K) \to C_q(K)$ is zero, because that will imply that $\lambda(f) = 0$, a contradiction.

Let σ be an oriented q-simplex of K and let τ be an oriented q-simplex of $K^{(m)}$ which occurs in the chain $\mu_q(\sigma)$. Thus, τ is contained in σ. Now, if $x \in \tau$, $d(s(x), g(x)) < \delta/3$ because s is a simplicial approximation of g. Hence, by $d(x, g(x)) \geq \delta$, we find that

$$d(x, s(x)) \geq d(x, g(x)) - d(g(x), s(x)) > \delta - \delta/3 = 2\delta/3.$$

Next, let $y \in \sigma$, then $d(x, y) < \delta/3$, and by what we proved just now,

$$d(y, s(x)) \geq d(s(x), x) - d(x, y) > 2\delta/3 - \delta/3 = \delta/3.$$

This means $s(x)$ and y do not lie in the same simplex of K and, therefore, $s(\tau) \cap \sigma = \phi$ for each face $\tau \subset \sigma$. Hence, in the chain $s_q \circ \mu_q(\sigma)$, the coefficient of σ is zero. In other words, we have proved that trace $s_q \circ \mu_q = 0$. This completes the proof. ■

It may be pointed out that the Hopf's Trace Formula stated earlier for a simplicial map $f\colon K \to K$ is a generalization of the Euler-Poincaré Theorem 4.5.8 when $f = I_K$. We also observe that for a simplicial map $g\colon |K| \to |K|$, which has the property that for each simplex $\sigma \in K$, $\sigma \cap g(\sigma) = \emptyset$, clearly, the Hopf Trace Formula will say that $\lambda(g) = 0$. Thus, the whole point of the Lefschetz Fixed-Point Theorem was to prove that under the given conditions, if f has no fixed points, then f can be approximated by a simplicial map g which has the property stated above.

Corollary 4.9.10. (Brouwer's Fixed-Point Theorem) *Let X be a compact polyhedron which is contractible. Then any continuous map $f\colon X \to X$ will have a fixed point.*

Proof. We just note that because X is contractible, $H_0(X; \mathbb{Q}) \cong \mathbb{Q}$ and $H_q(X) = 0$, $\forall\, q > 0$. Moreover, the induced homomorphism $f_*\colon H_0(X; \mathbb{Q}) \to H_0(X; \mathbb{Q})$ is the identity map. Therefore, $\lambda(f) = 1 \neq 0$, and f has a fixed point by Lefschetz Fixed-Point Theorem. ■

Corollary 4.9.11. *Let* $f\colon \mathbb{S}^n \to \mathbb{S}^n$ *be a continuous map. Then* $\lambda(f) = 1 + (-1)^n \deg f$. *In particular, if* $\deg f \neq \pm 1$, *then it has a fixed point.*

Proof. We know that $H_0(\mathbb{S}^n;\mathbb{Q}) \cong \mathbb{Q} \cong H_n(\mathbb{S}^n;\mathbb{Q})$ and $H_q(\mathbb{S}^n;\mathbb{Q}) = 0$ for all other values of q. Since $f_*\colon H_0(\mathbb{S}^n;\mathbb{Q}) \to H_0(\mathbb{S}^n;\mathbb{Q})$ is the identity map and $f_*\colon H_n(\mathbb{S}^n;\mathbb{Q}) \to H_n(\mathbb{S}^n;\mathbb{Q})$ has trace equal to the $\deg f$, the result is now immediate. ■

Exercises

1. Let $f\colon X \to X$ be a continuous map where X is a compact polyhedron. Recall that the definition of Lefschetz number $\lambda(f)$ is given using homology with the rational coefficients $\mathbb{Q}$. Prove that $\lambda(f)$ must always be an integer. Show by an example that this is not true if X is not a compact polyhedron.

2. Let X be a compact polyhedron and $f\colon X \to X$ be a constant map, i.e., $f(x) = x_0, \forall x \in X$ and for some $x_0 \in X$. Determine the induced homomorphism $f_*\colon H_q(X) \to H_q(X)$ in homology for all $q \geq 0$. Hence or otherwise, prove that a null homotopic map $f\colon X \to X$ must have a fixed point.

3. Let the additive group $\mathbb{R}$ of real numbers act on a compact polyhedron X with $\chi(X) \neq 0$ under the action $\theta\colon \mathbb{R} \times X \to X$ (see Appendix). For each $t \in \mathbb{R}$, let us define a map $\theta_t\colon X \to X$ by putting $\theta_t(x) = \theta(t,x)$. Prove that

   (a) Any two maps $\theta_t, \theta_{t'}$ are homotopic to each other.

   (b) Each θ_t has a fixed point in X.

   (c) For each $n \geq 1$, any fixed point of $\theta_{1/2^{n+1}}$ is a fixed-point of $\theta_{1/2^n}$.

   (d) There is a point $x_0 \in X$ which is a fixed point of θ_t for all diadic rational $t = (m/2^n)$.

   (e) There is a point $x_0 \in X$ which is fixed under the group $\mathbb{R}$.

4. Let G be a path-connected topological group. Show that any left translation $L_g\colon G \to G$ defined by $L_g(x) = g\cdot x$ is homotopic to the identity map. Hence or otherwise deduce that the Euler characteristic of a compact connected triangulable topological group G must be zero. (**Hint**: Let ω_g be a path in G joining the identity element with g and define $F\colon G \times I \to G$ by $F(x,t) = \omega_{x^{-1}}(t)\cdot x$)

5. Prove that an even-dimensional sphere $\mathbb{S}^{2n}$ cannot be a topological group.

6. Prove that the only compact surface which can be a topological group is the torus T. Why not the Klein bottle?

4.9.5 The Borsuk-Ulam Theorem

We now come to another very interesting result about the maps from spheres $\mathbb{S}^m$ to spheres $\mathbb{S}^n$, where m and n are arbitrary nonnegative integers. Here, we consider the n-sphere $\mathbb{S}^n$ as a subspace of $\mathbb{R}^{n+1}$ defined by

$$\mathbb{S}^n = \{(x_1, \ldots, x_{n+1}) \mid \sum_{i=1}^{n+1} x_i^2 = 1\}.$$

Two points $x, y \in \mathbb{S}^n$ are said to be antipodal points if $y = -x$, i.e., these are opposite to each other with respect to the origin. The map $A\colon \mathbb{S}^n \to \mathbb{S}^n$ defined by $A(x) = -x$ is called the antipodal map, and we have already come across this map earlier (see Definition 4.8.7). A continuous map $f\colon \mathbb{S}^m \to \mathbb{S}^n$, m, n arbitrary positive integers, is said to "preserve" antipodal points if for all $x \in \mathbb{S}^m$, $f(A(x)) = A(f(x))$. The question that we would like to ask is: What are the values of m and n so that there is a continuous map $f\colon \mathbb{S}^m \to \mathbb{S}^n$ preserving the antipodal points? The answer is trivial if $m \leq n$. To see this, recall that for $m \leq n$, $\mathbb{S}^m$ is embedded into $\mathbb{S}^n$ in such a manner that antipodal points of $\mathbb{S}^m$ are also the antipodal points of $\mathbb{S}^n$. For example, we can place $\mathbb{S}^1 = \{(x, y) \in \mathbb{R}^2 \mid x^2 + y^2 = 1\}$ into $\mathbb{S}^2 = \{(x, y, z) \in \mathbb{R}^3 \mid x^2 + y^2 + z^2 = 1\}$ simply by taking the z-coordinate to be zero. More generally, take $\mathbb{S}^m$ as a subset of $\mathbb{S}^n$ where the last $n - m$ coordinates are zero. Then the inclusion map $i\colon \mathbb{S}^m \to \mathbb{S}^n$ defined by $i(x_1, \ldots, x_{m+1}) = (x_1, \ldots, x_{m+1}, 0, \ldots, 0)$ clearly preserves the antipodal points. In particular, note that the identity map of $\mathbb{S}^n$ preserves the antipodal points. It is the case when $m > n$, however, which is not at all clear! In fact, the Borsuk-Ulam Theorem asserts that if $m > n$, then there can be no continuous map $f\colon \mathbb{S}^m \to \mathbb{S}^n$ preserving the antipodal points. This is what we will prove in this section. It may be mentioned that this result was conjectured by S. Ulam and first proved by K. Borsuk in 1933.

Once we have proved the above result, several interesting consequences, some of which are in fact equivalent to the Borsuk-Ulam Theorem, follow very quickly. For example,

Theorem 4.9.12. *Any continuous map $f\colon \mathbb{S}^n \to \mathbb{R}^n$ must map a pair of antipodal points to the same point.*

Proof. If this theorem is not true, then we find that for each $x \in \mathbb{S}^n$, $f(-x) \neq f(x)$, i.e., $f(x) - f(-x) \neq 0$. Hence we can define a map $g\colon \mathbb{S}^n \to \mathbb{S}^{n-1}$ by the formula

$$g(x) = \frac{f(x) - f(-x)}{\| f(x) - f(-x) \|}.$$

Then, clearly, g is continuous and for all $x \in \mathbb{S}^n$, $g(-x) = -g(x)$, i.e., g preserves antipodal points, a contradiction to Borsuk-Ulam Theorem. ∎

Corollary 4.9.13. *The n-sphere $\mathbb{S}^n$, $n \geq 1$, cannot be embedded into $\mathbb{R}^n$. In particular, the 2-sphere $\mathbb{S}^2$ cannot be put into the plane $\mathbb{R}^2$.*

This follows at once because any embedding $i \colon \mathbb{S}^n \to \mathbb{R}^n$ is continuous, and so by the preceding theorem, must map at least one pair of points to a single point of $\mathbb{R}^n$, i.e., it cannot be an injective map.

Here is an interesting illustration of the above form of Borsuk-Ulam Theorem: Suppose our Earth is a 2-sphere and T, P are functions on the earth defining temperature and barometric pressure at any point of time and place on the Earth. Then these are continuous functions and, hence, so is the map $h \colon \mathbb{S}^2 \to \mathbb{R}^2$ defined by $h(x) = (T(x), P(x))$. The Borsuk-Ulam Theorem says that at any point of time there exists a pair of opposite points on the earth where the temperature and barometric pressure are identical!

Corollary 4.9.14. (Invariance of Dimension). *If $m \neq n$, then the Euclidean space $\mathbb{R}^m$ cannot be homeomorphic to $\mathbb{R}^n$.*

Proof. One can assume that $m > n$. If possible, let $f \colon \mathbb{R}^m \to \mathbb{R}^n$ be a homeomorphism. Then f will map the n-sphere $\mathbb{S}^n \subset \mathbb{R}^m$ into $\mathbb{R}^n$ as a continuous injective map, i.e., f restricted to $\mathbb{S}^n$ is an embedding into $\mathbb{R}^n$, contradicting the earlier corollary. ■

There is yet another interesting consequence of the Borsuk-Ulam Theorem which must be mentioned: Let $\mathbb{S}^1$ be the unit circle in the plane $\mathbb{R}^2$. If we cover $\mathbb{S}^1$ by two sets $A_1 = \{e^{i\theta} \mid 0 \leq \theta < \pi\}$ and $A_2 = \{e^{i\theta} \mid \pi \leq \theta < 2\pi\}$, then note that none of these sets contains a pair of antipodal points. None of these sets, however, is closed. If we take one of them to be closed, then observe that A_1 or A_2 will definitely contain a pair of antipodal points. A similar situation is true for $\mathbb{S}^2$ also, viz., if we cover $\mathbb{S}^2$ by three closed sets, then at least one of them will definitely contain a pair of antipodal points. The next result says that these observations can be proved for $\mathbb{S}^n$ for any n. We have

Theorem 4.9.15. (Lusternik-Schnirelmann Theorem). *If $\{A_1, A_2, \ldots, A_{n+1}\}$ is a covering of $\mathbb{S}^n$, $n \geq 1$, by closed sets, then at least one of these sets must contain a pair of antipodal points.*

Proof. Let us define a map $f \colon \mathbb{S}^n \to \mathbb{R}^n$ by the formula $f(x) = (d(x, A_1), d(x, A_2), \ldots, d(x, A_n))$, where $d(x, A_i)$ denotes the distance of point x from the closed set A_i, $i = 1, 2, \ldots, n$. Since each of these distance functions is continuous in x, the map f itself is continuous. Hence, by the Borsuk-Ulam Theorem, there must be an $x \in \mathbb{S}^n$ such that $(d(x, A_1), \ldots, d(x, A_n)) = (d(-x, A_1), \ldots, d(-x, A_n))$, which says that $d(x, A_i) = d(-x, A_i)$ for every i. Now, if $d(x, A_i) > 0$ for each i, then $d(-x, A_i) > 0$ for each i also, and so both x, $-x$ must be in the remaining set A_{n+1} because $\{A_1, A_2, \ldots, A_{n+1}\}$ covers $\mathbb{S}^n$. On the other hand, if $d(x, A_i) = 0$ for some i then, since A_i is closed, both x and $-x$ must be in A_i. ■

Remark 4.9.16. In the proof of the above theorem we used the fact that out of $n+1$ sets only n are closed.

Note that the antipodal map $A\colon \mathbb{S}^n \to \mathbb{S}^n$ is a homeomorphism of $\mathbb{S}^n$ of order 2. This says that the group $G = \{1, -1\}$ of order two acts on $\mathbb{S}^n$ for each n via the antipodal maps. Now, if $f\colon \mathbb{S}^m \to \mathbb{S}^n$ is a continuous map preserving the antipodal points then, in terms of group action, this means the map f preserves the above G-action. Such a map induces a continuous map $\bar{f}\colon \mathbb{S}^m/G \to \mathbb{S}^n/G$ on orbit spaces, i.e., $\bar{f}$ is a continuous map from the real projective space $\mathbb{RP}^m$ to $\mathbb{RP}^n$. Thus, the Borsuk-Ulam Theorem says that if $m > n$, then there cannot be a continuous map from $\mathbb{RP}^m$ to $\mathbb{RP}^n$ induced by any equivariant map from $\mathbb{S}^m$ to $\mathbb{S}^n$.

Now, we return to the proof of the Borsuk-Ulam Theorem, viz., **there cannot be a continuous map $f\colon \mathbb{S}^m \to \mathbb{S}^n$ preserving the antipodal points if** $m > n$. (The case when $m = n+1$ is Borsuk-Ulam Theorem). Suppose, on the contrary, that $m > n$ and there is a continuous map $f\colon \mathbb{S}^m \to \mathbb{S}^n$ preserving antipodal points. Then, to prove the result we will, on the basis of this assumption, arrive at a contradiction. In fact, we will prove (Theorem 4.9.18) that any $g\colon \mathbb{S}^n \to \mathbb{S}^n$ mapping antipodal points to antipodal points must have odd degree. On the other hand, since $f\colon \mathbb{S}^m \to \mathbb{S}^n$, $m > n$, preserves antipodal points, it restricts to a map $g\colon \mathbb{S}^n \to \mathbb{S}^n$ preserving antipodal points. We will show that this restriction map has degree zero, giving a contradiction. The latter assertion follows rather very easily. For, the restriction $g\colon \mathbb{S}^n \to \mathbb{S}^n$ of $f\colon \mathbb{S}^m \to \mathbb{S}^n$ clearly has an extension to $\mathbb{D}^{n+1} \subset \mathbb{S}^{n+1} \subseteq \mathbb{S}^m$ and, therefore, g is homotopic to a constant map. Since homotopic maps have same degrees and the constant map has degree zero, we find that $\deg g = 0$. Thus, it is the first part which requires a good amount of work. We now proceed to prove that result next.

Note that for a given map $g\colon \mathbb{S}^n \to \mathbb{S}^n$, its degree $\deg(g) = d$ is an integer having the property that if $h\colon |K| \to \mathbb{S}^n$ is any triangulation of $\mathbb{S}^n$, then the induced homomorphism $(h^{-1}gh)_*\colon H_n(K) \to H_n(K)$ is simply multiplication by d. We have already seen that degree d is independent of the triangulation. Therefore, we can choose any triangulation of $\mathbb{S}^n$ which is convenient to us and can work with that only. We choose a special triangulation $\pi\colon |\Sigma^n| \to \mathbb{S}^n$ described below: For each integer $i = 1, 2, \ldots, n+1$, let v_i be the point of $\mathbb{R}^{n+1}$ having all its coordinates zero except the i-th coordinate which is 1. Let $J = \{1, 2, \ldots, n+1\}$ and let $a\colon J \to \{1, -1\}$ be any map. For each a, we define an n-simplex $\sigma_a^n = \langle a(1)v_1, a(2)v_2, \ldots, a(n+1)v_{n+1}\rangle$. Then, note that the union of the geometric carriers of all these simplexes is the following subset of $\mathbb{R}^{n+1}$:

$$Q^n = \{(x_1, \ldots, x_{n+1}) \in \mathbb{R}^{n+1} \mid \sum_{i=1}^{n+1} |x_i| = 1\}.$$

The case $n = 1$ is illustrated in the following figure:

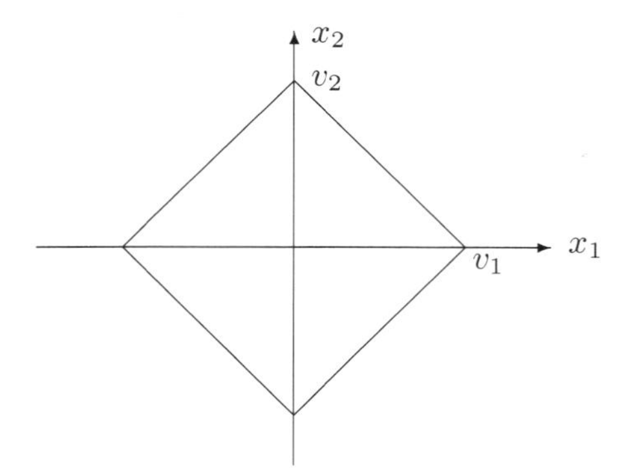

Note that the set of all n-simplexes described above, along with all of their faces, gives a triangulation Σ^n of $\mathbb{S}^n$ – the homeomorphism $h\colon |\Sigma^n| \to \mathbb{S}^n$ is just the "radial projection" from the origin of $\mathbb{R}^{n+1}$: any ray starting from origin meets $Q^n = |\Sigma^n|$ and $\mathbb{S}^n$ in exactly one point, say, x and y, respectively, and we define $h(x) = y$. This triangulation is specially suited for dealing with antipodal map A because both $\mathbb{S}^n$ and Q^n are invariant under A. The simplicial complex Σ^n can also be regarded as invariant with respect to the antipodal map because A maps an n-simplex σ^n_a of Σ^n to σ^n_{-a}. Thus, the antipodal map A is also a simplicial map from Σ^n to itself and, therefore, induces a chain map in the simplicial chain complex $C(\Sigma^n)$ of Σ^n. First, we have

Theorem 4.9.17. *Let $f\colon \mathbb{S}^n \to \mathbb{S}^n$ be a continuous map preserving the antipodal points. Then the Lefschetz number $\lambda(f)$ of f is an even integer.*

Proof. We use the above triangulation $\pi\colon |\Sigma^n| \to \mathbb{S}^n$ of $\mathbb{S}^n$ and recall that $\lambda(f) = \lambda(\pi^{-1} f \pi)$. Let $\Sigma^{(k)}$ be a barycentric subdivision of Σ^n, which is star related to $\pi^{-1} f \pi$, and let $\tau\colon |\Sigma^{(k)}| \to |\Sigma^n|$ be a simplicial approximation of $\pi^{-1} f \pi$. We assert that τ can be chosen so that τ preserves the antipodal points of $\Sigma^{(k)}$ and Σ^n. We divide the vertices of $\Sigma^{(k)}$ into two parts B and the image $A(B)$ of B under A so that no points of B are antipodal. For each vertex v in B, choose a vertex $\tau(v)$ in Σ so that $\pi^{-1} f \pi(\text{st } (v) \subseteq \text{st } (\tau(v))$. Then define $\tau(v)$ for elements of $A(B)$ by putting $\tau A(x) = A\tau(x)$. Then it is straightforward to see, using the fact that $\pi^{-1} f \pi$ maps antipodal points to antipodal points, that the simplicial map τ so defined is a simplicial approximation of $\pi^{-1} f \pi$, and τ preserves antipodal points.

Next, we consider the subdivision chain map $\mu\colon C(\Sigma^n) \to C(\Sigma^{(k)})$ and the orientation of $\Sigma^{(k)}$. From the definition of μ, each q-simplex σ^q_j of Σ^n, $\mu(\sigma^q_j) = \sum \epsilon^i_j \alpha^q_i$, where the sum is over all q-simplexes of $\Sigma^{(k)}$ contained in σ^q_j and $\epsilon^r_j = +1$ or -1. We say that a simplex α^q_i of $\Sigma^{(k)}$ contained in σ^q_j is oriented like σ^q_j if $\epsilon^i_j = +1$ and is unlike σ^q_j if $\epsilon^i_j = -1$. We may orient $\Sigma^{(k)}$ so that whenever α^q_i is contained in σ^q_j, then α^q_i is oriented like σ^q_j if and only if $A(\alpha^q_j)$ is oriented like $A(\sigma^q_j)$. The purpose of this particular orientation is to ensure that μ commutes with $A_\#$. This is easily verified since for each simplex

σ_j^q of Σ^n, we have

$$A_\#\mu(\sigma_j^q) = A_\#(\sum \epsilon_j^i \alpha_i^q) = \sum \epsilon_j^i A(\alpha_i^q) = \mu A_\#(\alpha_j^q).$$

Next, we consider the Lefschetz number $\lambda(\pi^{-1}f\pi) = \sum(-1)^i tr\ (\tau_\#\mu)$. Let us consider the $tr\ (\tau_\#\mu)$ in a particular dimension, say, q. Let $M = (m_j^i)$ be the matrix representing $\tau_\#\mu$ with respect to the basis of $C_q(\Sigma^n)$ consisting of elementary integral q-chains $\sigma_j^q \in B$ and $A\sigma_j^q \in A(B)$. Then the integer m_j^j is the coefficient of σ_j^q in the chain $\tau_\#\mu(\sigma_j^q)$ and the sum of all these numbers is the *trace* $(\tau_\#\mu)$ in dimension q. Now, we will prove that such a coefficient in this sum appears twice. Let us denote the coefficient of σ_j^q in $\tau_\#\mu(\sigma_j^q)$ by the symbol $g(\sigma_j^q)$, i.e., $m_j^j = g(\sigma_j^q)$. It is enough to show that $g(A(\sigma_j^q)) = g(\sigma_j^q)$. We have,

$$\tau_\#\mu(\sigma_j^q) = g(\sigma_j^q)\sigma_j^q + \textstyle\sum_{i\neq j} m_j^i \sigma_i^q,$$

so that

$$\begin{aligned}
\tau_\#\mu(A(\sigma_j^q)) &= \tau_\#\mu(A_*(\sigma_j^q)) = A_*\tau_\#\mu(\sigma_j^q) \\
&= A_*(g(\sigma_j^q)\sigma_j^q + \sum_{i\neq j} m_j^i \sigma_j^q) \\
&= g(\sigma_j^q)(A(\sigma_j^q)) + \sum_{i\neq j} m_j^i A(\sigma_i^q).
\end{aligned}$$

Since the coefficient of $A(\sigma_j^q)$ in $\tau_\#\mu(A(\sigma_j^q))$ is $g(\sigma_j^q)$, we conclude that $g(A(\sigma_j^q)) = g(\sigma_j^q)$. This shows that trace of $\tau_\#\mu$ is even in each dimension, and so $\lambda(f) = \lambda(\pi^{-1}f\pi)$ is even. ■

Now, we have

Theorem 4.9.18. *Let $f\colon \mathbb{S}^n \to \mathbb{S}^n$, $n \geq 1$, be a continuous map preserving antipodal points. Then* $\deg f$ *is odd.*

Proof. We know that $H_0(\mathbb{S}^n) \cong \mathbb{Z} \cong H_n(\mathbb{S}^n)$, and $H_q(\mathbb{S}^n) = 0$ for other values of q. We also know, because $\mathbb{S}^n$ is path-connected, that $f_*\colon H_0(\mathbb{S}^n) \to H_0(\mathbb{S}^n)$ is the identity map. Hence, by the preceding theorem, $\lambda(f) = 1 + (-1)^n \deg f$ is even, which implies that $\deg f$ must be odd. ■

Theorem 4.9.19. (Borsuk-Ulam). *If $m > n \geq 0$, then there cannot be a continuous map $f\colon \mathbb{S}^m \to \mathbb{S}^n$ preserving the antipodal points.*

Proof. Suppose, on the contrary, there is a continuous map $f\colon \mathbb{S}^m \to \mathbb{S}^n$ which preserves the antipodal points. Let $i\colon \mathbb{S}^n \to \mathbb{S}^m$ denote the standard inclusion map, i.e., $\mathbb{S}^n$ is obtained from $\mathbb{S}^m$ by putting last $m - n$ coordinates equal to zero. Then the composite $i \circ f\colon \mathbb{S}^m \to \mathbb{S}^m$ also preserves the antipodal points. Hence, by Theorem 4.9.18 , $\deg(i \circ f)$ must be odd. On the other hand, the composite $(i \circ f)_* = i_* \circ f_* \colon H_m(\mathbb{S}^m) \to H_m(\mathbb{S}^n) \to H_m(\mathbb{S}^m)$ factors through $H_m(\mathbb{S}^n) = 0$, which means $(i \circ f)_*$ is the zero map, i.e., $\deg(i \circ f)$ is zero. This is a contradiction, proving the theorem. ■

Corollary 4.9.20. *Let $f \colon \mathbb{S}^n \to \mathbb{R}^n$ be a continuous map preserving antipodal points. Then there exists a point $x \in \mathbb{S}^n$ such that $f(x) = 0$.*

Proof. If $f(x) \neq 0$, $\forall\ x \in \mathbb{S}^n$, then we can define a map $g \colon \mathbb{S}^n \to \mathbb{S}^{n-1}$ by the formula $g(x) = \frac{f(x)}{\|f(x)\|}$. Then g is a continuous map preserving antipodal points, a contradiction to Borsuk-Ulam Theorem. ■

Exercises

1. Prove that the Borsuk-Ulam theorem is equivalent to the following statement: if $f : \mathbb{S}^n \to \mathbb{R}^n$ is a continuous map, then there exists a point $x_0 \in \mathbb{S}^n$ such that $f(x_0) = f(-x_0)$.

2. Show that $\mathbb{S}^n$ is not homeomorphic to a proper subspace of itself.

3. Show that there is no continuous map $f \colon \mathbb{S}^m \to \mathbb{S}^n$, for all $m > n \geq 0$, preserving antipodal points if and only if there is no continuous map $g \colon \mathbb{S}^{n+1} \to \mathbb{S}^n$, preserving antipodal points for all $n \geq 0$.

4. Suppose a map $f \colon \mathbb{S}^n \to \mathbb{S}^n$ satisfies the condition that $f(-x) = f(x)$, $\forall\ x \in \mathbb{S}^n$. Then prove that $\deg f$ is even. If n is even, show that $\deg f = 0$.

5. If $f \colon \mathbb{S}^n \to \mathbb{S}^n$ has odd degree, then prove that $f(-x) = -f(x)$ for some $x \in \mathbb{S}^n$. (**Hint:** Suppose the contrary. Consider the map $h(x) = (f(x) + f(-x))/ \parallel f(x) + f(-x) \parallel$.)

6. Let A_1, A_2, A_3 be three bounded convex subsets of $\mathbb{E}^3$, and define a function $f : \mathbb{S}^3 \to \mathbb{E}^3$ w.r.t. A_1, A_2, A_3 as follows: For $x \in S^3$ take a hyperplane $P(x)$ perpendicular to x passing through (0,0,0,1/2). Let $f_i(x)$ be the volume of that part of A_i which lies on the same side of the hyperplane $P(x)$ as the point x, $i = 1, 2, 3$. Put

$$f(x) = (f_1(x), f_2(x), f_3(x))$$

   Prove that f is continuous.
   (Hence, by the Borsuk-Ulam Theorem there exists an $x_0 \in \mathbb{S}^3$ such that $f(x_0) = f(-x_0)$, i.e., $f_1(x_0) = f_1(-x_0)$, $f_2(x_0) = f_2(-x_0)$, $f_3(x_0) = f_3(-x_0)$. This result is called the **Ham-Sandwich Theorem** because it proves the following surprising result: Given a three-layered Ham-sandwich it is possible to cut it by a single stroke of knife so that each of the three pieces is divided into two equal parts!)

4.10 Homology of General Simplicial Complexes

So far we have defined and studied the finite simplicial complexes K and their homology groups $H_q(K), q \geq 0$. We know that for such complexes, the induced topological space $|K|$ are compact polyhedra. Using this fact, we defined and studied the homology groups of a compact polyhedron X. The homology of a noncompact polyhedron, however, has not been defined yet. We note that the Euclidean space $\mathbb{R}^n, n \geq 1$, is a polyhedron and a triangulation of $\mathbb{R}^n$ is an infinite simplicial complex. This means we have not defined homology of even such nice spaces as $\mathbb{R}^n$. In this section, we will extend the definition of homology to any general simplicial complex K, which will enable us to define the simplicial homology of any polyhedron X, compact or noncompact. In particular, we will define $H_q(\mathbb{R}^n), q \geq 0$, for all Euclidean spaces, and the homology of many other nice noncompact polyhedra.

Let K be a general simplicial complex as in Section 3.5. We give an orientation to K by orienting each simplex of K arbitrarily and then define the simplicial chain complex $C_*(K)$ of K as follows:

$$C_*(K) : \ldots \longrightarrow C_{q+1}(K) \xrightarrow{\partial_{q+1}} C_q(K) \xrightarrow{\partial_q} C_{q-1}(K) \ldots \xrightarrow{\partial_1} C_0(K) \longrightarrow 0.$$

Here $C_q(K)$ is the free abelian group whose basis is the set of all q-simplexes of K, and $\partial_q \colon C_q(K) \to C_{q-1}(K)$ is defined as before viz.,

$$\partial_q(\langle v_0, \ldots, v_q \rangle) = \textstyle\sum_0^q (-1)^i \langle v_0, \ldots, \hat{v}_i, \ldots, v_q \rangle$$

where $\langle v_0, \ldots, v_q \rangle$ is a basis element of $C_q(K)$. We know that $\partial_q \circ \partial_{q+1} = 0, \forall\, q \geq 0$. Since the simplicial complex K is arbitrary, the above chain complex will be of infinite length. Each $C_q(K)$ could be a finitely generated or infinitely generated abelian group depending on whether the number q-simplexes is finite or infinite. Then we define the homology groups $H_q(K)$ of K by

$$H_q(K) = \frac{\ker \partial_q}{\operatorname{Im} \partial_{q+1}}, \quad q \geq 0.$$

We observe that, in contrast with finite simplicial complexes, now $H_q(K)$ may be nonzero for infinitely many values of q and each of these groups could be finitely generated or infinitely generated.

Next, let $f \colon K \to L$ be a simplicial map. Then we can define the induced chain map $f_\# \colon C_q(K) \to C_q(L)$ exactly as in finite simplicial complexes. This will induce a homomorphism $f_* \colon H_q(K) \to H_q(L), \forall\, q \geq 0$, in homology, and they will satisfy the functorial properties, viz., (i) the identity map $I_K \colon K \to K$ will induce identity map $(I_K)_* \colon H_q(K) \to H_q(K) \forall\, q \geq 0$, and (ii) the induced map will preserve the composite maps, i.e., if $f \colon K \to L$ and $g \colon L \to M$ are simplicial maps, then $(g \circ f)_* = g_* \circ f_* \colon H_q(K) \to H_q(M)\ \forall\, q \geq 0$.

Homomorphism induced by a continuous map

Let X be a polyhedron, compact or noncompact. We now wish to define the simplicial homology $H_q(X), q \geq 0$ of X. Let us take a triangulation K of X and define

$$H_q(X) = H_q(K) \text{ for all } q \geq 0.$$

In order that the above is well defined up to isomorphism, we need to prove that if K' is any other triangulation of X, then for all $q \geq 0, H_q(K) \approx H_q(K')$. As in the case of finite simplicial complexes, this raises the question as to what is the homomorphism in homology induced by a continuous map $f\colon |K| \to |L|$. Let us suppose that $f\colon |K| \to |L|$ is a continuous map which need not arise from a simplicial map from K to L. How this map f will induce a homomorphism $f_*\colon H_q(K) \to H_q(L) \forall\ q \geq 0$, in their simplicial homology? For this, we need a simplicial approximation theorem for the continuous map $f\colon |K| \to |L|$ for general simplicial complexes K and L.

As in the case of finite complexes, for a simplicial approximation theorem, we require the concept of barycentric subdivision of a general simplicial complex K. If we try to proceed as in the case of finite simplicial complexes, we confront a basic problem viz., for general K, $|K|$ has the coherent topology and so it need not be a metric space. More than that, even if $|K|$ is a metric space, it need not lie in an Euclidean space $\mathbb{R}^n$ and so we cannot assert that there is a simplex of K having the largest diameter. In order words, it is not possible to define the mesh of a simplicial complex K. Therefore, we cannot have subdivisions of K having arbitrary small mesh. This problem has been resolved by the concept of **generalized barycentric subdivision** of K. This is a technical generalization of the usual barycentric subdivision where we consider barycentric subdivision of higher dimension skeleton $K^{(n)}$ of K without disturbing the subdivision of $K^{(n-1)}$. We refer the reader to the book of J.R. Munkres [16] for all details of the generalized barycentric subdivision. Using this subdivision of an arbitrary simplicial complex K, one can prove the following (see [16]).

Theorem 4.10.1. (Generalized Simplicial Approximation Theorem). *Let $f\colon |K| \to |L|$ be a continuous map. Consider the open cover $\mathcal{V} = \{\text{st } w : w \text{ is a vertex of } L\}$ of $|L|$. Then there exists a generalized barycentric subdivision $\text{Sd}^N(K) = K'$ of K such that the cover $\mathcal{U} = \{\overline{\text{st } v} : v \text{ is a vertex of } K'\}$ by closed vertex stars is a refinement of $\{f^{-1}(v) : v \in \mathcal{V}\}$. This means $|K|$ is star-related to $|L|$ w.r.t. f.*

Assuming the above simplicial approximation theorem for general simplicial complexes, now we can define the homomorphism induced by a continuous map.

Definition 4.10.2. *Let $f\colon |K| \to |L|$ be a continuous map and let $g\colon |K'| \to |L|$ be a generalized simplicial approximation of f. Suppose $\lambda\colon C_*(K) \to C_*(K')$ is* *the subdivision operator. Then we define* $f_*: H_q(K) \to H_q(L)$ *as the composite* $f_* = g_* \circ \lambda_*$.

It can now be proved, exactly as in the case of finite simplicial complexes, that the induced homomorphism $f_*: H_q(K) \to H_q(L)$ is really independent of the simplicial approximation. Hence f_* is well-defined. Then one can easily prove the following:

Theorem 4.10.3. (i) *Let* $I: |K| \to |K|$ *be the identity map. Then the induced homomorphism* $I_*: H_q(K) \to H_q(K)$ *is also the identity map* $\forall\, q \geq 0$.

(ii) *Let* $f: |K| \to |L|$ *and* $g: |L| \to |M|$ *be two continuous maps. Then,* $\forall q \geq 0$

$$(g \circ f)_* = g_* \circ f_*: H_q(K) \to H_q(M).$$

Corollary 4.10.4. *Let* $h: |K| \to |L|$ *be a homeomorphism. Then* $h_*: H_q(K) \to H_q(L)$ *is an isomorphism* $\forall\, q \geq 0$.

Proof. Let $k: |L| \to |K|$ be the inverse of h. Then $k \circ h: |K| \to |K|$ is the identity on $|K|$ and $h \circ k: |L| \to |L|$ is the identity on $|L|$. Hence by (i) and (ii) above, we have $\forall\, q \geq 0$,

$$(k \circ h)_* = k_* \circ h_* = I_{H_q(K)}$$

$$(h \circ k)_* = h_* \circ k_* = I_{H_q(L)}.$$

This means $h_*: H_q(K) \to H_q(L)$ is an isomorphism $\forall\, q \geq 0$. ■

Simplicial Homology of a Polyhedron

Now we can define the simplicial homology of a polyhedron which has been the objective of this section.

Definition 4.10.5. *Let* X *be a general polyhedron. We define* $H_q(X) = H_q(K) \forall\, q \geq 0$, *where* K *is any triangulation of* X.

We remark that the above definition of homology of X is unambiguous (up to isomorphism) because if K' is any other triangulation of X, then we have homeomorphisms $h: |K| \to X$, $k: |K'| \to X$, which means $k^{-1} \circ h: |K| \to |K'|$ is a homeomorphism. Therefore, by the above corollary it induces an isomorphism $(k^{-1} \circ h)_*: H_q(K) \to H_q(L) \forall\, q \geq 0$.

Thus we have now defined the simplicial homology of any polyhedron X (of course up to group isomorphism). Let us be clear what we mean by saying that '$H_q(X)$ is defined up to isomorphism'. This means if we consider the class of all simplicial complexes K which triangulate the polyhedron X, then the homology groups of all such complexes are isomorphic to each other and we define $H_q(X)$ to be simply this isomorphism class. Hence if we identify all these isomorphic groups, then $H_q(X)$ is uniquely defined, and this is what we do in the sequel.

Next, let X, Y be two polyhedra and $f\colon X \to Y$ be a continuous map. To define the induced homomorphism $f_*\colon H_q(X) \to H_q(Y)$, we take triangulations K and L of X and Y respectively. Then we have a commutative diagram

$$\begin{array}{ccc} X & \xrightarrow{f} & Y \\ {\scriptstyle h}\uparrow & & \uparrow{\scriptstyle k} \\ |K| & \xrightarrow[g]{} & |L| \end{array}$$

Here the maps h and k are homeomorphism and the map g is being defined as $g = k^{-1}fh\colon |K| \to |L|$. Now we define f_* to be simply $g_*\colon H_q(K) \to H_q(L)$ $\forall q \geq 0$. The definition of f_* depends not only on the map f and topology of $|K|$, $|L|$, but it also depends on the simplicial complexes K and L. If we identify $H_q(X)$ with $H_q(K)$ and $H_q(Y)$ with $H_q(L)$ where K and L are triangulations of X and Y, then we can as well identify f_* with g_*. This means f_* is also uniquely defined up to the isomorphism.

We can also prove the following:

Proposition 4.10.6. (i) *Let $I\colon X \to X$ be the identity map. Then $\forall\, q \geq 0$ the induced homomorphism $I_*\colon H_q(X) \to H_q(X)$ is also the identity map.*

(ii) *If $f\colon X \to Y$ and $g\colon Y \to Z$ are continuous map of polyhedra, then $\forall\, q \geq 0$*

$$(g \circ f)_* = g_* \circ f_*\colon H_q(X) \to H_q(Z).$$

The above proposition tells us that the homology groups of a polyhedron X are topological invariants.

Example 4.10.7. Let us compute the homology groups of the real line $\mathbb{R}$. We know that $K = \{\langle n\rangle,\ \langle n, n+1\rangle : n \in N\}$ is a triangulation of $\mathbb{R}$. Hence the simplicial chain complex $C_*(\mathbb{R})$ is given by

$$\begin{array}{lccccccc} C_*(\mathbb{R}): & 0 & \longrightarrow & \oplus\mathbb{Z}\langle n, n+1\rangle & \xrightarrow{\partial_1} & \oplus\mathbb{Z}\cdot\langle n\rangle & \longrightarrow & 0 \\ & & & \| & & \| & & \\ & & & C_1(\mathbb{R}) & & C_0(\mathbb{R}) & & \end{array}$$

Now note that there is not a single non-zero 1-cycle since $\partial_1\langle n, n+1\rangle = \langle n+1\rangle - \langle n\rangle = 0$ means $n = n+1$, a contradiction. Hence $H_1(\mathbb{R}) = 0 = H_2(\mathbb{R}) = \ldots$.

To compute $H_0(\mathbb{R})$, note that the 0-cycle $\langle 1\rangle$ is not a boundary. Moreover if we take any other 0-cycle, say for example $\langle 3\rangle$, then

$$\partial_1(\langle 1,2\rangle + \langle 2,3\rangle) = \langle 3\rangle - \langle 1\rangle,$$

which means the cycle $\langle 3\rangle$ is homologous to $\langle 1\rangle$. Therefore, there is exactly one nontrivial homology class $\{\langle 1\rangle\}$. Hence $H_0(\mathbb{R}) \approx \mathbb{Z}\cdot\{\langle 1\rangle\}$.

By a similar consideration using a triangulation of $\mathbb{R}^n$, we can prove that

$$H_0(\mathbb{R}^n) = \begin{cases} \mathbb{Z}, & q = 0, \\ 0, & \text{otherwise.} \end{cases}$$

We can also prove the following.

Theorem 4.10.8. *Suppose X and Y are two polyhedra. If $f,\ g\colon X \to Y$ are two continuous maps which are homotopic, then for all $q \geq 0$*

$$f_* = g_*\colon H_q(X) \to H_q(Y).$$

The above theorem implies that homology groups $H_q(X)$ of a polyhedron are not only topological invariants, but they are also homotopy invariants, i.e., if X and Y are two polyhedra of the same homotopy type, then $H_q(X) \approx H_q(Y)\ \forall\ q \geq 0$.

In order to prove the above theorem we require a number of results which we summarize below. Since their proofs are technically involved, we refer to ([16]) for details. First, we have

Theorem 4.10.9. *If X is a polyhedron, then the product space $X \times I$ is also a polyhedron.*

Proof of Theorem 4.10.8: In fact, one can prove the following sharper result: For any polyhedron $|K|$, we can always find a triangulation M of $|K| \times I$ such that for each simplex $\sigma \in K$, $\sigma \times \{0\}$ and $\sigma \times \{1\}$ are both simplexes of M. Moreover the subset $\sigma \times I$ is an acyclic subcomplex of M. Now let $F\colon |K| \times I \to |L|$ be a homotopy from a continuous map $f\colon |K| \to |L|$ to the continuous map $g\colon |K| \to |L|$. Suppose $f_*\colon H_q(|K|) \to H_q(|L|)$ and $g_*\colon H_q(|K|) \to H_q(|L|)$ are the induced homomorphisms in homology. We now have the following diagram of continuous maps and polyhedra.

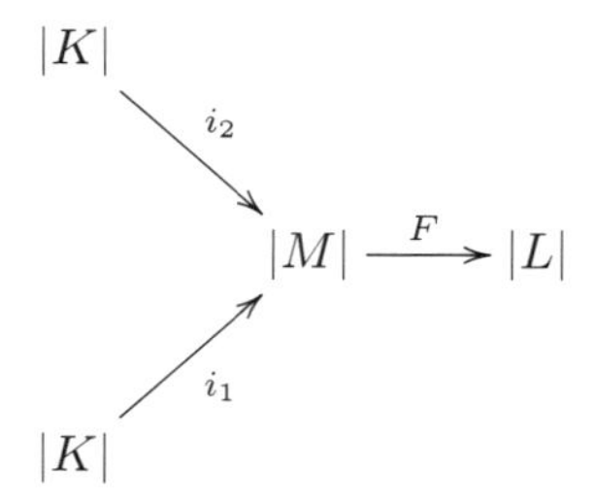

where $i_1(\sigma) = \sigma \times \{0\}$ and $i_2(\sigma) = \sigma \times \{1\}$ are the canonical injections. Then $F \circ i_1 = f$ and $F \circ i_2 = g$. Using the 'acyclic model' theorem, one can prove that the induced chain maps $i_{1\#}, i_{2\#} \colon C_*(|K|) \to C_*(|L|)$ are chain homotopic. This means that the induced maps in homology

$$f_* = F_* \circ i_{1*} = F_* \circ i_{2*} = g_*. \qquad \blacksquare$$

Next, let $(|K|, |L|)$ be a polyhderal pair, i.e., L is a subcomplex of K. Then we can define the simplicial homology $H_q(|K|, |L|)$ of the pair to be simply the homology of the quotient chain complex of the following exact sequence of chain complexes

$$0 \to C_*(L) \to C_*(K) \to \frac{C_*(K)}{C_*(L)} \to 0$$

In other words, we define $H_q(|K|, |L|) = H_q\left(\frac{C_*(K)}{C_*(L)}\right) \; \forall\, q \geq 0$. We can also define the homomorphism induced by a continuous map of pairs $f \colon (|K|, |L|) \to (|K'|, |L'|)$ by considering the chain map $f_\#$ induced by the following diagram of chain complexes.

$$\begin{array}{ccccccccc} 0 & \longrightarrow & C_*(|L|) & \longrightarrow & C_*(|K|) & \longrightarrow & C_*(|K|, |L|) & \longrightarrow & 0 \\ & & \big\downarrow (f|_{|L|})_\# & & \big\downarrow (f|_{|K|})_\# & & \big\downarrow f_\# & & \\ 0 & \longrightarrow & C_*(|L'|) & \longrightarrow & C_*(|K'|) & \longrightarrow & C_*(|K'|, |L'|) & \longrightarrow & 0 \end{array}$$

Then, clearly $f_\#$ induces the homomorphism

$$f_* \colon H_q(|K|, |L|) \to H_q(|K'|, |L'|) \; \forall\, q \geq 0$$

The above result now can be used to define the simplicial homology theory on the category of polyhedral pairs and their morphisms (see the chapter on singular homology). One can now easily prove the following general homotopy theorem for pairs:

Theorem 4.10.10. *Let $f,\ g \colon (X, A) \to (Y, B)$ be two maps of polyhedral pairs which are homotopic as maps of pairs. Then the induced homomorphism in homology of pairs are equal, i.e., $f_* = g_* \colon H_q(X, A) \to H_q(Y, B) \; \forall\, q \geq 0$.*

Proof. Let $F \colon (X \times I, A \times I) \to (Y, B)$ be a homotopy from the map f to g. Define $\bar{h_0},\ \bar{h_1} \colon (X, A) \to (X \times I, \bar{A \times I})$ by $h_0(x) = (x, 0),\ h_1(x) = (x, 1)$. If $h_0,\ h_1 \colon X \to X \times I$ are defined by $\bar{h_0},\ \bar{h_1}$, then we know that there is a natural chain homotopy

$$D : (h_0)_\# \cong (h_1)_\# : S_*(X) \to S_*(X \times I).$$

This chain homotopy restricts to a chain homotopy

$$D : (h_0)_\# \cong (h_1)_\# : S_*(A) \to S_*(A \times I).$$

Hence we can pass on to the quotient chain homotopy

$$\bar{D} : (\bar{h_0})_{\#} \cong (\bar{h_1})_{\#} : S_*(X, A) \to S_*(X \times I, A \times I).$$

Therefore, that the induced homomorphism in homology are equal, i.e.,

$$(\bar{h}_0)_* = (\bar{h}_1)_* : H_q(X, A) \to H_q(X \times I, A \times I).$$

Hence $f_* = F_* \circ (\bar{h}_0)_* = F_* \circ (\bar{h}_1)_* = g_*$. ■

Exercises

1. Let (X, A) be a polyhedral pair. Show that there is a long exact sequence in simplicial homology:

$$\ldots \to H_q(A) \xrightarrow{i_*} H_q(X) \xrightarrow{j_*} H_q(X, A) \xrightarrow{\partial_*} H_{q-1}(A) \to \ldots,$$

   where $i \colon A \to X$, $j \colon X \to (X, A)$ are the inclusion maps of pairs and $\partial_* \colon H_q(X, A) \to H_{q-1}(A)$ is the connecting homomorphism.

2. Let L be a subcomplex of K. Prove that
   (a) if there is a retraction $r \colon |K| \to |L|$, then $H_q(K) \approx H_q(L) \oplus H_q(K, L)$.
   (b) if the identity map $i \colon K \to K$ is homotopic to a map $f \colon |K| \to |L|$, then
   $$H_q(L) \approx H_q(K) \oplus H_{q+1}(K, L).$$
   (c) if the inclusion $j \colon |L| \to |K|$ is homotopic to some constant map, then
   $$H_q(K, L) \approx \tilde{H}_q(K) \oplus \tilde{H}_{q-1}(L)$$

3. Give an example to show that there are polyhedral pairs (X, A) and (Y, B) such that for all q, we have $H_q(X) \approx H_q(Y)$, $H_q(A) \approx H_q(B)$ but $H_q(X, A) \not\approx H_q(Y, B)$.

4. Suppose we have a sequence of finitely generated abelian groups $G_0, G_1, \ldots$, where G_0 is free and nontrivial. Then show that there is a simplicial complex K such that $H_q(K) \approx G_i \; \forall \, i \geq 0$.

Chapter 5

Covering Projections

5.1 Introduction

Let $\exp\colon \mathbb{R} \to \mathbb{S}^1$ denote the normalized exponential map defined by $\exp(t) = e^{2\pi it}$, $t \in \mathbb{R}$. Recall (Chapter 2, Theorem 2.6.5, 2.6.6) that this map has the Path Lifting Property as well as the Homotopy Lifting Property. We now wish to generalize these results to a wider class of continuous maps $p\colon \tilde{X} \to X$, called **covering projections**. The theory of covering projections is of great importance not only in topology, but also in other branches of mathematics like complex analysis, differential geometry and Lie groups, etc. The concept of fundamental group, discussed in Chapter 2, is intimately related to covering projections. Among other things, we will see that the problem of "lifting" a continuous map $f\colon A \to X$ to a continuous map $\tilde{f}\colon A \to \tilde{X}$, where $p\colon \tilde{X} \to X$ is a covering projection, has a complete solution in terms of the fundamental groups of the spaces involved and the induced homomorphisms among them.

Let $p\colon \tilde{X} \to X$ be a continuous map. Then an open set U of X is said to be **evenly covered** by p if $p^{-1}(U)$ is a disjoint union of open subsets of $\tilde{X}$ each of which is mapped homeomorphically onto U by the map p. In such a case, we say that U is an **admissible open** set of X. It is possible that $p^{-1}(U)$ can be expressed as a disjoint union of open sets satisfying the above condition in more than one way. However, if U is a path-connected open set, then there is only one way of expressing $p^{-1}(U)$ as a disjoint union of open sets of $\tilde{X}$ such that each open set is homeomorphic to U under the map p. These open sets are simply path-components of $p^{-1}(U)$ – these are path-components because each one is an open as well as a closed subset of $p^{-1}(U)$, and is path-connected as each one is homeomorphic to the path-connected space U. It is for this reason that, in this chapter, **all of our spaces are assumed to be path-connected and locally path-connected unless stated otherwise.**

We also mention a topological fact to be kept in mind (see Proposition 1.6.10). Every connected and locally path-connected space is path connected. Thus, for locally path-connected spaces, connectedness is equivalent to path-connectedness and so all of our spaces are assumed to be locally path-connected and connected (rather than path-connected).

The following is our basic concept:

Definition 5.1.1. *A continuous map $p\colon \tilde{X} \to X$ is said to be a* **covering projection** *or* **a covering map** *if each point $x \in X$ has a path connected open neighbourhood U which is evenly covered by p, i.e., U is* **an admissible neighbourhood** *of $x \in X$. The space $\tilde{X}$ is called a* **covering space** *of X and X itself is called the* **base space** *of the covering projection p.*

Evidently, if an open set U is admissible, then any path connected open subset of U is also admissible.

Example 5.1.2. Let $\exp\colon \mathbb{R} \to \mathbb{S}^1$ be the exponential map defined by $\exp(t) = e^{2\pi i t}$. Then exp is a covering projection. To see this, recall how the above map wraps the real line $\mathbb{R}$ onto the unit circle $\mathbb{S}^1$. See Fig. 5.1.

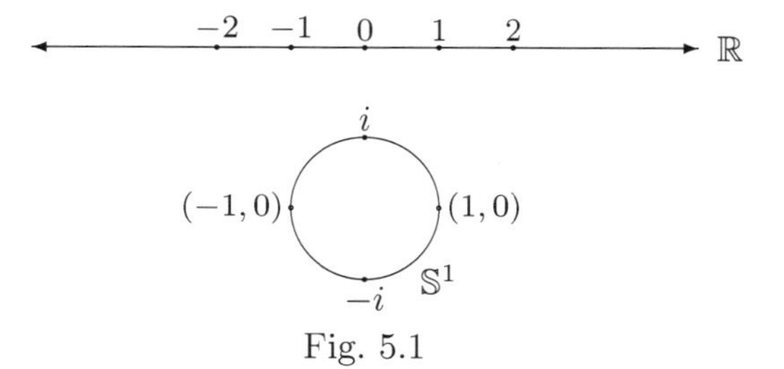

Fig. 5.1

Note that the complex number $1 = 1 + i \cdot 0$ and $-1 = -1 + i \cdot 0$ lie on $\mathbb{S}^1$. Consider the open set $U_1 = \mathbb{S}^1 - \{-1\}$ and observe that

$$(\exp)^{-1}(U_1) = \bigcup_{-\infty}^{\infty} (n - \frac{1}{2}, n + \frac{1}{2})$$

is a disjoint union of open intervals, each of which is homeomorphic to U_1 under the map exp. Similarly, one can see that the open set $U_2 = \mathbb{S}^1 - \{1\}$ is also evenly covered by exp map. Thus, each point of $\mathbb{S}^1$ has an admissible open neighbourhood in $\mathbb{S}^1$. As a matter of fact, any proper connected arc of $\mathbb{S}^1$ is easily seen to be evenly covered by the given exponential map. The same argument shows that the map $p\colon \mathbb{R} \to \mathbb{S}^1$ defined by $p(t) = e^{i\alpha t}$, where $\alpha \in \mathbb{R}$ is a fixed nonzero number, is also a covering projection; the case $\alpha = 1$ is the usual exponential map and will be frequently used later on.

Example 5.1.3. Any homeomorphism $p\colon \tilde{X} \to X$ is trivially a covering projection. More generally, let Y be a discrete space and X be any space. If $p\colon X \times Y \to X$ denotes the projection map on the first factor then p is a covering projection. In fact, here the space X itself is evenly covered because

$$p^{-1}(X) = X \times Y = \bigcup_{y \in Y} X \times \{y\}$$

is a disjoint union and X is homeomorphic to $X \times \{y\}$ for each $y \in Y$ under the map p.

Example 5.1.4. For any nonzero integer n, let us define a map $q_n\colon \mathbb{S}^1 \to \mathbb{S}^1$ by

$$q_n(z) = z^n,$$

where $z \in \mathbb{S}^1$ is the unit complex number and z^n is the nth power of z. Then q_n is a covering projection. As an illustration, consider the case when $n = 4$. If we divide the domain

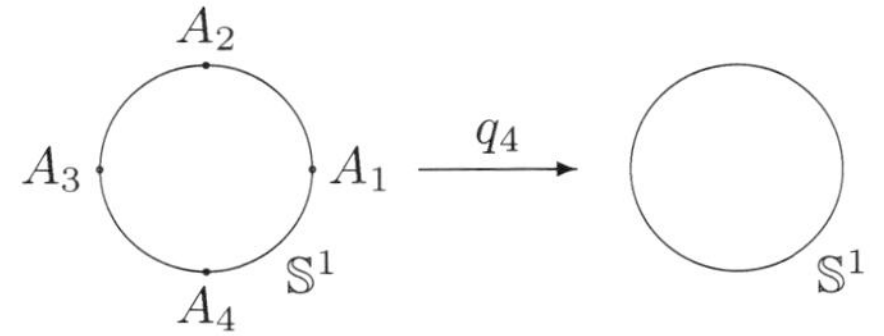

Fig. 5.2: q_4 wraps first circle four times onto second

circle into four equal arcs (Fig. 5.2), then q_4 maps each arc $[A_i, A_{i+1}]$, $i = 1, 2, 3, 4$ and $A_5 = A_1$ onto the whole of range circle $\mathbb{S}^1$, so that the open arc (A_i, A_{i+1}) gets mapped to $\mathbb{S}^1 - \{A_1\}$ homeomorphically under the map q_4. This means that each point of the range circle $\mathbb{S}^1 - \{A_1\}$ has an open neighbourhood which is evenly covered by q_4. If we consider mid-points B_i of arcs $[A_i, A_{i+1}]$, $B_5 = B_1$, then each open arc (B_i, B_{i+1}) gets mapped homeomorphically under q_4 onto the open set $\mathbb{S}^1 - \{A_3\}$ of the range circle. Thus it follows that each point of the range circle has an open neighbourhood which is evenly covered by $q_n\colon \mathbb{S}^1 \to \mathbb{S}^1$. We may also observe here that q_n wraps the domain $\mathbb{S}^1$ onto the range circle exactly n times either clockwise or anticlockwise depending on whether n is negative or positive.

Example 5.1.5. If $p\colon \tilde{X} \to X$ and $q\colon \tilde{Y} \to Y$ are covering projections, then the product map $p \times q\colon \tilde{X} \times \tilde{Y} \to X \times Y$ defined by $(p \times q)(x, y) = (p(x), q(y))$ is also a covering projection. To see this, let $(x_0, y_0) \in X \times Y$. Choose an open neighbourhood U of x_0 and an open neighbourhood V of y_0 which are evenly covered by p and q, respectively. Then $U \times V$ is a neighbourhood of (x_0, y_0) in $X \times Y$ which is easily verified to be evenly covered by $p \times q$.

In particular we get: the map $(\exp, \exp)\colon \mathbb{R} \times \mathbb{R} \to \mathbb{S}^1 \times \mathbb{S}^1$ is a covering projection. In fact, for every $n \in \mathbb{N}$, the product map $(\exp)^n\colon \mathbb{R}^n \to T^n$, where $T^n = \prod_1^n \mathbb{S}^1$ is the n-dimensional torus, is a covering projection.

Example 5.1.6. Let $X = [0,4]$, $Y = \mathbb{S}^1$ and $p = \exp|_X$, i.e., p is the restriction of the usual exponential map to the space X. Then p wraps X four times on Y, but is not a covering projection. To see this, note that each point of $\mathbb{S}^1 - \{(1,0)\} = U$ has an open neighbourhood, viz., U itself, which is evenly covered but the point (1,0) does not have any small neighbourhood in $\mathbb{S}^1$ which is evenly covered by p, since the inverse image of any small open neighbourhood N around (1,0) in $\mathbb{S}^1$ will have five pieces out of which three will be homeomorphic to N and the remaining two, being semi-closed intervals, will not be homeomorphic to N under any map and so, in particular, under p.

Example 5.1.7. Note that if $\tilde{X} \to X$ is any covering map, then it follows from definitions that there must be an open set $\tilde{U}$ of $\tilde{X}$ and an open set U of X which are homeomorphic. Now, let $I = [0,1]$ denote the closed unit interval and $p\colon I \times I \to I$ be the projection on one of the factors. Then p cannot be a covering projection because no open set of $I \times I$ can be homeomorphic to any open set of I; this last fact is clear since if that was the case then an open rectangle in $\mathbb{R}^2$ must be homeomorphic to an open interval of $\mathbb{R}$, a contradiction. The same argument proves that **no** map from $I \times I$ to I can be a covering projection.

Example 5.1.8. Let $\mathbb{C}$ denote the complex plane. We claim that the exponential map $\exp\colon \mathbb{C} \to \mathbb{C} - \{0\}$ defined by $\exp(z) = e^z = \sum_0^\infty \frac{z^n}{n!}$ is a covering map. To show this, we begin by observing that the map $k\colon \mathbb{C} \to \mathbb{R} \times \mathbb{R}$ defined by $k(x+iy) = (x,y)$ is a homeomorphism. Next, we see that any element w of $\mathbb{C} - \{0\}$ can be uniquely written (in terms of polar coordinates) as $w = re^{i\theta}$, $0 < r$ and $0 \leq \theta < 2\pi$. Since the map $\exp\colon \mathbb{R} \to \mathbb{R}_+ = \{x \in \mathbb{R} \mid x > 0\}$ is a homeomorphism, we can write $w = e^x \cdot e^{iy}$, where x is uniquely determined by r and $y = \theta$. Hence it is clear that the map $h\colon \mathbb{C} - \{0\} \to \mathbb{R}_+ \times \mathbb{S}^1$ defined by $h(w) = (e^x, e^{iy})$ is a homeomorphism. Since $p\colon \mathbb{R} \to \mathbb{R}_+$ defined by $p(x) = e^x$ is a homeomorphism and $q\colon \mathbb{R} \to \mathbb{S}^1$ defined by $q(y) = e^{iy}$ is a covering projection, it follows from Example 5.1.5 that the product map $p \times q\colon \mathbb{R} \times \mathbb{R} \to \mathbb{R}_+ \times \mathbb{S}^1$ is a covering projection. Now, one can easily see that $e^z = h^{-1} \circ (p \times q) \circ k$ and hence is a covering projection from $\mathbb{C}$ to $\mathbb{C} - \{0\}$. We must point out here that in this example, we have used the obvious fact that if $\tilde{X} \to X$ is a covering map and $\tilde{X}$ is replaced by a homeomorphic space $\tilde{Y}$ or X is replaced by a homeomorphic space Y, then $\tilde{Y}$ is a covering space of Y through a suitable map.

Example 5.1.9. Let us observe that for any integer $n \neq 0$, the map $f\colon \mathbb{R}_+ \to \mathbb{R}_+$ defined by $f(x) = x^n$ is a homeomorphism and so is a covering projection. Also, the map $q\colon \mathbb{S}^1 \to \mathbb{S}^1$ defined by $q(e^{iy}) = e^{iny}$ is a covering projection. It follows, therefore, that $p \times q\colon \mathbb{R}_+ \times \mathbb{S}^1 \to \mathbb{R}_+ \times \mathbb{S}^1$ is a covering projection. Then composing with the homeomorphisms h and h^{-1} defined in the previous example, we conclude that the map $g\colon \mathbb{C} - \{0\} \to \mathbb{C} - \{0\}$ defined by $g(z) = z^n$ is a covering projection for any $n \neq 0$. Also, one can easily see that for any $n \neq 0, 1$, the map g is not a covering map from $\mathbb{C}$ to $\mathbb{C}$.

Example 5.1.10 (Figure Eight)**.** Consider the following subset X of the product space $\mathbb{S}^1 \times \mathbb{S}^1$:

$$X = \{(z, w) \in \mathbb{S}^1 \times \mathbb{S}^1 \mid z = (1,0) \text{ or } w = (1,0)\}.$$

Here, $\mathbb{S}^1$ should be thought of as the standard unit circle in the complex plane. The space X can be thought of as two circles which are touching each other at one point - this is the space usually known as the **figure of eight**. Now consider the following subspace $\tilde{X}$ of the plane $\mathbb{R}^2$ (see Fig. 5.3);

$$\tilde{X} = \{(x, y) \in \mathbb{R}^2 \mid x \text{ or } y \text{ is an integer}\}.$$

Define $p\colon \tilde{X} \to X$ by

$$p(x, y) = (e^{2\pi i x}, e^{2\pi i y}).$$

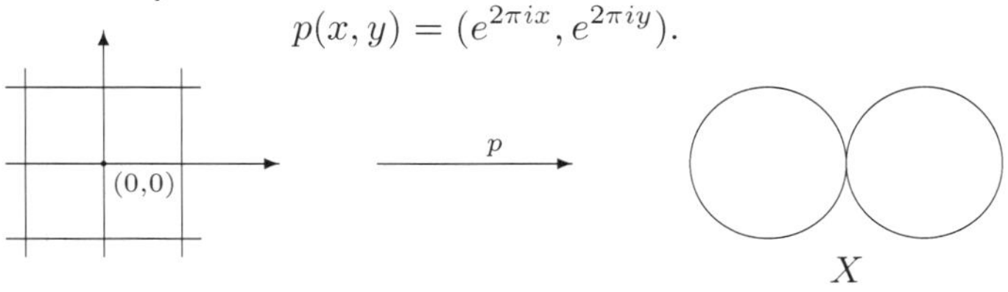

Fig. 5.3

For each integer x, p is the usual covering map; similarly, for each integer y, the map p is the same exponential covering map. Hence it is not difficult to see that p is indeed a covering projection.

A mapping $p\colon E \to B$ from a space E to a space B is said to be a **local homeomorphism** if each point $e \in E$ has an open neighbourhood U such that $p(U)$ is open in B, and the map $p|_U : U \to p(U)$ is a homeomorphism. If $p\colon \tilde{X} \to X$ is a covering projection, then clearly it is a local homeomorphism: For any $\tilde{x} \in \tilde{X}$, let U be an admissible open neighbourhood of $p(\tilde{x})$ and $\tilde{U}$ be the component of $p^{-1}(U)$ which contains $\tilde{x}$ and is homeomorphic to U. Then $\tilde{U}$ is the required open neighbourhood of $\tilde{x}$. The converse, however, need not be true. Let us consider the following

Example 5.1.11. Suppose $p\colon \mathbb{R} \to \mathbb{S}^1$ is the exponential map $p(x) = e^{2\pi i x}$. Consider the restriction $q = p|_{(0,8)}$ of p to the open interval (0,8). Then q is a local homeomorphism but cannot be a covering map because the point $(1,0) \in \mathbb{S}^1$ does not have an admissible neighbourhood in $\mathbb{S}^1$ - there will be nine components of the inverse image of a small open neighbourhood of (0, 1) in which seven will be homeomorphic to the small neighbourhood whereas the other two will not be homeomorphic to that neighbourhood under the map p.

5.2 Properties of Covering Projections

In this section, we will prove some of the basic properties of a covering map $p\colon \tilde{X} \to X$. We have already seen that a covering map is a local homeomorphism. It follows from the definition itself that a covering map is an onto map. Recall that a space X is locally path connected if and only if each path component of an open subset of X is open.

Proposition 5.2.1. *A covering projection $p\colon \tilde{X} \to X$ is always an open map.*

Proof. Let V be any open subset of $\tilde{X}$. We must show that $p(V)$ is open in X. For that, let $x \in p(V)$ and choose $\tilde{x} \in V$ such that $p(\tilde{x}) = x$. Let U be an admissible neighbourhood of x and W be the path component of $p^{-1}(U)$ which contains $\tilde{x}$. Since $\tilde{X}$ is locally path connected, W is open in $p^{-1}(U)$ and hence in $\tilde{X}$. Since p maps W homeomorphically onto U, p must map the open set $W \cap V$ onto an open subset $p(W \cap V)$ of U and hence of X. Clearly, $x \in p(V \cap W) \subseteq p(V)$, i.e., $p(V)$ is a neighbourhood of each of its points and hence is open in X. ■

Note that the above result combines two facts together, viz., a covering projection is a local homeomorphism and a local homeomorphism is an open mapping.

Proposition 5.2.2. *The cardinality of all the fibres of a covering projection $p\colon \tilde{X} \to X$ are same.*

Proof. By the definition of a covering projection, each point $p \in X$ has an open neighbourhood U such that $p^{-1}(U)$ is homeomorphic to the disjoint union $\bigcup\{U_\alpha\colon \alpha \in I\}$ such that U_α is homeomorphic to U under the map p for each α. The space $\bigcup_{\alpha \in I} U_\alpha$ is clearly homeomorphic to the product $U \times I$, where I has the discrete topology. Hence fibres over U have the same cardinality as that of the indexing set I. In other words, the cardinality of fibres over the base space X is locally constant. Since X is connected, the cardinality of the fibres must be constant over X, which proves the result. ■

Because a covering map $p\colon \tilde{X} \to X$ is always an open onto continuous map, we conclude that the topology of the base space X is the quotient topology induced from $\tilde{X}$ by the map p.

We have seen (Example 5.1.11) that the restriction of a covering map $p\colon \tilde{X} \to X$ to an open set U of $\tilde{X}$ need not be a covering map even though it may be onto X. However, the following is a useful result:

Proposition 5.2.3. *Let $p\colon \tilde{X} \to X$ be a covering map and A be a connected and locally path connected subset of X. If $\tilde{A}$ is a path component of $p^{-1}(A)$, then the restriction map $p|_{\tilde{A}} : \tilde{A} \to A$ is a covering projection.*

Proof. For each $a \in A$, choose an admissible open neighbourhood U of a (in X) and let $\{U_i\}$ be the path components of $p^{-1}(U)$ such that each U_i is homeomorphic to U under p. Then, clearly, $U \cap A$ is evenly covered by $U_i \cap p^{-1}(A)$. Since A is locally path connected, we can find a path-connected open (in A) neighbourhood V of a contained in $U \cap A$. Then, clearly, V is evenly covered by

p. If any path component V_i of $p^{-1}(V)$ intersects $\tilde{A}$, then it must be contained in $\tilde{A}$ as $\tilde{A}$ is path connected. Therefore, $p|_{\tilde{A}} : \tilde{A} \to A$ is a covering projection. ∎

Path Lifting Property

The following is an important topological problem (see Chapter 1, last section) which one usually comes across in various situations, particularly in topology.

Lifting Problem. Let $p\colon \tilde{X} \to X$ be any continuous, onto map, not necessarily a covering projection. Suppose a continuous map $f\colon A \to X$ is given. Can we find a continuous map $\tilde{f}\colon A \to \tilde{X}$ such that the following triangle is commutative?

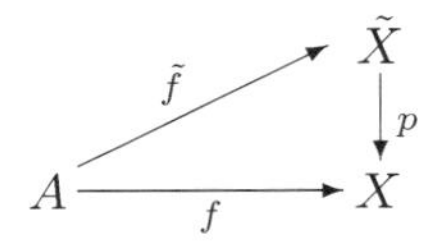

In other words, the problem is to find a map $\tilde{f}\colon A \to \tilde{X}$ such that $p \circ \tilde{f} = f$. This question may or may not have an affirmative answer; furthermore, if there is one $\tilde{f}$ satisfying the above condition, then there could be more than one such map; any such $\tilde{f}$ is called a **lift** of f. We will be able to answer the above question "quite satisfactorily" when $p\colon \tilde{X} \to X$ is a covering projection. However, first we will consider the case when the map f is a path in X. But, before that, we can easily decide the uniqueness of lifts. We have

Theorem 5.2.4. *Let $p\colon \tilde{X} \to X$ be a covering map and A be any space (not necessarily connected or locally path connected). If $f,\ g\colon A \to \tilde{X}$ are any two continuous maps such that $p \circ f = p \circ g$, then the set of points of A, where f and g agree, is a clopen subset of A.*

Proof. Let $S = \{a \in A \mid f(a) = g(a)\}$. To prove that S is open we will show that S contains a neighbourhood of each of its points. Let $x_0 \in S$. Choose an admissible neighbourhood U of $pf(x_0) = pg(x_0)$, and let V be the path component of $p^{-1}(U)$ containing $f(x_0) = g(x_0)$. Then V is open in $\tilde{X}$ and hence $f^{-1}(V) \cap g^{-1}(V)$ is an open set of A containing x_0. We will show that $f^{-1}(V) \cap g^{-1}(V)$ is a subset of S. Let $x \in f^{-1}(V) \cap g^{-1}(V)$. This means $f(x)$ and $g(x)$ are in V. Since p restricted to V is a homeomorphism, and we are given that $pf(x) = pg(x)$, it follows that $f(x) = g(x)$ which proves the required assertion.

Next, we prove that S is a closed subset of A. Suppose b is a limit point of S which is not in S. Then $f(b) \neq g(b)$. Since $pf(b) = pg(b)$, we can find an admissible neighbourhood W of $pf(b)$ such that the points $f(b)$ and $g(b)$ must be in distinct path components V_1 and V_2 of $p^{-1}(W)$. Now, $b \in f^{-1}(V_1) \cap g^{-1}(V_2)$ which is an open set containing b and since b is a limit point of S, there should be point $a \in S$ such that $a \in f^{-1}(V_1) \cap g^{-1}(V_2)$. But this a contradiction, since $V_1 \cap V_2 = \phi$ and $a \in f^{-1}(V_1) \cap g^{-1}(V_2)$ implies $f(a) = g(a) \in V_1 \cap V_2$. Hence all limit points of S must be in S and so S is closed. ■

The following gives the uniqueness result for the lifts of a map $f\colon A \to X$ when A is a connected space.

Corollary 5.2.5. *Let $p\colon \tilde{X} \to X$ be a covering projection. Suppose $g_1, g_2\colon A \to \tilde{X}$ are lifts of the same map $f\colon A \to X$ such that g_1 and g_2 agree at one point of A. Then $g_1 = g_2$.*

Proof. By the given condition, $pg_1 = f = pg_2$. Hence, by the above theorem, the set $S = \{a \in A \mid g_1(a) = g_2(a)\}$ is a clopen subset of A. The fact that g_1, g_2 agree at some point of A implies that S is nonempty. Since A is connected, we conclude that $S = A$, i.e., $g_1 = g_2$. ■

The following theorem shows that any path in the base space X of a covering projection $p\colon \tilde{X} \to X$ can always be lifted to a path in $\tilde{X}$, and is unique if the initial point is already prescribed.

Theorem 5.2.6. (The Path Lifting Property). *Let $p\colon \tilde{X} \to X$ be a covering projection and $\alpha\colon I \to X$ be a path beginning at some point $x_0 \in X$. Then given any point $\tilde{x}_0$ over x_0 (i.e., $p(\tilde{x}_0) = x_0$), there exists a unique path $\tilde{\alpha}$ in $\tilde{X}$ beginning at $\tilde{x}_0$ which lifts α, i.e., $p \circ \tilde{\alpha} = \alpha$.*

Proof. Suppose $\alpha(I)$ is contained in some admissible open neighbourhood U of X. Choose a path component V of $p^{-1}(U)$. Then $\tilde{\alpha} = (p|_V)^{-1} \circ \alpha\colon I \to \tilde{X}$ is a path in $\tilde{X}$ which has the property that

$$p \circ \tilde{\alpha} = p \circ (p|_V)^{-1} \circ \alpha = \alpha.$$

Hence the path α has been lifted to a path $\tilde{\alpha}$. If $\alpha(I)$ is not contained in a single admissible neighbourhood of X, choose an open cover $\{V_i\}$ of X consisting of admissible open neighbourhoods. Let $\epsilon > 0$ be a Lebesgue number for the open covering $\{\alpha^{-1}(V_i)\}$ of I. We can now select a partition

$$0 = t_0 < t_1 < \ldots < t_n = 1$$

of I so that $|t_{i+1} - t_i| \leq \epsilon$, for all $i = 0, 1, \ldots, n-1$. Then, obviously $\alpha[t_i, t_{i+1}]$ is contained in some admissible open neighbourhood, say, V_{i+1}, $i = 0, 1, \ldots, n-1$. Let V_1 denote the path component of $p^{-1}(V_1)$ to which the desired point $\tilde{x}_0$ belongs and then define $\tilde{\alpha}(t) = (p|_{V_1})^{-1} \circ \alpha(t)$, for all $t \in [t_0, t_1]$. Then $\tilde{\alpha}$ is a path defined on $[t_0, t_1]$ in $\tilde{X}$ which lifts $\alpha|_{[t_0,t_1]}$. Now, inductively, suppose $k > 1$ and we have defined a lift $\tilde{\alpha}\colon [t_0, t_k] \to \tilde{X}$ of $\alpha|_{[t_0,t_k]}$. Then $\alpha[t_k, t_{k+1}] \subset U_{k+1}$. Now, we choose a path component V_{k+1} of $p^{-1}(U_{k+1})$ which contains $\tilde{\alpha}(t_k)$.

Since $\alpha|_{V_{k+1}}$ is a homeomorphism, define the extension $\bar{\alpha}$ of $\tilde{\alpha}$ by putting $\bar{\alpha}(t) = (p|_{V_{k+1}})^{-1} \circ \alpha(t)$, $t \in [t_k, t_{k+1}]$. Then $\bar{\alpha}$ is a path in $\tilde{X}$ defined on $[t_0, t_{k+1}]$ which lifts $\alpha|_{[t_0,t_{k+1}]}$. Taking $k = n-1$, we have a path in $\tilde{X}$ defined on I which lifts the given path α in X and has the initial point $\tilde{x}_0$.

The uniqueness of $\tilde{\alpha}$, which lifts α and starts at the prescribed point $\tilde{x}_0$ in X, follows at once from the preceding corollary because I is connected and, by assumption, any two lifts of α agree at the point $a = 0 \in I$. ■

Homotopy Lifting Property

Having proved the path lifting property of a covering projection $p\colon \tilde{X} \to X$, now we ask whether or not any homotopy $F\colon A \times I \to X$ between two maps $f, g\colon A \to X$ can be lifted to a homotopy $\tilde{F}\colon A \times I \to \tilde{X}$. Since any $f\colon A \to X$ is homotopic to itself, lifting of this homotopy $F\colon A \times I \to X$ will automatically imply that any map f can always be lifted to a map $\tilde{f}\colon A \to X$, which evidently need not be true. Hence the general answer to the above question cannot be in affirmative. However, the answer is "yes" provided a lift of f or any map which is homotopic to f under F is already given. This result is known as the homotopy lifting property of the covering projection $p\colon \tilde{X} \to X$ and the map $\tilde{F}$ is also known as covering homotopy. We have

Theorem 5.2.7. (The Homotopy Lifting Property). *Let $p\colon \tilde{X} \to X$ be a covering projection and A be a compact space. Suppose $\tilde{f}\colon A \to \tilde{X}$ is a continuous map and $F\colon A \times I \to X$ is a homotopy starting from $p \circ \tilde{f}$. Then there is a homotopy $\tilde{F}\colon A \times I \to \tilde{X}$ which starts from $\tilde{f}$ and lifts F. Furthermore, if F is a homotopy relative to a subset S of A, then so is $\tilde{F}$.*

Proof. Since F is continuous and A is compact, the subset $F(A \times I)$ of X must be compact. Therefore, we can cover $F(A \times I)$ by finite number of admissible open sets, say, $U_1, \ldots, U_r$. Then the open sets $F^{-1}(U_1), \ldots, F^{-1}(U_r)$ cover $A \times I$. For each point $(a, t) \in A \times I$, find a rectangular neighbourhood $V \times (-\epsilon, \epsilon)$ which is contained in $F^{-1}(U_i)$ for some i. Then, using compactness of $A \times I$, one can find a finite number of open connected sets $\{V_\alpha\}_1^n$ of A covering A and a finite partition $0 = t_0 < t_1 < \ldots < t_n = 1$ of I such that for every α, $1 \le \alpha \le n$, and every i, $0 \le i \le k-1$, $F(V_\alpha \times [t_i, t_{i+1}]) \subset V_j$ for some j, $1 \le j \le r$.

First, we construct a continuous map $G\colon A \times [0, t_1] \to \tilde{X}$ which agrees with f on $A \times \{0\}$ and lifts $F|_{A \times [0,t_1]}$. For every α, we know that $F(V_\alpha \times [0, t_1]) \subset U_j$ for some j, and also we know that $f(V_\alpha \times \{0\})$, being connected, must be contained in one of the components, say, W_α of $p^{-1}(U_j)$. Hence we can define a map

$$G_1^\alpha = (p|_{W_\alpha})^{-1} \circ F\colon V_\alpha \times [0, t_1] \to \tilde{X}.$$

Then, obviously, G_1^α is continuous, $G_1^\alpha(a, 0) = f(a)$ for all $a \in A$ and $p \circ G_1^\alpha = F|_{V_\alpha \times [0,t_1]}$. Next, we prove that for every pair α, β for which $V_\alpha \cap V_\beta \neq 0$,

G_1^α, G_1^β will match on the common domain $(V_\alpha \cap V_\beta) \times [0, t_1]$. For this, note that on $(V_\alpha \cap V_\beta) \times \{0\}$, $G_1^\alpha = f = G_1^\beta$. Moreover, for any t, $0 \leq t \leq t_1$ and any $a \in V_\alpha \cap V_\beta$, the point (a, t) can be joined to $(a, 0)$ by an obvious path lying in $(V_\alpha \cap V_\beta) \times [0, t_1]$. This implies that G_1^α will map $(V_\alpha \cap V_\beta) \times [0, t_1]$ into W_β also and so into $W_\alpha \cap W_\beta$. Similarly G_1^β will also map it into $W_\alpha \cap W_\beta$. Thus, on $(V_\alpha \cap V_\beta) \times [0, t_1]$,

$$G_1^\alpha = (p|_{W_\alpha \cap W_\beta})^{-1} \circ F = G_1^\beta.$$

Since $\{V_\alpha \times [0, t_1]\}$ is an open cover of $A \times [0, t_1]$, all the G_1^α combine to give a continuous map $G_1 \colon A \times [0, t_1] \to \tilde{X}$ such that $G_1(a, 0) = f(a)$ for all $a \in A$ and $p \circ G_1 = F|_{A \times [0, t_1]}$. Now, using the map $G_i|_{A \times \{t_1\}}$ in place of f, we can construct a continuous map $G_2 \colon A \times [t_1, t_2] \to \tilde{X}$ such that $G_2(a, t_1) = G_1(a, t_1)$ for all $a \in A$ and $p \circ G_2 = F|_{A \times [t_1, t_2]}$. Proceeding inductively, we can now construct continuous functions $G_i \colon A \times [t_{i-1}, t_i] \to \tilde{X}$, $i = 2, 3, \ldots, k-1$, such that for all $a \in A$, $G_i(a, t_i) = G_{i+1}(a, t_i)$ and $p \circ G_i = F|_{A \times [t_{i-1}, t_i]}$. Applying continuity lemma for functions G_i on closed sets $A \times [t_{i-1}, t_i]$, we have a continuous function $\tilde{F} \colon A \times I \to \tilde{X}$ such that $p \circ \tilde{F} = F$ and $\tilde{F}(a, 0) = f(a)\ \forall\ a \in A$. This completes the proof of the existence of $\tilde{F}$ satisfying the stated conditions.

Finally, it is clear from the construction of G_i's that if for any $a \in A$, $F(a, t)$ remains constant for all $t \in I$, then so are G_i's. Therefore, if the homotopy F is relative to a subset S of A, then so is the homotopy $\tilde{F}$. ■

We may remark that the above result remains true even when A is not compact, but the proof would be different.

5.3 Applications of Homotopy Lifting Theorem

We will now present some applications of the path lifting property and homotopy lifting property of covering projections. First of all, we must point out that the path lifting property is a consequence of the homotopy lifting property as shown below.

Corollary 5.3.1. *If $p \colon \tilde{X} \to X$ is a covering projection and $\tilde{x}_0 \in \tilde{X}, x_0 \in X$ are such that $p(\tilde{x}_0) = x_0$, then for any path $w \colon I \to X$ such that $w(0) = x_0$, there exists a unique path $\tilde{w} \colon I \to \tilde{X}$ such that $\tilde{w}(0) = \tilde{x}_0$ and $p \circ \tilde{w} = w$.*

Proof. The uniqueness follows from the fact that I is connected. Let $P = \{a\}$ be a singleton space and consider the map $f \colon P \to X$ defined by $f(a) = x_0$. The path w defines a homotopy $F \colon P \times I \to X$ on P by putting $F(a, t) = w(t)$. By the Homotopy Lifting Property, we have a map $\tilde{F} \colon P \times I \to \tilde{X}$ such that $\tilde{F}(a, 0) = \tilde{x}_0$ and $p \circ \tilde{F} = F$. Then $\tilde{w} \colon I \to \tilde{X}$ defined by $\tilde{w}(t) = \tilde{F}(a, t)$, $t \in I$, is a path in $\tilde{X}$ which starts from $\tilde{x}_0$ and has the property that for $t \in I$

$$p\tilde{w}(t) = p\tilde{F}(a, t) = F(a, t) = w(t). \qquad ■$$

Recall that any continuous map $f\colon X \to Y$ induces a group homomorphism $f_*\colon \pi_1(X,x) \to \pi_1(Y, f(x))$ for all $x \in X$, defined by $f_*[\alpha] = [f \circ \alpha]$. If $p\colon \tilde{X} \to X$ is a covering map, then we know that it is onto. But the induced map, in general, need not be onto. However, it follows from the following theorem that $p_*\colon \pi_1(\tilde{X}, \tilde{x}_0) \to \pi_1(X, x_0)$ is always an injective map.

Theorem 5.3.2. (The Monodromy Theorem). *Let $p\colon \tilde{X} \to X$ be a covering projection. Suppose $\tilde{\alpha}$, $\tilde{\beta}$ are two paths in $\tilde{X}$ having the same initial point $\tilde{x}_0$. Then $\tilde{\alpha}$, $\tilde{\beta}$ are equivalent if and only if $p\tilde{\alpha}$, $p\tilde{\beta}$ are equivalent paths in X.*

Remark. By definition, two paths in the space are equivalent means they have the same initial and terminal points. Thus, if $p\tilde{\alpha}$ and $p\tilde{\beta}$ are equivalent, then by above theorem $\tilde{\alpha}$, $\tilde{\beta}$ are equivalent. We are given that they have the same initial point. Hence it is a consequence (not a hypothesis) of the above theorem that $\tilde{\alpha}$, $\tilde{\beta}$ must have the same terminal point.

Proof. If $F\colon I \times I \to \tilde{X}$ is a continuous map such that

$$F(s,0) = \tilde{\alpha}(s),\ F(s,1) = \tilde{\beta}(s), s \in I$$

$$F(0,t) = \tilde{x}_0,\ F(1,t) = \tilde{x}_1, t \in I,$$

then $p \circ F\colon I \times I \to X$ is a continuous map such that for all $s \in I$, $p \circ F(s,0) = p\tilde{\alpha}(s)$, $p \circ F(s,1) = p\tilde{\beta}(s)$ and for all $t \in I$, $p \circ F(0,t) = p(\tilde{x}_0)$, $p \circ F(1,t) = p(\tilde{x}_1)$. This proves the direct part.

Conversely, suppose $F\colon I \times I \to X$ is a homotopy between $p\tilde{\alpha}$ and $p\tilde{\beta}$ relative to $\{0,1\} \subset I$. By the Homotopy Lifting Property, there is a unique homotopy $\tilde{F} : I \times I \to \tilde{X}$ such that $\tilde{F}(0,0) = \tilde{x}_0$ and $p\tilde{F} = F$. Restricting $\tilde{F}$ on $(s,0)$, $s \in I$, we find a path $s \to \tilde{F}(s,0)$ starting from $\tilde{x}_0$ and lifting $p\tilde{\alpha}$. Clearly, $s \rightsquigarrow \tilde{\alpha}(s)$ is also a path in $\tilde{X}$ starting from $\tilde{x}_0$ and lifting $p\tilde{\alpha}$. Hence, by the uniqueness property of the covering paths, $\tilde{F}(s,0) = \tilde{\alpha}(s)$ for all $s \in I$. By a similar argument, we have $\tilde{F}(s,1) = \tilde{\beta}(s)$. Restricting $\tilde{F}$ on $(0,t)$, $t \in I$, we get a path $t \rightsquigarrow \tilde{F}(0,t)$ which projects under p to the constant path at x_0; on the other hand, the constant path $t \rightsquigarrow \tilde{x}_0$ in $\tilde{X}$ also starts at $\tilde{x}_0$ and projects under p to the constant path $t \rightsquigarrow x_0$ in X. Hence, again by the uniqueness theorem, $t \rightsquigarrow \tilde{F}(0,t)$ must be a constant path based at $\tilde{x}_0$. Similarly, the path $t \rightsquigarrow \tilde{F}(1,t)$ must be a constant path based at some point $\tilde{x}_1$ over x_1. This proves that $\tilde{F}$ is a homotopy between $\tilde{\alpha}$ and $\tilde{\beta}$ relative to $\{0,1\}$. Therefore, $\tilde{\alpha}$ is equivalent to $\tilde{\beta}$. ■

Now we can state the remark made earlier as a useful

Corollary 5.3.3. *If $p\colon \tilde{X} \to X$ is a covering projection and $p(\tilde{x}) = x$, then the induced homomorphism $p_* : \pi_1(\tilde{X}, \tilde{x}) \to \pi_1(X, x)$ is a one-one map.*

As yet another application of the path lifting property of a covering projection $p\colon \tilde{X} \to X$, we prove the next result regarding the cardinality of fibres

$p^{-1}\{x\}$, $x \in X$ once again; the earlier proof used only set topology whereas the next one uses the concept of algebraic topology with a little gain, viz., a specific bijection can be given between any two fibres.

Proposition 5.3.4. *Let $p\colon \tilde{X} \to X$ be a covering map. Then for any two points a, $b \in X$, the fibres $p^{-1}\{a\}$ and $p^{-1}\{b\}$ have the same cardinality.*

Proof. It is enough to produce a bijection from $p^{-1}\{a\}$ to $p^{-1}\{b\}$: Fix a path w in X with $w(0) = a$ and $w(1) = b$, and let $x \in p^{-1}\{a\}$. Using the unique path lifting property, we can find a path $\tilde{w}_x$ in $\tilde{X}$ such that $\tilde{w}_x(0) = x$ and $p\tilde{w}_x = w$. Then $\tilde{w}_x(1) \in p^{-1}\{b\}$ is uniquely determined by x. We define a map $f\colon p^{-1}\{a\} \to p^{-1}\{b\}$ by putting $f(x) = \tilde{w}_x(1)$. We assert that f is one-one because $f(x_1) = f(x_2)$ implies $\tilde{w}_{x_1}(1) = \tilde{w}_{x_2}(1)$, which means $\tilde{w}_{x_1}$ and $\tilde{w}_{x_2}$ both agree at a point of I and hence, by uniqueness of the lifted paths, $\tilde{w}_{x_1} = \tilde{w}_{x_2}$. Therefore, $x_1 = \tilde{w}_{x_1}(0) = \tilde{w}_{x_2}(0) = x_2$. Next, we claim that f is onto. For that, let $z \in p^{-1}\{b\}$ and consider the path w^{-1} in X defined by $w^{-1}(t) = w(1-t)$. Let $\tilde{w}_z$ be the path in $\tilde{X}$ such that $\tilde{w}_z(0) = z$ and $p\tilde{w}_z = w^{-1}$. Then

$$\begin{aligned} p(\tilde{w}_z)^{-1}(t) &= p(\tilde{w}_z)(1-t) \\ &= w^{-1}(1-t) \\ &= w(t), \end{aligned}$$

and $(\tilde{w}_z)^{-1}(0) = \tilde{w}_z(1)$. Hence $(\tilde{w}_z)^{-1}$ is a lift of w starting at $\tilde{w}_z(1)$. Putting $\tilde{w}_z(1) = x$, we see that $x \in p^{-1}\{a\}$, since both $(\tilde{w}_z)^{-1}$ and $\tilde{w}_x$ start at x and both lift w, $w_x = (\tilde{w}_z)^{-1}$. It follows that $\tilde{w}_x(1) = (\tilde{w}_z)^{-1}(1) = \tilde{w}_z(0) = z$. ■

From the above result, we see that for any covering projection $p\colon \tilde{X} \to X$, all the fibres have same cardinality. This common cardinal number is called the **number of sheets** of $\tilde{X}$ over X. If this number is a natural number n, then we also express this fact by saying that $\tilde{X}$ is a n**-sheeted** covering of X. For example, the covering projection $p\colon \mathbb{S}^1 \to \mathbb{S}^1$ defined by $p(z) = z^4$ is a four-sheeted covering of $\mathbb{S}^1$ by itself. Similarly, the quotient map $q\colon \mathbb{S}^2 \to \mathbb{P}^2$ identifying the antipodal points of $\mathbb{S}^2$ is a two-sheeted covering of the projective plane $\mathbb{P}^2$.

The figure 5.4 depicts a three-sheeted covering of the **figure-eight**. Note that the figure-eight space is a closed curve with one self-intersection, where as the three-sheeted covering is a closed curve with three self-intersections.

The Right Action of the Fundamental Group

The number of sheets of a covering projection has another interesting interpretation given by the following result:

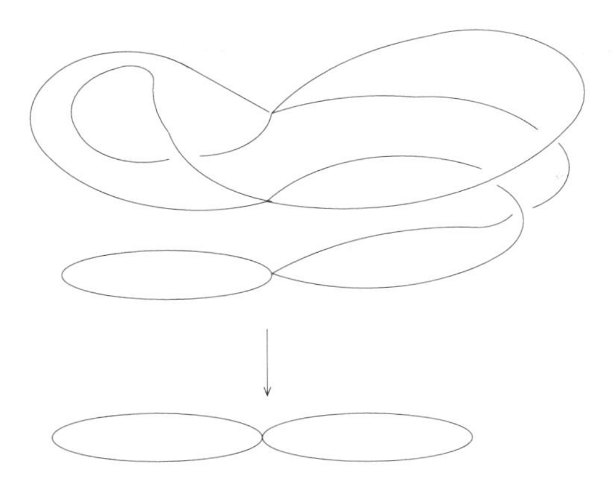

Fig. 5.4: The three-sheeted covering of the figure-eight.

Proposition 5.3.5. *Let $p\colon \tilde{X} \to X$ be a covering projection. Then for any $x \in X$ and for any $\tilde{x} \in p^{-1}\{x\}$, there is a canonical one-one correspondence between the set $p^{-1}\{x\}$ (number of sheets) and the set $\pi_1(X,x)/p_*(\pi_1(\tilde{X},\tilde{x}))$ of all cosets of the subgroup $p_*(\pi_1(\tilde{X},\tilde{x}))$ in the group $\pi_1(X,x)$.*

Proof. We will define a natural right action of the group $\pi_1(X,x)$ on the set $p^{-1}\{x\}$ and show that the action is transitive with $p_*(\pi_1(\tilde{X},\tilde{x}))$ as the isotropy group at the point $\tilde{x} \in p^{-1}\{x\}$. Take an element $[\alpha] \in \pi_1(X,x)$ and point $\tilde{x} \in p^{-1}\{x\}$. By the unique path lifting property, there is unique path $\tilde{\alpha}_x$ in $\tilde{X}$ which starts from $\tilde{x}$ and lifts α. We define

$$\tilde{x} \cdot [\alpha] = \tilde{\alpha}_x(1). \tag{5.3.1}$$

We must show that the point $\tilde{\alpha}_x(1)$ depends on the path class $[\alpha]$, not on any representative of $[\alpha]$. For that, choose another loop β in X based at x so that $[\alpha] = [\beta]$. Let $\tilde{\beta}_x$ be a path in $\tilde{X}$ which starts at $\tilde{x}$ and lifts β. Since the paths α, β are homotopic and $\tilde{\alpha}_x, \tilde{\beta}_x$, respectively, are paths in $\tilde{X}$ lifting them and each starting at the same point $\tilde{x}$, we can apply the Monodromy Theorem to conclude that $\tilde{\alpha}_x$ and $\tilde{\beta}_x$ must have the same terminal points, i.e, $\tilde{\alpha}_x(1) = \tilde{\beta}_x(1)$.

To prove that (5.3.1) defines an action, let ϵ_x be the constant loop in X based at x. Then, for any $x \in X$, the constant loop $\tilde{\epsilon}_x$ will be a path in $\tilde{X}$ which lifts ϵ_x and starts at $\tilde{x}$. It follows that

$$\tilde{x} \cdot [\epsilon_x] = \tilde{\epsilon}_x(1) = \tilde{x}. \tag{1}$$

Next, let $[\alpha], [\beta] \in \pi_1(X,x)$. For any $\tilde{x} \in p^{-1}\{x\}$, first choose a path $\tilde{\alpha}_x$ in $\tilde{X}$ starting at $\tilde{x}$ and lifting α and then choose a path $\tilde{\beta}_x$ in $\tilde{X}$ which lifts β and starts at $\tilde{\alpha}_x(1)$. Then, by definition,

$$(\tilde{x} \cdot [\alpha]) \cdot [\beta] = \tilde{\alpha}_x(1) \cdot [\beta] = \tilde{\beta}_x(1).$$

On the other hand, $\tilde{\alpha}_x * \tilde{\beta}_x$ is a path in $\tilde{X}$ starting at $\tilde{x}$ and lifting $\alpha * \beta$. This means

$$\tilde{x} \cdot ([\alpha] \cdot [\beta]) = \tilde{x} \cdot [\alpha * \beta] = (\tilde{\alpha}_x * \tilde{\beta}_x)(1) = \tilde{\beta}_x(1).$$

Consequently, for each $\tilde{x} \in p^{-1}\{x\}$,

$$\tilde{x} \cdot ([\alpha] \cdot [\beta]) = (\tilde{x} \cdot [\alpha]) \cdot [\beta]. \tag{2}$$

Thus, in view of (1) and (2), the group $\pi_1(X, x)$ acts on right on the set $p^{-1}\{x\}$.

Finally, we prove that the above action is also transitive. For that, let $\tilde{x}_0, \tilde{x}_1$ be any two points of $p^{-1}\{x\}$. Since $\tilde{X}$ is path connected, there is a path $\tilde{w}$ in $\tilde{X}$ such that $\tilde{w}(0) = \tilde{x}_0$ and $\tilde{w}(1) = \tilde{x}_1$. Evidently, then $p\tilde{w} = w$ is a loop in X based at x, i.e, $[w] \in \pi_1(X, x)$, and by definition, $x_0 \cdot [w] = \tilde{w}(1) = x_1$. The isotropy group of the above action at any $\tilde{x}$ is obviously $p_*(\pi_1(\tilde{X}, \tilde{x}))$ because $\tilde{x} \cdot [\alpha] = \tilde{x}$ if and only if there is a closed loop $\tilde{\alpha}$ in $\tilde{X}$ which lifts α and starts at the point $\tilde{x}$, i.e., $[\alpha] = p_*([\tilde{\alpha}])$, which means $[\alpha] \in p_*(\pi_1(\tilde{X}, \tilde{x}))$. Hence it follows from the property of the isotropy groups and the orbits that $p^{-1}\{x\}$ will be in 1-1 correspondence with the set $\pi_1(X, x)/p_*(\pi(\tilde{X}, \tilde{x}))$. ■

Corollary 5.3.6. *If $p_* \colon \pi_1(\tilde{X}, \tilde{x}) \to \pi_1(X, x)$ is onto, then the map $p \colon \tilde{X} \to X$ is a homeomorphism.*

5.4 Lifting of an arbitrary map

Now, we return to the "lifting problem" that we mentioned earlier: Given a covering projection $p \colon \tilde{X} \to X$ and a continuous map $f \colon A \to X$, can we find a continuous map $\tilde{f} \colon A \to \tilde{X}$ such that $p \circ \tilde{f} = f$? If f is a path or a homotopy between paths, we have seen that the answer to the lifting problem is in affirmative. However, for any continuous map f, the answer is negative. We have the following:

Example 5.4.1. Consider the exponential map $\exp \colon \mathbb{R} \to \mathbb{S}^1$ defined by $\exp(t) = e^{2\pi i t}$. We assert that the identity map $I_d \colon \mathbb{S}^1 \to \mathbb{S}^1$ cannot be lifted to a continuous map $\phi \colon \mathbb{S}^1 \to \mathbb{R}$ making the triangle

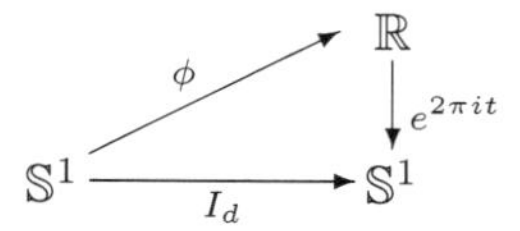

commutative. If there was such a ϕ, then ϕ must be one-one. But then, $\mathbb{S}^1$ being compact will mean ϕ is an embedding of $\mathbb{S}^1$ into $\mathbb{R}$. This is impossible because any compact connected subset of $\mathbb{R}$ must be a closed interval and that can never be homeomorphic to $\mathbb{S}^1$.

In view of the above example, the lifting problem for covering projections now amounts to asking for a necessary and sufficient condition so that any map $f: A \to X$ can be lifted. It is very interesting to remark that the above question can be beautifully answered in terms of fundamental groups of spaces A, X, $\tilde{X}$ and the homomorphisms amongst them induced by f and p. This is one of the several fine illustrations of the methods of algebraic topology where the topological problem of lifting a continuous map is faithfully translated into an easier algebraic problem regarding fundamental groups. This result is stated in the next theorem.

Theorem 5.4.2. (Lifting Theorem). *Let $p: \tilde{X} \to X$ be a covering projection and $f: A \to X$ be any continuous map. Then, given any three points $a_0 \in A$, $x_0 \in X$ and $\tilde{x}_0 \in \tilde{X}$ such that $f(a_0) = x_0$ and $p(\tilde{x}_0) = x_0$, there exists a continuous map $\tilde{f}: A \to \tilde{X}$ satisfying $\tilde{f}(a_0) = \tilde{x}_0$ such that $p \circ \tilde{f} = f$ if and only if $f_*(\pi_1(A, a_0)) \subset p_*(\pi_1(\tilde{X}, \tilde{x}_0))$.*

Proof. First, suppose there exists a continuous map $\tilde{f}: A \to \tilde{X}$ satisfying the given conditions. Then the following diagram of continuous maps is commutative:

$$\begin{array}{ccc} & & (\tilde{X}, \tilde{x}_0) \\ & \overset{\tilde{f}}{\nearrow} & \downarrow p \\ (A, a_0) & \xrightarrow[f]{} & (X, x_0) \end{array}$$

Passing on to the fundamental groups and induced homomorphisms, we see that the following diagram must be commutative:

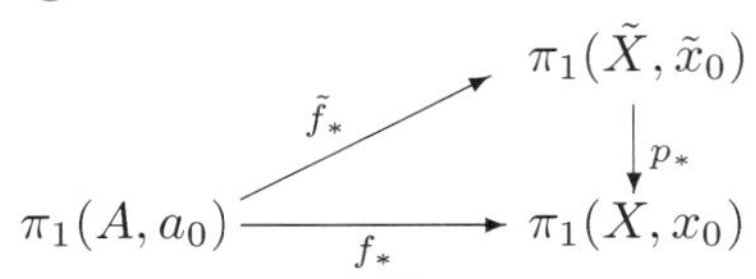

This shows that $f_*(\pi_1(A, a_0)) = p_*(\tilde{f}_*(\pi_1(A, a_0)) \subset p_*(\pi_1(\tilde{X}, \tilde{x}_0))$, and proves the necessity of the given condition.

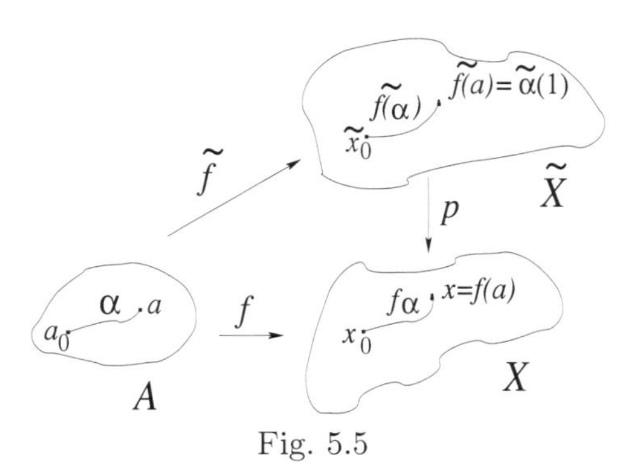

Fig. 5.5

Conversely, suppose the given algebraic condition is satisfied. We define $\tilde{f}: A \to \tilde{X}$ as follows: Let $a \in A$ and choose a path α in A such that $\alpha(0) = a_0$ and $\alpha(1) = a$. Then $f\alpha$ is a path in X starting at x_0. By the Path Lifting Property, there exists a path $\tilde{\alpha}$ in $\tilde{X}$ starting at $\tilde{x}_0$ which projects to $f\alpha$. We define $\tilde{f}(a) = \tilde{\alpha}(1)$. In order that $\tilde{f}$ is well-defined, we must show that $\tilde{f}(a)$ is independent of the path α that joins a_0 to a. So, let β be any other path in A joining a_0 to a and suppose $\tilde{\beta}$ is a path in $\tilde{X}$ starting from $\tilde{x}_0$ such that $p\tilde{\beta} = f\beta$. It suffices to show that $\tilde{\alpha}(1) = \tilde{\beta}(1)$. Now, observe that $\alpha * \beta^{-1}$ is a loop in A based at a_0. Hence, by the given algebraic condition, there exists a loop δ in $\tilde{X}$ based at $\tilde{x}_0$ such that the loop $p\delta$ is homotopic to $f(\alpha * \beta^{-1}) = f\alpha * (f\beta)^{-1}$. But then by the Homotopy Lifting Theorem, there is a loop w based at $\tilde{x}_0$ such that $pw = f\alpha * (f\beta)^{-1}$. We now break the loop w into the product $w_1 * w_2^{-1}$ of paths w_1, w_2 in $\tilde{X}$ by defining $w_1(t) = w(t/2)$, $w_2(t) = w(1+t/2)$, $t \in I$. Then for each $t \in I$, $pw_1(t) = pw(t/2) = (f\alpha) * (f\beta)^{-1}(t/2) = f\alpha(t)$ and, similarly, $pw_2(t) = f\beta(t)$. Since w_1 and $\tilde{\alpha}$ agree at $\tilde{x}_0$ and $pw_1 = f\alpha = p\tilde{\alpha}$, we can apply the Monodromy Theorem to conclude that $w_1(1) = \tilde{\alpha}(1)$ and $w_2(1) = \tilde{\beta}(1)$. However, since $w_1(1) = w_2(1) = w(\frac{1}{2})$, we find that $\tilde{\alpha}(1) = \tilde{\beta}(1)$. It is clear from the definition that $p\tilde{f} = f$.

Finally, we prove that $\tilde{f}$ is continuous. Let $a \in A$ and U be an open neighbourhood of $\tilde{f}(a)$. Choose an admissible neighbourhood U' of $p\tilde{f}(a) = f(a)$ such that $U' \subset p(U)$. Let W be the path component of $p^{-1}(U')$ which contains the point $\tilde{f}(a)$, and let U'' be an admissible neighbourhood of $f(a)$ such that $U'' \subset p(U \cap W)$. Then the path component of $p^{-1}(U'')$ containing $\tilde{f}(a)$ must be contained in U. Since f is continuous and A is locally path connected, we can find a path connected neighbourhood V of a such that $f(V) \subset U'$. Then one can readily verify that $\tilde{f}(V) \subset U$. ■

5.5 Covering Homomorphisms

We have already proved several important results concerning a covering projection $p\colon \tilde{X} \to X$. In this section we want to consider the following question: Given a space X, how many distinct covering projections of X one can find? In other words, in how many distinct ways a space X can be covered? Before answering this question, we must specify clearly as to what is meant by distinct covering projections of a given space X. As an example, suppose $p\colon \tilde{X} \to \mathbb{R}$ is a covering projection. Then, by the Monodromy Theorem, the induced homomorphism $p_*\colon \pi_1(\tilde{X}) \to \pi_1(\mathbb{R})$ must be one-one and since $\pi_1(\mathbb{R}) = 0$, it must also be onto. Therefore, by Corollary 5.3.6, $p\colon \tilde{X} \to \mathbb{R}$ must be a homeomorphism, i.e, any covering projection p of $\mathbb{R}$ must be a homeomorphism. The same argument shows that if X is simply connected, then any covering map of X must be a homeomorphism. We emphasize that all of our covering spaces here are assumed to be connected as well as locally path connected.

As another example, the map $q_n\colon \mathbb{S}^1 \to \mathbb{S}^1$ defined by $q_n(z) = z^n$ is a n-sheeted covering projection of the circle $\mathbb{S}^1$ by itself. Hence, by any plausible definition, the coverings q_n and q_m must be treated as distinct if $m \neq n$. It must be mentioned here that the space $\tilde{X}$ and the map $p\colon \tilde{X} \to X$ both are essential parts of a covering projection – the same space $\tilde{X}$ can cover X in so many distinct ways. These considerations lead us to the following well-established definition:

Definition 5.5.1. *Let $p_1 \colon \tilde{X}_1 \to X$, $p_2 \colon \tilde{X}_2 \to X$ be two covering projections of a space X. A continuous map $h \colon \tilde{X}_1 \to \tilde{X}_2$ is said to be a* **homomorphism** *of covering spaces if the diagram*

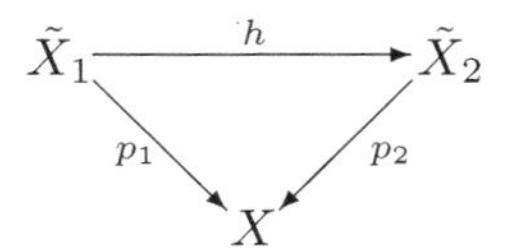

is commutative. If, in addition, the map h is a homeomorphism, then h is called an **isomorphism** *of the two covering projections.*

If we consider the class of all covering projections of a space X, then one can easily verify that (i) the identity map $I_{\tilde{X}} \colon \tilde{X} \to \tilde{X}$ is a homomorphism, and (ii) the composite of two homomorphisms is again a homomorphism. It follows that covering projections $p \colon \tilde{X} \to X$ as objects and homomorphism amongst them as morphisms, form a category, and the isomorphisms in this category are just the isomorphisms of covering projections as defined above.

The covering spaces $p_1 \colon \tilde{X}_1 \to X, p_2 \colon \tilde{X}_2 \to X$ will be called **comparable** if one can find a homomorphism from either one of them to the other; they will be considered **same** if there exists an isomorphism from one onto the other. An isomorphism h from a covering space $p \colon \tilde{X} \to X$ to itself is called an **automorphism or a deck transformation**. In this section, we will answer the following two questions very nicely in terms of the fundamental groups and the induced homomorphisms amongst them: When are two covering projections of a space X (a) comparable? (b) same?

Group of Deck Transformations

Let $p \colon \tilde{X} \to X$ be a covering projection. Then the set of all automorphisms of this covering projection is closed under the obvious operation of composition of maps. Furthermore, since the identity map of $\tilde{X}$ to itself is an automorphism and the inverse of an automorphism is again an automorphism, the set of all automorphisms of a covering projection forms a group with respect to the composition operation. This group, usually denoted by $A(\tilde{X}, p)$, is called the **automorphism group** of the covering projection $p \colon \tilde{X} \to X$. The automorphisms are also known as the **covering transformations** or **deck transformations** of the covering projection p.

Let $p_1 \colon \tilde{X}_1 \to X$, $p_2 \colon \tilde{X}_2 \to X$ be covering projections. Suppose g, $h \colon \tilde{X}_1 \to \tilde{X}_2$ are any two homomorphisms. Then one can think of each of g

and h as lifts of the map $p_1: \tilde{X}_1 \to X$ with respect to the covering projection $p_2: \tilde{X}_2 \to X$.

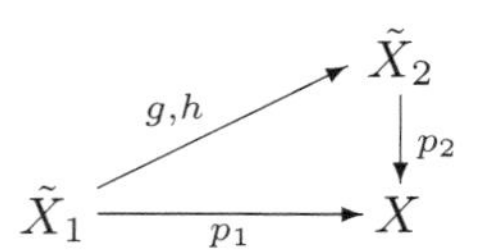

Consequently, if g and h both agree on a single point of $\tilde{X}_1$, then $g = h$ because $\tilde{X}_1$ is connected (Corollary 5.2.1). This yields the next

Proposition 5.5.2. *Let g, $h: \tilde{X}_1 \to \tilde{X}_2$ be two homomorphisms of the covering projections. If $g(\tilde{x}_1) = h(\tilde{x}_1)$ for some $\tilde{x}_1 \in \tilde{X}_1$, then $g = h$.*

We already know that the group Homeo(Y) of all homeomorphisms of a space Y always acts on the set Y by the action defined by

$$h \cdot y = h(y),$$

where $h \in$ Homeo(Y) and $y \in Y$. Since the group $A(\tilde{X}, p)$ of all automorphisms of a covering projection $p: \tilde{X} \to X$ is a subgroup of the group of all homeomorphisms of $\tilde{X}$, $A(\tilde{X}, p)$ also acts on the space $\tilde{X}$ by the above action. In this terminology, the following result is a special case of the foregoing proposition:

Corollary 5.5.3. *The group $G = A(\tilde{X}, p)$ of all automorphisms of the covering projections $p: \tilde{X} \to X$ acts on $\tilde{X}$ freely, i.e., if $g \in G$ and $g(\tilde{x}) = \tilde{x}$ for any $\tilde{x} \in \tilde{X}$, then $g = I_{\tilde{X}}$.*

Since an automorphism $h: \tilde{X} \to \tilde{X}$ is a homeomorphism making the diagram

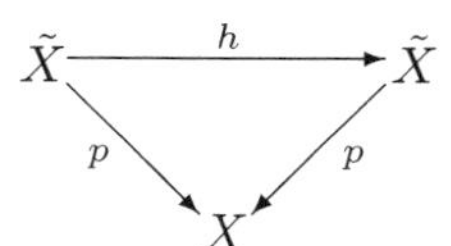

commutative, h must map each point of the fibre $p^{-1}\{x\}$ to a point of itself, i.e, h simply permutes the points of each fibre. We now wish to answer the following question: Given any two points $\tilde{x}_1$, $\tilde{x}_2$ of $p^{-1}\{x\}$, can we find an automorphism $h: \tilde{X} \to \tilde{X}$ of the covering projection $p: \tilde{X} \to X$ such that $h(\tilde{x}_1) = \tilde{x}_2$? In the terminology of group actions, the question amounts to asking whether the group $G = A(\tilde{X}, p)$ acts on $\tilde{X}$ so that it is transitive on each fibre $p^{-1}\{x\}$, $x \in X$. We will shortly see that the question has a very simple answer in algebraic terms involving fundamental groups. Applying the Lifting Theorem to the map $p_1: \tilde{X}_1 \to X$ with respect to the covering projection $p_2: \tilde{X}_2 \to X$, we have

Proposition 5.5.4. *Let $p_1: \tilde{X}_1 \to X$, $p_2: \tilde{X}_2 \to X$ be covering projections over X and $\tilde{x}_1 \in \tilde{X}_1$, $\tilde{x}_2 \in \tilde{X}_2$ be such that $p_1(\tilde{x}_1) = p_2(\tilde{x}_2)$. Then there exists a homomorphism $h: \tilde{X}_1 \to \tilde{X}_2$ such that $h(\tilde{x}_1) = \tilde{x}_2$ if and only if*

$$p_{1*}(\pi_1(\tilde{X}_1, \tilde{x}_1)) \subseteq p_{2*}(\pi_1(\tilde{X}_2, \tilde{x}_2)).$$

If $h\colon \tilde{X}_1 \to \tilde{X}_2$ above is a homeomorphism then not only $p_1\colon \tilde{X}_1 \to X$ has a lift h but the map $p_2\colon \tilde{X}_2 \to X$ is also lifted by $h^{-1}\colon \tilde{X}_2 \to \tilde{X}_1$ for the covering projection $p_1\colon \tilde{X}_1 \to X$. Hence, in the above proposition, the reverse inclusion also holds, i.e, $p_{2*}(\pi_1(\tilde{X}_2, \tilde{x}_2)) \subseteq p_{1*}(\pi_1(\tilde{X}_1, \tilde{x}_1))$. Conversely, suppose the equality holds. Then, again, by the Lifting Theorem, there are homomorphisms $h_1\colon \tilde{X}_1 \to \tilde{X}_2$ and $h_2\colon \tilde{X}_2 \to \tilde{X}_1$ such that $p_2 h_1 = p_1$, $p_1 h_2 = p_2$, $h_1(\tilde{x}_1) = \tilde{x}_2$ and $h_2(\tilde{x}_2) = \tilde{x}_1$. This means $h_2 \circ h_1\colon \tilde{X}_1 \to \tilde{X}_1$ is a homomorphism which maps $\tilde{x}_1$ to itself, and so by Proposition 5.5.2, $h_2 \circ h_1 = I_{\tilde{X}_1}$. Similarly, we have $h_1 \circ h_2 = I_{\tilde{X}_2}$. Thus, each of h_1 and h_2 is an isomorphism and we have

Proposition 5.5.5. *Let $p_1\colon \tilde{X}_1 \to X$, $p_2\colon \tilde{X}_2 \to X$ be covering projections and let $\tilde{x}_1 \in \tilde{X}_1$, $\tilde{x}_2 \in \tilde{X}_2$ be points such that $p_1(\tilde{x}_1) = p_2(\tilde{x}_2)$. Then there exists an isomorphism $h\colon \tilde{X}_1 \to \tilde{X}_2$ satisfying $h(\tilde{x}_1) = \tilde{x}_2$ if and only if $p_{1*}(\pi_1(\tilde{X}_1, \tilde{x}_1)) = p_{2*}(\pi_1(\tilde{X}_2, \tilde{x}_2))$.*

This immediately yields

Corollary 5.5.6. *Let $p\colon \tilde{X} \to X$ be a covering projection and $\tilde{x}_1$, $\tilde{x}_2 \in p^{-1}\{x\}$. Then there exists an automorphism h of $\tilde{X}$ such that $h(\tilde{x}_1) = \tilde{x}_2$ if and only if $p_*(\pi_1(\tilde{X}, \tilde{x}_1)) = p_*(\pi_1(\tilde{X}, \tilde{x}_2))$. In particular, the group $G = A(\tilde{X}, p)$ acts transitively on the fibre $p^{-1}\{x\}$ if $p_*(\pi_1(\tilde{X}, \tilde{x}_1)) = p_*(\pi_1(\tilde{X}, \tilde{x}_2))$ for every $\tilde{x}_1, \tilde{x}_2 \in p^{-1}\{x\}$.* ■

Next, we note the following:

Proposition 5.5.7. *Let $p\colon \tilde{X} \to X$ be a covering projection. Then the set $\{p_*(\pi_1(\tilde{X}, \tilde{x})) \mid \tilde{x} \in p^{-1}\{x\}\}$ forms a complete conjugacy class of subgroups of the group $\pi_1(X, x)$.*

Proof. Let $\tilde{x}_1$, $\tilde{x}_2 \in p^{-1}\{x\}$. Since $\tilde{X}$ is path connected, there is a path, say, w in $\tilde{X}$ joining $\tilde{x}_1$ to $\tilde{x}_2$. Then the map $f_w\colon \pi_1(\tilde{X}, \tilde{x}_1) \to \pi_1(\tilde{X}, \tilde{x}_2)$ defined by $f_w[\alpha] = [w^{-1} * \alpha * w]$ is clearly an isomorphism. Note that pw is a closed path at x. Since the diagram

$$\begin{array}{ccc} \pi_1(\tilde{X}_1, \tilde{x}_1) & \xrightarrow{\ f_w\ } & \pi_1(\tilde{X}_1, \tilde{x}_2) \\ {\scriptstyle p_*}\downarrow & & \downarrow{\scriptstyle p_*} \\ \pi_1(X, x) & \xrightarrow[\ f_{pw}\ ]{} & \pi_1(X, x) \end{array}$$

is easily seen to be commutative, we find that

$$p_*(\pi_1(\tilde{X}, \tilde{x}_2)) = p_* f_w(\pi_1(\tilde{X}, \tilde{x}_1)) = f_{pw}(p_*(\pi_1(\tilde{X}, \tilde{x}_1))).$$

Because f_{pw} is simply a conjugation in $\pi_1(X,x)$ by the element $[pw]$, we conclude that $p_*(\pi_1(X,\tilde{x}_1))$ and $p_*(\pi_1(X,\tilde{x}_2))$ are conjugate in $\pi_1(X,x)$ for any two points $\tilde{x}_1$, $\tilde{x}_2$ in $p^{-1}\{x\}$. On the other hand, suppose H is a subgroup of $\pi_1(X,x)$ which is conjugate to $p_*(\pi_1(\tilde{X},\tilde{x}_0))$ for some $\tilde{x}_0 \in p^{-1}\{x\}$. Then there is a closed path δ in X based at x such that

$$H = [\delta]^{-1} p_*(\pi_1(\tilde{X},\tilde{x}_0))[\delta].$$

Let $\tilde{\delta}$ be a lift of δ starting from $\tilde{x}_0$ and put $\tilde{x}_1 = \tilde{\delta}(1)$. Then we assert that $H = p_*(\pi_1(\tilde{X},\tilde{x}_1))$. To see this, let β be any loop in $\tilde{X}$ based at $\tilde{x}_0$. Then $\delta^{-1} * p\beta * \delta$ is a loop in X based at x. Since $p(\tilde{\delta}^{-1}\beta\tilde{\delta}) = \delta^{-1}p\beta\delta$ and $\tilde{\delta}^{-1}\beta\tilde{\delta}$ is a loop based at $\tilde{x}_1$ in $\tilde{X}$,

$$[\delta^{-1}]p_*(\pi_1(\tilde{X},\tilde{x}_0))[\delta] \subseteq p_*(\pi_1(\tilde{X},\tilde{x}_1)).$$

Conversely, if α is a loop in $\tilde{X}$ based at the point $\tilde{x}_1$, then $p\alpha \sim \delta^{-1}\delta p\alpha\delta^{-1}\delta \sim \delta^{-1}(p\tilde{\delta} * p\alpha * (p\tilde{\delta})^{-1})\delta \sim \delta^{-1}p(\tilde{\delta} * \alpha * \delta^{-1})\delta$. Because $\tilde{\delta} * \alpha * \tilde{\delta}^{-1}$ is a loop in $\tilde{X}$ based at $\tilde{x}_0$, $p_*(\pi_1(\tilde{X},\tilde{x}_1)) \subseteq [\delta^{-1}]p_*(\pi_1(\tilde{X},\tilde{x}))[\delta]$. ■

Corollary 5.5.8. *The group* $G = A(\tilde{X},p)$ *acts transitively on a fibre* $p^{-1}\{x\}$ *if and only if the subgroup* $p_*(\pi_1(\tilde{X},\tilde{x}))$ *is normal in* $\pi_1(X,x)$ *for every* $\tilde{x} \in p^{-1}\{x\}$.

Now, we can characterize as to when two covering projections of X are comparable. We have

Proposition 5.5.9. *Let* $p_1\colon \tilde{X}_1 \to X$ *and* $p_2\colon \tilde{X}_2 \to X$ *be two covering projections of the space* X. *Then there exists a homomorphism* $h\colon \tilde{X}_1 \to \tilde{X}_2$ *of covering projections if and only if for any* $\tilde{x}_1 \in p_1^{-1}\{x\}$ *and* $\tilde{x}_2 \in p_2^{-1}\{x\}$, $p_{1*}(\pi_1(\tilde{X}_1,\tilde{x}_1))$ *is conjugate to a subgroup of* $p_{2*}(\pi_1(\tilde{X}_2,\tilde{x}_2))$.

Proof. First, assume that there is a homomorphism $h\colon \tilde{X}_1 \to \tilde{X}_2$. Then, by the Lifting Theorem, we must have

$$p_{1*}(\pi_1(\tilde{X}_1,\tilde{x}_1)) \subseteq p_{2*}(\pi_1(\tilde{X}_2,h(\tilde{x}_1))).$$

By the previous proposition, since the subgroup $p_{2*}(\pi_1(\tilde{X}_2,h(\tilde{x}_1)))$ is conjugate to $p_{2*}(\pi_1(\tilde{X}_2,\tilde{x}_2))$ in $\pi_1(X,x)$, it follows that $p_{1*}(\pi_1(\tilde{X}_1,\tilde{x}_1))$ is conjugate to a subgroup $p_{2*}(\pi_1(\tilde{X}_2;\tilde{x}_2))$.

Conversely, if

$$[\alpha]^{-1}p_{1*}(\pi_1(\tilde{X},\tilde{x}_1))[\alpha] \subseteq p_{2*}(\pi_1(\tilde{X}_2,\tilde{x}_2))$$

then

$$\begin{aligned} p_{1*}(\pi_1(\tilde{X}_1,\tilde{x}_1)) &\subseteq [\alpha]p_{2*}(\pi_1(\tilde{X}_2,\tilde{x}_2))[\alpha]^{-1} \\ &= p_{2*}(\pi_1(\tilde{X}_2,\tilde{x}_3)) \end{aligned}$$

for some $\tilde{x}_3 \in p_2^{-1}\{x\}$. Hence, again by the Lifting Theorem, there exists a homomorphism $h\colon \tilde{X}_1 \to \tilde{X}_2$ (such that $h(\tilde{x}_1) = \tilde{x}_3$). ■

The question as to when are two covering spaces of a space "same" is completely answered in terms of the fundamental groups by the following result:

Theorem 5.5.10. *Let $p_1\colon \tilde{X}_1 \to X$, and $p_2\colon \tilde{X}_2 \to X$ be two covering projections. Then $\tilde{X}_1$ is isomorphic to $\tilde{X}_2$ if and only if for any $\tilde{x}_1 \in \tilde{X}_1$, $\tilde{x}_2 \in \tilde{X}_2$ such that $p_1(\tilde{x}_1) = p_2(\tilde{x}_2)$, $p_{1*}(\pi_1(\tilde{X}_1, \tilde{x}_1))$ is conjugate to $p_{2*}(\pi_1(\tilde{X}_2, \tilde{x}_2))$.*

Proof. First, suppose $h\colon \tilde{X}_1 \to \tilde{X}_2$ is an isomorphism of the two covering projections. Let $\tilde{x} \in \tilde{X}_1$ be such that $p_1(\tilde{x}) = p_1(\tilde{x}_1)$. Then, by the Lifting Theorem,

$$p_{1*}(\pi_1(\tilde{X}_1, \tilde{x})) \subseteq p_{2*}(\pi_1(\tilde{X}_2, h(\tilde{x}))).$$

Since $h^{-1}\colon \tilde{X}_2 \to \tilde{X}_1$ is also an isomorphism and maps $h(\tilde{x})$ to $\tilde{x}$, we have, by the same theorem,

$$p_{2*}(\pi_1(\tilde{X}_2, h(\tilde{x}))) \subseteq p_{1*}(\pi_1(\tilde{X}_1, \tilde{x})).$$

Hence the above inclusions reduce to equality. Since, by Proposition 5.5.7, $p_{1*}(\pi_1(\tilde{X}_1, \tilde{x}_1))$ is conjugate to $p_{1*}(\pi_1(\tilde{X}_1, h(\tilde{x})))$ and $p_{2*}(\pi_1(\tilde{X}_2, h(\tilde{x})))$ is conjugate to $p_{2*}(\pi_1(\tilde{X}_2, h(\tilde{x})))$, and thus we have proved the direct part.

Conversely, suppose the given groups are conjugate. Then, by Proposition 5.5.9, there is a homomorphism $h_1\colon \tilde{X}_1 \to \tilde{X}_2$ such that $h_1(\tilde{x}_1) = \tilde{x}_2$. Similarly, there is a homomorphism $h_2\colon \tilde{X}_2 \to \tilde{X}_1$ such that $h_2(\tilde{x}_2) = \tilde{x}_1$. It follows that $h_2 h_1\colon \tilde{X}_1 \to \tilde{X}_1$ is a homomorphism fixing the element $\tilde{x}_1 \in \tilde{X}_1$, which, in turn, says that $h_2 h_1 = I_{\tilde{X}_1}$. Likewise $h_1 h_2 = I_{\tilde{X}_2}$. Hence each of h_1 and h_2 is an isomorphism. ■

Remark. Note that for any covering projection $p\colon \tilde{X} \to X$, the subgroups $\{p_*(\pi_1(\tilde{X}, \tilde{x})) \mid \tilde{x} \in p^{-1}\{x\}\}$ form a conjugacy class of subgroups of $\pi_1(X, x)$. The above theorem says that a conjugacy class of subgroups of $\pi_1(X, x)$ completely determines the covering projection p up to isomorphism. Here, we are not saying that given any conjugacy class of subgroups of $\pi_1(X, x)$, there always exists a covering projection $p\colon \tilde{X} \to X$ which yields that conjugacy class, though it is true. This is an important question which we have not considered so far (see exercises at the end).

5.6 Universal Covering Space – Applications

The concept of universal covering space of a given space X is extremely useful in several situations. Before we give its definition and examples, let us prove the following result.

Proposition 5.6.1. *Let $p_1 \colon \tilde{X}_1 \to X$ and $p_2 \colon \tilde{X}_2 \to X$ be two covering projections and suppose $h \colon \tilde{X}_1 \to \tilde{X}_2$ is a homomorphism of covering spaces. Then h itself is a covering projection.*

Proof. First, we show that h is onto. Let $z \in \tilde{X}_2$. Since p_1 is onto and $p_2(z) \in X$, there is a point $x_1 \in \tilde{X}_1$ such that $p_1(x_1) = p_2(z)$. Let $x_2 = h(x_1)$. Since $\tilde{X}_2$ is path connected, there is a path w_2 in $\tilde{X}_2$ which starts from x_2 and terminates into z. Then, evidently, $p_2 w_2$ is a loop based at $p_1(x_1)$. Hence, by the Path Lifting Property, there is a path w_1 in $\tilde{X}_1$ starting at x_1 such that $p_1 w_1 = p_2 w_2$. Let $\tilde{z} = w_1(1)$. Note that hw_1 and w_2 are both paths in $\tilde{X}_2$ starting from $h(x_1) = x_2$ such that

$$p_2 h w_1 = p_1 w_1 = p_2 w_2.$$

Hence, by the uniqueness of lifted paths, $hw_1 = w_2$. This means

$$h(\tilde{z}) = h(w_1(1)) = w_2(1) = z.$$

Next, we prove that each point of $\tilde{X}_2$ has an open admissible neighbourhood. Let $z \in \tilde{X}_2$. Observe that there is an open admissible neighbourhood U_1 of $p_2(z)$ for the covering p_1, and an open admissible neighbourhood U_2 for the covering p_2. Hence, if we take U to be the path component of $U_1 \cap U_2$ which contains $p_2(z)$, then U is an admissible neighbourhood for p_1 and p_2 both. Now, let V be the path component of $p_2^{-1}(U)$ which contains z. We claim that V is an admissible neighbourhood for $h \colon \tilde{X}_1 \to \tilde{X}_2$. To see this, let x_1 be any point in $\tilde{X}_1$ such that $h(x_1) = z$. Then $x_1 \in p_1^{-1}(p_2(z))$. Since p_1 is a covering projection, the path component V' of $p_1^{-1}(U)$, which contains x_1, is homeomorphic to U under p_1. But this means $h|_{V'} : V' \to V$ is a homeomorphism because

$$h|_{V'} = (p_2|_U)^{-1} \circ (p_1|_{V'}). \qquad \blacksquare$$

Now, we have

Definition 5.6.2. *A covering projection $q \colon U \to X$ in which the covering space U is simply connected is called the* **Universal Covering** *space of X.*

Let us observe that, since U is simply connected, $\pi_1(U) = 0$ and so for any covering projection $p \colon \tilde{X} \to X$, $q_*(\pi_1(U)) = \{0\} \subseteq p_*(\pi_1(\tilde{X}))$. Therefore, by the Lifting Theorem, there exists a map $h \colon U \to \tilde{X}$ such that $p \circ h = q$. By the previous proposition, this means $h \colon U \to \tilde{X}$ is a covering projection. It follows that the universal covering space U of a space X is big enough so that it covers any covering space of X. It is for this reason that U is called **Universal**. It also follows from above that any two universal covering spaces of X are isomorphic and hence the universal covering space U of X is unique, provided it exists.

Example 5.6.3. (a) Since $\mathbb{R}$ is simply connected, the exponential map $p\colon \mathbb{R} \to \mathbb{S}^1$, defined by $p(t) = e^{2\pi i t}$, is the universal covering of $\mathbb{S}^1$.

(b) Since $\mathbb{R}\times\mathbb{R}$ is also simply connected, the map $q\colon \mathbb{R}\times\mathbb{R} \to \mathbb{S}^1\times\mathbb{S}^1$, defined by $q(t_1, t_2) = (e^{2\pi i t_1}, e^{2\pi i t_2})$, is the universal covering space of torus $T = \mathbb{S}^1 \times \mathbb{S}^1$.

(c) The quotient map $f\colon \mathbb{S}^2 \to \mathbb{P}^2$ which identifies the antipodal points is a covering projection which is universal since $\mathbb{S}^2$ is simply connected. More generally, since $\mathbb{S}^n$ $(n \geq 2)$ is simply connected, the quotient map which identifies the antipodal points is a universal covering projection of the projective space $\mathbb{P}^n$, $(n \geq 2)$.

The next result gives an interesting interpretation of the fundamental group of a space X, which admits the universal covering space.

Proposition 5.6.4. *Let $q\colon U \to X$ be the universal covering of a space X. Then the automorphism group $A(U, q)$ is isomorphic to the fundamental group $\pi_1(X)$ of X. Moreover, the order of $\pi_1(X)$ is equal to the number of sheets of the universal covering.*

Proof. The last part follows from Proposition 5.3.5, since $\pi_1(U) = 0$. To prove the main result, let $u_0 \in U$ and $q(u_0) = x_0$. We define a map $T\colon A(U, q) \to \pi_1(X, x_0)$ as follows: If $f \in A(U, q)$, then since f permutes the points of the fibre $q^{-1}(x_0)$, the point $f(u_0)$ is in this fibre. Let α be a path in U joining u_0 with $f(u_0)$. Then $q\alpha$ is a loop in X based at x_0. We define

$$T(f) = [q\alpha].$$

If β is any other path joining u_0 with $f(u_0)$ then, since U is simply connected, α is equivalent to β which means $[q\alpha] = [q\beta]$. Then the map T is well defined.

To prove that T is a homomorphism, let $f,\ g \in A(U, q)$ and suppose $\alpha,\ \beta$ are paths in U joining u_0 to $f(u_0)$ and to $g(u_0)$, respectively. Then $T(f) = [q\alpha]$ and $T(g) = [q\beta]$. Observe that $f\beta$ is a path joining $f(u_0)$ to $f(g(u_0))$ and so $\alpha * f\beta$ is a path in U joining u_0 with $f(g(u_0))$. Therefore,

$$T(f \circ g) = [q(\alpha * f\beta)] = [q\alpha * qf\beta] = [q\alpha] \circ [qf\beta].$$

Since $qf = q$, we see that $T(f \circ g) = Tf \circ Tg$. Next, we see that T is one-one because $T(f) = T(g)$ implies $[q\alpha] = [q\beta]$ where $\alpha,\ \beta$ are paths in U starting from u_0. This means $q_*[\alpha] = q_*[\beta]$ and so, by the Monodromy Theorem, $\alpha,\ \beta$ must have the same terminal point also, i.e, $f(u_0) = g(u_0)$. Since any two automorphisms of $q\colon U \to X$ which agree at a single point must agree everywhere (Proposition 5.5.2), we must have $f = g$.

Finally, to show that T is onto, let $\alpha \in \pi_1(X, x_0)$. Let $\tilde{\alpha}$ be a lift of the path α in U such that $\tilde{\alpha}(0) = u_0$. Consider the diagram

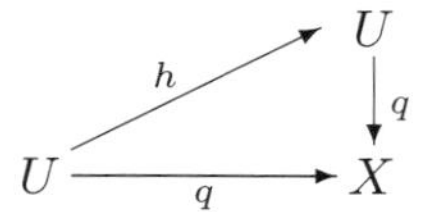

Since U is simply connected covering projection of X, there exists a homomorphism $h\colon U \to U$ such that $h(u_0) = \tilde{\alpha}(1)$. By the same argument, there is also a homomorphism $k\colon U \to U$ such that $k(\tilde{\alpha}(1)) = u_0$. It follows that the homomorphism $k \circ h\colon U \to U$ must map u_0 to itself and so, by Proposition 5.5.2, it must be the identity map; similarly $h \circ k$ too must be identity. This shows that $h \in A(U, q)$ and, by definition, $T(h) = [q\tilde{\alpha}] = [\alpha]$. ■

Example 5.6.5. The above result immediately yields the fundamental group $\pi_1(\mathbb{RP}^n)$ of the projective space $\mathbb{RP}^n$ for each $n \geq 2$. For this, consider the universal covering $q\colon \mathbb{S}^n \to \mathbb{RP}^n$, where $\mathbb{S}^n$ is the standard n-sphere and the map q identifies the antipodal points of $\mathbb{S}^n$. Since q is a 2-sheeted covering projection, we can apply the above proposition to assert that the group $\pi_1(\mathbb{RP}^n)$ must be of order two. Hence, $\pi_1(\mathbb{RP}^n) \cong \mathbb{Z}/2\mathbb{Z}$. The other alternative is to verify directly that the group of automorphisms of $q\colon \mathbb{S}^n \to \mathbb{RP}^n$ has only two elements.

Example 5.6.6. The covering $\mathbb{R} \to \mathbb{S}^1$ by exponential map is a universal covering. Consider the set of all translations of $\mathbb{R}$ by all integers. Each of this is easily seen to be a deck transformation of the covering and so that group of deck transformations is isomorphic to the additive group of integers. This says that the fundamental group of the circle is the group $\mathbb{Z}$ of integers – a result already proved directly.

Example 5.6.7. The product covering $\mathbb{R} \times \mathbb{R} \to \mathbb{S}^1 \times \mathbb{S}^1$ is again a universal covering and the group of Deck transformations is easily seen to be isomorphic to $\mathbb{Z} \oplus \mathbb{Z}$. It follows that the fundamental group of torus is $\mathbb{Z} \oplus \mathbb{Z}$, which is also already computed.

Recall the concepts of 'group action' and 'properly discontinuous action' as discussed in Appendix 7.3.

Proposition 5.6.8. *Let G act freely and properly discontinuously on X. Then the quotient map $\eta\colon X \to X/G$ is a covering projection.*

Proof. Any element of X/G will be of the form $\eta(x)$ for some $x \in X$. Since the action is free, it follows from the definition of properly discontinuous action that there exists an open set V of X with $x \in V$ and $g \cap gV = \phi$ for any $g \neq e$ in G. Then $W = \eta(V)$ is an open set containing $\eta(x)$ in X/G. Moreover, $\eta^{-1}(W) = \bigcup_{g \in G} gV$ a disjoint union and $\eta|_{gV} : gV \to W$ is a homeomorphism for each $g \in G$. ■

For covering projections which arise, as above, from a free and properly discontinuous action of a group G on a simply connected space X, it is interesting to observe (see Bredon [2] p.149) that G is isomorphic to the fundamental group of the orbit space X/G. This follows from Proposition 5.6.4 and the fact that if X is simply connected, then the group of deck transformations of the covering map $X \to X/G$ is isomorphic to G itself. (See Exercise 14 at the end of the chapter for a general result.)

As an application of the last observation, we can determine the fundamental group of lens spaces $L(p, q)$.

Example 5.6.9. (Fundamental Group of lens spaces). Recall (see Appendix 7.3) that if p, q are any two positive integers, which are relatively prime, then we have defined an action of the finite group $\mathbb{Z}_p$ on the space $\mathbb{S}^3$. The quotient space $\mathbb{S}^3/\mathbb{Z}_p$ is called the **lens space** $L(p, q)$. Since the action of $\mathbb{Z}_p$ on $\mathbb{S}^3$ is free and properly discontinuous, and $\mathbb{S}^3$ is simply connected, it follows from the above observation that $\pi_1(L(p, q)) \cong \mathbb{Z}_p$ for each q.

Covering projections and fundamental groups, as we have seen so far, are very closely related. Fundamental groups were used not only in proving the existence of covering projections, but also in classifying them. There are situations when both can be combined to produce interesting results of deeper significance. Recall that, using homology groups and degree concept of a map from $\mathbb{S}^n$ to $\mathbb{S}^n$, we proved the Borsuk-Ulam theorem, viz., there cannot be a continuous map from $\mathbb{S}^n$ to $\mathbb{S}^{n-1}$ preserving antipodal points. This classical result can also be proved using the ideas of covering projections and homotopy groups. We prove below the case $n = 2$ of Borsuk-Ulam theorem using covering projections and fundamental groups; the case $n \geq 3$ requires, in addition, the computation of the cohomology algebra of higher-dimensional projective spaces which we have not dealt with so far.

Example 5.6.10. (Borsuk-Ulam Theorem for $n = 2$). *There cannot be a continuous map $\mathbb{S}^2 \to \mathbb{S}^1$ which preserves antipodal points.* Suppose, on the contrary, there is a continuous map $f \colon \mathbb{S}^2 \to \mathbb{S}^1$ satisfying the condition that $f(-x) = -f(x)$ for any $x \in \mathbb{S}^2$. We show that this leads to a contradiction. We pass on to the quotient spaces $\mathbb{P}^2$ and $\mathbb{P}^1$ of $\mathbb{S}^2$ and $\mathbb{S}^1$, respectively, by identifying antipodal points. Note that the projective space $\mathbb{P}^1$ is homeomorphic to $\mathbb{S}^1$ itself. Since f preserves antipodal points, it induces a continuous map $g \colon \mathbb{P}^2 \to \mathbb{P}^1$ which makes the following diagram commutative:

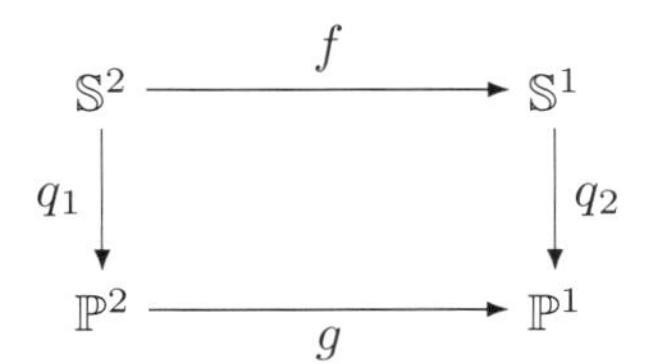

We know that both q_1 and q_2 are two sheeted covering projections. Note that $\pi_1(\mathbb{P}^2) \cong \mathbb{Z}/2\mathbb{Z}$ and $\pi_1(\mathbb{P}^1) \cong \mathbb{Z}$ and the latter group has no elements of finite order other than the identity. Hence the induced homomophism $g_*\colon \pi_1(\mathbb{P}^2) \to \pi_1(\mathbb{P}^1)$ must be a trivial homomorphism. Next, we show that g_* is indeed nontrivial. Let α be a path in $\mathbb{S}^2$ joining a pair of antipodal points, say, x_0 and $-x_0$. Then $q_1 \circ \alpha$ is a closed path in $\mathbb{P}^2$ and $f \circ \alpha$ is a path joining the pair of antipodal points $f(x_0), f(-x_0)$. Therefore, $q_2 \circ (f \circ \alpha)$ is also a closed path in $\mathbb{P}^1$, and we assert that this is not null-homotopic. To see this, suppose it is null-homotopic and let $C_{f(x_0)}$ be the constant path in $\mathbb{S}^1$ based at $f(x_0)$. Then

$$q_{2*}[C_{f(x_0)}] = [q_2 \circ (f \circ \alpha)] = q_{2*}[f \circ \alpha].$$

Hence, by the Monodromy Theorem, $f \circ \alpha$ must have the same terminal point as $C_{f(x_0)}$, i.e, $f \circ \alpha(1) = f(-x_0) = f(x_0)$, a contradiction. Now, since

$$g_*[q_1 \circ \alpha] = [g \circ (q_1 \circ \alpha)] = [q_2 \circ f \circ \alpha] = q_{2*}[f \circ \alpha] \neq 0,$$

we conclude that g_* is nontrivial.

Exercises

1. Let $p\colon \tilde{X} \to X$ be a covering projection. Prove that the family of admissible open neighbourhoods is a basis for the topology of X.

2. Suppose X is connected and locally path-connected and $p\colon \tilde{X} \to X$ is a covering projection. Prove that
   (a) X is Hausdorff implies $\tilde{X}$ is Hausdorff.
   (b) X is completely regular if and only if $\tilde{X}$ is completely regular.
   (c) X is locally compact if and only if $\tilde{X}$ is locally compact.

3. Give an example of a Hausdorff space X and a group G acting on X freely and properly discontinuously so that the orbit space X/G is not Hausdorff. This shows that the converse part of 2(a) is not true. (**Hint:** See Massey [13] pp. 167–169.)

4. Let $p\colon \tilde{X} \to X$ be a covering projection where X is connected and locally path-connected. Prove that $\tilde{X}$ is compact if and only if X is compact and p is finite sheeted.

5. Let $p_n\colon \tilde{X}_n \to X_n$ be the covering projection $\exp : \mathbb{R} \to \mathbb{S}^1$, for all $n \geq 1$. Show that the product $\prod_1^k p_n\colon \prod_1^k \tilde{X}_n \to \prod_1^k X_n$ is a covering projection.

6. Let $L(p, q)$ denote the lens space, where p, q are relatively prime positive integers. Prove that if $p_1 \neq p_2$, then $L(p_1, q)$, $L(p_2, q)$ are not homeomorphic.

7. Construct a 6-sheeted covering projection of the figure of eight.

8. Let $\mathbb{RP}^n (n \geq 2)$ denote the n-dimensional real projective space. Prove that any map $f\colon \mathbb{RP}^n \to \mathbb{S}^1$, the unit circle, is null-homotopic. Is it true when $\mathbb{RP}^n$ is replaced by $\mathbb{S}^n$?

9. Let G, $\tilde{G}$ be connected and locally path-connected topological groups and $p\colon \tilde{G} \to G$ be a covering projection which is also a group homomorphism. Prove that $\ker p \cong \mathrm{Aut}(\tilde{G}, p)$.

Existence of universal coverings

10. A space X is said to be **semi-locally simply connected** if for each $x \in X$, there exists an open neighbourhood U of x in X such that the inclusion map $U \subset X$ induces trivial homomorphism $\pi_1(U) \to \pi_1(X)$. Prove that a connected, locally path-connected space X has a universal covering if and only if it is semi-locally simply connected.

11. Let X be a connected, locally path-connected and semi-locally simply connected space. Show that for any given conjugacy class $\{H\}$ of subgroups of $\pi_1(X)$, there exists a covering $p\colon \tilde{X} \to X$ such that the conjugacy class of the subgroup $p_*(\pi_1(\tilde{X}))$ agrees with the class of H.

12. Determine the fundamental group of an infinite product of circles. Deduce that the latter has no universal covering space, whereas any product of n-spheres $(n \geq 2)$ always has an obvious universal covering. (**Hint**: Use Exercise 10 above.)

13. Let X be the subset of the real plane consisting of circles having radius $1/n$ and center $(1/n, 0)$ for $n = 1, 2, \ldots$. Show that X has no universal covering. (**Hint**: Use Ex 10 above.)

14. For a covering projection $p\colon \tilde{X} \to X$, let $p(\tilde{x}_0) = x_0$ and $H = p_\#(\pi_1(\tilde{X}, \tilde{x}_0))$. Suppose $N(H)$ is the normalizer of H in $\pi_1(X, x_0)$. Prove that $N(H)/H$ is isomorphic to the group $\mathrm{Aut}(\tilde{X}, p)$. In particular, if $\tilde{X}$ is simply connected, then $\pi_1(X, x_0) \cong \mathrm{Aut}(\tilde{X}, p)$.

Chapter 6

Singular Homology

6.1 Introduction

Having defined the simplicial homology groups of a polyhedron (space of a general simplicial complex) now we come to another kind of homology groups, called **singular homology groups** of a topological space. The first interesting feature of these homology groups is that they are defined for all topological spaces X, not only for polyhedra. Singular homology groups were first defined by S. Lefschetz in 1933, and were perfected in their present form by S. Eilenberg (1913–1998) in the beginning of the 1940's. These turn out to be the most important and natural generalization of simplicial homology groups, and are most suited for the study of topological manifolds. We may recall (Theorem 4.9.6) that it took a good amount of hard work and new ideas to prove the topological invariance of simplicial homology in Chapter 4. This was first done by J.W. Alexander (1888–1971). In a contrast to this, we will see in this chapter that the topological invariance of singular homology follows almost obviously – this is another attractive feature of singular homology.

Historically, several kinds of homology groups, e.g., Vietoris homology, Čech homology, etc., were defined, but each was defined for a restricted class of topological spaces. However, their basic properties were found to be the same or at least very similar to each other. Each of these basic properties was considered a deep theorem for the particular homology under consideration. It was only in 1945 that S. Eilenberg, jointly with N. Steenrod (1910–1971), succeeded in a very ingenious way, in clarifying and unifying all these homology groups by defining a **homology theory** axiomatically. They recognized the basic data underlying different homology groups and identified a set of seven axioms to define a homology theory. Each of these seven axioms was a basic theorem for different homologies proposed at the time. Eilenberg and Steenrod also proved the interesting fact that any two homology theories, for which the zero-dimensional groups of a point-space (viz., the coefficient group) were isomorphic, gave the same result for any compact polyhedron. This was indeed their important **uniqueness theorem.** The compact polyhedra, whose geometry and topology we studied in Chapter 3, therefore, gained even better status than before.

The most remarkable aspect which emerged out of Eilenberg-Steenrod axiomatization was that several results, valid for other homology theories, were derived using only these axioms, not the specific definitions of those homologies. As a matter of fact, this axiomatization brought about a sort of revolution in algebraic topology and the subject grew thereafter very fast cutting deep into other areas of mathematics. It continues unabated ever since. One of the greatest outcomes of this revolution was the birth of "generalized homology and cohomology theories", which include fascinating topics like K-theory, bordism and cobordism theories, etc. The concept of relative homology of a pair (X, A), which will be explained later, is of crucial importance in axiomatizing a homology theory. Since our scope is quite limited, we will not go into other aspects of singular homology except for giving its brief introduction. A curious reader, however, may referred to any of the advanced books such as [2], [6], [15], [18], for a fuller and comprehensive study of this subject.

In this chapter, we explain the definition of singular homology of a topological pair, and establish that it is a homology theory in the sense of Eilenberg-Steenrod. Our approach is quite brief and pointed towards proving those theorems about singular homology, which go to show that the singular homology satisfies the seven axioms of Eilenberg-Steenrod. It is true that the real purpose and importance of singular homology lies in proving theorems of topology and solving open problems of topology, geometry and other areas of mathematics. But that can be done after accomplishing this major goal. Numerous results about singular homology become available to us as a consequence of singular homology being a homology theory. We don't have to prove these results as something special to singular homology. Eilenberg-Steenrod also defined the concept of a "cohomology theory" as a dual of a homology theory, satisfying seven axioms analogous to those of a homology theory. After defining singular homology, we introduce singular cohomology valid for the category of all topological pairs and continuous maps. It turns out that a cohomology theory has more structure than a homology theory, viz., there is a natural product called "cup product" between any two cohomology classes of a given pair (X, A). As a result, the direct sum of cohomology groups of (X, A) of all dimensions becomes not only a graded-ring but a graded R-algebra, where R is the ground ring. Again, the singular cohomology is the most general example of a "cohomology theory" in the sense of Eilenberg-Steenrod. Their classic book (see [6]) deals with all of these aspects almost exhaustively.

6.2 Singular Chain Complex

Let $\mathbb{R}^\omega$ denote a countably infinite product of copies of the real line $\mathbb{R}$. We fix the following points of $\mathbb{R}^\omega$:

$$\begin{aligned}
e_0 &= (0, 0, \ldots) \\
e_1 &= (1, 0, 0, \ldots) \\
e_2 &= (0, 1, 0, \ldots) \\
&\vdots \\
e_n &= (0, 0, \ldots, 1, 0, \ldots), \text{ 1 at the } n\text{th place, etc.}
\end{aligned}$$

Let $\mathbb{R}^n$ denote the n-dimensional Euclidean space. Then $\mathbb{R}^n$ can be identified with a subspace of $\mathbb{R}^\omega$ under the map $(x_1, x_2, \ldots, x_n) \to (x_1, x_2, \ldots, x_n, 0, 0, \ldots)$. With this identification, $\mathbb{R} \subset \mathbb{R}^2 \subset \cdots \subset \mathbb{R}^n \subset \cdots$ Let $\Delta_n = \langle e_0, e_1, \ldots, e_n \rangle$ be the standard geometric n-simplex spanned by the points $e_0, e_1, \ldots, e_n$ in $\mathbb{R}^\omega$.

Given a set of $q + 1$ points $x_0, x_1, \ldots, x_q$ in $\mathbb{R}^n$, there is a unique affine map from Δ_q to $\mathbb{R}^n$ which maps e_i to x_i for $i = 0, 1, \ldots, q$. We denote this map by the symbol $(x_0, x_1, \ldots, x_q)$. Now we have our basic concept,

Definition 6.2.1. *Let X be a topological space. A continuous map $\sigma \colon \Delta_q \to X$ is called a* **singular q-simplex** *in X.*

Note that since Δ_q is a compact connected space, the image set $\sigma(\Delta_q)$ in X must be compact and connected. Thus, a singular 1-simplex in X is simply a path in X, whereas a singular 0-simplex in X is just a point in X; a singular 2-simplex in X will be a curved triangle or region (see Fig. 6.1) with its interior.

Fig. 6.1: A singular 1-chain and a singular 2-chain

Now, let R be a commutative PID (principal ideal domain) with $1 \neq 0$, and let $S_q(X)$ be the free R-module generated by the set of all singular q-simplexes in X. The elements of $S_q(X)$, which are formal linear combinations of singular q-simplexes in X with coefficients in R, are called **singular q-chains** in X. We now proceed to define boundary homomorphisms $S_q(X) \to S_{q-1}(X)$, for all $q \geq 1$, so that we get a chain complex. For this, let $F_q^i \colon \Delta_{q-1} \to \Delta_q$,

$0 \le i \le q$, be the unique affine maps which map the vertices $(e_0, e_1, \ldots, e_{q-1})$ to the set of points $(e_0, e_1, \ldots, \hat{e}_i, \ldots, e_q)$, where $\hat{e}_i$ means the point e_i is omitted. In other words, let

$$F_q^i(e_j) = \begin{cases} e_j, & j < i \\ e_{j+1}, & j \ge i \end{cases}$$

The map F_q^i is called the ith **face operator** of the q-simplex Δ_q, and it simply maps the $(q-1)$-simplex Δ_{q-1} into the ith face of the q-simplex Δ_q, i.e., to the face of Δ_q opposite to the vertex e_i. Thus, F_2^i will map the standard 1-simplex $\langle e_0, e_1\rangle$ to $\langle e_1, e_2\rangle$, $\langle e_0, e_2\rangle$ and $\langle e_0, e_1\rangle$ under the face operators F_2^0, F_2^1, F_2^2, respectively. For a given singular q-simplex σ, the composite $\sigma^{(i)} = \sigma \circ F_q^i$ is called the ith face of σ, $i = 0, 1, \ldots, q$. Now, we can define the boundary $\partial(\sigma)$ of a singular q-simplex σ by the formula

$$\partial(\sigma) = \sum_0^q (-1)^i \sigma^{(i)}.$$

The right hand side is the sum of $(q+1)$ singular $(q-1)$- simplexes. Since $S_q(X)$ is the free R-module generated by the set of all singular q-simplexes in X, we can extend ∂ by linearity to an R-homomorphism $\partial_q \colon S_q(X) \to S_{q-1}(X)$, for all $q \ge 1$, i.e., for any singular chain $c = \sum r_\sigma \sigma$, we define

$$\partial_q(c) = \sum r_\sigma \partial(\sigma), \quad r_\sigma \in R.$$

We note that if in the standard q-simplex $\Delta_q = \langle e_0, e_1, \ldots, e_q\rangle$, we first drop the ith vertex followed by the jth vertex, $j < i$, then the result is same as first dropping the jth vertex and then dropping $(i-1)$th vertex. Reversing this observation and using the definition of face operators, we get the following:

Lemma 6.2.2. *The face operators F_q^i, F_{q-1}^j, $0 \le j < i \le q$, satisfy the following condition:*

$$F_q^i \circ F_{q-1}^j = F_q^j \circ F_{q-1}^{i-1}.$$

Using the above lemma, we have

Proposition 6.2.3. *The composite $\partial_{q-1} \circ \partial_q = 0$, for all $q \ge 1$.*

Proof. It suffices to prove that the composite $\partial \circ \partial$ vanishes on generators $\sigma \in S_q(X)$. We have

$$
\begin{aligned}
\partial\partial(\sigma) &= \sum(-1)^i\partial(\sigma^{(i)}) \\
&= \sum(-1)^i(\sum(-1)^j((\sigma\circ F_q^i)\circ F_{q-1}^j) \\
&= \sum_{j<i=1}^{q}(-1)^{i+j}\sigma\circ(F_q^iF_{q-1}^j)+\sum_{0\le i\le j}^{q-1}(-1)^{i+j}\sigma\circ(F_q^iF_{q-1}^j) \\
&= \sum_{j<i=1}^{q}(-1)^{i+j}\sigma\circ(F_q^jF_{q-1}^{(i-1)})+\sum_{0\le i\le j}^{q-1}(-1)^{i+j}\sigma\circ(F_q^iF_{q-1}^j) \\
&= \sum_{0\le k\le l}^{q}(-1)^{k+l+1}\sigma\circ(F_q^kF_{q-1}^l)+\sum_{0\le i\le j}^{q-1}(-1)^{i+j}\sigma\circ(F_q^iF_{q-1}^j) \\
&= 0,
\end{aligned}
$$

since, by the foregoing lemma, everything cancels: we simply applied the lemma in the last but two step and replaced j by k, $i-1$ by l in the last but one step, which shows that the two expressions are identical except for signs. ■

Remark 6.2.4. Whenever one has to prove a result of the above kind, it is always helpful to verify the first few cases when σ is a 2-simplex, and a 3-simplex, and notice the result how the terms cancel with each other. Then one can take up the general case to prove it carefully.

It follows from the above proposition that $S_*(X) = \{S_q(X), \partial_q\}_{q\ge0}$ is a chain complex of free R-modules. This chain complex is called the **singular chain complex** of the space X. Its homology $\{H_q(S_*(X))\}_{q\ge0}$ is called the **singular homology** of X; each of these groups is an R-module. If we take $\mathbb{Z}$ for R, the resulting homology groups are called **integral singular homology groups** of the space X.

Next, we are going to prove some basic properties of the singular homology. We begin by computing the singular homology of a point-space.

Example 6.2.5 (Dimension Axiom). Let $X = \{x\}$ be a single-point space. What is $S_q(X)$ for each $q \ge 0$? The answer is obvious, viz., for each q, there is only one singular q-simplex $\sigma\colon \Delta_q \to X$ which is just the constant map taking all of Δ_q to the single point x. Therefore, $S_q(X) = R\cdot\sigma^q$, where σ^q denotes the constant map $\sigma\colon \Delta_q \to X$. Note that $\partial(\sigma_1) = \sigma_0 - \sigma_0 = 0$, $\partial(\sigma_2) = \sigma_1 - \sigma_1 + \sigma_1 = \sigma_1$, etc. In fact, more generally, we have

$$
\partial(\sigma_q) = \begin{cases} \sigma_{q-1}, & \text{if } q \text{ is even} > 0 \\ 0, & \text{otherwise.} \end{cases}
$$

Thus, the singular chain complex, in this case is

$$\cdots \xrightarrow{I} R\sigma_3 \xrightarrow{0} R\sigma_2 \xrightarrow{I} R\sigma_1 \xrightarrow{0} R\sigma_0 \longrightarrow 0 \cdots,$$

where 0 denotes the zero map and I denotes the identity map. It follows that $H_0(X) \cong R$ under the map $r \cdot \sigma_0 \mapsto r$. $H_1(X) = Z_1/B_1 \cong R/R = 0$. $H_2(X) \cong 0/0 = 0, \ldots$ Thus, we conclude that

$$H_q(X) \cong \begin{cases} R, & \text{if } q = 0 \\ 0, & \text{otherwise.} \end{cases}$$

Instead of taking the coefficient module as the ring R, we could have taken, more generally, any R-module G as the coefficient module. Then what we proved above can be stated, more generally, as

Proposition 6.2.6. *If X is a point-space, then*

$$H_q(X; G) \cong \begin{cases} G, & \text{if } q = 0 \\ 0, & \text{otherwise.} \end{cases}$$

Zero-Dimensional Homology

Next, we compute the zero-dimensional homology module of a space X with coefficients in R itself. The results are true with coefficients in any general R-module G, and will also be suppressed unless we state it otherwise. The zero-dimensional homology is obtained in terms of its path-components. First, we have

Proposition 6.2.7. *If X is path-connected, then $H_0(X) \cong R$.*

Proof. Note that, by definition, $Z_0(X) = S_0(X)$. To determine $B_0(X)$, observe that if $c = \sum r_i\sigma_i$ is a 1-chain in X, then each σ_i is a path in X and $\partial(c) = \sum r_i(\sigma(1) - \sigma(0)) \in B_0(X)$ and it has the property that the sum of the coefficients of $\partial(c)$ is zero. Note that the converse of this observation is also true. For this, let $x_0 \in X$ be a fixed point and suppose $c = \sum_{i=1}^k r_i x_i$ is a 0-chain which has the property that $\sum_{i=1}^k r_i = 0$. Since X is path-connected, we can find a path, say, σ_{x_i}, joining x_0 to x_i for all $i = 1, 2, \ldots, k$. Then

$$\begin{aligned} c = \sum r_i x_i &= \sum r_i x_i - (\sum r_i) x_0 \\ &= \sum_{i=1}^k r_i(x_i - x_0) \\ &= \sum r_i \partial(\sigma_{x_i}) \\ &= \partial(\sum_{i=1}^k r_i \sigma_{x_i}). \end{aligned}$$

Thus, we find that $B_0(X)$ consists of all those 0-chains $c = \sum r_i x_i$ in X which have the property that the sum $\sum r_i$ of their coefficients in the ring R is zero. Now, we define an R-homomorphism $\epsilon\colon S_0(X) \to R$ called **augmentation map**, by putting $\epsilon(\sum r_i x_i) = \sum r_i$. The map ϵ is clearly onto and its kernel is $B_0(X)$. This says that

$$H_0(X) = \frac{Z_0(X)}{B_0(X)} = \frac{S_0(X)}{\ker \epsilon} \cong R. \qquad \blacksquare$$

More generally, we have

Proposition 6.2.8. *Let $\{X_i, i \in I\}$ be the family of all path-components of a space X. Then, for all $q \geq 0$,*

$$H_q(X) \cong \oplus_{i \in I} H_q(X_i).$$

Proof. In fact, since Δ_q is a path-connected space, any singular q-simplex $\sigma\colon \Delta_q \to X$ is indeed a singular simplex in X_i for some $i \in I$. This implies that

$$S_q(X) = \bigoplus_{i \in I} S_q(X_i)$$

$$Z_q(X) = \bigoplus_{i \in I} Z_q(X_i)$$

$$B_q(X) = \bigoplus_{i \in I} B_q(X_i).$$

Therefore, it follows at once that for all $q \geq 0$, $H_q(X) \cong \bigoplus_{i \in I} H_q(X_i)$. $\blacksquare$

Corollary 6.2.9. *For any space X, $H_0(X)$ is the free R-module consisting of as many copies of R as there are path-components of X.*

Proof. Let $x_i \in X_i$, $i \in I$, be a fixed point of the path component X_i of X. Define a singular 0-simplex $c_{x_i}\colon \Delta_0 \to X$ by putting $c_{x_i}(e_0) = x_i$. We claim that the 0-cycles $\{x_i\}$, $i \in I$, form a basis for $H_0(X)$. Note that any singular 0-simplex c_x, $x \in X$ is homologous to c_{x_i} for some i. This is because if x lies in the path component X_i, we have a path $\omega_x\colon I \to X$ such that $\omega_x(0) = x_i$ and $\omega_x(1) = x$. Then, clearly $\partial(\omega_x) = c_x - c_{x_i}$ is a boundary. It follows that any singular 0-chain will be homologous to a linear combination of c_{x_i}, $i \in I$. To see that these are also linearly independent, let us suppose $\sum r_i c_{x_i}$ is a 0-cycle, i.e., it is a boundary. Then there exists a singular 1-chain $\sum d$ such that $\partial(\sum d) = \sum r_i c_{x_i}$. Note that any singular 1-simplex lies in a path component of X. Collecting all 1-chains d_i which are in the path component X_i, we find that for all i, $\partial(\sum d_i) = r_i c_{x_i}$. Applying the augumentation map ϵ on both sides we find that $0 = r_i$. $\blacksquare$

Remark 6.2.10. It may be mentioned here, just for information, that all of the foregoing results about zero-dimensional homology modules are valid for singular homology only. In general, this need not be true for other homologies.

6.3 One-Dimensional Homology and the Fundamental Group

We have already seen in Corollary 6.2.1 that zero-dimensional singular homology of a space X is closely related to the path-components $\pi_0(X)$ of X. Next, we are going to show that the one-dimensional integral singular homology $H_1(X)$ of a space X is also nicely related to the fundamental group of X. In fact, the exact relationship for a path-connected space X is that $H_1(X)$ is isomorphic to the factor group $\frac{\pi_1}{[\pi_1, \pi_1]}$, where $\pi_1 = \pi_1(X, x_0)$ is the fundamental group of the space X at some point $x_0 \in X$ and $[\pi_1, \pi_1]$ denotes the commutator subgroup of π_1.

For the time being, therefore, we assume that X is a path-connected space and we let $H_1(X)$ denote the singular homology of X with integer coefficients. We fix a point $x_0 \in X$, and for each $x \in X$, we let w_x be a path joining x_0 to x; there are several such paths and we choose one of them to be w_x. Recall that a path w in X is a continuous map $w\colon I \to X$, where $I = [0, 1]$, whereas a singular 1-simplex σ is a continuous map from the standard 1-simplex $[e_0, e_1]$ into X. Since there is an affine homeomorphism between I and any 1-simplex in $\mathbb{R}^n$, a path in X is a singular 1-simplex in X and a continuous map from any closed interval in $\mathbb{R}^n$ can be regarded as a path in X via the affine homeomorphism.

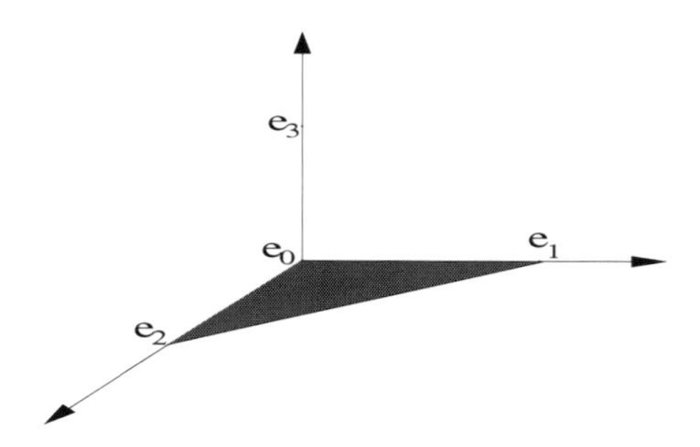

We begin with the observation that if w is a loop in X based at any point, then the corresponding singular 1-simplex is a cycle in the singular chain complex $S_*(X)$. Also, note that a constant path w in X mapping the whole domain to the point $x_0 \in X$ is not only a cycle, but is a boundary. To see this, consider a singular 2-simplex $\sigma\colon \Delta_2 \to X$ which maps the whole of Δ_2 to that point x_0. Then $\partial(\sigma_2)$ is the constant path w. In fact, much more is true, viz., if w is a path joining x_0 to x_1, then the 1-chain $w + w^{-1}$ is a boundary. To see this also, just consider the singular 2-simplex $\sigma\colon \Delta_2 \to X$ which maps $[e_1, e_2]$ as w and is constant on every line of Δ_2 parallel to $[e_0, e_2]$. Then the restriction of σ on $[e_0, e_1]$ gives w^{-1}. Since restriction of σ on $[e_0, e_2]$ is constant, we find that $\partial(\sigma) = w - \text{constant} + w^{-1}$. Because the constant path is already a boundary, we conclude that $w + w^{-1}$ is a boundary. Thus, we have proved the following:

Lemma 6.3.1. *Any constant path in X is a singular 1-boundary. If w is a path in X, then the singular 1-chain $w + w^{-1}$ is also a boundary.*

Now, let us define a map from $\pi_1(X, x_0)$ to $H_1(X)$ just by sending a loop w based at x_0 to the homology class $\{w\}$ of the singular 1-cycle w in X. First, we prove that such a map is well-defined. We have

Lemma 6.3.2. *Suppose $w \sim w'$ rel $\{0, 1\}$. Then the 1-chain $w - w'$ in X is a boundary.*

Proof. Let $F: I \times I \to X$ be a path-homotopy from w to w'. Because F is constant on the left vertical side of $I \times I$, the above map factors through the map $I \times I \to \Delta_2$, which maps the left vertical side to the vertex e_0 of Δ_2. Hence we get a singular 2-simplex $\sigma: \Delta_2 \to X$ such that $\sigma = w$ on $[e_0, e_1]$, $\sigma = w'$ on $[e_0, e_2]$ and σ is the constant path w_{x_0} on $[e_1, e_2]$. Then we have

$$\partial(\sigma) = w + w_{x_0} - w'.$$

Since the constant path w_{x_0} is already a boundary, we find that $w - w'$ is a boundary. ■

In view of the above lemma, we now have a well-defined map $\phi: \pi_1(X, x_0) \to H_1(X)$ given by $\phi[w] = \{w\}$. Next, we prove that ϕ is a homomorphism. This follows from the following result.

Proposition 6.3.3. *If w, w' are two loops based at $x_0 \in X$, then the 1-chain $w * w' - w - w'$ is a boundary in X, i.e., $\{w * w'\} = \{w\} + \{w'\}$ in $H_1(X)$.*

Proof. We may assume that w is defined on the edge $[e_0, e_1]$ and w' is defined on $[e_1, e_2]$. Now we define a singular 2-simplex $\sigma: \Delta_2 \to X$ by declaring it to be constant on lines parallel to the line joining e_1 with mid-point of $[e_0, e_2]$. Then σ restricted to $[e_0, e_2]$ clearly gives the product path $w * w'$. Hence, keeping the definition of boundary map ∂ in mind, we get

$$\partial(\sigma) = w + w' - (w * w'). \quad ■$$

Having shown that $\phi: \pi_1(X, x_0) \to H_1(X)$ is a homomorphism in which $H_1(X)$ is abelian, we find that ϕ maps the commutator subgroup $[\pi_1, \pi_1]$ to zero, and so we get the induced homomorphism

$$\bar{\phi}: \frac{\pi_1}{[\pi_1, \pi_1]} \to H_1(X).$$

We next prove that the map $\bar{\phi}$ is indeed an isomorphism. We have

Theorem 6.3.4. (Poincaré-Hurewicz Theorem) *Assume that X is a path-connected space. Then the map $\bar{\phi}$ defined above is an isomorphism.*

We have only to show that the map ϕ is onto and has the kernel equal to $[\pi_1, \pi_1]$. We will prove both these results by producing a homomorphism from $H_1(X)$ to $\frac{\pi_1}{[\pi_1,\pi_1]}$ which will be the inverse of $\bar{\phi}$. For this, let $\sigma\colon \Delta_1 \to X$ be a singular 1-simplex in X. Putting $\sigma(e_0) = \sigma(0)$ and $\sigma(e_1) = \sigma(1)$, we get a loop $\hat{\sigma} = w_{\sigma(0)} * \sigma * w_{\sigma(1)}^{-1}$ based at x_0. Recall that w_x is a path joining x_0 to x and there are several choices for w_x. If μ is another path joining x_0 to $\sigma(0)$, then $w_{\sigma(0)} * \sigma * w_{\sigma(1)}^{-1}$ and $\mu_{\sigma(0)} * \sigma * \mu_{\sigma(1)}^{-1}$ are conjugate. This means the loop $\hat{\sigma}$ is not uniquely defined by σ, but note that its coset in $\frac{\pi_1}{[\pi_1,\pi_1]}$ is uniquely determined. This follows from the fact that two conjugate elements in a group G determine the same elements of $\frac{G}{[G,G]}$. We define $\psi(\sigma) = [\hat{\sigma}]$, where $[\hat{\sigma}]$ is a well-defined element of $\frac{\pi_1}{[\pi_1,\pi_1]}$. Since $S_1(X)$ is the free abelian group generated by all such 1-simplexes σ in X, we get a homomorphism

$$\psi\colon S_1(X) \to \frac{\pi_1}{[\pi_1, \pi_1]}.$$

The next result shows that ψ maps the group $B_1(X)$ of 1-boundaries of X into the trivial element of $\frac{\pi_1}{[\pi_1,\pi_1]}$. We have

Lemma 6.3.5. Im $\psi\partial = \{1\}$, *the identity element of* $\frac{\pi_1}{[\pi_1,\pi_1]}$.

Proof. Since $B_1(X)$ is generated by all elements of the type $\partial(\sigma)$ where $\sigma\colon \Delta_2 \to X$ is a singular 2-simplex, it suffices to prove that for any such σ, $\psi\partial(\sigma)$ is the identity element. Let us put $\sigma(e_i) = y_i$; $i = 0, 1, 2$ and $f = \sigma^{(2)}, g = \sigma^{(0)}, h = (\sigma^{(1)})^{-1}$ denote the ith face of σ (see Fig.6.2).

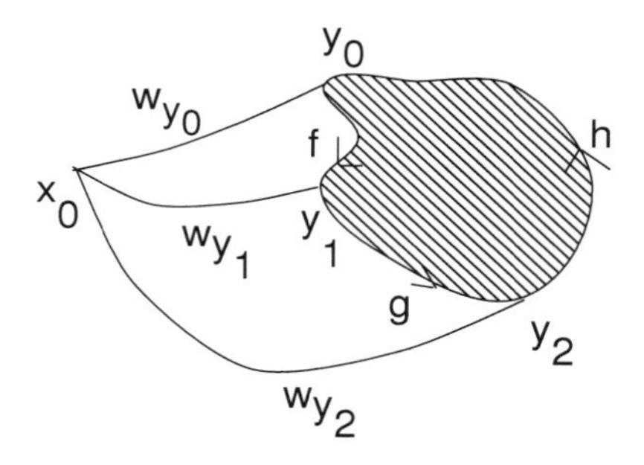

Fig. 6.2: Path $f * g * h$ is null-homotopic

We have

$$\begin{aligned}\psi(\partial\sigma) &= \psi(\sigma^{(0)} - \sigma^{(1)} + \sigma^{(2)}) \\ &= \psi(f + g - h^{-1}) \\ &= \psi(f)\cdot\psi(g)\cdot(\psi(h^{-1}))^{-1} \\ &= [[\hat{f}][\hat{g}][(\hat{h}^{-1})^{-1}]] \\ &= [\hat{f} * \hat{g} * (\hat{h}^{-1})^{-1}] \\ &= [[w_{y_0} * f * w_{y_1}^{-1} * w_{y_1} * g * w_{y_2}^{-1} * (w_{y_0} * h^{-1} * w_{y_2}^{-1})^{-1}]] \\ &= [[w_{y_0} * f * g * h * w_{y_0}^{-1}]].\end{aligned}$$

Since the loop $f * g * h$ has an extension to the 2-disk, it is homotopic to the constant path at y_0, which implies that the loop in bracket is path-homotopic to the constant loop at x_0. ∎

We need one more notation. The map $x \to w_x$ is, in fact, a map from the generators of $S_0(X)$ into $S_1(X)$ and so it can be extended to a homomorphism $w\colon S_0(X) \to S_1(X)$. We use the following notation to write $w_{(\sum n_i x_i)} = \sum n_i w_{x_i}$. We have

Proposition 6.3.6. *If σ is a singular 1-simplex in X, then $\bar{\phi}\psi(\sigma)$ is represented by the cycle $\sigma + w_{\sigma(0)} - w_{\sigma(1)} = \sigma - w_{\partial(\sigma)}$. If c is a 1-chain, then $\bar{\phi}\psi(c) = \{c - w_{\partial(c)}\}$, and $\bar{\phi}\psi(c) = \{c\}$ if c is a cycle.*

Proof. We have already noted that for any path w, $w + w^{-1}$ is a boundary, i.e., $\{w\} = -\{w^{-1}\}$ in $\frac{S_1(X)}{B_1(X)}$. Now, for any singular 1-simplex σ, we have

$$\begin{aligned}\bar{\phi}\psi(\sigma) &= \bar{\phi}[w_{\sigma(0)} * \sigma * w_{\sigma(1)}^{-1}] \\ &= \{w_{\sigma(0)}\} + \{\sigma\} + \{w_{\sigma(1)}^{-1}\} \\ &= \{w_{\sigma(0)} + \sigma - w_{\sigma(1)}\} \\ &= \{\sigma - w_{\partial(\sigma)}\}.\end{aligned}$$

The next part is obvious from the definition whereas the last part follows from the fact that a constant path is a boundary. ∎

Proof of the Poincaré-Hurewicz Theorem. We have seen that $\bar{\phi}\colon \frac{\pi_1}{[\pi_1,\pi_1]} \to H_1(X)$ defined earlier is a homomorphism. The last lemma clearly defines a homomorphism $\bar{\psi}\colon H_1(X) \to \frac{\pi_1}{[\pi_1,\pi_1]}$ by putting $\bar{\psi}\{z\} = \psi(z)$. Moreover, for each path-class $[\alpha]$ based at x_0, we have

$$\begin{aligned}\bar{\psi}\bar{\phi}[[\alpha]] &= \bar{\psi}(\{\alpha\}) \\ &= [[w_{x_0} * \alpha * w_{x_0}^{-1}]] \\ &= [[\alpha]]\end{aligned}$$

in $\frac{\pi_1}{[\pi_1,\pi_1]}$. Conversely, for each $\{z\} \in H_1(X)$, the last proposition shows that

$$\bar{\phi}\bar{\psi}\{z\} = \bar{\phi}(\psi(z)) = \{z\}.$$

This completes the proof that $\bar{\phi}$ is an isomorphism. ■

The geometric idea of cycles and boundaries in the case of simplicial homology, as we saw in Chapter 4, is quite apparent. In the case of singular homology, however, the pictures underlying cycles, boundaries, etc., get highly complicated and very often even lost. The proof of the main result elaborated above certainly gives us a glimpse of how 1-cycles and 1-boundaries look like in singular theory. In this respect, the above relation between the 1-dimensional singular homology and the fundamental group of a space X is quite revealing. Sometimes, the Poincaré-Hurewicz isomorphism theorem is expressed by saying that the 1-dimensional integral singular homology of a space is "its fundamental group abelianized".

Example 6.3.7. Since the fundamental groups of several spaces are already determined (see Chapter 2), using Poincaré-Hurewicz Theorem, we can quickly tell their one-dimensional singular homology with integer coefficients. As an illustration of this approach, we get the following results where symbols have their usual meaning:

$$\begin{aligned} H_1(\mathbb{S}^1) &\cong \mathbb{Z} \\ H_1(\mathbb{S}^n) &= 0 \text{ for all } n \geq 2 \\ H_1(T) &\cong \mathbb{Z} \oplus \mathbb{Z} \\ H_1(\mathbb{R}P^n) &\cong \mathbb{Z}/2\mathbb{Z},\ n \geq 2 \\ H_1(L(p,q))) &\cong \mathbb{Z}_p. \end{aligned}$$

If X denotes the figure of eight (see Example 5.1.10), then we know that its fundamental group is the free group on two generators (nonabelian). Hence, abelianizing this free group, we conclude that $H_1(X) \cong \mathbb{Z} \oplus \mathbb{Z}$.

Homomorphism Induced by a Continuous Map

Let $f\colon X \to Y$ be a continuous map. If $\sigma\colon \Delta_q \to X$ is a singular simplex in X then, clearly, $f \circ \sigma\colon \Delta_q \to Y$ is a singular simplex in Y. Thus, for all $q \geq 0$, we can define a homomorphism $f_{\#}\colon S_q(X) \to S_q(Y)$ simply by putting $f_{\#}(\sigma) = f \circ \sigma$ on generators of $S_q(X)$ and then extending it to $S_q(X)$ by linearity. Now, we have

Proposition 6.3.8. *The sequence $f_{\#}\colon S_q(X) \to S_q(Y)$, $q \geq 0$, of homomorphisms is a chain map.*

Proof. We must prove that for all $q \geq 1$, the following squares are commutative:

$$\begin{array}{ccccccc}
\longrightarrow & S_q(X) & \xrightarrow{\partial} & S_{q-1}(X) & \longrightarrow \\
 & \downarrow f_\# & & \downarrow f_\# & \\
\longrightarrow & S_q(Y) & \xrightarrow{\partial} & S_{q-1}(Y) & \longrightarrow
\end{array}$$

However, this is immediate because for a generator $\sigma \in S_q(X)$, we have:

$$\begin{aligned}
f_\#\partial(\sigma) &= f_\#(\sum(-1)^i\sigma^i) \\
&= \sum(-1)^i f_\#(\sigma \circ F_q^i) \\
&= \sum(-1)^i (f \circ \sigma) \circ F_q^i \\
&= \partial(f \circ \sigma) \\
&= \partial f_\#(\sigma).
\end{aligned}$$

■

It follows from the above proposition that a continuous map $f\colon X \to Y$ induces an R-homomorphism $f_*\colon H_q(X) \to H_q(Y)$ for all $q \geq 0$, in singular homology. The actual definition of f_* is the following: if $z \in S_q(X)$ is a q-cycle, then $f_*\{z\} = \{f_\#(z)\}$. With this definition, we can easily verify the following functorial properties of induced homomorphisms:

Proposition 6.3.9. (Identity Axiom). *If $I\colon X \to X$ is the identity map, then the induced homomorphism $I_*\colon H_q(X) \to H_q(X)$ is also identity map, for all $q \geq 0$.*

Proposition 6.3.10. (Composition Axiom). *If $f\colon X \to Y$, $g\colon Y \to Z$ are continuous maps then, for all $q \geq 0$, the induced homomorphisms in homology satisfy*

$$(g \circ f)_* = g_* \circ f_*\colon H_q(X) \to H_q(Z).$$

The above two propositions yield the following topological invariance of singular homology groups:

Corollary 6.3.11. *If $f\colon X \to Y$ is a homeomorphism then, for all $q \geq 0$, $f_*\colon H_q(X) \to H_q(Y)$ is an isomorphism.*

Proof. Let f^{-1} denote the inverse of the homeomorphism f. Then, clearly $f^{-1} \circ f = I_X$ and $f \circ f^{-1} = I_Y$. Considering the induced homomorphisms $f_*\colon H_q(X) \to H_q(Y)$ and $f_*^{-1}\colon H_q(Y) \to H_q(X)$ in homology, we find from Proposition 6.3.10 and Proposition 6.3.9, that f_* is the inverse of f_*^{-1}. This means f_* is an isomorphism. ■

Remark 6.3.12. The above corollary says that the singular homology modules are topological invariants. At this point, we must recall what we presented earlier in Chapter 4 (Theorem 4.9.6), viz., the proof of the topological invariance of the simplicial homology. That proof was indeed quite long and surprisingly difficult. Even the proof of the functorial property for simplicial homology, analogous to the composition axiom for singular homology proved above, took considerable effort. This is in great contrast with the corresponding results for the singular homology where they have simple and obvious proofs.

Proposition 6.3.13. *Let X, Y be path-connected spaces and $f\colon X \to Y$ be a continuous map. Then the induced homomorphism $f_*\colon H_0(X) \to H_0(Y)$ in singular homology is an isomorphism. In particular, when $X = Y$, any map $f\colon X \to X$ induces the identity map $f_*\colon H_0(X) \to H_0(X)$.*

Proof. To see this, note that the augmentation map $\epsilon\colon S_0(X) \to R$ defined by $\epsilon(\sum r_i x_i) = \sum r_i$ gives us the isomorphism $H_0(X) = \frac{S_0(X)}{\mathrm{Im}\,\partial_1} \to R$. The same argument for the space Y gives that $H_0(Y) = \frac{S_0(Y)}{\mathrm{Im}\,\partial_1} \to R$ is an isomorphism. The chain map $f_\#\colon S_0(X) \to S_0(Y)$ clearly maps $B_0(X)$ into $B_0(Y)$ making the following diagram commutative:

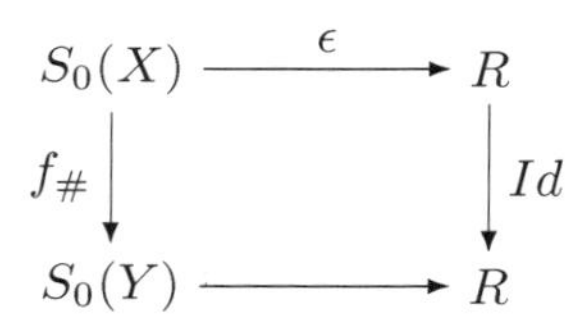

This clearly proves that $f_*\colon H_0(X) \to H_0(Y)$ is an isomorphism. For the case when $X = Y$, recall from Proposition 6.2.3 that a singular 0-chain $\sum r_i x_i$ in X is a boundary if and only if $\sum r_i = 0$. For any 0-chain $c = \sum r_i x_i$, we find that $f_\#(c) = \sum r_i f(x_i)$. But the 0-chain $c - f_\#(c) = \sum r_i x_i - \sum r_i f(x_i)$ has the property that the sum of their coefficients is zero. This means $c - f_\#(c)$ is a boundary, i.e., $\{c\} = \{f_\#(c)\} = f_*\{c\}$ in $H_0(X)$. This proves that f_* is the identity map. ■

Proposition 6.3.14. *If $f\colon X \to Y$ is a constant map, then the induced homomorphism $f_*\colon H_q(X) \to H_q(Y)$ in homology is zero, for all $q > 0$.*

Proof. To see this, suppose $f(x) = y_0 \in Y$ for every $x \in X$. Then the map $f\colon X \to Y$ factors as $X \xrightarrow{c} \{y_0\} \xrightarrow{i} Y$ where c is the obvious constant map and i is the inclusion. Hence the induced homomorphism $f_*\colon H_q(X) \to H_q(Y)$ factors as $H_q(X) \xrightarrow{c_*} H_q(\{y_0\}) \xrightarrow{i_*} H_q(Y)$, for all $q \geq 0$. Since $H_q(\{y_0\}) = 0$, for all $q > 0$, $f_* = i_* \circ c_* = 0$. ■

Reduced Homology Groups

Let $X \neq \emptyset$ and let $S_*(X)$ denote the singular chain complex of the space X. We consider an abstract chain complex C_q such that $C_0 = R$, the R-module R, and $C_q = 0$ for all $q \neq 0$. Then there is a chain map $\epsilon\colon S_*(X) \to C_*$ defined by $\epsilon(\sum r_i x_i) = \sum r_i$ at the zero level and $\epsilon = 0$ in positive dimensions. This chain map ϵ is clearly an onto map. The $\ker \epsilon$, which is a sub chain complex of $S_*(X)$, is denoted by $\tilde{S}_*(X)$ and is called the **reduced singular chain complex** of X. Since $S_*(X)$ is a chain complex of free R-modules, the augmentation map ϵ splits, and we find that $S_0(X) = \tilde{S}_0(X) \oplus R$ and $S_q(X) = \tilde{S}_q(X)$ for all $q > 0$. We define the **reduced singular homology modules** of X, denoted by $\tilde{H}_q(X)$, to be the homology of the chain complex $\tilde{S}_*(X)$. Since $\epsilon(\text{Im } \partial_1) = 0$, we find that $\text{Im } \partial_1 \subseteq \ker \epsilon = \tilde{S}_0(X)$. Therefore,

$$H_0(X) = \frac{S_0(X)}{\text{Im } \partial_1} = \frac{\tilde{S}_0(X)}{\text{Im } \partial_1} \oplus R = \tilde{H}_0(X) \oplus R$$

and $H_q(X) = \tilde{H}_q(X)$ for all $q > 0$. This last result says that the reduced singular homology modules are completely determined in terms of the usual singular homology modules of X and vice-versa. The reduced groups are very useful in computations of homology groups.

What is a basis for the reduced group $\tilde{H}_0(X)$ with integer coefficients? By definition, all the elements of $\tilde{S}_0(X)$ are reduced 0-cycles, and any element of $\tilde{S}_0(X)$ is a singular 0-chain $c = \sum n_i x_i$ in X which satisfies the condition that $\epsilon(\sum n_i x_i) = 0$. Here we are identifying a singular 0-simplex $\Delta_0 \to \{x_i\}$ by the point x_i of X.

Let us fix some point $x_0 \in X$. Then for any $x \in X$, $x - x_0$ is reduced 0-cycle, since the sum of their coefficients is zero. If x and x_0 both lie in the same path components of X, then there is a path $\omega\colon I \to X$ such that $\omega(0) = x$ and $\omega(1) = x_0$. But this means $\partial_1 \omega = x_0 - x$ is a boundary, and hence it represents the zero element of $\tilde{H}_0(X)$. It is also now clear that if x and x_0 lie in different path components of X, then the reduced 0-cycle $x - x_0$ is not a zero element of $\tilde{H}_0(X)$. Let us choose one point x_i $(i \neq 0)$ in each path component X_i of X. Then we claim that the set $\{x_i - x_0\}$ forms a basis of $\tilde{H}_0(X)$. To see this, let $c = \sum n_i x_i$ be a reduced 0-cycle. Then $\sum n_i = 0$. Hence

$$\begin{aligned} c = \sum n_i x_i &= \sum n_i x_i - (\sum n_i) x_0 \\ &= \sum n_i (x_i - x_0), \end{aligned}$$

shows that the set $\{x_i - x\}$ generates the group $\tilde{H}_0(X)$. Also, this set is linearly independent, since

$$\sum n_i (x_i - x_0) = 0$$

in $\tilde{H}_0(X)$ means $\sum n_i(x_i - x_0)$ is a boundary, i.e., there is a chain $\sum d$ such that $\partial(\sum d) = \sum n_i(x_i - x_0)$. Since a singular 1-simplex is simply a path and hence it lies in a single path component, say X_i of X. Collecting all paths in X_i, we find that $\partial(\sum d_i) = (x_i - x_0)$ for each i. Applying ϵ on both sides, we get $0 = n_i$. Hence, the set $\{x_i - x\}$ is also linearly independent. Thus, these elements of $\tilde{H}_0(X)$ form a basis for $\tilde{H}_0(X)$.

Let $f\colon X \to Y$ be a continuous map between two nonempty spaces and consider the induced homomorphism $f_\#\colon S_0(X) \to S_0(Y)$. Since $f_\#$ maps $B_0(X)$ into $B_0(Y)$, we find that $f_\#$ induces a homomorphism $\tilde{f}_\#\colon \tilde{S}_0(X) \to \tilde{S}_0(Y)$ which, in turn, induces a homomorphism $f_*\colon \tilde{H}_0(X) \to \tilde{H}_0(Y)$. Now, it can be easily verified that these induced homomorphisms in reduced singular homology satisfy the two functorial properties analogous to Propositions 6.3.9 and 6.3.10.

Definition 6.3.15. *A space X is said to be* **acyclic** *if $\tilde{H}_q(X) = 0$ for all $q \geq 0$.*

A point-space is evidently acyclic. After proving the homotopy axiom for singular homology, we will see that all contractible spaces are also acyclic. Let us compute the singular homology of convex subsets of the Euclidean space, like an n-disc, n-simplex and a cone etc. First we have a little general concept.

Definition 6.3.16. *A subset X of some Euclidean space $\mathbb{R}^n$ is said to be* **star-convex with respect to a point** *$x_0 \in X$ if each point $x \in X$ can be joined to x_0 by a line segment which lies in X.*

For example, a disc $\mathbb{D}^n$ is star convex with respect to any of its points. The following subspace of the plane $\mathbb{R}^2$ is not convex, but is star convex with respect to $x_0 \in X$.

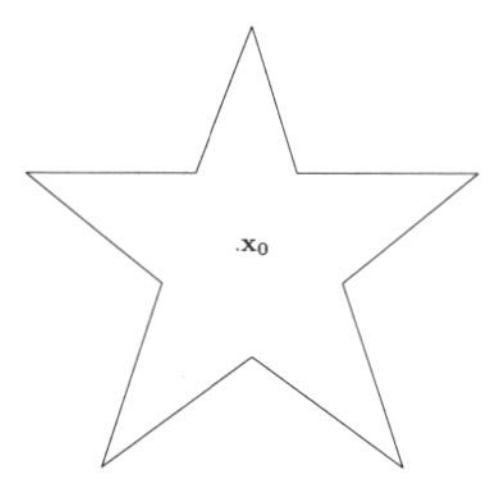

Let us have the following

Definition 6.3.17. *Let $X \subset \mathbb{R}^n$ be star convex relative to $x_0 \in X$. We define a* **bracket operation** *$[\,\cdot\,,\, x_0]\colon S_q(X) \to S_{q+1}(X)$ as follows: Let σ be a singular q-simplex in X. We define $[\sigma, x_0] \in \Delta_{q+1}(X)$ to be the linear map which carries the line segment joining $x \in \Delta_q$ to $e_{q+1} \in \Delta_{q+1}$ to the line segment joining*

$\sigma(x)$ to x_0. Then we extend this map linearly, i.e., if $c \in S_q(X)$ and $c = \sum r_i\sigma_i$, then we put

$$[c, x_0] = \sum r_i[\sigma_i, x_0].$$

Note that the map $[c, x_0]$ when restricted to the face Δ_q of Δ_{q+1} gives the chain c itself. Also, notice that if σ is a linear map

$$\sigma(e_0, \ldots, e_q) \to (y_0, \ldots, y_q)$$

then $[\sigma, x_0]$ is also the linear map taking $(e_0, \ldots, e_q, e_{q+1})$ to $(y_0, \ldots, y_q, x_0)$. We can now prove the following

Lemma 6.3.18. *The map $[\sigma, x_0]\colon \Delta_{q+1} \to X$ is continuous.*

Proof. Clearly, there is the quotient map $\Delta_q \times I \to \Delta_{q+1}$ defined by $H(x,t) = (1-t)x + t \cdot e_{q+1}$ which collapses the top $\Delta_1 \times \{1\}$ to a single point e_{q+1} and is 1-1 otherwise. The continuous map $f\colon \Delta_q \times I \to X$ defined by

$$f(x,t) = (1-t)\sigma(x) + tx_0$$

is constant on $\Delta_q \times \{1\}$ and hence induces a continuous map $\Delta_{q+1} \to X$, which is nothing but the map $[\sigma, x_0]$. ■

Now we have the following

Proposition 6.3.19. *Suppose X is a star convex with respect to $x_0 \in X$ and c is a singular q-chain in X. Then*

$$\partial[c, x_0] = \begin{cases} [\partial c, x_0] + (-1)^{q+1}c, & q > 0 \\ \epsilon(c)\sigma_{x_0} - c, & q = 0 \end{cases}$$

where σ_{x_0} is the singular 0-simplex mapping Δ_0 to x_0.

Proof. If σ_x is a singular 0-simplex, then $[\sigma_x, x_0]$ maps the simplex Δ_1 linearly onto the line segment from $\sigma_x(\Delta_0)$ to x_0. Then $\partial[\sigma_x, x_0] = \sigma_{x_0} - \sigma_x$, and the second part of the formula is verified.

Suppose $q > 0$. It suffices to check the formula for singular q-simplexes σ. Let us compute

$$\partial[\sigma, x_0] = \sum_0^{q+1} (-1)^i [\sigma, x_0] \circ F^i,$$

where $F^i\colon \Delta_q \to \Delta_{q+1}$ denotes the linear map

$$(e_0, \ldots, \hat{e_i}, \ldots, e_{q+1}) \mapsto (e_0, \ldots, 0, \ldots, e_{q+1}).$$

The map F^{q+1} is simply the inclusion map $\Delta_q \to \Delta_{q+1}$. Since restriction of $[\sigma, x_0]$ to Δ_q equals σ itself, the last term in above summation is $(-1)^{q+1}\sigma$.

To complete the proof, let us consider $[\sigma, x_0] \circ F^i$ where $i < q+1$. The map F^i carries Δ_q homeomorphically onto the ith face of Δ_{q+1}, it carries $(e_0, \dots, e_q)$ to $(e_0, \dots, e_{i-1}, e_{i+1}, \dots, e_{q+1})$ respectively. Therefore, the restriction of F^i to $\Delta_{q-1} = (e_0, \dots, e_{q-1})$ carries this simplex by a linear map onto the simplex spanned by $(e_0, \dots, e_{i-1}, e_{i+1}, \dots, e_q)$ since $i < q+1$. Thus

$$F^i\big|_{\Delta_{q-1}} = F^i(e_0, \dots \hat{e_i}, \dots, e_q)$$

Now we compute $[\sigma, x_0] \circ F^i \colon \Delta_q \to X$. Let x be a general point of Δ_{q-1}. Since $F^i \colon \Delta_q \to \Delta_{q+1}$ is a linear map, it carries the line segment from x to e_q linearly onto the line segment from $F^i(x)$ to e_{q+1}. Since $F^i(x) \in \Delta_q$, the map $[\sigma, x_0] \colon \Delta_{q+1} \to X$ carries the line segment linearly onto the line segment joining $\sigma(F^i(x))$ to x_0. Therefore, by definition,

$$[\sigma, x_0] \circ F^i = [\sigma \circ (F^i\big|_{\Delta_{q-1}}), x_0]$$

Putting the value of $F^i\big|_{\Delta_{q-1}}$ obtained earlier, we get

$$\begin{aligned} \partial[\sigma, x_0] &= \sum_0^q (-1)^i [\sigma_0 F(e_0, \dots, \hat{e_i}, \dots, e_q), x_0] + (-1)^{q+1}\sigma \\ &= [\partial\sigma, x_0] + (-1)^{q+1}\sigma. \end{aligned}$$

■

Theorem 6.3.20. *Let X be a star convex subset of $\mathbb{R}^n$ with respect to some point $x_0 \in X$. Then X is acyclic in singular homology.*

Proof. First we prove that $\tilde{H}_0(X) = 0$. Recall that a singular 0-chain in X is a 0-cycle in reduced singular homology $\tilde{H}_0(X)$ iff the augumentation map ϵ carries it to zero. Let c be a singular 0-chain such that $\epsilon(c) = 0$. Then by previous proposition

$$\partial[c, x_0] = \epsilon(c)\sigma_{x_0} - c = -c.$$

Hence c is a boundary, i.e., $\tilde{H}_0(X) = 0$. To show that $H_q(X) = 0$ for $q > 0$, let z be a singular q-cycle in X. By the previous proposition again

$$\partial[z, x_0] = [\partial z, x_0] + (-1)^{q+1}z = (-1)^{q+1}z,$$

which is a boundary. ■

Corollary 6.3.21. *Any simplex Δ_q, $q \geq 0$ is acyclic in singular homology. The product $\Delta_q \times I$ is also acyclic.*

We will compute the singular homology of n-spheres $\mathbb{S}^n, n \geq 0$ and of several other interesting spaces later. In order to compute the singular homology of various other topological spaces, it will be quite cumbersome if we apply the definition of singular homology and make the computation. Hence, first we should prove some general results or theorems which could be effectively used to make the actual computation a pleasant task. It is very similar to computing the differential coefficients of various differentiable functions in differential calculus. With this objective in our mind, we will now prove some basic theorems about singular homology – in fact these theorems will later become basic axioms of 'a homology theory' so that we can compute the singular homology using only these axioms and avoid the definition completely. We call these theorems as axioms so that their importance is indicated right in the beginning.

Exercises

1. Prove that the n-simplex Δ^n is acyclic for all $n \geq 0$. (**Hint:** Regard $\mathbb{Z}$ as a chain complex C_* where $C_0 = \mathbb{Z}$ and $C_q = 0$, for all $q \neq 0$. Prove that the augmentation homomorphism $\epsilon\colon S_*(\Delta^n) \to \mathbb{Z}$ is a chain equivalence. To do so, fix a point $v_0 \in \Delta^n$ and define a homomorphism $\tau\colon \mathbb{Z} \to S_0(\Delta_n)$ by putting $\tau(1) = v_0$. Then show that $\epsilon \circ \tau = I_{\mathbb{Z}}$ and $\tau \circ \epsilon$ is chain homotopic to the identity map $I_{S_*(\Delta_n)}$.)

2. Compute the singular homology with integer coefficients of the Topologist's Sine curve.

3. Compute the singular homology groups of the space X if
(a) X is the discrete space having countable number of points.
(b) X is the space of all rational numbers.
(c) X is the space of all irrational numbers.
(d) X is the classical Cantor set.

6.4 Homotopy Axiom for Singular Homology

Here, we first prove a special case (the absolute version) of the homotopy axiom. The general case (the relative version) will be proved later. Note that if $f\colon X \to Y$, $g\colon Z \to W$ are two continuous maps, then the product map $f \times g\colon X \times Z \to Y \times W$ defined by $(f \times g)(x, z) = (f(x), g(z))$, is continuous.

Theorem 6.4.1. (Homotopy Axiom). *Let f, $g\colon X \to Y$ be two maps which are homotopic. Then for all $q \geq 0$, $f_* = g_*\colon H_q(X) \to H_q(Y)$, i.e., the homotopic maps induce identical maps in singular homology.*

Proof. Let $F\colon X \times I \to Y$ be a homotopy from f to g so that $F(x, 0) = f(x)$ and $F(x, 1) = g(x)$ for all $x \in X$. Let i_0, $i_1\colon X \to X \times I$ be the bottom and top inclusion maps. Then, clearly, $f = F \circ i_0$ and $g = F \circ i_1$. Therefore, it is enough to prove that $i_{0*} = i_{1*}\colon H_q(X) \to H_q(X \times I)$ for all $q \geq 0$. Since chain homotopic maps induce identical maps in homology, we will prove that the two chain maps $i_{0\#}$, $i_{1\#}\colon S(X) \to S(X \times I)$ are chain-homotopic: For this, we will define chain-homotopy simultaneously for all spaces X and by induction on q:

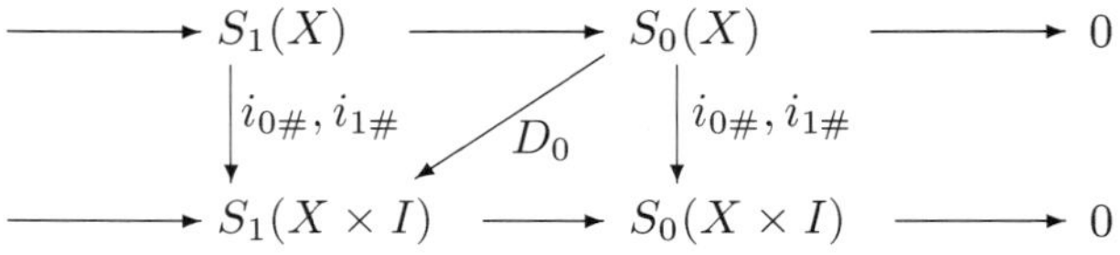

$$
\begin{array}{ccccccc}
\longrightarrow & S_1(X) & \longrightarrow & S_0(X) & \longrightarrow & 0 \\
& \Big\downarrow i_{0\#}, i_{1\#} & \swarrow D_0 & \Big\downarrow i_{0\#}, i_{1\#} & & \\
\longrightarrow & S_1(X \times I) & \longrightarrow & S_0(X \times I) & \longrightarrow & 0
\end{array}
$$

To start, let us define $D_0 \colon S_0(X) \to S_1(X \times I)$ by $D_0(\sigma) = \sigma \times I$, where σ is a singular 0-simplex in X and $\sigma \times I$ denotes the singular 1-simplex in $X \times I$. Extend D_0 to all 0-chains by linearity. Then, for all $x \in X$, we have

$$
\begin{aligned}
\partial D_0(x) &= \partial(x \times I) \\
&= (x, 1) - (x, 0) \\
&= i_1(x) - i_0(x).
\end{aligned}
$$

This implies $\partial D_0 = i_{1\#} - i_{0\#}$.

Note that D_0 is a natural map, i.e., if $h \colon X \to Y$ is a continuous map, then the diagram

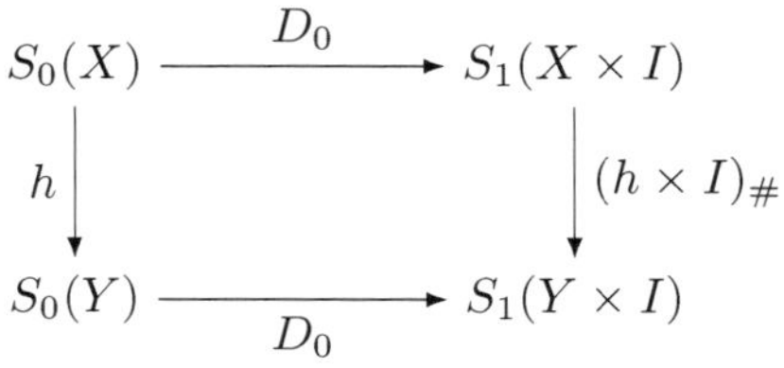

$$
\begin{array}{ccc}
S_0(X) & \xrightarrow{D_0} & S_1(X \times I) \\
h \Big\downarrow & & \Big\downarrow (h \times I)_\# \\
S_0(Y) & \xrightarrow[D_0]{} & S_1(Y \times I)
\end{array}
$$

is commutative because for any singular 0-simplex σ in X, we have

$$
\begin{aligned}
(h \times I)_\# \circ D_0(\sigma) &= (h \times I) \circ (\sigma \times I) \\
&= (h \circ \sigma) \times I \\
&= D_0(h\sigma) \\
&= D_0 h_\#(\sigma).
\end{aligned}
$$

Having defined D_0 for all spaces, we now assume that $D_j \colon S_j(X) \to S_{j+1}(X \times I)$ has been defined for all $j \leq q - 1$ and for all spaces having the following two properties:

(a) $\partial_{j+1} D_j + D_{j-1} \partial_j = i_{1\#} - i_{0\#}$.

(b) For each map $h \colon X \to Y$, $D_j h_\# = (h \times I)_\# D_j$.

Let us observe that in view of (a) for $j = q - 1$, we have

$$
\begin{aligned}
\partial_q (i_{1\#} - i_{0\#}) &= (i_{1\#} - i_{0\#}) \partial_q \\
&= (\partial_q D_{q-1} - D_{q-2} \partial_{q-1}) \partial_q \\
&= \partial_q D_{q-1} \partial_q.
\end{aligned}
$$

This means

$$\partial_q(i_{1\#} - i_{0\#} - D_{q-1}\partial_q) = 0$$

for all spaces X.

To construct D_q for all spaces X, let us assume that $\delta_q \colon \Delta_q \to \Delta_q$ denotes the identity map. Then taking $X = \Delta_q$, we find that

$$(i_{1\#} - i_{0\#} - D_{q-1}\partial_q)(\delta_q)$$

is a q-cycle in $S_q(\Delta_q \times I)$. Since the space $\Delta_q \times I$ is acyclic, there is a $w \in S_{q+1}(\Delta_q \times I)$ such that

$$\partial_{q+1}(w) = (i_{1\#} - i_{0\#} - D_{q-1}\partial_q)(\delta_q).$$

We define $D_q(\delta_q) = w$. This yields

$$\partial_{q+1}D_q(\delta_q) + D_{q-1}\partial_q(\delta_q) = (i_{1\#} - i_{0\#})(\delta_q).$$

Then, for a singular q-simplex σ in $S_q(X)$, we define

$$D_q(\sigma) = (\sigma \times I)_\#(D_q(\delta_q)),$$

and extend it by linearity to $S_q(X) \to S_{q+1}(X \times I)$. Now, we will show that D_q also satisfies both (a) and (b). Condition (b) follows at once because

$$\begin{aligned} D_q h_\#(\sigma) &= D_q(h\sigma) \\ &= (h\sigma \times I)_\# D_q(\delta_q) \\ &= (h \times I)_\#(\sigma \times I)_\#(D_q(\delta_q)) \\ &= (h \times I)_\# D_q(\sigma). \end{aligned}$$

To verify (a), we compute

$$\begin{aligned} \text{(i) } \partial_{q+1}D_q(\sigma) &= \partial_{q+1}(\sigma \times I)_\# D_q(\delta_q) \\ &= (\sigma \times I)_\# \partial_{q+1} D_q(\delta_q). \end{aligned}$$

Writing $\sigma = \sigma_\#(\delta_q)$ and applying ∂_q yields

$$\partial_q \sigma = \sigma_\# \partial_q(\delta_q).$$

Therefore,

$$\begin{aligned} \text{(ii) } D_{q-1}\partial_q\sigma &= D_{q-1}\sigma_\#\partial_q(\delta_q) \\ &= (\sigma \times I)_\# D_{q-1}\partial_q(\delta_q), \qquad \text{by (b)}. \end{aligned}$$

Combining (i) and (ii) together, we have

$$\begin{aligned} & (\partial_{q+1}D_q + D_{q-1}\partial_q)(\sigma) \\ = \; & (\sigma \times I)_\#(\partial_{q+1}D_q + D_{q-1}\partial_q)(\delta_q) \\ = \; & (\sigma \times I)_\#(i_{1\#} - i_{0\#})(\delta_q) \\ = \; & (i_{1\#} - i_{0\#})\sigma_\#(\delta_q), \quad \text{because } (\sigma \times I)i_j = i_j\sigma; \quad j = 0, 1. \\ = \; & (i_{1\#} - i_{0\#})(\sigma). \end{aligned}$$

This completes the induction and hence the proof. ■

Corollary 6.4.2. *Let f, $g\colon X \to Y$ be two homotopic maps. Then the induced chain maps $f_\#$, $g_\#\colon S_*(X) \to S_*(Y)$ are chain-homotopic.*

Proof. We have already shown that the bottom and top inclusion maps i_0, $i_1\colon X \to X \times I$ induce chain-homotopic maps $S_*(X) \to S_*(X \times I)$. Since $f = F \circ i_0$, $g = F \circ i_1$, we find $f_\# = F_\# \circ i_{0\#}$ is chain homotopic to $F_\# \circ i_{1\#} = g_\#$, because composites of chain-homotopic maps are again chain-homotopic. ■

The following result is of deeper consequence and establishes the homotopy invariance of singular homology.

Corollary 6.4.3. *If $f\colon X \to Y$ is a homotopy equivalence, then for all $q \geq 0$, $f_*\colon H_q(X) \to H_q(Y)$ is an isomorphism.*

Proof. Let $g\colon Y \to X$ be a continuous map such that $g \circ f$ is homotopic to I_X and $f \circ g$ is homotopic to I_Y. Then, by Proposition 6.3.9, Proposition 6.3.10 and Theorem 6.4.1, the induced homomorphisms $f_*\colon H_q(X) \to H_q(Y)$ and $g_*\colon H_q(Y) \to H_q(X)$ have the property that $g_* \circ f_* = (g \circ f)_* = (I_X)_* = I_{H_q(X)}$ and $f_* \circ g_* = I_{H_q(Y)}$, for all $q \geq 0$. This proves the desired result. ■

Remark 6.4.4. The proof of homotopy invariance theorem is indeed a special case of a very powerful theorem called "acyclic model theorem". We refer the interested reader to [18] p. 164–167 for details. Also, the above corollary has a very interesting consequence: it says that any two spaces having the same homotopy type will have isomorphic homology groups. **In particular, if we find that two spaces X, Y are such that $H_q(X)$ is not isomorphic to $H_q(Y)$ for some $q \geq 0$, then the spaces X, Y cannot be of the same homotopy type, and so they cannot be homeomorphic in the first place.** We frequently use this observation to distinguish between two non-homeomorphic spaces.

6.5 Relative Homology and the Axioms

By a **topological pair** (X, A) we mean a topological space X along with a subspace A of X. When $A = \emptyset$, we identify the pair $(X, \emptyset)$ with the space X and thus the concept of a pair can be considered as generalization of topological spaces. If (X, A) and (Y, B) are two such pairs, then a continuous map $f\colon X \to Y$ is called a **map of pairs** if $f(A) \subseteq B$; we write such maps as $f\colon (X, A) \to (Y, B)$. It is clear that the composite of two maps of pairs is again a map of pairs. It follows that the class of all pairs, with morphisms as maps of pairs, is a category which includes the category of all topological spaces as a subcategory.

We will now define the singular homology of a pair (X, A). Note that the singular chain complex $S_*(A)$ of the space A is a subcomplex of the singular chain complex $S_*(X)$. Hence we get the quotient chain complex $S_*(X)/S_*(A)$, which we denote by $S_*(X, A)$. This chain-complex is called the singular-chain complex of the pair (X, A) and its homology modules, denoted by $H_q(X, A)$, $q \geq 0$, are called the **relative singular homology** modules of the pair (X, A). In the case $A = \phi$, the relative homology modules are called **absolute homology** modules, distinguishing it from the relative case when $A \neq \phi$.

Let us consider the relative chain complex $S_*(X, A)$ a bit further. The kernel of the homomorphism $\frac{S_q(X)}{S_q(A)} \to \frac{S_{q-1}(X)}{S_{q-1}(A)}$, denoted by $Z'_q(X, A)$, is the module of all relative q-cycles. The image under the homomorphism $\frac{S_{q+1}(X)}{S_{q+1}(A)} \to \frac{S_q(X)}{S_q(A)}$, denoted by $B'_q(X, A)$, is the module of all relative q-boundaries of the pair (X, A). Each of these is a submodule of the quotient module $S_q(X)/S_q(A)$ and so their elements are cosets $c + S_q(A)$, where $c \in S_q(X)$. The relative homology modules $H_q(X, A) = Z'_q(X, A)/B'_q(X, A)$ are, therefore, quotients of the quotient modules. We make these simpler by defining

$$Z_q(X, A) = \{c \in S_q(X) \mid \partial(c) \in S_{q-1}(A)\},$$

$$B_q(X, A) = \{b \in S_q(X) \mid b - \partial(c') \in S_q(A), \text{ for some } c' \in S_{q+1}(X)\}.$$

Then it is clear that $Z'_q(X, A) = Z_q(X, A)/S_q(A)$ and $B'_q(X, A) = B_q(X, A)/S_q(A)$. Hence it follows that $H_q(X, A) = Z'_q(X, A)/B'_q(X, A) \cong Z_q(X, A)/B_q(X, A)$. With this interpretation, we have a better hold on the homology classes of the pair (X, A).

Induced Homomorphism for Pairs

Let $f\colon (X, A) \to (Y, B)$ be a map of pairs. The map $f\colon X \to Y$ induces a chain map $f_\#\colon S_*(X) \to S_*(Y)$ and its restriction $f|_A : A \to B$ induces a chain map $(f|_A)_\#\colon S_*(A) \to S_*(B)$. Hence there is an induced chain map in their quotients, i.e., we get a chain map $f_\#\colon S_*(X, A) \to S_*(Y, B)$. We are using the same notation $f_\#$ for the induced map in the relative as well as the absolute chain complexes, but this should cause no confusion. The above chain map clearly induces a homomorphism $f_*\colon H_q(X, A) \to H_q(Y, B)$ for all $q \geq 0$, in their singular homology modules. Strictly speaking, f_* should also have the index q attached to it, but we have suppressed it because homology modules always indicate the correct dimension. Now, the following functorial properties of these induced homomorphisms can be easily verified:

Proposition 6.5.1. (i) **(Identity Axiom):** *If $I\colon (X, A) \to (X, A)$ is the identity map, then the induced homomorphism $I_*\colon H_q(X, A) \to H_q(X, A)$ is also the identity map for all $q \geq 0$.*

(i) **(Composition Axiom):** *If $f\colon (X,A) \to (Y,B)$ and $g\colon (Y,B) \to (Z,C)$ are maps of pairs then, for all $q \geq 0$, the induced homomorphisms in homology satisfy the following:*

$$(g \circ f)_* = g_* \circ f_* \colon H_q(X,A) \to H_q(Z,C).$$

Next, we want to show that homotopic maps from (X,A) to (Y,B) induce identical maps in relative homology. To see this, recall that the two continuous maps $f,\ g\colon (X,A) \to (Y,B)$ are said to be homotopic if there exists a map $F\colon (X \times I, A \times I) \to (Y,B)$ of pairs such that $F(x,0) = f(x),\ F(x,1) = g(x)$ for all $x \in X$. With this definition, we have

Theorem 6.5.2. (Homotopy Axiom). *Let $f,\ g\colon (X,A) \to (Y,B)$ be homotopic maps. Then the induced homomorphisms in relative homology are identical, i.e., for all $q \geq 0$,*

$$f_* = g_* \colon H_q(X,A) \to H_q(Y,B).$$

Proof. To prove this theorem we go back to the proof of this result (see Theorem 6.4.1) in the absolute case. Let $i_0, i_1\colon (X,A) \to (X \times I, A \times I)$ be the bottom and top inclusion maps and let F be a homotopy from f to g. As in the absolute case, it evidently suffices to show that these bottom and top inclusion maps induce chain-homotopic maps $S_*(X,A) \to S_*(X \times I, A \times I)$. We have already proved in the previous section that there is a chain homotopy $D_q\colon S_q(X) \to S_{q+1}(X \times I)$, $q \geq 0$, between the chain maps induced in the absolute case and that the same is natural. We now consider the following part of the diagram of chain complexes:

$$\begin{array}{ccccccccc}
0 & \longrightarrow & S_q(A) & \longrightarrow & S_q(X) & \longrightarrow & S_q(X,A) & \longrightarrow & 0 \\
 & & \downarrow D_q & & \downarrow D_q & & \downarrow \bar{D}_q & & \\
0 & \longrightarrow & S_{q+1}(A \times I) & \longrightarrow & S_{q+1}(X \times I) & \longrightarrow & S_{q+1}(X \times I, A \times I) & \longrightarrow & 0
\end{array}$$

By the naturality of the chain homotopy $\{D_q\}$, the first square is commutative for all $q \geq 0$. Therefore, we get induced homomorphisms $\bar{D}_q\colon S_q(X,A) \to S_{q+1}(X \times I, A \times I)$ for all $q \geq 0$. This gives us the chain homotopy between the chain maps $i_{0\#},\ i_{1\#}$ induced in the relative chain complexes $S_*(X,A) \to S_*(X \times I, A \times I)$. Since $f = F \circ i_0,\ g = F \circ i_1$, we have $f_* = F_* \circ i_{0*}$ and $g_* = F_* \circ i_{1*}$, by the functorial properties of induced homomorphisms in homology. Now the theorem follows because $i_{0*} = i_{1*}$. ■

Next, let us consider a pair (X,A). Let $i\colon A \to X$ and $j\colon (X,\phi) \to (X,A)$ denote the inclusion maps. These maps induce homomorphisms $i_*\colon H_q(A) \to H_q(X)$ and $j_*\colon H_q(X) \to H_q(X,A)$ for all $q \geq 0$. With these in mind, we have the following important result:

Theorem 6.5.3. (Exactness Axiom). *For any pair (X, A), there is the following long exact sequence in homology:*

$$(*) \quad \cdots \longrightarrow H_q(A) \xrightarrow{i_*} H_q(X) \xrightarrow{j_*} H_q(X,A) \xrightarrow{\partial} H_{q-1}(A) \xrightarrow{i_*} \cdots,$$

where $\partial \colon H_q(X,A) \to H_{q-1}(A)$ is the connecting homomorphism and other homomorphisms are induced by inclusion maps.

Proof. The proof follows from the fundamental theorem of homological algebra (Theorem 7.2.22). Note that, for the given pair (X, A), the two inclusion maps $i \colon A \to X$ and $j \colon X \to (X, A)$ give rise to the following exact sequence of singular chain complexes

$$0 \longrightarrow S_*(A) \xrightarrow{i_\#} S_*(X) \xrightarrow{j_\#} S_*(X,A) \longrightarrow 0.$$

Therefore, by the fundamental theorem mentioned above, we get the long exact sequence $(*)$ in homology, where ∂ is the connecting homomorphism. ■

We also have the following

Theorem 6.5.4. (Commutativity Axiom). *The connecting homomorphism $\partial \colon H_q(X,A) \to H_{q-1}(A)$ is natural in the pair (X, A), i.e., if $f \colon (X,A) \to (Y,B)$ is a map of pairs, then the squares in the second diagram below are commutative for every $q \geq 0$.*

Proof. The map $f \colon (X,A) \to (Y,B)$ of pairs clearly gives continuous maps $f|_A : A \to B$ and $f \colon X \to Y$. Hence we get induced homomorphisms in their singular chain complexes making the following diagram of chain-complexes commutative in which, by definition, the rows are exact:

$$\begin{array}{ccccccccc}
0 & \longrightarrow & S_*(A) & \xrightarrow{i_\#} & S_*(X) & \xrightarrow{j_\#} & S_*(X,A) & \longrightarrow & 0 \\
 & & \downarrow (f|_A)_\# & & \downarrow f_\# & & \downarrow f_\# & & \\
0 & \longrightarrow & S_*(B) & \xrightarrow[i_\#]{} & S_*(Y) & \xrightarrow[j_\#]{} & S_*(Y,B) & \longrightarrow & 0
\end{array}$$

Now, if we write the two long exact homology sequences for pairs (X, A) and (Y, B) then, because the connecting homomorphism is natural in short exact sequences of chain complexes, the following diagram must be commutative for all $q \geq 0$:

$$\begin{array}{ccccccccc}
\longrightarrow & H_q(X) & \xrightarrow{j_*} & H_q(X,A) & \xrightarrow{\partial} & H_{q-1}(A) & \xrightarrow{i_*} & H_{q-1}(X) & \longrightarrow \\
 & \downarrow f_* & & \downarrow f_* & & \downarrow (f|_A)_* & & \downarrow f_* & \\
\longrightarrow & H_q(Y) & \xrightarrow[j_*]{} & H_q(Y,B) & \xrightarrow[\partial]{} & H_{q-1}(B) & \xrightarrow[i_*]{} & H_{q-1}(Y) & \longrightarrow
\end{array}$$

■

The exact homology sequence of a pair (X, A) yields some immediate consequences which are worth mentioning. We have

Proposition 6.5.5. *For any space X, $H_q(X, X) = 0$ for all $q \geq 0$.*

Proof. This follows at once when we write the exact homology sequence for the pair (X, X) and note that the inclusion map $i\colon X \to X$, which is identity, induces the identity homomorphism $i_*\colon H_q(X) \to H_q(X)$ for all $q \geq 0$. ■

Proposition 6.5.6. *If $A \subset X$ is a retract of X, then for all $q \geq 0$,*

$$H_q(X) \cong H_q(A) \oplus H_q(X, A),$$

and so, $H_q(A)$ is a direct summand of $H_q(X)$ for all $q \geq 0$.

Proof. Let $r\colon X \to A$ be a retraction map. Then $r \circ i\colon A \to A$ is the identity map. This means for all $q \geq 0$, $r_* \circ i_*\colon H_q(A) \to H_q(X) \to H_q(A)$ is also identity. Now, consider the exact homology sequence of the pair (X, A)

$$\to H_q(A) \underset{r_*}{\overset{i_*}{\rightleftarrows}} H_q(X) \xrightarrow{j_*} H_q(X, A) \xrightarrow{\partial} H_{q-1}(A) \to$$

Since $r_* \circ i_*$ is identity, i_* is a 1-1 map, i.e., $\ker i_* = 0 = \operatorname{Im} \partial$. But this says $j_*\colon H_q(X) \to H_q(X, A)$ is onto for all $q \geq 0$. In other words, for all $q \geq 0$, we have a short exact sequence

$$0 \longrightarrow H_q(A) \underset{r_*}{\overset{i_*}{\rightleftarrows}} H_q(X) \xrightarrow{j_*} H_q(X, A) \longrightarrow 0,$$

which splits. Therefore, for all $q \geq 0$, $H_q(X) \cong H_q(A) \oplus H_q(X, A)$. ■

Proposition 6.5.7. *If $A \subset X$ is a deformation retract, then $H_q(X, A) = 0$ for all $q \geq 0$.*

Proof. If A is a deformation retract, then the inclusion map $i\colon A \to X$ is a homotopy equivalence and the retraction $r\colon X \to A$ is its homotopy inverse. Hence, by Corollary 6.4.2, $i_*\colon H_q(A) \to H_q(X)$ is an isomorphism for all $q \geq 0$. Now it follows from the exact sequence of the pair (X, A) that $j_*\colon H_q(X) \to H_q(X, A)$ must be a zero map, and so, by exactness, $0 = \operatorname{Im} j_* = \ker \partial = H_q(X, A)$. ■

6.6 The Excision Theorem

Now, we come to the last theorem about singular homology which is, like the homotopy axiom, a bit difficult to prove. Since the proof involves several interesting ideas about the singular homology, it is all worthwhile. The final result, known as the **excision theorem for singular homology**, will complete our main objective of showing that the singular homology is a homology theory in the sense of Eilenberg-Steenrod. The theorem says that if $U \subseteq X$ is such that $\bar{U} \subseteq \text{int } A$, then the inclusion map $\rho\colon (X - U, A - U) \to (X, A)$, called **excision**, induces isomorphism $\rho_*\colon H_q(X - U, A - U) \to H_q(X, A)$ for all $q \geq 0$ in homology. To 'excise' means to 'cut' or 'remove' the subset U from the pair (X, A) making the pair smaller without affecting the homology groups of the pair (X, A). This theorem is a powerful tool not only in making computations of relative homology groups of the pair (X, A), but also in other contexts like homology manifolds, orientation of manifolds, etc.

Before we proceed any further, let us show that the foregoing hypothesis on the set U, which is to be excised from the pair (X, A), is really needed. We give an example to show that, in general, an excision map need not induce isomorphism in singular homology. We have

Example 6.6.1. Let us take the 2-sphere $\mathbb{S}^2$ and let $A = \{(x, y, 0) \in \mathbb{S}^2 \mid x \geq 0\}$, $U = P = (1, 0, 0)$. We will later show that $H_2(\mathbb{S}^2) \cong \mathbb{Z}$ for singular homology also, though we already know that this is true for the simplicial homology. Note that A is a part of the great circle of $\mathbb{S}^2$ lying in the (x, y)-plane and so it is homeomorphic to the closed interval [0,1]. Since the sequence

$$\cdots \to H_2(\mathbb{S}^2 - U) \to H_2(\mathbb{S}^2 - U, A - U) \to H_1(A - U) \to \cdots$$

for the pair $(\mathbb{S}^2 - U, A - U)$ is exact, and since $\mathbb{S}^2 - U$ is homeomorphic to $\mathbb{R}^2$, we find that $H_2(\mathbb{S}^2 - U, A - U) = 0$. On the other hand, the exact sequence

$$\cdots \to H_2(A) \to H_2(\mathbb{S}^2) \to H_2(\mathbb{S}^2, A) \to H_1(A) \to \cdots$$

shows that $H_2(\mathbb{S}^2, A) \cong H_2(\mathbb{S}^2) \cong \mathbb{Z}$. Thus we find that $H_2(\mathbb{S}^2 - U, A - U)$ is not isomorphic to $H_2(\mathbb{S}^2, A)$, i.e., the excision theorem is not valid in this case. The reason is also obvious, viz., int $A = \phi$ and so the hypothesis that $\bar{U} \subseteq \text{int } A$ is not satisfied.

Singular Chains of Small Sizes

Though the proof of the excision theorem is, in fact, motivated by the results of simplicial homology, it uses new ideas about singular homology not discussed so far. One important new result which we will prove says, roughly speaking, that singular homology of a space X can be computed using only chains of smaller sizes, rather than all chains. To make this point precise, let $\mathcal{U}$ be an open covering of X. A singular simplex $\sigma\colon \Delta_q \to X$ is said to be **small of order** $\mathcal{U}$ if $\sigma(\Delta_q) \subset U$ for some $U \in \mathcal{U}$. Now, observe that if σ is small of order $\mathcal{U}$, then so is every face $\sigma^{(i)}$ of σ. A singular chain c in X is said to be small of order $\mathcal{U}$ if every singular simplex occuring in c is small of order $\mathcal{U}$. Let $S_q(X, \mathcal{U})$ denote the free abelian group generated by all singular q-simplexes of X which are small of order $\mathcal{U}$. Then it is now clear that $\{S_q(X, \mathcal{U})\}_{q \geq 0}$ forms a sub chain complex of the singular chain complex $\{S_q(X)\}_{q\geq 0}$. The interesting result that we will prove is that the inclusion chain map $S_*(X,\mathcal{U}) \subset S_*(X)$ induces isomorphism in homology $H_q(X,\mathcal{U}) \to H_q(X)$ for all $q \geq 0$. We also observe that both the chain complexes $S_*(X,\mathcal{U})$ and $S_*(X)$ consist of free abelian groups, it follows from a result in homological algebra (see E.H. Spanier [18] Theorem 10 on p. 192) that the inclusion map $S_*(X,\mathcal{U}) \to S_*(X)$ is a chain homotopy equivalence.

Now, let us consider singular simplexes in a fixed n-simplex Δ_n. A singular q-simplex $\sigma\colon \Delta_q \to \Delta_n$ is said to be **affine** if $\sigma(\sum t_i p_i) = \sum t_i \sigma(p_i)$, where $t_i \in I$ and $\sum t_i = 1$. Note that if σ is affine then each of its faces $\sigma^{(i)}$ is also affine, $i = 0, 1, \ldots, q$. The set of all affine simplexes in Δ_n generate a subcomplex $S'_*(\Delta_n) \subset S_*(\Delta_n)$.

It is evident that an affine simplex σ in Δ_n is completely determined by the images of its vertices: If $x_0, x_1, \ldots, x_q$ are points in Δ_n, then let us denote by $(x_0, x_1, \ldots x_q)$ the affine simplex which maps the vertex p_i of Δ_q into the point x_i, $0 \leq i \leq q$. Then it is clear that

$$\partial(x_0, x_1, \ldots, x_q) = \sum(-1)^i(x_0, \ldots, \hat{x}_i, \ldots, x_q).$$

Also, notice that the identity map $\xi_n\colon \Delta_n \to \Delta_n$ is a linear simplex $(p_0, \ldots, p_n)$; this ξ_n will play a fundamental role.

Now, let b_n denote the barycentre of Δ_n, i.e., $b_n = \sum \frac{1}{n+1} p_i$. For $q \geq 0$, we define a homomorphism

$$\beta_n\colon S'_q(\Delta_n) \to S'_{q+1}(\Delta_n)$$

by the formula

$$\beta_n(x_0, \ldots, x_q) = (b_n, x_0, \ldots, x_q).$$

Then, as we have seen in the case of simplicial homology, β_n gives a **chain contraction** of the chain complex $S'(\Delta_n)$. An epimorphism $\epsilon\colon C_0 \to \mathbb{Z}$, such that the composite $C_1 \xrightarrow{\partial_1} C_0 \xrightarrow{\epsilon} \mathbb{Z}$ is trivial, is called an **augmentation** for a chain complex C. A nonnegative chain complex C with an augmentation is called an **augmented chain complex**.

Now, if $\tau\colon \mathbb{Z} \to S'_0(\Delta_n)$ is defined by $\tau(1) = (b_n)$, then one can easily see by direct computation that $\beta_n\colon I_{S'(\Delta_n)} \cong \tau \circ \epsilon$ is a chain homotopy equivalence, i.e., $\partial\beta_n + \beta_n\partial = I_d$ on $S'_*(\Delta_n)$. In particular, $\partial\beta_n = I_d$ on all cycles.

Using the chain contraction β_n, we now define an augmentation preserving **subdivision chain map** $Sd\colon S(X) \to S(X)$ in **singular theory**, simultaneously along with a chain deformation $D\colon S(X) \to S(X)$ from Sd to $I_{S(X)}$, which will be natural in X, i.e., for each continuous map $f: X \to Y$, we will have commutative squares:

$$\begin{array}{ccc} S_q(X) & \xrightarrow{Sd} & S_q(X) \\ f_\# \downarrow & & \downarrow f_\# \\ S_q(Y) & \xrightarrow[Sd]{} & S_q(Y) \end{array} \qquad \begin{array}{ccc} S_q(X) & \xrightarrow{D} & S_{q+1}(X) \\ f_\# \downarrow & & \downarrow f_\# \\ S_q(Y) & \xrightarrow[D]{} & S_{q+1}(Y) \end{array}$$

Both will be defined by induction on q-chains, $q \geq 0$. If c is a 0-chain, we put $Sd(c) = c$ and $D(c) = 0$. Now, let $n \geq 1$ and assume that Sd and D both have been defined for $0 \leq q < n$. Then we define both of these on the singular n-simplex $\xi_n: \Delta_n \to \Delta_n$ by

$$Sd(\xi_n) = \beta_n(Sd(\partial\xi_n)) \tag{a}$$

$$D(\xi_n) = \beta_n(Sd(\xi_n) - \xi_n - D\partial(\xi_n)). \tag{b}$$

Then, for every singular n-simplex $\sigma: \Delta^n \to X$, let us define

$$Sd(\sigma) = \sigma_\#(Sd(\xi_n))$$

$$D(\sigma) = \sigma_\#(D(\xi_n)).$$

Now we have the following relations:

Proposition 6.6.2. (i) $\partial\, Sd = Sd\, \partial$, (ii) $\partial D = Sd - I_d - D\partial$.

Proof. (i) We have

$$\begin{aligned} Sd\,\partial(\sigma) &= Sd\,\partial(\sigma_\#(\xi_n)) \\ &= Sd\,\sigma_\#(\partial\xi_n) \\ &= \sigma_\# Sd(\partial\xi_n) \text{ (by naturality)} \\ &= \sigma_\#\partial Sd(\xi_n) \text{ (by (a) and the fact that } Sd(\partial\xi_n) \text{ is a cycle)} \\ &= \partial\, Sd(\sigma) \text{ (by definition)}. \end{aligned}$$

(ii) Here, we have

$$\begin{aligned} D\partial\sigma &= D\partial\sigma_\#(\xi_n) \\ &= D\sigma_\#(\partial\xi_n) \\ &= \sigma_\# D\partial(\xi_n) \end{aligned}$$

and

$$\begin{aligned} \partial D\sigma &= \partial\sigma_\#(D\xi_n) \\ &= \sigma_\#\partial D(\xi_n), \end{aligned}$$

so that for each σ,

$$\begin{aligned}(D\partial + \partial D)(\sigma) &= \sigma_{\#}((D\partial + \partial D)(\xi_n)) \\ &= \sigma_{\#}((Sd - I_d)(\xi_n)) \text{ by (b)} \\ &= (Sd - I_d)(\sigma).\end{aligned}$$

■

Note that the above properties of the chain homotopy have been established using the already known similar properties for affine simplexes. Yet another point which must be observed is that image of any σ under the maps Sd and D lie in $\sigma(\Delta_n)$.

Now, if X is a metric space and $c = \sum n_i\sigma_i$ is a singular q-chain in X, then we can define

$$\text{mesh } c = \sup\{\text{diam } \sigma_i(\Delta_q) \mid n_i \neq 0\}.$$

With this definition one can easily prove the following by induction on q:

Lemma 6.6.3. *Let Δ_n be the Euclidean n-simplex and c be an affine q-chain in Δ_n. Then*

$$\text{mesh } (Sd(c)) \leq \frac{q}{q+1} \text{ mesh } (c).$$

We can also define, by induction on m, the subdivision chain maps $Sd^m\colon S(X) \to S(X)$ of higher order m. We put $Sd^0 = I_{S(X)}$ and $Sd^m = Sd(Sd^{m-1})$, $m \geq 1$. From the remarks made earlier, it also follows that the subdivision operator Sd^m is also chain homotopic to the identity map $I_{S(X)}$. Since Δ_n is a metric space, it follows from the preceding lemma that if c is any affine q-chain in Δ_n, then for all $m \geq 1$, mesh $(Sd^m(c)) \leq (\frac{q}{q+1})^m$ mesh (c). Next, we have

Theorem 6.6.4. *Let $\{z\} \in H_n(X, A)$. Then, given any open covering $\mathcal{V}$ of X, we can find an $r > 0$ such that $Sd^r(z)$ is a linear combination of singular n-simplexes in X which are small of the order of $\mathcal{V}$.*

Proof. Let σ be a singular n-simplex occurring in the relative cycle z. It suffices to show that there exists an $r > 0$ such that each singular simplex occurring in $Sd^r(\sigma)$ is contained in V for some $V \in \mathcal{V}$. Note that $\sigma\colon \Delta_n \to X$ is continuous and Δ_n is a compact metric space. Therefore, the open cover $\{\sigma^{-1}(V) \mid V \in \mathcal{V}\}$ will have a Lebesgue number, say, λ. Now, we choose r so large that $(\frac{q}{q+1})^r \text{ diam}(\Delta_n) < \lambda$, which means mesh $(Sd^r(\xi_n)) < \lambda$, i.e., each singular simplex occurring in $\sigma_{\#}(Sd^r(\xi_n)) = Sd^r(\sigma)$ will be contained in V for some $V \in \mathcal{V}$. ■

Now, we prove what we stated in the beginning.

Theorem 6.6.5. *Let $\mathcal{U}$ be any open covering of X. Then the inclusion map $S(X,\mathcal{U}) \to S(X)$ induces isomorphism $H_q^{\mathcal{U}}(X) \to H_q(X)$ for all $q \geq 0$.*

Proof. To show that the above map is a monomorphism, let $\{z\} \in H_q^{\mathcal{U}}(X)$ which goes to zero under the above homomorphism. This means $z = \partial(e)$ for some $e \in S_{q+1}(X)$. It suffices to show that $z = \partial(e')$ for some $e' \in S_{q+1}(X,\mathcal{U})$. Note that we can find an integer k such that $Sd^k(e) \in S_{q+1}(X,\mathcal{U})$ and

$$\begin{aligned} Sd^k(e) - e &= D_k\partial(e) + \partial D_k(e) \\ &= D_k(z) + \partial D_k(e), \end{aligned}$$

which means

$$\partial Sd^k(e) - \partial(e) = \partial D_k(z)$$

and, therefore,

$$z = \partial(Sd^k(e) - D_k(z)) \in \partial(S_{q+1}(X,\mathcal{U})),$$

proving the result.

To prove the onto part, let $z \in S_q(X)$ such that $\partial(z) = 0$. We find an integer k such that $Sd^k(z) \in S_q(X,\mathcal{U})$. Then

$$\begin{aligned} Sd^k(z) - z &= D_k\partial(z) + \partial D_k(z) \\ &= \partial D_k(z). \end{aligned}$$

This means $Sd^k(z)$, being a cycle in $S_q(X,\mathcal{U})$, is homologous to z in $S_q(X)$, i.e., the class $\{Sd^k(z)\}$ goes to the class $\{z\}$ under the above homomorphism. ■

Now, we can prove

Theorem 6.6.6. (Excision Theorem). *Let (X, A) be any pair and $U \subseteq X$ such that $\bar{U} \subseteq \text{int } A$. Then the excision map $\rho\colon (X - U, A - U) \to (X, A)$ induces an isomorphism*

$$\rho_*\colon H_q(X - U, A - U) \to H_q(X, A) \textit{ for all } q \geq 0.$$

Proof. Let $\{z\} \in H_q(X, A)$, where z is a q-cycle in X mod A. Note that $\{X - \bar{U},\ \text{int } A\}$ is an open covering of X (see Fig. 6.3).

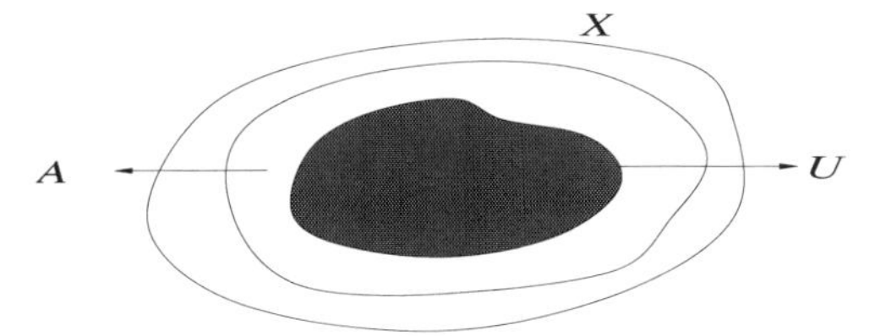

Fig. 6.3: U can be excised

Hence we can write $z = \sum_1^n n_i\sigma_i$, where each σ_i is small of the order of the covering $\{X - \bar{U},\ \text{int } A\}$. This means any σ_i which is not in int A will lie in $X - \bar{U} \subseteq X - U$. Dropping all those σ_i's which lie in int A, we find that $z' = \sum_{i=1}^m n_i\sigma_i$, where $m \leq n$, is a q-cycle in $X - U$ mod $A - U$. Since the inclusion map takes z' to itself and $\{z\} = \{z'\}$ in $H_q(X, A)$, we find that $\rho_*\{z'\} = \{z\}$. This shows ρ_* is onto.

To prove that it is one-one, let $\{z\} \in H_q(X - U, A - U)$ such that $z \sim 0$ in $H_q(X, A)$. Thus, $z = z' + \partial(w)$, where z' is a q-chain in A and w is $(q+1)$-chain in X. Now, we choose r so large that $Sd^r(w)$ is a linear combination of simplexes which are small of order $\{X - \bar{U},\ \text{int } A\}$. Thus, we can write $Sd^r(w) = w_1 + w_2$, where all singular simplexes occurring in w_1 map in $X - U$ and those occurring in w_2 map in A. Then we get

$$Sd^r(z) - \partial(w_1) = Sd^r(z') + \partial(w_2).$$

Since the left hand side is a chain in $X - U$, the right hand side is a chain in A, we find that each is a chain in $A - U$. This shows that $Sd^r(z) \sim 0$ in $(X - U)$ mod $(A - U)$. But this means $z \sim 0$ in $X - U$ mod $A - U$, i.e., ρ_* is a 1-1 map. ■

6.7 Homology and Cohomology Theories

Let us now summarize what we have done so far. Our ground ring R is a commutative principal ideal domain. For every pair (X, A), we defined a sequence $H_q(X, A)$, $q \geq 0$, of R-modules called **relative** homology module of the pair (X, A); the topological space X is identified with the pair $(X, \emptyset)$ in which case the relative homology modules $H_q(X, \emptyset) = H_q(X)$ are called **absolute** homology modules of X. Given a map $f\colon (X, A) \to (Y, B)$ of pairs, we showed that this induces a sequence $f_*\colon H_q(X, A) \to H_q(Y, B)$ of R-homomorphisms. Finally, we showed that for each pair (X, A) there exists a sequence $\partial\colon H_q(X, A) \to H_{q-1}(A)$ of R-homomorphisms, called **connecting** homomorphisms. Among other things, we proved that all of the above satisfy the following properties:

1. If $f\colon (X, A) \to (X, A)$ is the identity map, then the induced homomorphism $f_*\colon H_q(X, A) \to H_q(X, A)$ is also the identity map for all $q \geq 0$.

2. If $f\colon (X, A) \to (Y, B)$ and $g\colon (Y, B) \to (Z, C)$ are maps of pairs then, for all $q \geq 0$,

$$(g \circ f)_* = g_* \circ f_*\colon H_q(X, A) \to H_q(Z, C).$$

3. For any map $f\colon (X, A) \to (Y, B)$ of pairs, the square

$$\begin{array}{ccc} H_q(X,A) & \xrightarrow{\partial_{(X,A)}} & H_{q-1}(A) \\ {\scriptstyle f_*}\downarrow & & \downarrow{\scriptstyle (f|_A)_*} \\ H_q(Y,B) & \xrightarrow[\partial_{(Y,B)}]{} & H_{q-1}(B) \end{array}$$

is commutative for all $q \geq 1$. This says that for each pair (X, A), the connecting homomorphism $\partial_{(X,A)} \colon H_q(X,A) \to H_{q-1}(A)$ is a natural transformation from the functor $H_q(X, A)$ to the functor $H_{q-1}(A)$ for all $q \geq 1$.

4. For each pair (X, A), there is a long exact sequence of homology modules, viz.,

$$\cdots \longrightarrow H_q(A) \xrightarrow{i_*} H_q(X) \xrightarrow{j_*} H_q(X, A) \overset{\partial_{(X,A)}}{\longrightarrow} H_{q-1}(A) \longrightarrow \cdots .$$

5. If $f,\ g\colon (X,A) \to (Y,B)$ are homotopic maps of pairs then, for all $q \geq 0$,

$$f_* = g_* \colon H_q(X,A) \to H_q(Y,B).$$

6. Given a pair (X, A), let $U \subset X$ be such that $\bar{U} \subseteq \operatorname{int} A$. Then the excision map $\rho\colon (X-U, A-U) \to (X, A)$ induces isomorphism $\rho_* \colon H_q(X-U, A-U) \to H_q(X,A)$ for all $q \geq 0$.

7. If $X = \{p\}$ is a point-space then $H_q(X) = 0$ for all $q \neq 0$. The R-module $H_0(X) = G$ is the **coefficient** module.

All the results summarized above express the deep fact that the singular homology, defined for all pairs (X, A) of topological spaces, is a "homology theory" in the sense of Eilenberg and Steenrod. The seven properties, proved earlier and stated above, simply say that singular homology satisfies all the seven axioms of Eilenberg and Steenrod on the category of all topological pairs with morphisms as maps of these pairs. In fact, Eilenberg and Steenrod defined a homology theory, more generally, on a suitable admissible category. According to them, a **Homology Theory** (H, ∂) on an admissible category $\mathcal{C}$ of pairs consists of the following:

- A sequence $H_q(X, A)$, $q \in \mathbb{Z}$, of R-modules for each pair (X, A) in category $\mathcal{C}$.
- For each morphism $f\colon (X, A) \to (Y, B)$ in $\mathcal{C}$, there is a sequence of R-homomorphisms $f_*\colon H_q(X, A) \to H_q(Y, B)$, $q \in \mathbb{Z}$.
- For each pair (X, A), there is a natural transformation $\partial_{(X,A)} \colon H_q(X, A) \to H_{q-1}(A)$, $q \in \mathbb{Z}$.

In addition, these are required to satisfy the seven axioms stated above. Similarly, a **Cohomology Theory** (H, δ) on an admissible category $\mathcal{C}$ consists of the following:

- A sequence $H^q(X, A)$, $q \in \mathbb{Z}$, of R-modules for each pair (X, A) in category $\mathcal{C}$.
- For each morphism $f\colon (X, A) \to (Y, B)$ in $\mathcal{C}$, there is a sequence of R-homomorphisms $f^*\colon H^q(Y, B) \to H^q(X, A)$, $q \in \mathbb{Z}$.
- For each pair (X, A), there is a natural transformation $\delta_{(X,A)}\colon H^q(A) \to H^{q+1}(X, A)$, $q \in \mathbb{Z}$.

And these are required to satisfy the seven analogous axioms of Eilenberg and Steenrod (see [6] for details) for a cohomology, viz., they satisfy the identity axiom, composition axiom, exactness axiom, commutativity axiom, homotopy axiom, excision axiom and the dimension axiom.

On the basis of these axioms, Eilenberg and Steenrod derived interesting properties of any homology theory. As a simple illustration of these ideas, we now compute the homology groups of the n-sphere $\mathbb{S}^n$ using only these axioms. This will mean, in particular, that these are also the singular homology modules of $\mathbb{S}^n$. We state the results for the coefficient $R = \mathbb{Z}$, the additive group of integers.

First, let us prove the following version of the excision property for a homology theory.

Proposition 6.7.1. *Let $A \subset X$ and $V \subset U \subset A$. Suppose V can be excised and $(X - U, A - U)$ is a deformation retract of $(X - V, A - V)$. Then U can also be excised.*

Proof. Since $(X - U, A - U)$ is a deformation retract of $(X - V, A - V)$, the inclusion map $i\colon (X - U, A - U) \to (X - V, A - V)$ is a homotopy equivalence. This means $i_*\colon H_q(X - U, A - U) \to H_q(X - V, A - V)$ is an isomorphism by the functorial properties and the homotopy axiom of the given homology theory. Since $(X - U, A - U) \to (X, A)$ is just the composition of the inclusion maps i and the given excision map $(X - V, A - V) \to (X, A)$, it follows that U can be excised from the pair (X, A). ■

Homology of Spheres

Next, we are going to compute the integral homology groups of an n-sphere, $n \geq 0$, using only the axioms of a homology theory.

Proposition 6.7.2. *Let E_n^+ and E_n^- denote the northern and southern hemispheres of the n-spheres $\mathbb{S}^n$, $n \geq 1$, so that $\mathbb{S}^{n-1} = E_n^+ \cap E_n^-$ is their common* *boundary. Then the excision map* $(E_n^+, \mathbb{S}^{n-1}) \to (\mathbb{S}^n, E_n^-)$ *induces an isomorphism* $H_q(E_n^+, \mathbb{S}^{n-1}) \to H_q(\mathbb{S}^n, E_n^-)$ *for all* $q \geq 0$.

Proof. Note that if we could excise the open set $U = \text{Int}(E_n^-)$ from the pair $(\mathbb{S}^n, E_n^-)$, then the result would be obvious (see Fig. 6.4).

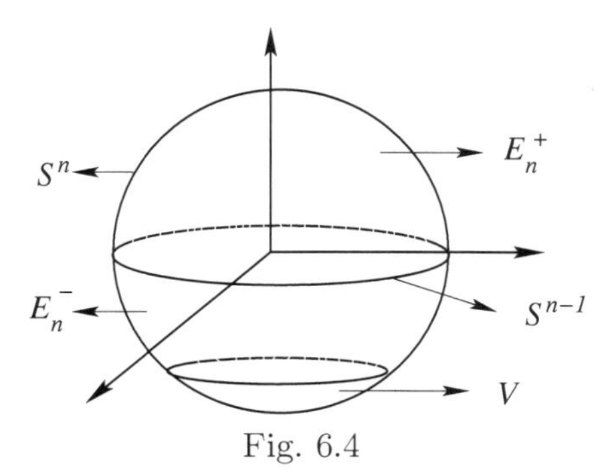

Fig. 6.4

Since $\bar{U}$ is not contained in $\text{Int}(E_n^-)$, the excision property cannot be applied directly. Let us define $V = \{x \in \mathbb{S}^n \mid x_{n+1} < -1/2\}$. Then V satisfies the hypothesis of the excision axiom. On the other hand one can readily verify that $(\mathbb{S}^n - U, E_n^- - U)$ is a deformation retract of $(\mathbb{S}^n - V, E_n^- - V)$ and so, by the preceding proposition, U can be excised proving the assertion. ■

Proposition 6.7.3. *Let* $\mathbb{S}^0 = \{1, -1\}$ *denote the 0-sphere. Then*

$$H_q(\mathbb{S}^0) = \begin{cases} \mathbb{Z} \oplus \mathbb{Z}, & q = 0 \\ 0, & \text{otherwise.} \end{cases}$$

Proof. Let $P_1 = \{1\}, P_2 = \{-1\}$ be two point-spaces. Then $\mathbb{S}^0 = P_1 \cup P_2$ is a discrete space in which P_1 and P_2 both are clopen sets. This means we can excise the set P_1 from the pair $(\mathbb{S}^0, P_1)$ so that the excision $k\colon P_2 \to (\mathbb{S}^0, P_1)$ induces an isomorphism $k_*\colon H_q(P_2) \to H_q(\mathbb{S}^0, P_1)$, for all $q \geq 0$. Now consider the exact homology sequence for the pair $(\mathbb{S}^0, P_1)$:

$$\cdots \longrightarrow H_q(P_1) \xrightarrow{i_{1*}} H_q(\mathbb{S}^0) \xrightarrow{j_*} H_q(\mathbb{S}^0, P_1) \to H_{q-1}(P_1) \longrightarrow \cdots,$$

where $i_1\colon P_1 \to \mathbb{S}^0$, $i_2\colon P_2 \to \mathbb{S}^0$, $j : \mathbb{S}^0 \to (\mathbb{S}^0, P_1)$ denote the inclusion maps. Note that $k = j \circ i_2$. Since k_* is an isomorphism, $(j \circ i_2)_* k_*^{-1}$ is identity on $H_q(\mathbb{S}^0, P_1)$, i.e., the sequence

$$H_q(P_1) \xrightarrow{i_{1*}} H_q(\mathbb{S}^0) \xrightarrow{j_*} H_q(\mathbb{S}^0, P_1)$$

splits for all $q \geq 0$. This means $H_q(\mathbb{S}^0) \cong H_q(P_1) \oplus H_q(P_2)$ for all $q \geq 0$. By dimension axiom, we know that for $i = 1, 2$

$$H_q(P_i) = \begin{cases} \mathbb{Z}, & \text{if } q = 0 \\ 0, & \text{otherwise} \end{cases}$$

and that proves the proposition. ■

Remark 6.7.4. If $X = \{a_1, a_2, \ldots, a_n\}$ is a discrete space having n-points, then it follows from the above result that

$$H_q(X) = \begin{cases} \bigoplus_1^n \mathbb{Z}, & q = 0 \\ 0, & \text{otherwise.} \end{cases}$$

Remark 6.7.5. The above consideration also yields the result that if a space $X = X_1 \cup X_2$ is the disjoint union of two clopen sets then, for all $q \geq 0$,

$$H_q(X) \cong H_q(X_1) \oplus H_q(X_2)$$

for any homology theory (H_q, ∂).

Definition 6.7.6. *If $X \neq \emptyset$ then we can define the reduced homology groups $\tilde{H}_q(X)$ for any homology theory as follows: Let $\epsilon\colon X \to P$ be the constant map, where P is a point-space. Then, if $i\colon P \to X$ is any map, we find that $\epsilon \circ i = I_P$ so that the induced homomorphism $\epsilon_*\colon H_q(X) \to H_q(P)$ is onto. We define $\tilde{H}_q(X) = \ker \epsilon_*$ and call it the* **reduced homology group** *of the space X.*

It follows from the above definition that the following sequence is exact:

$$0 \to \tilde{H}_0(X) \to H_0(X) \to H_0(P) \to 0.$$

Since the group $H_0(P) \cong \mathbb{Z}$ is just the coefficient group of the homology theory, we find that $H_0(X) \cong \tilde{H}_0(X) \oplus \mathbb{Z}$. We also have $\tilde{H}_q(X) = H_q(X)$ for all $q \neq 0$. Moreover, if $A \neq \emptyset$, we define $\tilde{H}_q(X, A) = H_q(X, A)$ for all $q \geq 0$. Now, it can be verified that the reduced homology groups of a nonempty pair (X, A) also satisfies the exactness axiom. This last fact is of frequent occurrence.

From the above remarks, we have

Theorem 6.7.7. *If X is a nonempty contractible space then $\tilde{H}_q(X) = 0$ for all $q \geq 0$.*

Now we have:

Theorem 6.7.8. *The reduced homology group of the n-sphere $\mathbb{S}^n$, $n \geq 0$, are given by*

$$\tilde{H}_q(\mathbb{S}^n) = \begin{cases} \mathbb{Z}, & \text{if } q = n \\ 0, & \text{otherwise.} \end{cases}$$

Proof. We observe that the theorem is true for $n = 0$ by Proposition 6.7.3. Also, we note that each of the hemispheres E_n^+ and E_n^- defined earlier is homeomorphic to the n-disk $\mathbb{D}^n$, which is contractible. Thus, by the homotopy and the dimension axiom, $\tilde{H}_q(E_n^+) = 0 = \tilde{H}_q(E_n^-)$ for all $q \geq 0$. Writing out the exact homology sequence for the pair $(\mathbb{S}^n, E_n^-)$, we have

$$\cdots \to \tilde{H}_q(E_n^-) \to \tilde{H}_q(\mathbb{S}^n) \to \tilde{H}_q(\mathbb{S}^n, E_n^-) \to \tilde{H}_{q-1}(E_n^-) \to \cdots.$$

From this exact sequence, we conclude that $\tilde{H}_q(\mathbb{S}^n) \cong \tilde{H}_q(\mathbb{S}^n, E_n^-)$. However, by the excision axiom, we have already proved that $\tilde{H}_q(\mathbb{S}^n, E_n^-) \cong \tilde{H}_q(E_n^+, \mathbb{S}^{n-1})$ for all $q \geq 0$. Then writing the following exact sequence

$$\cdots \to \tilde{H}_q(E_n^+) \to \tilde{H}_q(E_n^+, \mathbb{S}^{n-1}) \to \tilde{H}_{q-1}(\mathbb{S}^{n-1}) \to \tilde{H}_q(E_n^+) \to \cdots$$

for the pair $(E_n^+, \mathbb{S}^{n-1})$, we find that $\tilde{H}_q(E_n^+, \mathbb{S}^{n-1}) \cong \tilde{H}_{q-1}(\mathbb{S}^{n-1})$. Combining the two results, we see that, for all $q \geq 0$, $\tilde{H}_q(\mathbb{S}^n) \cong \tilde{H}_{q-1}(\mathbb{S}^{n-1})$. Taking $q = n$, we find that

$$\tilde{H}_n(\mathbb{S}^n) \cong \tilde{H}_{n-1}(\mathbb{S}^{n-1}) \cong \cdots \cong \tilde{H}_0(\mathbb{S}^0) = \mathbb{Z},$$

and taking $q \neq n$, we find that

$$\tilde{H}_q(\mathbb{S}^n) \cong \tilde{H}_{n-q}(\mathbb{S}^0) = 0. \qquad \blacksquare$$

Corollary 6.7.9. *For any homology theory (H_q, ∂),*

$$H_q(\mathbb{S}^n) \cong \begin{cases} \mathbb{Z}, & \text{if } q = 0 \text{ or } n \\ 0, & \text{otherwise.} \end{cases}$$

Exercises

1. Use the axioms of the homology theory to show that the relative homology of the pair $(\mathbb{R}^n, \mathbb{S}^{n-1})$, $n \geq 2$, is given by

$$H_q(\mathbb{R}^n, \mathbb{S}^{n-1}) = \begin{cases} \mathbb{Z}, & q = n \\ 0, & \text{otherwise.} \end{cases}$$

2. If (X, A) is a pair such that both X and A are path-connected, then show that $H_0(X, A) = 0$, and mention an example where under the same conditions, $H_q(X, A)$ need not vanish in higher dimensions.

3. Give examples to show that, in general, for a pair (X, A)
   (i) $H_q(X, A) \neq H_q(X - A)$
   (ii) $H_q(X, A) \neq H_q(X/A)$, where X/A is the quotient space of X obtained by collapsing A to a point.

4. If $i\colon A \subset X$ is such that for all $q \geq 0$ the induced map $i_*\colon H_q(A) \to H_q(X)$ is an isomorphism, then compute the homology groups $H_q(X, A), q \geq 0$.

5. Let X be a topological n-manifold, $n \geq 1$. For each point $x \in X$, show that

$$H_q(X, X - x) = \begin{cases} \mathbb{Z}, & \text{if } q = n \\ 0, & \text{otherwise.} \end{cases}$$

6.8 Singular Homology with Coefficients

Let R be a PID and G be an R-module. We are now going to discuss the singular homology with coefficients in the module G. The two special cases of this coefficient module G are: (i) when R is the ring $\mathbb{Z}$ of integers so that G is simply an abelian group, and (ii) when R is a field so that G is a vector space over R. The first case practically covers everything whatever is true for an R-module G with R a principal ideal domain. In fact, this is one reason why we always tend to call "homology modules" simply as "homology groups". This happens because of the well-known algebraic result which says that a submodule of a free R-module is free provided R is a principal ideal domain. On the other hand, the case of a field gives the result in the simplest possible form. We consider the case of a general R-module G so that both the cases, mentioned above, are covered.

For a pair (X, A), recall that there is the following short exact sequence of singular chain complexes of free R-modules, where chain maps are induced by the inclusion maps $i\colon A \to X$, $j\colon X \to (X, A)$:

$$0 \longrightarrow S_*(A) \xrightarrow{i_\#} S_*(X) \xrightarrow{j_\#} S_*(X, A) \longrightarrow 0.$$

If we take the tensor product of the above exact sequence by the R-module G (on left or right, it makes no difference since tensor product is commutative), then since the chain complexes are free, we get the new short sequence of chain complexes and chain maps

$$0 \longrightarrow S_*(A) \otimes G \xrightarrow{i_\# \otimes 1} S_*(X) \otimes G \xrightarrow{j_\# \otimes 1} S_*(X, A) \otimes G \longrightarrow 0 \quad (*)$$

which remains exact. The chain complex $S_*(X, A) \otimes G$, denoted by $S_*(X, A; G)$, is called the singular chain complex of the pair (X, A) **with coefficients in the R-module** G. Its homology modules $H_q(S_*(X, A; G))$, denoted by $H_q(X, A; G)$, $q \geq 0$, are called the **singular homology modules** of (X, A) **with coefficients in** G. The case, when G is the R-module R, is the original case of singular homology modules $H_q(X, A)$ of the pair (X, A). The special case, when $R = \mathbb{Z}$ and the group $G \cong \mathbb{Z}/2\mathbb{Z}$, is known as the (mod 2) **homology** of the pair (X, A) and it has been historically very important.

It may be remarked that the homology with coefficients is really a concept derived from the integral homology. To see this, note that if $S_*(X)$ denotes the integral singular chain complex of X then the elements of the group $S_q(X)$ are integral linear combinations like $k_1 f_1 + \cdots + k_n f_n$ of singular q-simplexes $f_1, \ldots, f_n$ in X. We already know that an abelian group G is isomorphic to the group $\mathbb{Z} \otimes G$ under the inverse of the map $k \otimes a \to k.a$, $a \in G$. Hence any element of $G \otimes S_q(X)$ will be of the type $a_1 \otimes (k_1 f_1) + a_2 \otimes (k_2 f_2) + \cdots + a_n \otimes (k_n f_n) = b_1 f_1 + b_2 f_2 + \cdots + b_n f_n$, where $b_i = k_i a_i \in G$, $i = 1, \ldots, n$. Observe that the result of taking tensor product could become different, e.g., $k_i f_i$ may be nonzero while $(a_i.k_i)f_i$ could very well be zero, depending on the abelian group G. Furthermore, if G is an R-module, then G is an abelian group with a ring R of endomorphisms of G acting on G. Each of these endomorphisms induces an abelian group endomorphism of $G \otimes S_q(X)$, and so the R-module $S_*(X, G)$ is simply the abelian group $G \otimes S_*(X)$ with R acting on it as a ring of endomorphisms. Thus, a typical element of the R-module $G \otimes S_*(X)$ is again of the type $c_1 f_1 + \cdots + c_n f_n$, where $c_i \in G$ for all i.

Bockstein Homomorphism

Homology with coefficients has several interesting properties. First of all, let us note that if

$$0 \longrightarrow G' \xrightarrow{f} G \xrightarrow{g} G'' \longrightarrow 0$$

is an exact sequence of R-modules then, since $S_*(X, A)$ is a free chain complex, the following sequence of chain complexes with different coefficients remains exact:

$$0 \longrightarrow S_*(X, A) \otimes G' \xrightarrow{1 \otimes f} S_*(X, A) \otimes G \xrightarrow{1 \otimes g} S_*(X, A) \otimes G'' \longrightarrow 0.$$

Hence, by the fundamental theorem of homological algebra, we get the following long exact sequence connecting the singular homology with different coefficients:

$$\to H_q(X, A; G') \xrightarrow{f_*} H_q(X, A; G) \xrightarrow{g_*} H_q(X, A; G'') \xrightarrow{\partial_q} H_{q-1}(X, A; G') \to$$

The connecting homomorphism $\partial_q \colon H_q(X, A; G'') \to H_{q-1}(X, A; G')$, occurring in the above homology exact sequence, is quite important and it is known as the **Bockstein homology operator**. The special case of Bockstein operator, arising from the exact sequence $0 \to \mathbb{Z} \to \mathbb{Z} \to \mathbb{Z}_p \to 0$, where p is a prime, has found an interesting interpretation and several applications so much so that a symbol $\beta_q : H_q(X, \mathbb{Z}_p) \to H_{q-1}(X, \mathbb{Z})$ has been almost reserved for it.

We may recall that homology groups (with integer coefficients) are the simplest tools for distinguishing between two non-homeomorphic spaces. In this context, we will see later that by varying the coefficients, we will not be able to distinguish between those spaces which are not already distinguished by their integral homologies and yet the homology with coefficients is very interesting – the main reason for this is the fact that very often homology with particular coefficients arises very naturally. Properties of tensor products are crucial to the study of homology with coefficients. We have

Proposition 6.8.1. *On the category of all topological pairs, the singular homology with coefficients in a fixed R-module G is a homology theory in the sense of Eilenberg-Steenrod.*

Proof. The proof consists of easy verifications. Let $f\colon (X,A) \to (Y,B)$ be a map of pairs. Suppose $f_{\#}\colon S_*(X,A) \to S_*(Y,B)$ denote the induced chain map in singular chain complexes. Then the tensor product $f_{\#} \otimes 1\colon S_*(X,A) \otimes G \to S_*(Y,B) \otimes G$, where $1\colon G \to G$ is the identity map on the R-module G, induces homomorphism $f_*\colon H_q(X,A;G) \to H_q(Y,B;G)$ for all $q \geq 0$. The verification of the identity axiom and the composition axiom now is plain from the properties of tensor product. The exactness axiom follows from the short exact sequence $(*)$ defining the singular homology of the pair (X,A) with coefficients in G. The commutativity axiom follows from the following commutative diagram of chain complexes and chain maps in which the rows are exact:

$$\begin{array}{ccccccccc}
0 & \longrightarrow & S_*(A) \otimes G & \xrightarrow{i_{\#}} & S_*(X) \otimes G & \xrightarrow{j_{\#}} & S_*(X,A) \otimes G & \longrightarrow & 0 \\
 & & \downarrow{\scriptstyle (f|_A)_{\#} \otimes 1} & & \downarrow{\scriptstyle f_{\#} \otimes 1} & & \downarrow{\scriptstyle f_{\#} \otimes 1} & & \\
0 & \longrightarrow & S_*(B) \otimes G & \xrightarrow[i_{\#}]{} & S_*(Y) \otimes G & \xrightarrow[j_{\#}]{} & S_*(Y,B) \otimes G & \longrightarrow & 0
\end{array}$$

To prove the homotopy axiom, let $f,\ g\colon (X,A) \to (Y,B)$ be homotopic maps of pairs. These induce chain maps $f_{\#},\ g_{\#}\colon S_*(X,A) \to S_*(Y,B)$ which are chain homotopic by the homotopy axiom of singular theory. Taking the tensor product by the R-module G, we find that the chain maps $f_{\#} \otimes 1,\ g_{\#} \otimes 1\colon S_*(X,A) \otimes G \to S_*(Y,B) \otimes G$ are also chain homotopic, by Proposition 7.2.21. Therefore, the induced homomorphism in homology are equal, i.e., $f_* = g_*\colon H_q(X,Y;G) \to H_q(Y,B;G)$ for all $q \geq 0$. Finally, to prove the excision axiom, let $U \subseteq A$ such that $\bar{U} \subset \text{int } A$. The excision map $\rho\colon (X-U, A-U) \to (X,A)$ evidently induces chain map $\rho_{\#}\colon S_*(X-U, A-U) \to S_*(X,A)$. We know that the induced homomorphism $\rho_*\colon H_q(X-U, A-U) \to H_q(X,A)$ in singular homology is isomorphism for all $q \geq 0$, by the excision theorem for singular homology. Since singular chain complexes $S_*(X-U, A-U)$ and $S_*(X,A)$ are free and $\rho_{\#}$ induces isomorphism in homology, it follows from Exercise 4, Section 7.2.6 that $\rho_{\#}$ is indeed a chain equivalence. Taking tensor product with G, we find that

$$\rho_{\#} \otimes 1\colon S_*(X-U, A-U) \otimes G \to S_*(X,A) \otimes G$$

is also a chain equivalence. Hence the induced homomorphism $\rho_*\colon H_q(X-U, A-U; G) \to H_q(X,A;G)$ in homology is an isomorphism for all $q \geq 0$. Since the dimension axiom is evidently true, this completes the proof. ■

In Chapter 4, we defined simplicial homology of a polyhedral pair. Now, we prove the following theorem.

Theorem 6.8.2. *Simplicial homology, defined on the category of all polyhedral pairs, is a homology theory in the sense of Eilenberg-Steenrod.*

Proof. Recall that a topological pair (X, A) is a polyhedral pair if there exists a pair (K, L) of polyhedra and a homeomorphism $h\colon (|K|, |L|) \to (X, A)$ of topological pairs giving a triangulation of (X, A). We define the simplicial homology $H_q(X, A)$ of the pair (X, A) by putting $H_q(X, A) := H_q(C_*(K, L))$ for all $q \geq 0$; here, $C_*(K, L)$ is the quotient chain complex $C_*(K)/C_*(L)$ defined by the following exact sequence of simplicial chain-complexes:

$$0 \to C_*(L) \to C_*(K) \to C_*(K, L) \to 0.$$

We have already proved in Chapter 4 that the groups $H_q(X, A)$ defined above are independent of the choice of a triangulation (K, L) of the pair (X, A). Now, let $f\colon (X, A) \to (Y, B)$ be a map of polyhedral pairs. Suppose $h\colon (|K|, |L|) \to (X, A)$ and $k : (|K_1|, |L_1|) \to (Y, B)$ are triangulations of (X, A) and (Y, B). Let $s\colon (K^{(m)}, L^{(m)}) \to (|K_1|, |L_1|)$ be a simplicial approximation of the map $k^{-1}fh\colon (|K|, |L|) \to (|K_1|, |L_1|)$. Then we define the induced homomorphism $f_*\colon H_q(X, A) \to H_q(Y, B)$ for all $q \geq 0$ to be the composite homomorphism $(k^{-1}fh)_* \circ \mu\colon H_q(K, L) \to H_q(K^{(m)}, L^{(m)}) \to H_q(K_1, L_1)$, where $\mu\colon H_q(K, L) \to H_q(K^{(m)}, L^{(m)})$ is the subdivision homomorphism. Following Proposition 4.10.6, one can prove that these induced homomorphisms in homology satisfy the functorial properties. For any pair (K, L) of simplicial complexes, the exact sequence

$$0 \to C_*(L) \to C_*(K) \to C_*(K, L) \to 0$$

of simplicial chain-complexes yields the exactness axiom for a polyhedral pair $(X, A) \approx (|K|, |L|)$. The commutativity axiom follows easily from the Simplicial Approximation Theorem and the results about chain complexes of R-modules. The homotopy axiom follows from Section 4.10.10. For excision, let (X, A) be a polyhedral pair and $U \subset A$ such that $\bar{U} \subset \text{int } A$ and that the inclusion map $(X - U, A - U) \to (X, A)$ is a map of polyhedral pairs. Then we must show that this inclusion induces isomorphism in simplicial homology. Then we must show that this inclusion induces isomorphism in simplicial homology. Let (K_1, L_1) be a triangulation of the pair $(X - U, A - U)$ and (K, L) be a triangulation of (X, A). There is a subtle point involved here. The simplicial pair (K_1, L_1) need not be a subcomplex of the pair (K, L). First we assume that this is the case. Then consider the chain map $C_*(K_1) \to C_*(K) \to C_*(K)/C_*(L)$ which is the composite of the inclusion and the quotient maps. This is clearly an onto map since $U \subset |L|$. Its kernel is obviously $C_*(L_1)$. Therefore we have the following isomorphism of chain complexes: $C_*(K)/C_*(L) \cong C_*(K_1)/C_*(L_1)$. This implies that

$$C_*(|K| - U, |L| - U) = C_*(K_1, L_1) \cong C_*(K, L) = C_*(|K|, |L|).$$

Since chain maps commute with the boundary maps, we find that the excision map induces isomorphism in homology $H_q(X - U, A - U) \to H_q(X, A), \ \ \forall q \geq 0$. In case the simplicial pair (K_1, L_1) is not a subpair of (K, L), we refer to [16] for details. ∎

Remark 6.8.3. Note that when we excise the subset M from the simplicial pair (K, L), we really excise, geometrically speaking, not only $|M|$ but also the open cones over simplexes of M whose vertices may be in L. Thus, in terms of geometric carriers, we may be excising more than what is permitted in singular homology - this is an advantage of simplicial homology over singular homology. For example, from the double cone $(\mathbb{S}^2, \mathbb{D}^2_-)$ over the boundary of a triangle, when we excise the lower vertex v_4 (see Fig. 6.5) from the pair $(\mathbb{S}^2, \mathbb{D}^2)$, we really excise the open lower hemisphere, viz., $\text{Int}(\mathbb{D}^2_-)$ to get the pair $(\mathbb{D}^2_+, \mathbb{S}^1)$.

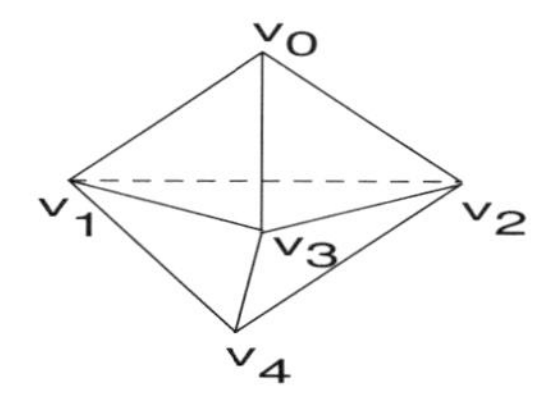

Fig. 6.5: The bottom vertex to be excised

Uniqueness of Homology Theory

As mentioned in the introduction, Eilenberg-Steenrod also proved that any two homology theories, which are isomorphic for point-spaces, are actually isomorphic on all compact polyhedra (see Spanier [18], p. 203). By now we have shown that on the category of compact polyhedra, both simplicial homology as well as the singular homology are homology theories. Therefore, by the Eilenberg and Steenrod Uniqueness Theorem, we find the nice result that *if X is a compact polyhedron, then singular homology of X is the same as its simplicial homology.* For instance, if K is a triangulation of Sphere, Torus, Klein bottle or Möbius band, then simplicial homology of $|K|$, viz., $H_q(K)$ is isomorphic to the singular homology of the space $|K|$. *We will use this result from now onwards without mentioning it explicitly.*

Universal Coefficient Theorem for Singular Homology

We continue our assumption that R is a PID and G is an R-module. The Universal Coefficient Theorem for singular homology gives an exact relationship between the homology modules $H_q(X, A)$ and the homology modules $H_q(X, A; G)$ with coefficients in G. The exact relationship, given later, is determined in terms of the tensor product and torsion product of G and the modules $H_q(X, A)$.

For a pair (X, A), we consider the singular chain complex $S_*(X, A)$ defined earlier. This is a chain complex of free R-modules. Let $Z_q(X, A)$ and $B_q(X, A)$ denote the submodules of $S_q(X, A)$ consisting of relative q-cycles and relative q-boundaries. Now, note that if $z \in Z_q(X, A)$ is a q-cycle then for any $g \in G$,

$z \otimes g$ is a q-cycle in $S_q(X, A) \otimes G$. Similarly, if $b \in B_q(X, A)$ is a boundary then $b \otimes g \in S_q(X, A) \otimes G$ is also a boundary. Now, we can define a map $f \colon H_q(X, A) \times G \to H_q(X, A; G)$ by putting $f(\{z\}, g) = \{z \otimes g\}$. It is quickly seen that this map is an R-bilinear map and hence it defines an R-homomorphism $\mu \colon H_q(X, A) \otimes G \to H_q(X, A; G)$ by the formula $\mu(z \otimes g) = \{z \otimes g\}$ on generators. Using this homomorphism, one can prove the following Universal Coefficient Theorem for singular homology (see Spanier [18] for a proof) which says that the above homomorphism μ is a monomorphism with cokernel $\mathrm{Tor}(H_{q-1}(X, A); G)$:

Theorem 6.8.4. *For a topological pair (X, A) and for any R-module G, there is the following split exact sequence of R-modules for all $q \geq 0$:*

$$0 \longrightarrow H_q(X, A) \otimes G \xrightarrow{\mu} H_q(X, A; G) \longrightarrow Tor(H_{q-1}(X, A); G) \longrightarrow 0.$$

The exact sequence is natural in (X, A) and in the coefficient module G, but the splitting is not natural.

The splitting of the above sequence implies that $H_q(X, A; G)$ is the direct sum of $H_q(X, A) \otimes G$ and $\mathrm{Tor}(H_{q-1}(X, A); G)$. Note that the naturality of the sequence in (X, A) implies, in particular, that if $f \colon (X, A) \to (Y, B)$ is a map which induces isomorphism in integral singular homology, then it induces isomorphism in singular homology with coefficients in an arbitrary abelian group G. This follows from the Five Lemma because, by naturality, the following diagram is commutative:

$$\begin{array}{ccccccccc}
0 & \longrightarrow & H_q(X, A) \otimes G & \longrightarrow & H_q(X, A; G) & \longrightarrow & \mathrm{Tor}(H_{q-1}(X, A); G) & \longrightarrow & 0 \\
 & & \cong \downarrow f_* \otimes 1 & & \downarrow f_* & & \cong \downarrow \mathrm{Tor}(f_*, 1) & & \\
0 & \longrightarrow & H_q(Y, B) \otimes G & \longrightarrow & H_q(Y, B; G) & \longrightarrow & \mathrm{Tor}(H_{q-1}(Y, B); G) & \longrightarrow & 0
\end{array}$$

Thus, if two spaces cannot be distinguished by integral homology groups then they cannot be distinguished by homology with any coefficients.

Example 6.8.5. We already know the integral singular homology groups of the projective plane $\mathbb{P}^2$ as follows: $H_0(\mathbb{P}^2) \cong \mathbb{Z}$, $H_1(\mathbb{P}^2) \cong \mathbb{Z}_2$ and $H_q(\mathbb{P}^2) = 0$ for all other values of q. Hence, by the Universal Coefficient Theorem, the nontrivial (mod 2) homology of $\mathbb{P}^2$ is given by:

$$H_0(\mathbb{P}^2; \mathbb{Z}_2) \cong \mathbb{Z} \otimes \mathbb{Z}_2 \cong \mathbb{Z}_2,$$

$$H_1(\mathbb{P}^2; \mathbb{Z}_2) \cong \mathbb{Z}_2 \otimes \mathbb{Z}_2 \oplus \mathrm{Tor}(\mathbb{Z}, \mathbb{Z}_2) \cong \mathbb{Z}_2,$$

$$H_2(\mathbb{P}^2; \mathbb{Z}_2) = 0 \oplus \mathrm{Tor}(\mathbb{Z}_2, \mathbb{Z}_2) \cong \mathbb{Z}_2.$$

Example 6.8.6. If K is the Klein bottle then we know the integral singular homology as $H_0(K) \cong \mathbb{Z}$, $H_1(K) \cong \mathbb{Z} \oplus \mathbb{Z}_2$ and $H_q(K) = 0$ for all other values of q. Hence, by the Universal Coefficient Theorem, the (mod 2) homology is given by $H_0(K; \mathbb{Z}_2) \cong \mathbb{Z}_2$, $H_1(K; \mathbb{Z}_2) \cong (\mathbb{Z} \oplus \mathbb{Z}_2) \otimes \mathbb{Z}_2 \cong \mathbb{Z}_2 \oplus \mathbb{Z}_2$, $H_2(K; \mathbb{Z}_2) = \operatorname{Tor}(\mathbb{Z} \oplus \mathbb{Z}_2, \mathbb{Z}_2) \cong \mathbb{Z}_2$.

Note that (mod 3) homology of the projective plane $\mathbb{P}^2$ is given by $H_0(\mathbb{P}^2; \mathbb{Z}_3) \cong \mathbb{Z}_3$, $H_q(\mathbb{P}^2; \mathbb{Z}_3) = 0$, $q \neq 0$, i.e., the projective plane is like a 2-disk with respect to the (mod 3) homology. Similarly, the Klein bottle K has the same homology as torus with respect to the $\mathbb{Z}_2$-coefficients. Thus, though the integral homology distinguishes the Klein bottle from torus, the two spaces cannot be distinguished by the (mod 2) homology.

Example 6.8.7. Let $L(p, q)$ denote the lens space, where $(p, q) = 1$. Then the integral homology is given by $H_0(L(p, q)) \cong \mathbb{Z}$, $H_1(L(p, q)) \cong \mathbb{Z}_p$, $H_3(L(p, q)) = \mathbb{Z}$, $H_i(L(p, q)) = 0$ for other values of i. Hence, by the Universal Coefficient Theorem, $H_0(L(p, q); \mathbb{Z}_q) \cong \mathbb{Z}_q$, $H_1(L(p, q); \mathbb{Z}_q) = 0$ and $H_3(L(p, q); \mathbb{Z}_q) \cong \mathbb{Z}_q$ whereas $H_0(L(p, q); \mathbb{Z}_p) \cong \mathbb{Z}_p$, $H_1(L(p, q); \mathbb{Z}_p) \cong \mathbb{Z}_p$, $H_3(L(p, q); \mathbb{Z}_p) \cong \mathbb{Z}_p$. Thus, with coefficients in $\mathbb{Z}_q$, $L(p, q)$ has the same homology as 3-sphere, but with respect to $\mathbb{Z}_p$-coefficients, the two are different.

Künneth Formula for Singular Homology

We are now going to state a result which not only generalizes the Universal Coefficient Theorem for singular homology, but also gives the singular homology of the product of two spaces X and Y in terms of the singular homologies of the spaces X and Y. For convenience and simplicity, we state the result for integral coefficients but the theorem is true for arbitrary coefficients over a PID. The special case, when the ground ring is a field, must always be noted.

Let $S_*(X)$ and $S_*(Y)$ denote the integral singular chain complexes of the spaces X and Y, respectively. First of all, let us state the important theorem due to Eilenberg-Zilber (see Spanier [18] p. 232), which asserts that the singular chain complex $S_*(X \times Y)$ of the product space $X \times Y$ is indeed chain equivalent to the tensor product $S_*(X) \otimes S_*(Y)$ of the singular chain complexes of X and Y. In fact, this result of Eilenberg-Zilber shows how the geometric situation of the cartesian product has a nice algebraic representation in terms of the tensor product of singular chain complexes of factor spaces. As a result, we immediately see that for all $q \geq 0$, $H_q(X \times Y) \cong H_q(S_*(X) \otimes S_*(Y))$. Thus, the problem of computing the homology groups of the product space $X \times Y$ is really the problem of computing the homology of the tensor product of two free chain complexes. However, we have discussed this question for two abstract chain complexes in Appendix, and the same can be applied to obtain the following basic result:

Theorem 6.8.8. (Künneth). *Let X, Y be any two spaces. Then, for all $n \geq 0$, there is the following split short exact sequence of abelian groups:*

$$0 \to \sum_{p+q=n} H_p(X) \otimes H_q(Y) \xrightarrow{\mu} H_n(X \times Y) \to$$

$$\sum_{p+q=n-1} \mathrm{Tor}(H_p(X), H_q(Y)) \to 0.$$

The above sequence is natural in X, Y but the splitting is not natural.

The monomorphism μ above is called the **homology cross product**. Let us remark that if G is an abelian group, we can define an abstract chain complex C_* as follows: We put $C_q = 0$ for all $q \neq 0$ and $C_0 = G$ with obvious homomorphisms. Since the Künneth theorem is true for abstract chain complexes, we can replace $S_*(Y)$ by the above chain complex C_*. Then the tensor product $S_*(X) \otimes C_*$ is just the chain complex $S_*(X) \otimes G$. Since $H_q(C_*) = G$ for $q = 0$, and zero otherwise, the above Künneth Theorem reduces to the following exact sequence for all $n \geq 0$:

$$0 \to H_n(X) \otimes G \to H_n(X, G) \to Tor(H_{n-1}(X), G) \to 0.$$

This is exactly the Universal Coefficient Theorem for singular homology.

If we take the singular homologies of X and Y with coefficients in abelian groups G and G', respectively, we find the following theorem:

Theorem 6.8.9. (Künneth). *For any two spaces X and Y and for any two abelian groups G and G' such that $\mathrm{Tor}(G, G') = 0$, there is the following split exact sequence, for all $n \geq 0$, of singular homology groups:*

$$0 \to \sum_{p+q=n} H_p(X, G) \otimes H_q(Y, G') \to H_n(X \times Y, G \otimes G')$$

$$\to \sum_{p+q=n-1} \mathrm{Tor}(H_p(X, G), H_q(Y, G')) \to 0.$$

Moreover, the sequence is natural in spaces X, Y as well as in the coefficients G and G', but the splitting is not natural.

Example 6.8.10. The integral singular homology of the n-sphere $\mathbb{S}^n$ is given by: $H_0(\mathbb{S}^n) \cong \mathbb{Z} \cong H_n(\mathbb{S}^n)$, and $H_q(\mathbb{S}^n) = 0$ for other values of q. Hence, by the Künneth Theorem, we get for $m \neq n$,

$$H_q(\mathbb{S}^m \times \mathbb{S}^n) = \begin{cases} \mathbb{Z}, & q = 0, m, n \text{ or } m+n \\ 0, & \text{otherwise.} \end{cases}$$

But, for $m = n$, we have

$$H_q(\mathbb{S}^n \times \mathbb{S}^n) = \begin{cases} \mathbb{Z}, & q = 0, 2n \\ \mathbb{Z} \oplus \mathbb{Z}, & q = n \\ 0, & \text{otherwise.} \end{cases}$$

Note that $\mathbb{P}^2$ is a two-dimensional polyhedron which means $\mathbb{P}^2 \times \mathbb{P}^2$ is 4-dimensional polyhedron. As mentioned below, the integral homology of $\mathbb{P}^2 \times \mathbb{P}^2$ in dimensions higher than three vanishes but, as can be seen easily, the (mod 2) homology of $\mathbb{P}^2 \times \mathbb{P}^2$ does not vanish up to dimension 4.

Example 6.8.11. Let $\mathbb{P}^2$ denote the projective plane. Then, by the Künneth Formula, we have

$$\begin{aligned} H_0(\mathbb{P}^2 \times \mathbb{P}^2) &\cong \mathbb{Z}, \\ H_1(\mathbb{P}^2 \times \mathbb{P}^2) &\cong \mathbb{Z}_2 \oplus \mathbb{Z}_2, \\ H_2(\mathbb{P}^2 \times \mathbb{P}^2) &\cong \mathbb{Z}_2 \cong H_3(\mathbb{P}^2 \times \mathbb{P}^2), \text{ and} \\ H_q(\mathbb{P}^2 \times \mathbb{P}^2) &= 0 \text{ for all } q \geq 4. \end{aligned}$$

6.9 Mayer-Vietoris Sequence for Singular Homology

Let X_1, X_2 be two subspaces of a space X. We would like to know how the singular homology groups of $X_1 \cup X_2$ can be computed if we know the homology groups of X_1, X_2 and $X_1 \cap X_2$. The Mayer-Vietoris exact sequence provides a nice answer to this question. If X_1, X_2 both are closed (respectively open) in X and $X_1 \cap X_2 = \phi$ then, obviously, both X_1, X_2 are open (respectively, closed) in $X_1 \cup X_2$ and, in such a case, we have already seen (Remark 6.7.5) that for a general homology theory, $H_q(X_1 \cup X_2) \cong H_q(X_1) \oplus H_q(X_2)$ for all $q \geq 0$. However, if any one of the sets is only closed and the other is only open, then $X_1 \cap X_2 = \phi$ does not mean that such a relation will be valid. In case $X_1 \cap X_2 \neq \phi$, then Mayer-Vietoris exact sequence connects the homology modules of X_1, X_2, $X_1 \cap X_2$ and $X_1 \cup X_2$ under a suitable condition on X_1 and X_2. We explain it precisely: Let $i_1 \colon X_1 \cap X_2 \to X_1$, $i_2 \colon X_1 \cap X_2 \to X_2$, $j_1 \colon X_1 \to X_1 \cup X_2$ and $j_2 \colon X_2 \to X_1 \cup X_2$ be the obvious inclusion maps. These inclusion maps induce homomorphisms in singular chain complexes which are also inclusion maps. This evidently gives us the following sequence of free chain complexes and chain maps, which is easily seen to be exact:

$$0 \longrightarrow S_*(X_1 \cap X_2) \xrightarrow{(i_{1\#}, i_{2\#})} S_*(X_1) \oplus S_*(X_2)$$
$$\xrightarrow{j_{1\#} - j_{2\#}} S_*(X_1) + S_*(X_2) \longrightarrow 0.$$

The above chain maps are defined by:

$$(i_{1\#}, i_{2\#})(c) = (i_{1\#}(c), i_{2\#}(c)),$$
$$(j_{1\#} - j_{2\#})(c_1, c_2) = j_{1\#}(c_1) - j_{2\#}(c).$$

We can now apply the long exact homology sequence theorem to the above short exact sequence of chain complexes to get the following exact sequence in homolgy.

$$\cdots \to H_q(X_1 \cap X_2) \to H_q(X_1) \oplus H_q(X_2) \to H_q(S_*(X_1) + S_*(X_2)) \to$$

$$H_{q-1}(X_1 \cap X_2) \to \cdots .$$

In the above sequence we wish to replace $H_q(S_*(X_1)+S_*(X_2))$ by $H_q(X_1 \cup X_2)$ for each $q \geq 0$. To achieve this let us say that a pair (X_1, X_2) of subspaces of a space X is **excisive** if the inclusion map $S_*(X_1) + S_*(X_2) \subset S_*(X_1 \cup X_2)$ induces isomorphisms in homology. In view of Theorem 6.6.2, we know that the above inclusion map is a chain homotopy equivalence if $X = X_1 \cup X_2 = \text{int } X_1 \cup \text{int } X_2$. Thus, if we assume that $X = X_1 \cup X_2 = \text{int } X_1 \cup \text{int } X_2$, then the pair (X_1, X_2) will be excisive and the long exact homology sequence mentioned above gives us the following long exact sequence in singular homology:

$$(*) \quad \cdots \longrightarrow H_q(X_1 \cap X_2) \xrightarrow{(i_{1*}, i_{2*})} H_q(X_1) \oplus H_q(X_2) \xrightarrow{j_{1*} - j_{2*}}$$
$$H_q(X_1 \cup X_2) \xrightarrow{\partial} H_{q-1}(X_1 \cap X_2) \longrightarrow \cdots$$

The above sequence is known as the **Mayer-Vietoris Exact Sequence** for singular homology. This sequence is frequently useful in computing the homology of union of two subspaces if we know the homology of individual subspaces and their intersection.

Note that all the singular chain complexes involved in the preceding short exact sequence of chain complexes are free and hence the sequence will remain exact even after we tensor it by an R-module G. Thus, the resulting short exact sequence of chain complexes gives us a similar Mayer-Vietoris exact sequence for singular homology with coefficients.

For simplicial homology we have a stronger result. Observe that if K_1, K_2 are any two subcomplexes of a simplicial complex K then, by definition, $C_*(K_1 \cup K_2) = C_*(K_1) + C_*(K_2)$. This directly gives us the following short exact sequence of simplicial chain complexes:

$$0 \to C_*(K_1 \cap K_2) \xrightarrow{(i_{1\#}, i_{2\#})} C_*(K_1) \oplus C_*(K_2) \xrightarrow{j_{1\#} - j_{2\#}} C_*(K_1 \cup K_2) \to 0.$$

This, in turn, yields the following Mayer-Vietoris exact sequence in simplicial homology (for arbitrary K_1 and K_2):

$$\to H_q(K_1 \cap K_2) \to H_q(K_1) \oplus H_q(K_2) \to H_q(K_1 \cup K_2) \to H_{q-1}(K_1 \cap K_2) \to$$

Mayer-Vietoris Sequence for Reduced Homology

There is also a Mayer-Vietoris exact sequence for reduced singular homology for those subsets X_1, X_2 of a space X which satisfy the condition that $X_1 \cap X_2 \neq \emptyset$ and $X_1 \cup X_2 = \text{int } X_1 \cup \text{int } X_2$. To prove this note that the sequence $0 \longrightarrow \mathbb{Z} \xrightarrow{i} \mathbb{Z} \oplus \mathbb{Z} \xrightarrow{j} \mathbb{Z}$, where the maps i and j are defined by $i(n) = (n, n)$, $j(m, n) = m - n$, is clearly exact. Now, let $S_*(X_1) \to \mathbb{Z}$, $S_*(X_2) \to \mathbb{Z}$ and $S_*(X_1 \cap X_2) \to \mathbb{Z}$ be the augmentation homomorphisms. Then, with $\mathcal{U} = \{\text{int } X_1, \text{ int } X_2\}$, we have the following commutative diagram of chain complexes and chain maps in which the rows are exact and the vertical maps are onto.

$$\begin{array}{ccccccccc} 0 \to & S_*(X_1 \cap X_2) & \xrightarrow{(i_{1\#}, i_{2\#})} & S_*(X_1) \oplus S_*(X_2) & \xrightarrow{j_{1\#} - j_{2\#}} & S_*(X_1) + S_*(X_2) & \to 0 \\ & \downarrow & & \downarrow & & \downarrow & \\ 0 \longrightarrow & \mathbb{Z} & \xrightarrow[i]{} & \mathbb{Z} \oplus \mathbb{Z} & \xrightarrow[j]{} & \mathbb{Z} & \longrightarrow 0 \end{array}$$

It follows from the above diagram that the kernel sequence of the augmentation homomorphisms is also exact, viz., the sequence

$$0 \to \tilde{S}_*(X_1 \cap X_2) \xrightarrow{(i_{1\#}, i_{2\#})} \tilde{S}_*(X_1) \oplus \tilde{S}_*(X_2) \xrightarrow{j_{1\#} - j_{2\#}} \tilde{S}_*(X_1) + \tilde{S}_*(X_2) \to 0$$

is exact. Then, as in the proof of the unreduced case, this last exact sequence of chain complexes and chain maps gives us the following Mayer-Vietoris exact sequence in reduced singular homology under the conditions stated above:

$$\to \tilde{H}_q(X_1 \cap X_2) \to \tilde{H}_q(X_1) \oplus \tilde{H}_q(X_2) \to \tilde{H}_q(X_1 \cup X_2) \to \tilde{H}_{q-1}(X_1 \cap X_2) \to$$

We have a similar reduced Mayer-Vietoris exact sequence for simplicial homology and for homology with coefficients.

Mayer-Vietoris exact sequence is a very powerful and convenient tool for computing homology groups. We have

Example 6.9.1. (One-point union of n-circles). Let $X = X_1 \cup \cdots \cup X_n$ denote the one-point union of n-spaces X_i, each of which is homeomorphic to the unit circle. Then X (sometimes called n-leaved rose) has a triangulation, viz., it is homeomorphic to the union of n triangles $T_1, T_2, \ldots, T_n$ all having one vertex, say v, in common (see Fig. 6.6).

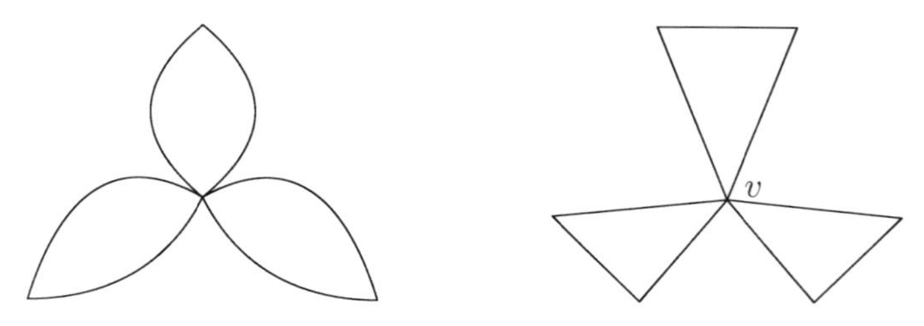

Fig. 6.6: The case $n = 3$ of n-leaved rose

Note that for each $i = 1, 2, \ldots, n$, $H_0(T_i) \cong \mathbb{Z} \cong H_1(T_i)$, $H_q(T_i) = 0$ for all $q > 1$. Since X is path-connected, $H_0(X) \cong \mathbb{Z}$. Applying reduced Mayer-Vietoris exact sequence to complexes K_1 and K_2, where $|K_1| \approx T_1$, $|K_2| \approx T_2$, $K_1 \cap K_2 = \{v\}$ and $|K_1 \cup K_2| = T_1 \cup T_2$, we find the nontrivial part of the exact sequence as

$$0 \to \tilde{H}_1(K_1) \oplus \tilde{H}_1(K_2) \to \tilde{H}_1(K_1 \cup K_2) \to 0.$$

Thus, we get $\tilde{H}_1(K_1 \cup K_2) = \mathbb{Z} \oplus \mathbb{Z}$ and for $q \geq 2$, $\tilde{H}_q(K_1 \cup K_2) = 0$. Applying induction and using the reduced Mayer-Vietoris sequence of simplicial homology, we find that $\tilde{H}_1(X_1 \cup X_2 \cup \cdots \cup X_n) = \oplus_1^n \mathbb{Z}$ and $H_q(X_1 \cup \cdots \cup X_n) = 0$ for all $q > 1$. Consequently, we have

$$H_q(X_1 \cup \cdots \cup X_n) \cong \begin{cases} \mathbb{Z}, & q = 0 \\ \bigoplus_1^n \mathbb{Z}, & q = 1 \\ 0, & \text{otherwise.} \end{cases}$$

Remark 6.9.2. We may mention here that in the above example, the Mayer-Vietoris sequence for singular homology does not apply directly – one has to extend the triangles at the vertices by a small amount, use homotopy axiom, etc., before applying the Mayer-Vietoris exact sequence. The above computation using simplicial homology is quite convenient; since X is a compact polyhedron, the final result is valid for any homology.

The above result also holds for the one-point union of a finite number of n-spheres for any $n \geq 1$. As a matter of fact, a close examination of the above example tells us that if we take a space X as the one-point union of n connected polyhedra $K_1, K_2, \ldots, K_n$, then the simplicial homology of X is clear: it is $\mathbb{Z}$ in dimension zero and is isomorphic to the direct sum $H_q(K_1) \oplus \cdots \oplus H_q(K_n)$ in dimension $q > 0$.

Remark 6.9.3. There is also a Mayer-Vietoris exact sequence for pairs of subspaces $\{(X_1, A_1), (X_2, A_2)\}$ of a topological pair (X, A) under suitable hypothesis on X_1, X_2 and A_1, A_2. The formulation of the statement, etc., is left as an exercise.

Remark 6.9.4. One can also compute the homology groups of the n-sphere $\mathbb{S}^n$ using Mayer-Vietoris exact sequence. For this, let N be the north pole and S be the south pole of $\mathbb{S}^n$ (see Fig. 6.7). Then $X_1 = \mathbb{S}^n - \{N\}$, $X_2 = \mathbb{S}^n - \{S\}$ are two contractible subspaces of $\mathbb{S}^n$ such that $\mathbb{S}^n = X_1 \cup X_2$, int $X_1 \cup$ int X_2 is of the same homotopy type as $\mathbb{S}^{n-1}$. Hence, applying the reduced Mayer-Vietoris sequence for X_1, X_2 for singular homology, we find that $\tilde{H}_q(\mathbb{S}^n) \cong \tilde{H}_{q-1}(\mathbb{S}^{n-1}) \cong \tilde{H}_{q-n}(\mathbb{S}^0) \cong \mathbb{Z}$ if $q = n$, and 0 otherwise. For simplicial homology, we can imagine $\mathbb{S}^n$ as the union of two cones X_1, X_2 over $\mathbb{S}^{n-1}$, i.e., $X_1 \cup X_2 = \mathbb{S}^n$, $X_1 \cap X_2 = \mathbb{S}^{n-1}$. Since cones are acyclic, we can apply the reduced Mayer-Vietoris sequence for simplicial homology to conclude that $\tilde{H}_q(\mathbb{S}^n) \cong \tilde{H}_{q-1}(\mathbb{S}^{n-1})$, etc. Note that, unlike the case of singular homology, homotopy axiom is not necessary in computing the simplicial homology of $\mathbb{S}^n$.

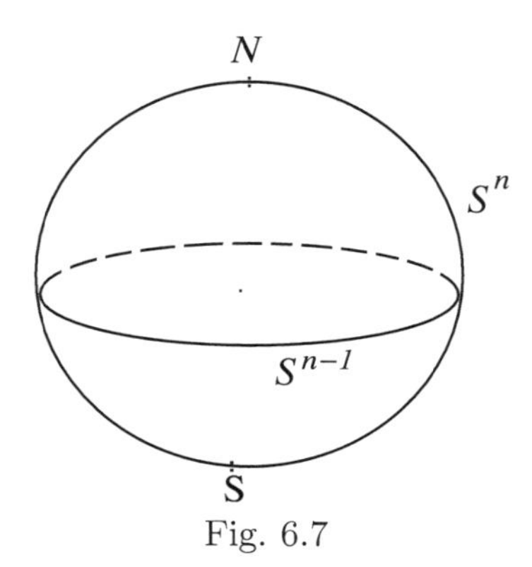

Fig. 6.7

Mayer-Vietoris Sequence for Abstract Homology

There is an important point about Mayer-Vietoris exact sequence which must be mentioned. This sequence is exact not only for simpicial or singular homology, but it is also valid for any homology theory. This appears rather too much in the first instance because a homology theory does not arise from any chain complex, and so the first question will be: how can a Mayer-Vietoris sequence be even formulated for an abstract homology theory? It is interesting to see that it is possible, though it does require some additional work to be done using only the axioms of the homology theory.

Let X_1, X_2 be two subspaces of a space X. We say that the pair (X_1, X_2) is an excisive couple or (X, X_1, X_2) is a **proper triad** if the inclusion maps $k_1: (X_1, X_1 \cap X_2) \to (X_1 \cup X_2, X_2)$ and $k_2: (X_2, X_1 \cap X_2) \to (X_1 \cup X_2, X_1)$ induce isomorphisms in homology, i.e., for all $q \geq 0, k_{1*}, k_{2*} : H_q(X_i, X_1 \cap X_2) \to H_q(X_1 \cup X_2, X_j)$ is an isomorphism for $i = 1, 2$ and $j = 2, 1$ respectively. Examples of proper triads are (i) any two open sets X_1, X_2 in a space X – this follows from the excision axiom and, (ii) any two subcomplexes $|K_1|, |K_2|$ of a simplicial complex $|K|$ – this is left as an exercise.

For a given proper triad (X, X_1, X_2), we put $A = X_1 \cap X_2$ and $X = X_1 \cup X_2$. Let us write the long exact sequences for the pairs (X_1, A) and (X, X_2) and then consider the vertical maps arising from the excision map $(X_1, A) \to (X, X_2)$. Then every square of the following diagram is easily seen to be commutative:

$$\begin{array}{ccccccccc}
\to H_q(A) & \xrightarrow{i_{1*}} & H_q(X_1) & \xrightarrow{j_{1*}} & H_q(X_1, A) & \xrightarrow{\partial} & H_{q-1}(A) & \to \\
\downarrow k_{1*} & & \downarrow k_{1*} & & \downarrow k_{1*} & & \downarrow k_{1*} & \\
\to H_q(X_2) & \xrightarrow{i'_{1*}} & H_q(X) & \xrightarrow{j'_{1*}} & H_q(X, X_2) & \xrightarrow{\partial} & H_{q-1}(X_2) & \to
\end{array}$$

Let us define maps ϕ_q, ψ_q and η_q as indicated in the following sequence, where η is given by $\eta(z) = \partial \circ k_*^{-1} \circ j'_{1*}(z)$:

$$(*) \cdots \quad \to H_q(A) \xrightarrow{\phi_q = (i_{1*}, k_{1*})} H_q(X_1) \oplus H_q(X_2) \xrightarrow{\psi_q = (k * - i_{1*})}$$

$$H_q(X) \xrightarrow{\eta_q} H_{q-1}(A) \quad \to \cdots .$$

Then a diagram chase shows that the above sequence is exact (this is left as an exercise). The above sequence is, in fact, the Mayer-Vietoris exact sequence of the proper triad (X, X_1, X_2). We emphasize again that the sequence has been derived using only the axioms of a homology theory.

6.10 Some Classical Applications

We have seen (see the chapter on Simplicial Homology) that using the simplicial homology of polyhedra, we could prove that $\mathbb{S}^m$ is not homeomorphic to $\mathbb{S}^n, (m \neq n)$; $\mathbb{R}^m$ is not homeomorphic to $\mathbb{R}^n, (m \neq n)$; the discs $\mathbb{D}^n$ have the fixed point property (Brouwer's fixed point theorem); the spheres $\mathbb{S}^n, n \geq 2$ do not have nonzero tangent vector fields, etc. All these results can also be proved, rather more easily, using the singular homology in place of simplicial homology. In this section we are going to present a few deeper applications of singular homology, proved by Brouwer as early as in 1911.

Proposition 6.10.1. *Let $\alpha \in H_q(X)$. Then there exists a compact subset K of X such that α lies in the image of $i_*\colon H_q(K) \to H_q(X)$, where $i\colon K \to X$ is the inclusion map.*

Proof. Note that if $\sigma\colon \Delta_q \to X$ is a singular simplex, then $\sigma(\Delta_q)$ is a compact subset of X, called the minimal carrier of σ. Clearly, the minimal carrier of a finite sum of singular simplexes is simply the union of their minimal carriers. Hence, if $c = \sum r_i \sigma_i \in S_q(X)$ is a singular q-chain in X such that $\alpha = \{c\}$, then the minimal carrier of c is a compact subset, say K of X. We can consider $c' = c$ as a singular q-chain in K. Then $i\colon K \to X$ maps c' to c. Since $\partial c' = \partial c$, if c is a cycle in X, it is also a cycle in K, i.e., $i_*\{c'\} = \{c\} = \alpha$, where $\{c'\} \in H_q(K)$.■

The above proposition is usually stated by saying that *singular homology has compact supports.*

Proposition 6.10.2. *Let $K \subset X$ be a compact subset and $\alpha \in H_q(K)$ is such that $i_*(\alpha) = 0$ in $H_q(X)$, where $i\colon K \to X$ is the inclusion map. Then, there exists a compact subset K'' containing K such that $j_*(\alpha) = 0$ in $H_q(K'')$, where $j\colon K \to K''$ is the inclusion map.*

Proof. If α is represented by the singular q-chain $c = \sum r_i\sigma_i$, where σ_i are singular q-simplexes, then $i(c) = c' = \sum r_i\sigma_i$ is a chain in X, which is a boundary by assumption. Therefore, there exists a $(q+1)$-chain d in X such that $\partial(d) = c'$. Let K' be the minimal carrier of d and put $K'' = K \cup K'$. Then c' becomes a boundary in K'', i.e., if $j\colon K \to K''$ is the inclusion map, then $j(c) = c'$ is a boundary in K''. This says that $j_*\{c\} = \{c'\} = 0$ in $H_q(K'')$. ■

The classical Jordan Curve theorem says that if we take any simple closed curve C (homeomorphic copy of the unit circle) in the plane $\mathbb{R}^2$, then C decomposes the plane into two components, one inside of the curve and the other outside of the curve, whose common boundary is the curve C itself.

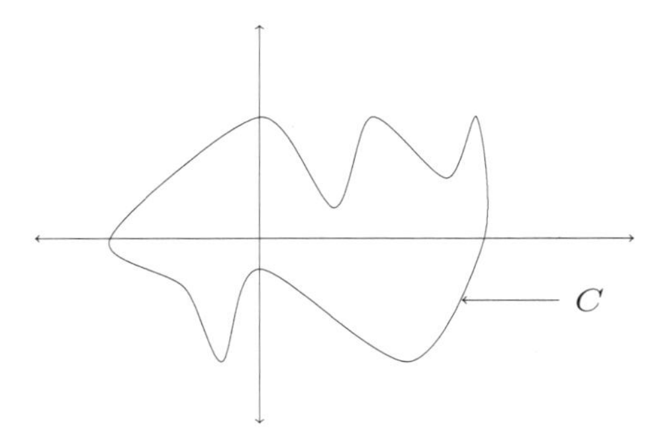

Fig. 6.5

In this section, we will prove this theorem and much more. We emphasize that we need a proof even in the case of plane since the shape of a simple closed curve in the plane could be very complicated, and beyond imagination; it need not look like the one in Fig. 6.5. Note also that if we take the one-point compactification $\mathbb{R}^2 \cup \{\infty\}$ (homeomorphic to $\mathbb{S}^2$) of $\mathbb{R}^2$, then the Jordan curve theorem says that any simple closed curve C in $\mathbb{S}^2$ decomposes the sphere into two components, whose common boundary is the curve C itself. We will prove a more general result: Let $n \geq 1$ be a positive integer and C be a subset of $\mathbb{S}^n$ which is homeomorphic to the sphere $\mathbb{S}^{n-1}$. Then $\mathbb{S}^n - C$ has two components and C is the common boundary of both the components. Here again, C need not look like $\mathbb{S}^{n-1}$ embedded in $\mathbb{S}^n$ in the canonical fashion; it could be visually unimaginable.

Let us start with the following observation: If we remove one point from the n-sphere $\mathbb{S}^n$, $n \geq 1$, the remaining space is homeomorphic to the Euclidean space $\mathbb{R}^n$. What happens if we remove a subset of $\mathbb{R}^n$ which looks like (is homeomorphic to) a closed interval (1-disc) or a 2-disc or more generally, a k-disc, $k \leq n$? The next theorem answers this question in terms of singular homology. We do not know about the homeomorphism type of the remaining space, but we prove below that the remaining space is homologically trivial, i.e., it is path connected and much more. Next, we also want to know what happens when we remove a subset A of $\mathbb{S}^n$ which is homeomorphic to a k-sphere $\mathbb{S}^k$, $0 \leq k < n$. Note that if we write $\mathbb{S}^n = \{(x_i, \dots, x_{n+1}) \in \mathbb{R}^{n+1} \mid \sum_i^{n+1} x_i^2 = 1\}$ and put $x_{n+1} = 0$, then we get $\mathbb{S}^{n-1}$ embedded in $\mathbb{S}^n$ canonically. Removing this $\mathbb{S}^{n-1}$ from $\mathbb{S}^n$ clearly leaves two open hemispheres $\mathbb{D}^n_+$ and $\mathbb{D}^n_-$ whose common boundary is $\mathbb{S}^{n-1}$ itself. However, if we remove a subset C of $\mathbb{S}^n$ which is homeomorphic to $\mathbb{S}^{n-1}$, it is not clear what is the remaining space. Jordan-Brouwer Separation theorem asserts that even in this case, we get the same result - this is indeed remarkable!

Let us prove the following :

Theorem 6.10.3. *Let D be a subset of $\mathbb{S}^n$ homeomorphic to the closed k-ball $\mathbb{D}^k$, $0 \leq k \leq n$. Then $\mathbb{S}^n - D$ is acyclic, i.e., $\tilde{H}_q(\mathbb{S}^n - D) = 0 \ \forall \ q \geq 0$.*

Proof. We prove the theorem by induction on k. When $k = 0$, $\mathbb{S}^n - D$ is simply a point in case $n = 0$, otherwise it is the punctured n-sphere and so is homeomorphic to $\mathbb{R}^n$. We know that $\mathbb{R}^n$ is acyclic and the theorem is true for $k = 0$. Now let us suppose that the theorem is true for $(k-1)$, $k \geq 1$, and D is a homeomorphic to a k-disc $\mathbb{D}^k$. Let $h\colon I^k \to D$ be a homeomorphism. Let $D_1 = h(I^{k-1} \times [0, \frac{1}{2}])$ and $D_2 = h(I^{k-1} \times [\frac{1}{2}, 1])$ be subsets of $\mathbb{S}^n$ and $C = h(I^{k-1} \times \{\frac{1}{2}\})$ be the disc of dimension $(k-1)$.

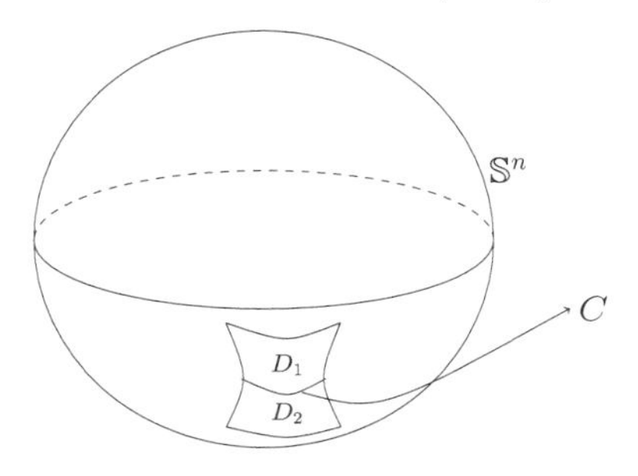

Fig. 6.6

Suppose $\alpha \in \tilde{H}_q(\mathbb{S}^n - D)$ is a nonzero element. There are inclusion maps $i_1\colon (\mathbb{S}^n - D) \to (\mathbb{S}^n - D_1)$ and $i_2\colon (\mathbb{S}^n - D) \to (\mathbb{S}^n - D_2)$. Let $X = \mathbb{S}^n - C$. Then by induction hypothesis, X is acyclic. Clearly, $X = X_1 \cup X_2$ where $X_1 = \mathbb{S}^n - D_1$ and $X_2 = \mathbb{S}^n - D_2$. Since X_1, X_2 are open in $\mathbb{S}^n$, these are also open in X. Hence by the reduced Mayer-Vietoris sequence, the following sequence is exact:

$$\tilde{H}_{q+1}(X) \to \tilde{H}_q(A) \to \tilde{H}_q(X_1) \oplus \tilde{H}_q(X_2) \to \tilde{H}_q(X).$$

Here $A = X_1 \cap X_2 = \mathbb{S}^n - D$. Since X is acyclic, the middle map is an isomorphism. We know the middle map which takes $\alpha \in H_q(A)$ to $(i_{1_*}(\alpha), -i_{2_*}(\alpha))$. Hence, at least one of the two elements $i_{1_*}(\alpha)$ or $i_{2_*}(\alpha)$ must be zero. Suppose $i_{1*}(\alpha) \neq 0$. Then, we write D_1 as the union of $h(I^{k-1} \times [0, \frac{1}{4}])$ and $h(I^{k-1} \times [\frac{1}{4}, \frac{1}{2}])$. By the same argument as above, we find that the image of α in $\tilde{H}_q(I^{k-1} \times [0, \frac{1}{4}])$ or $\tilde{H}_q(I^{k-1} \times [\frac{1}{4}, \frac{1}{2}])$ is nonzero. We continue like this and find a sequence of closed intervals $[a_1, b_1] \supset [a_2, b_2] \supset \cdots \supset [a_m, b_m]$ of lengths $\frac{1}{2}, \frac{1}{4}, \ldots$ such that the image of α in $\tilde{H}_q(I^{k-1} \times [a_m, b_m])$ is nonzero for each $m \geq 1$. Let x_0 be the unique point in the intersection of all these intervals and put $E = h(I^{k-1} \times \{x_0\})$. By induction hypothesis, the group $\tilde{H}_q(\mathbb{S}^n - E)$ is trivial. Hence the image of α in this group is also zero. Since the singular homology has compact supports (see Proposition 6.10.1), there is a compact subset A of $\mathbb{S}^n - E$ such that the image of α in $\tilde{H}_q(A)$ is zero. Since $\mathbb{S}^n - E$ is the union of open sets $\mathbb{S}^n - D_1, \mathbb{S}^n - D_2, \ldots,$ where $D_i = h(I^{k-1} \times [a_i, b_i])$ and A is compact, $A \subset \mathbb{S}^n - D_m$ for some m. This means image of α in $\tilde{H}_q(\mathbb{S}^n - D_m)$ must be zero, which contradicts the assertion proved earlier. Hence $\alpha \in \tilde{H}_q(\mathbb{S}^n - D)$ must be zero, i.e., $\mathbb{S}^n - D$ is acyclic.■

We have computed the singular homology of spaces obtained by removing k-discs from the n-sphere, $0 \leq k \leq n$. Now let us compute the singular homology of those spaces which are obtained by removing homeomorphic copies of k-spheres $\mathbb{S}^k$ from $\mathbb{S}^n$, $0 \leq k < n$. We have

Corollary 6.10.4. *Let $n > k \geq 0$ and $h\colon \mathbb{S}^k \to \mathbb{S}^n$ be an embedding. Then*

$$\tilde{H}_q(\mathbb{S}^n - h(S^k)) = \begin{cases} \mathbb{Z}, & q = n-1-k \\ 0, & \textit{otherwise.} \end{cases}$$

Proof. We will prove this Corollary again by induction on k. When $k = 0$, $h(\mathbb{S}^0)$ consists of two points, say a and b of $\mathbb{S}^n$. Since $\mathbb{S}^n - \{a\} - \{b\}$ is homeomorphic to $\mathbb{R}^n - \{0\}$, and the latter is of the same homotopy type as $\mathbb{S}^{n-1}$, we see that $\tilde{H}_q(\mathbb{S}^n - h(\mathbb{S}^0)) \approx \mathbb{Z}$ if $q = n-1$, and zero otherwise. Suppose now that the Corollary is true for $(k-1), k \geq 1$. Let D^k_+ and D^k_- be the upper and lower closed hemispheres of $\mathbb{S}^k$ and put $X_1 = \mathbb{S}^n - h(D^k_+)$, $X_2 = \mathbb{S}^n - h(D^k_-)$.

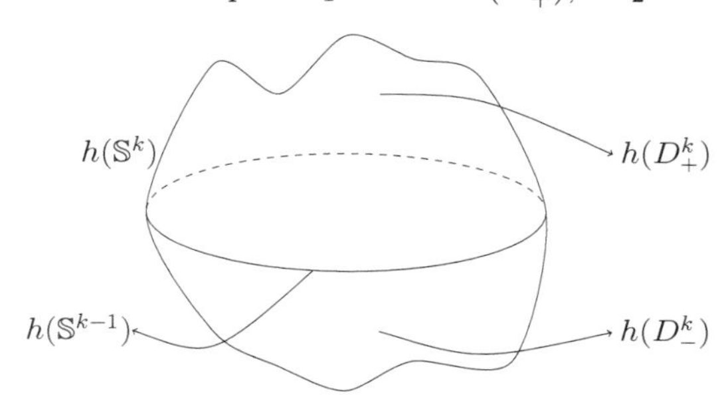

Fig. 6.7

Let $X = X_1 \cup X_2$ and $A = X_1 \cap X_2 = \mathbb{S}^n - h(\mathbb{S}^k)$. Since X_1, X_2 are open sets of X, we have the following part of the reduced Mayer-Vietoris exact sequence

$$\tilde{H}_{q+1}(X_1) \oplus \tilde{H}_{q+1}(X_2) \to \tilde{H}_{q+1}(X) \to \tilde{H}_q(A) \to \tilde{H}_q(X_1) \oplus \tilde{H}_q(X_2).$$

Both X_1, X_2 are acyclic by the previous theorem, and therefore $\tilde{H}_{q+1}(X) \to \tilde{H}_q(A)$ is an isomorphism, i.e., $\forall\ q$

$$H_{q+1}(\mathbb{S}^n - h(\mathbb{S}^{k-1})) \approx H_q(\mathbb{S}^n - h(\mathbb{S}^k)).$$

Hence, by induction hypothesis

$$\tilde{H}_q(\mathbb{S}^n - h(S^k)) = \begin{cases} \mathbb{Z}, & q+1 = n-1-(k-1) \\ 0, & \text{otherwise.} \end{cases}$$

■

Now we can prove our

Theorem 6.10.5. (Jordan-Brouwer Separation Theorem). *Let C be a subset of n-sphere $\mathbb{S}^n$, $n \geq 1$, which is homeomorphic to $\mathbb{S}^{n-1}$. Then $\mathbb{S}^n - C$ has exactly two components of which C is the common boundary.*

Proof. By the previous theorem we know that $\tilde{H}_0(\mathbb{S}^n - C) \approx \mathbb{Z}$. Hence $\mathbb{S}^n - C$ has exactly two path components. Since $\mathbb{S}^n - C$ is locally path connected open sets, path components and components are same. Now the use of algebraic topology in this proof is over. Let W_1 and W_2 be the two components. These are open in $\mathbb{S}^n$, since $\mathbb{S}^n - C$ is an open set. Hence the boundary of each of these is $\overline{W}_1 - W_1$ and $\overline{W}_2 - W_2$ respectively.

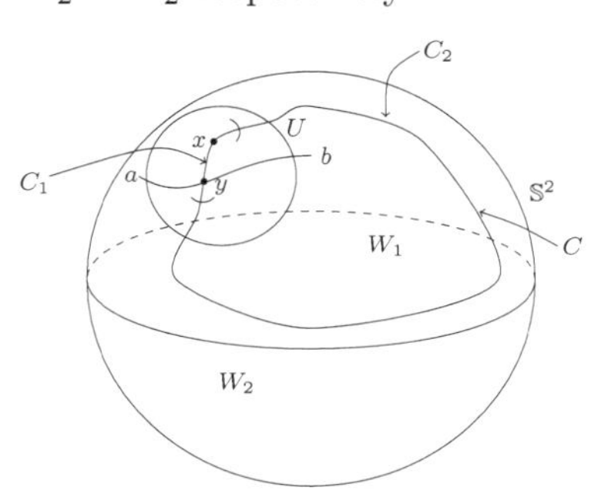

Fig. 6.8

Since no point of W_2 can be a limit point of W_1, $\overline{W}_1 - W_1 \subseteq C$. Similarly, $\overline{W}_2 - W_2 \subseteq C$. Hence it is enough to prove that $C \subseteq \overline{W}_i - W_i$, $i = 1, 2$. Let us show that $C \subseteq \overline{W}_1 - W_1$. Let $x \in C$. We show that x is a limit point of $\overline{W}_1 - W_1$, and since $\overline{W}_1 - W_1$ is closed, x will belong to $\overline{W}_1 - W_1$. To prove that x is a limit point of $\overline{W}_1 - W_1$, let U be an open neighbourhood of x in $\mathbb{S}^n$. We consider C as $\mathbb{S}^{n-1}$ and divide C into closed discs (upper hemisphere and lower hemisphere) C_1 and C_2. We can assume that C_1 is small enough to be contained in U. Now since $\mathbb{S}^n - C_2$ is connected, we take a path ω joining a point $a \in W_1$ and a point $b \in W_2$. This path ω will intersect $\overline{W}_1 - W_1$ because otherwise the image of ω (connected set) will lie in the union of W_1 and $\mathbb{S}^n - \overline{W}_1$, which are disjoint open sets, a contradiction. Let $y \in \overline{W}_1 - W_1$ be the point where the path ω intersects $\overline{W}_1 - W_1$. Since $y \in C$ and $y \notin C_2$, we find that $y \in C_1$. But $C_1 \subseteq U$ by our assumptions. Hence $y \in U$. Thus U intersects $\overline{W}_1 - W_1$ and the proof is complete. ■

The Jordan Curve Theorem is classically stated for simple closed curves in the plane $\mathbb{R}^2$, rather than for simple closed curves in $\mathbb{S}^2$. The same is also true for the more general Jordan Brouwer separation theorem. This can be proved as follows.

Theorem 6.10.6. *Let C be a subset of $\mathbb{R}^n$, $n \geq 2$, which is homeomorphic to $\mathbb{S}^{n-1}$. Then $\mathbb{R}^n - C$ has two path components of which C is the common boundary.*

Proof. We can consider $\mathbb{R}^n$ as $\mathbb{S}^n - \{N\}$, where N is the north pole of $\mathbb{S}^n$. Then C is a subset of $\mathbb{S}^n$ homeomorphic to $\mathbb{S}^{n-1}$. Hence by the Jordan-Brouwer separation theorem, $\mathbb{S}^n - C$ has two path components of which C is the common boundary. Let W_1 and W_2 be the two components and assume that $N \in W_1$. We claim that $W_1 - \{N\}$ is connected. This follows from a general fact that if U is an open connected set of $\mathbb{S}^n$ and a point $a \in U$ is removed, then $U - \{a\}$ remains connected: to see this suppose $U - \{a\}$ is not connected and consider its components (all of them are open in $\mathbb{S}^n$). There is a neighbourhood V of a which is homeomorphic to $\mathbb{R}^n$, and so $V - \{a\} \approx \mathbb{R}^n - \{0\}$ will be connected, since $n > 1$. Now let A be the component of $U - \{a\}$ which contains the connected set $V - \{a\}$ and B be the union of other components of $U - \{a\}$. Then $A \cup \{a\}$ and B give a decomposition of the connected set U, a contradiction. Therefore, it follows that $W_1 - \{N\}$ and W_2 are two components of $\mathbb{R}^n - C$, whose common boundary is C. ■

Next we are going to use the Jordan-Brouwer separation theorem to prove a remarkable result, called "invariance of domain" theorem. In set topology, we already know that a continuous bijective map need not be homeomorphism. However, it is interesting to see that if U is an open set of $\mathbb{R}^n$ or $\mathbb{S}^n$ and $f\colon U \to \mathbb{R}^n$ or $\mathbb{S}^n$ is an injective continuous map, then f is indeed a homeomorphism. We have

Theorem 6.10.7. (Invariance of Domain). *Let $U \subset \mathbb{R}^n$ be an open set and $f\colon U \to \mathbb{R}^n$ be a continuous injective map. Then $f(U)$ is an open set of $\mathbb{R}^n$ and $f\colon U \to f(U)$ is an embedding.*

Proof. Let $x \in U$ and $y = f(x)$. We will show that $f(U)$ is a neighbourhood of each of its points y and is, therefore, open in $\mathbb{R}^n$.

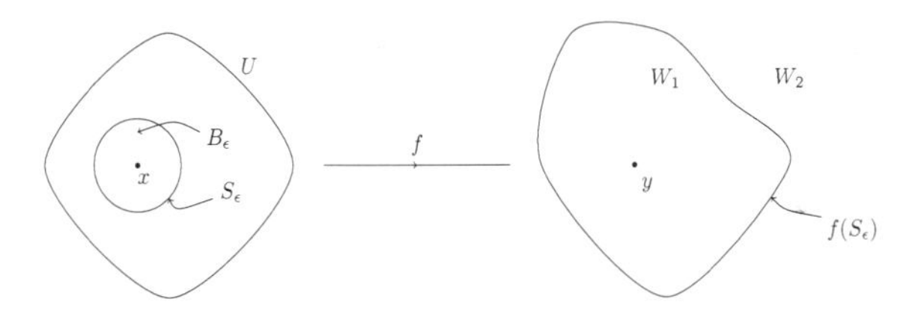

Fig. 6.9

Let us take an open ball B_ϵ of radius ϵ around x whose closure lies in U. Put $S_\epsilon = \overline{B}_\epsilon - B_\epsilon$ and note that S_ϵ is a $(n-1)$-sphere. Since $f\colon S_\epsilon \to f(S_\epsilon)$ is a continuous bijection from a compact space to a Hausdorff space, $f(S_\epsilon)$ is also an $(n-1)$-sphere. Since $f\colon S_\epsilon \to f(S_\epsilon)$ is an embedding, by Jordan-Brouwer separation theorem $f(S_\epsilon)$ decomposes $\mathbb{R}^n$ into two components say W_1 and W_2 which have $f(S_\epsilon)$ as the common boundary. Note that the set $f(B_\epsilon)$ is connected and is disjoint from $f(S_\epsilon)$. Hence it lies either in W_1 or in W_2. Suppose $f(B_\epsilon) \subseteq W_1$. We claim that $f(B_\epsilon) = W_1$ because otherwise $f(\overline{B}_\epsilon) = f(B_\epsilon) \cup f(S_\epsilon)$, being a disc, will separate $\mathbb{S}^n - f(\overline{B}_\epsilon)$ into two components viz., $W_1 - f(B_\epsilon)$ and W_2. This will give a contradiction to the fact that when a disc is removed from $\mathbb{S}^n$, the remaining space remains connected. Hence $f(B_\epsilon) = W_1$ and $f(U)$ is an open neighbourhood of y in $\mathbb{R}^n$. Now f is an embedding because the map $f\colon U \to f(U)$ is easily seen to be an open map, and it is already bijective. ■

6.11 Singular Cohomology and Cohomology Algebra

In this section, we define the singular cohomology of a topological pair (X, A) and for this we basically require the properties of the contravariant Hom(. , .) functor discussed in Appendix. For the sake of generality, we again assume that R is a PID. For a topological space X, let

$$S_*(X)\colon \longrightarrow S_{q+1}(X) \xrightarrow{\partial_{q+1}} S_q(X) \xrightarrow{\partial_q} S_{q-1}(X) \longrightarrow \cdots \longrightarrow S_0(X) \longrightarrow 0,$$

denote the singular chain complex of X. Here, each $S_q(X)$ is an R-module and $\partial_q \circ \partial_{q+1} = 0$ for all $q \geq 0$. If we apply the Hom(. , R) functor to the above chain complex, we get the following long sequence going in the opposite direction:

$$\mathrm{Hom}(S_*(X), R)\colon\ 0 \longrightarrow \mathrm{Hom}(S_0(X), R) \rightarrow \cdots \rightarrow \mathrm{Hom}(S_{q-1}(X), R)$$

$$\xrightarrow{\mathrm{Hom}(\partial_q, R)} \mathrm{Hom}(S_q(X), R) \xrightarrow{\mathrm{Hom}(\partial_{q+1}, R)} \mathrm{Hom}(S_{q+1}(X), R) \rightarrow$$

If, for all $q \geq 0$, we put $\mathrm{Hom}(S_q(X), R) = S^q(X)$ and $\mathrm{Hom}(\partial_q, R) = \delta^q$, then the above sequence of R-modules may be written as

$$S^*(X)\colon 0 \longrightarrow S^0(X) \longrightarrow \cdots \longrightarrow S^{q-1}(X) \xrightarrow{\delta^q} S^q(X) \xrightarrow{\delta^{q+1}} S^{q+1}(X) \longrightarrow$$

which has the property that for all $q \geq 0$,

$$\begin{aligned} \delta^{q+1} \circ \delta^q &= \operatorname{Hom}(\partial_{q+1}, R) \circ \operatorname{Hom}(\partial_q, R) \\ &= \operatorname{Hom}(\partial_q \circ \partial_{q+1}, R) \\ &= \operatorname{Hom}(0, R) = 0. \end{aligned}$$

Thus, $S^*(X)$ is indeed a cochain complex of R-modules, which is called the **singular cochain complex** of X. Its homology, viz., $H_q(S^*(X)) := \frac{\ker \delta^{q+1}}{\operatorname{Im} \delta^q}$, denoted by $H^q(X)$, $q \geq 0$, is called the **singular cohomology** module of the space X. For later use, let us recall that $\delta^q \colon S^{q-1}(X) \to S^q(X)$ is defined by $\delta^q(\alpha) = \alpha \circ \partial_q$ for any $\alpha \in \operatorname{Hom}(S_{q-1}(X), R)$. Hence, onwards we write δ^{q-1} for δ^q for consistency of notation.

Given a continuous map $f \colon X \to Y$, we have the induced chain map $f_\# \colon S_*(X) \to S_*(Y)$. Applying the functor $\operatorname{Hom}(\ .\ , R)$, we get a cochain map $\operatorname{Hom}(f_\#, R) \colon \operatorname{Hom}(S_*(Y), R) \to \operatorname{Hom}(S_*(X), R)$, which we denote by $f^\# \colon S^*(Y) \to S^*(X)$. Then, this cochain map obviously induces a homomorphism in homology, i.e., we get induced homomorphism $f^* \colon H^q(Y) \to H^q(X)$ for all $q \geq 0$, in singular cohomology modules of X and Y. With these definitions, one can easily verify the following functorial (contravariant) properties:

Proposition 6.11.1. (i) *If $I_X \colon X \to X$ is the identity map, then the induced homomorphism $(I_X)^* \colon H^q(X) \to H^q(X)$ is also the identity map for all $q \geq 0$.*

(ii) *If $f \colon X \to Y, g \colon Y \to Z$ are two maps then, for all $q \geq 0$,*

$$(g \circ f)^* = f^* \circ g^* \colon H^q(Z) \to H^q(X).$$

The following result should be compared with the corresponding homology result (see Proposition 6.2.8).

Proposition 6.11.2. *Let $\{X_i, i \in I\}$ be the path-components of a space X. Then, for all $q \geq 0$, $H^q(X) \cong \prod_{i \in I} H^q(X_i)$, i.e., the singular cohomology of X is isomorphic to the product of the singular cohomology of its path-components.*

Proof. Since the q-simplex Δ_q is path-connected, any singular q-simplex $\sigma \colon \Delta_q \to X$ maps Δ_q in some path-component X_i of X. Therefore, for all $q \geq 0$, $S_q(X) \cong \bigoplus_{i \in I} S_q(X_i)$. Since the Hom functor converts the direct sum into direct product, we find that

$$\operatorname{Hom}(S_q(X), R) \cong \operatorname{Hom}(\oplus S_q(X_i), R) \cong \prod_{i \in I} \operatorname{Hom}(S_q(X_i), R).$$

Since the group of q-cocycles and q-coboundaries of X are also the direct products of q-cocycles and coboundaries of $X_i, i \in I$, we get the desired result. ∎

The above proposition tells us, in particular, that the singular cohomology in dimension zero of a discrete space X will be isomorphic to the direct product of as many copies of R as there are points in X and will be zero for higher dimensions.

Reduced Cohomology

If $X \neq \phi$, then we can define an augmentation map $\eta \colon R \to \operatorname{Hom}(S_0(X), R)$ as the dual of the augmentation $\epsilon \colon S_0(X) \to R$ by putting $\eta(k)(\sum n_i x_i) = k \cdot \epsilon(\sum n_i x_i) = k \cdot (\sum n_i)$. Using this, we define the reduced singular cohomology $\tilde{H}^q(X)$ of the space X as the homology of the cochain complex:

$$0 \longrightarrow R \xrightarrow{\eta} S^0(X) \xrightarrow{\delta^0} S^1(X) \xrightarrow{\delta^1} \cdots \longrightarrow S^q(X) \xrightarrow{\delta^q} S^{q+1}(X) \longrightarrow \cdots,$$

where we have $\delta^0\eta = 0$. It follows from this definition that

$$\tilde{H}^0(X) \oplus R \cong H^0(X), \qquad \tilde{H}^q(X) = H^q(X), \text{ for all } q > 0.$$

Relative Cohomology and the Axioms

Let us now extend the definition of singular cohomology to topological pairs (X, A). Recall that

$$0 \to S_*(A) \to S_*(X) \to S_*(X, A) \to 0$$

is an exact sequence of singular chain complexes of R-modules in which each complex consists of free R-modules. Since the above sequence of chain complexes splits, by applying the $\operatorname{Hom}(., R)$ functor, we find that the induced sequence of singular cochain complexes

$$0 \longrightarrow S^*(X, A) \xrightarrow{j^\#} S^*(X) \xrightarrow{i^\#} S^*(A) \longrightarrow 0$$

is also exact. We define the cohomology of the cochain complex $S^*(X, A)$ to be the **singular cohomology of the pair** (X, A), which we denote by $H^q(X, A)$, $q \geq 0$. Since the preceding sequence of cochain complexes is exact, we get the following long exact sequence for singular cohomology:

$$\cdots \longrightarrow H^q(X, A) \xrightarrow{j_*} H^q(X) \xrightarrow{i_*} H^q(A) \xrightarrow{\partial} H^{q+1}(X, A) \longrightarrow \cdots.$$

This proves the **Exactness Axiom** for singular cohomology. The **Commutativity Axiom** for a map $f \colon (X, A) \to (Y, B)$ also follows by writing the long exact cohomology sequences for pairs (X, A) and (Y, B) arising from the following commutative diagram of singular cochain complexes and cochain maps:

$$\begin{array}{ccccccccc}
0 & \longrightarrow & S^*(X,A) & \longrightarrow & S^*(X) & \longrightarrow & S^*(A) & \longrightarrow & 0 \\
 & & \uparrow f^{\#} & & \uparrow f^{\#} & & \uparrow (f|_A)^{\#} & & \\
0 & \longrightarrow & S^*(Y,B) & \longrightarrow & S^*(Y) & \longrightarrow & S^*(B) & \longrightarrow & 0
\end{array}$$

To prove the **Homotopy Axiom** for singular cohomology, let $i_0,\ i_1\colon (X,A) \to (X \times I, A \times I)$ denote the bottom and top inclusion maps. Then we have already proved that the induced chain maps $i_{0\#}, i_{1\#}\colon S_*(X,A) \to S_*(X \times I, A \times I)$ are chain homotopic. Consequently, applying the Hom(. , R) functor, we find that the cochain maps $i_0^{\#},\ i_1^{\#}\colon S^*(X \times I, A \times I) \to S^*(X,A)$ are cochain homotopic. Therefore, the induced homomorphisms in singular cohomology $i_0^*,\ i_1^*\colon H^q(X \times I, A \times I) \to H^q(X,A)$ are identical. But this means, as in the case of singular homology, that if $f,\ g\colon (X,A) \to (Y,B)$ are homotopic maps of pairs then the induced homomorphisms in singular cohomology $f^*,\ g^*\colon H^q(Y,B) \to H^q(X,A)$ are identical for all $q \geq 0$.

By a similar argument, one can easily prove the Excision Axiom for singular cohomology. The Dimension Axiom being evidently true, we have the following result:

Theorem 6.11.3. *The singular cohomology defined on the category of all topological pairs is a cohomology theory in the sense of Eilenberg-Steenrod.*

One can similarly define the simplicial cohomology of a polyhedral pair (X,A), using the simplicial chain complex $C_*(X,A)$ in place of $S_*(X,A)$ everywhere. If G is an R-module, then singular cohomology of a pair (X,A), with coefficients in G, can also be easily defined. Instead of the functor Hom(. , R), we apply the functor Hom(. , G) to the singular chain complex $S_*(X,A)$ and get the cochain complex $S^*(X,A;G) := \mathrm{Hom}(S_*(X,A),G)$. The corresponding homology modules, viz., $H_q(S^*(X,A;G))$, denoted by $H^q(X,A;G)$, are called the **singular cohomology** modules of the pair (X,A) **with coefficients in** G. One can again easily prove that $H^q(X,A;G), q \geq 0$, which are contravariant in the pair (X,A), are indeed covariant functors in the coefficient module G. Thus, an exact sequence $0 \to G' \to G \to G'' \to 0$ of coefficient R-modules gives us the following long exact sequence

$$\to H^q(X,A;G') \to H^q(X,A;G) \to H^q(X,A;G'') \stackrel{\delta^q}{\to} H^{q+1}(X,A;G') \to$$

of cohomology of R-modules. The connecting homomorphism $\delta^q\colon H^q(X,A;G'') \to H^{q+1}(X,A;G')$ is called the **Bockstein cohomology operator** and they are known to have better applications in cohomology as against homology.

Universal Coefficient Theorem for Cohomology

Having defined the singular homology and singular cohomology of a space X, it is natural to ask as to how are the homology and cohomology groups of X related. If our ground ring R is a PID, there is a short exact sequence connecting the singular homology with singular cohomology of X showing that cohomology groups are, to a great extent, dual to the homology groups. To appreciate the exact relationship, we need to recall the properties of the Ext(. , .) functor discussed in Appendix. The precise result is given by the following theorem:

Theorem 6.11.4. (Universal Coefficient Theorem for Cohomology). *For an R-module G, where R is a PID, and for any pair (X, A) of topological spaces, there is the following split short exact sequence of R-modules for all $q \geq 0$:*

$$0 \to \text{Ext}(H_{q-1}(X, A), G) \to H^q(X, A, G) \to \text{Hom}(H_q(X, A), G) \to 0.$$

The above sequence is natural in (X, A) as well as in the coefficient module G, but the splitting is not natural with respect to (X, A).

The proof of the above theorem follows from the corresponding result of abstract chain and cochain complexes because here the singular chain complexes are free (see Spanier [18] p. 243). Note that, because the above sequence splits, we find that for all $q \geq 0$, $H^q(X, A; G) \cong \text{Hom}(H_q(X, A), G) \oplus \text{Ext}(H_{q-1}(X, A), G)$. This result is quite nice, at least for computational purposes. We must keep the two special cases of the above theorem in mind: first, when we take $R = \mathbb{Z}$ and secondly, when R is a field. When R is a field, the Ext-term above vanishes and we see that cohomology is simply the dual of homology. The following interesting result, which we state only for $R = \mathbb{Z}$, is also true for any PID.

Corollary 6.11.5. *Suppose a continuous map $f: X \to Y$ induces isomorphism $f_*: H_q(X) \to H_q(Y)$ in integral singular homology for all $q \geq 0$. Then the induced homomorphisms*

$$f_*: H_q(X; G) \to H_q(Y; G)$$

and

$$f^*: H^q(Y; G) \to H_q(X; G),$$

in homology and cohomology with coefficients in G, are isomorphisms for any abelian group G and every $q \geq 0$.

Proof. Let $S_*(X)$, $S_*(Y)$ denote the integral singular chain complexes of X and Y and let $f_\#: S_*(X) \to S_*(Y)$ denote the induced chain map. Since these are chain complexes of free abelian groups and the induced homomorphism $f_*: H_q(X) \to H_q(Y)$ in homology by the chain map $f_\#$ is an isomorphism, a well-known result of homological algebra (see Ex. 4, Section 7.2) says that

$f_\#$ is a chain equivalence. This means $f_\# \otimes 1\colon S_*(X) \otimes G \to S_*(Y) \otimes G$ is also a chain equivalence of R-modules. Hence the induced homomorphism $f_*\colon H_q(X;G) \to H_q(Y;G)$ in homology, with coefficients in G, must be an isomorphism. Next, we apply the Hom(. ,G) functor to the chain equivalence $f_\#\colon S_*(X) \to S_*(Y)$ to infer that the induced homomorphism $f^\#\colon S^*(Y,G) \to S^*(X,G)$ in cochain complexes is also a cochain equivalence. This implies that the induced homomorphism $f^*\colon H^q(Y;G) \to H^q(X;G)$ in cohomology is also an isomorphism. ∎

Note that the above result says that **if two spaces cannòt be distinguished by integral homology groups then they cannot be distinguished by homology or cohomology groups with any coefficient.**

Example 6.11.6. Applying the universal coefficient theorem for cohomology, one can now easily compute the integral singular cohomology groups of the n-sphere $\mathbb{S}^n$, the torus T, the Möbius band M, the projective plane $\mathbb{P}^2$ and the Klein bottle K, etc. Notice that for an abelian group G, the Universal Coefficient Theorem says that

$$H^q(\mathbb{S}^n;G) \cong \operatorname{Hom}(H_q(\mathbb{S}^n),G) \oplus \operatorname{Ext}(H_{q-1}(\mathbb{S}^n),G).$$

Since $H_q(\mathbb{S}^n) \cong \mathbb{Z}$ for $q = 0$ or n, and is zero otherwise, we find

$$H^q(\mathbb{S}^n;G) = \begin{cases} G, & q = 0, n \\ 0, & \text{otherwise.} \end{cases}$$

Similarly, for the projective plane $\mathbb{P}^2$, we have $H^0(\mathbb{P}^2) \cong \mathbb{Z}$, $H^1(\mathbb{P}^2) \cong \operatorname{Hom}(\mathbb{Z}_2,\mathbb{Z}) \oplus \operatorname{Ext}(\mathbb{Z},\mathbb{Z}) = 0$, $H^2(\mathbb{P}^2) = \operatorname{Ext}(\mathbb{Z}_2,\mathbb{Z}) \cong \mathbb{Z}_2$, $H^q(\mathbb{P}^2) = 0$ for all $q \geq 3$.

For coefficients in $\mathbb{Z}_2$, we have $H^0(\mathbb{P}^2;\mathbb{Z}_2) \cong \mathbb{Z}_2$, $H^1(\mathbb{P}^2;\mathbb{Z}_2) = \operatorname{Hom}(\mathbb{Z}_2,\mathbb{Z}_2) \cong \mathbb{Z}_2$, $H^2(\mathbb{P}^2;\mathbb{Z}_2) = \operatorname{Hom}(0,\mathbb{Z}_2) \oplus \operatorname{Ext}(\mathbb{Z}_2,\mathbb{Z}_2) \cong \mathbb{Z}_2$,, $H^q(\mathbb{P}^2) = 0$ for all $q \geq 3$.

Remark 6.11.7. We emphasize that cohomology with coefficients in G is obtained by taking the Hom and Ext of integral homology, not of the homology with coefficients in G. This can be checked by the example of projective plane $\mathbb{P}^2$ computed above.

The fact that $H^1(\mathbb{P}^2) = 0$ shows that the fundamental group of a space is not related to the 1-dimensional cohomology of the space, notwithstanding the fact that it is nicely related to its 1-dimensional homology, viz., $\pi_1(\mathbb{P}^2) \cong H_1(\mathbb{P}^2)$.

We also point out that the integral cohomology of $\mathbb{S}^n$ is exactly the dual of its homology whereas the integral cohomology of the projective plane $\mathbb{P}^2$ is not the dual of its integral homology. We observe at the same time that, though $\mathbb{P}^2$

is not orientable, yet its second integral cohomology is nonzero. This says that orientability, which is related to the nonvanishing of nth integral homology of a n-manifold, is not related to its cohomology in the same way.

We also have the following Mayer-Vietoris exact sequence for singular cohomology.

Theorem 6.11.8. *Let X_1, X_2 be two subspaces of a space X such that $X_1 \cup X_2 = \operatorname{int} X_1 \cup \operatorname{int} X_2$. Then the following Mayer-Vietoris sequence for singular cohomology is exact for any R-module G (coefficients are suppressed):*

$$\xrightarrow{\delta^{q-1}} H^q(X_1 \cup X_2) \xrightarrow{(j_1^*, j_2^*)} H^q(X_1) \oplus H^q(X_2) \xrightarrow{i_1^* - i_2^*} H^q(X_1 \cap X_2) \xrightarrow{\delta^q} H^{q+1}(X_1 \cup X_2) \rightarrow$$

Subject to some conditions, the Künneth Formula for singular cohomology is also valid and is given by the next theorem:

Theorem 6.11.9. (Künneth). *Let X, Y be any two spaces. Then, for all $n \geq 0$, there is the following split short exact sequence for any coefficient groups G and G' satisfying the condition that $\operatorname{Tor}(G, G') = 0$, and (a) both $H^*(X), H^*(Y)$ are of finite type or, (b) $H^*(Y)$ is finite type and G' is finitely generated,*

$$0 \rightarrow \sum_{p+q=n} H^p(X; G) \otimes H^q(Y; G') \rightarrow H^n(X \times Y, G \otimes G')$$

$$\rightarrow \sum_{p+q=n+1} \operatorname{Tor}(H^p(X; G), H^q(Y; G')) \rightarrow 0.$$

Furthermore, the above sequence is natural in spaces X, Y and the coefficient groups G, G', but the splitting is not natural.

It is interesting to observe the case of Künneth formula when $G = G' = R$ is a field. In this simplest situation, everything being free, we find that the cohomology of the product space agrees exactly with the tensor product of the cohomologies of the factor spaces because the term involving the Tor functor vanishes. The proofs of the preceding theorems are omitted and can be found in ([18] p. 239, 247) in all their details.

Cohomology Algebra

Let X be a topological space. We continue to assume that R is a PID and that all homology, cohomology modules in this section are with coefficients in the R-module R. Note that we have an obvious R-isomorphism $m\colon R \otimes R \rightarrow R$ defined by $m(r \otimes s) = r \cdot s$, where $r \cdot s$ is the multiplication of the ring R. The cohomology cross product $H^p(X) \times H^q(X) \to H^{p+q}(X)$ of the Künneth formula can be used to define a product in the graded cohomology module $H^*(X) := \bigoplus_{q \geq 0} H^q(X)$ as follows: Let $d\colon X \to X \times X$ denote the diagonal map given by $d(x) = (x, x)$. Clearly, d induces a homomorphism $d^*\colon H^n(X \times X) \to H^n(X), n \geq 0$, of cohomology modules. Observe that for any $u \in H^p(X)$, $v \in H^q(X)$, the cohomology product $u \times v \in H^{p+q}(X \times X, R \otimes R)$ and the multiplication isomorphism m defines an element $m_*(u \times v) \in H^{p+q}(X \times X, R)$. Therefore, the element $d^*(m_*(u \times v)) \in H^{p+q}(X, R)$. We define the **cup product** of elements $u \in H^p(X), v \in H^q(X)$, denoted by $u \cup v$ or just by the juxtaposition uv, by putting $u \cup v = d^*(m_*(u \times v))$. Then, using the bilinearity, it is extended to define the cup product in the whole graded cohomology module $H^*(X)$ of X. This product has several interesting properties; it is associative, it has an identity element, viz., $1 \in H^0(X, R)$, where 1 is represented by the singular 0-cochain of X which maps all points of X (singular 0-chains) to the unity element $1 \in R$. The cup product is also commutative up-to a sign, i.e., $uv = (-1)^{\deg(u)\cdot\deg(v)}vu$ for all $u, v \in H^*(X)$. Thus, $H^*(X)$ becomes an R-algebra with respect to this cup product. This algebra is called the **cohomology algebra** of the space X. The cup product has one more striking property, viz., it is natural in X, i.e., if $f\colon X \to Y$ is a map then the induced homomorphism $f^*\colon H^*(Y) \to H^*(X)$ in cohomology preserves the cup product. In other words, $f^*(u \cup v) = f^*(u) \cup f^*(v)$ for all $u, v \in H^*(Y)$. This says, in particular, that the **cohomology algebra $H^*(X)$ of X is a topological invariant of the space X.**

In order to give an idea of the structure of the cohomology algebra of a space, we will just state a few examples without going into details. The interested reader can find the full details in any of the advanced books on algebraic topology such as Bredon [2] or Spanier [18], etc.

Example 6.11.10. The cohomology algebra $H^*(\mathbb{S}^n)$ of the n-sphere $\mathbb{S}^n, n \geq 1$, is trivial, i.e., if $a \in H^n(\mathbb{S}^n) \cong \mathbb{Z}$ is the generator, then $a^2 = 0$ and this says that the product of any two elements of positive degree will be trivial. On the other hand, the cohomology algebra of $\mathbb{S}^n \times \mathbb{S}^n$ is nontrivial, i.e., if $u, v \in H^*(\mathbb{S}^n \times \mathbb{S}^n) \cong \mathbb{Z} \oplus \mathbb{Z}$ are the generators, then their cup product $uv \in H^{2n}(\mathbb{S}^n \times \mathbb{S}^n) \cong \mathbb{Z}$ is again a generator. In fact, it can be shown that $u^2 = 0 = v^2$ and $uv \neq 0$.

For any field F, let $F[x]$ denote the polynomial ring over F in one indeterminate x. Let (x^{n+1}) denote the ideal of the polynomial ring generated by the element x^{n+1}. Then the quotient ring $F[x]/(x^{n+1})$, which is an algebra over F, consists of polynomials in x of degree at most n. Such an algebra is called a **truncated polynomial algebra** of height n. Furthermore, if $F[x]$ is regarded as a graded ring with indeterminate x of degree k, then $F[x]/(x^{n+1})$ is called a truncated polynomial algebra of degree k and height n. This concept is useful in understanding the cohomology algebra of some familiar spaces.

Example 6.11.11. Let $\mathbb{RP}^n$ denote the real projective space of dimension n. This space is a compact, connected topological manifold of dimension n which is triangulable. Its cohomology with coefficients in $\mathbb{Z}_2$ is given by $H^q(\mathbb{RP}^n, \mathbb{Z}_2) \cong \mathbb{Z}_2$ for $q = 0, 1, 2, \ldots, n$ and is trivial for other values of q. It can be shown that the nonzero element $w \in H^1(\mathbb{RP}^n, \mathbb{Z}_2)$ has the property that $w, w^2, w^3, \ldots, w^n$ are all nonzero and $w^{n+i} = 0$ for all $i \geq 1$. Thus, the cohomology algebra of $\mathbb{RP}^n$ is a truncated polynomial algebra over $\mathbb{Z}_2$ of degree 1 and height n.

Example 6.11.12. Let $\mathbb{CP}^n$ denote the complex projective space. It is a simply connected, compact, triangulable manifold of real dimension $2n$. The integral cohomology of $\mathbb{CP}^n$ is given by $H^q(\mathbb{CP}^n) \cong \mathbb{Z}$, $q = 0, 2, \ldots, 2n$ and is trivial in other dimensions. If $w \in H^2(\mathbb{CP}^n, \mathbb{Z})$ is a generator then it can be proved that $w^2, w^3, \ldots, w^n$ are all nonzero, and they generate the corresponding cohomology groups and $w^{n+1} = 0$. It follows that the cohomology algebra of $\mathbb{CP}^n$ over $\mathbb{Z}$ is the truncated polynomial algebra over $\mathbb{Z}$ of degree 2 and height n.

It must be remarked here that the cup product can sometimes distinguish between two nonhomeomorphic spaces when they cannot be distinguished by homology or cohomology groups. We have

Example 6.11.13. Let $X = \mathbb{S}^p \times \mathbb{S}^q$, $p \neq q$. In the first example above, we computed the cohomology algebra of X. Let us now consider a different space $Y = \mathbb{S}^p \vee \mathbb{S}^q \vee \mathbb{S}^{p+q}$, which is the one-point union of the three spheres $\mathbb{S}^p$, $\mathbb{S}^q$, $\mathbb{S}^{p+q}$. All these spheres are triangulable and therefore Y can be so triangulated that all the spheres have a vertex v in common. Then, using the Mayer-Vietoris exact sequence for simplicial homology, we find that

$$H_i(Y, \mathbb{Z}) = \begin{cases} \mathbb{Z}, & i = 0, p, q, p+q \\ 0, & \text{otherwise.} \end{cases}$$

Using the universal coefficient theorem for homology and cohomology, one can easily verify that the two spaces X and Y have isomorphic homology as well as isomorphic cohomology groups with all coefficients. It is obvious that the space X is not homeomorphic to the space Y because, for instance, any point of $\mathbb{S}^p - Y$ has a neighbourhood which is homeomorphic to $\mathbb{R}^p$ whereas there is no such point in the space X. In fact, the space X is a $(p+q)$-manifold and Y is not a manifold. This example exhibits the limitation of the additive structure of the homology and cohomology groups. However, it is interesting to observe that the additional structure of the cup product in cohomology does distinguish between X and Y. If $a \in H^p(\mathbb{S}^p \times \{y\})$, $b \in H^q(\{x\} \times \mathbb{S}^q)$ are the generators, then the product $ab \in H^{p+q}(\mathbb{S}^p \times \mathbb{S}^q)$ is nonzero whereas the product $ab \in H^{p+q}(Y)$ is zero. In fact, this gives even a stronger result, viz., X, Y are not even of the same homotopy type – a result which is not at all obvious by considerations of set topology.

Exercises

1. Prove that the Euler characteristic $\chi(X) = \sum(-1)^i rkH_i(X, F)$ of a space X is independent of the coefficient field F used in defining the Euler characteristic. Show, in particular, that $\chi(X) = \sum(-1)^i rkH_i(X) = \sum(-1)^i rkH_i(X, \mathbb{Q})$.

2. Let (X, A) be a pair. If any two of the groups $H_*(A)$, $H_*(X)$ and $H_*(X, A)$ are finitely generated, then show that the third group is also finitely generated, and furthermore $\chi(X) = \chi(A) + \chi(X, A)$.

3. Let X_1, X_2 be a pair of excisive couple of subspaces in a space X such that $H_*(X_1)$, $H_*(X_2)$ are finitely generated. Prove that $H_*(X_1 \cup X_2)$ is finitely generated if and only if $H_*(X_1 \cap X_2)$ is finitely generated, and if that happens, show that $\chi(X_1 \cup X_2) = \chi(X_1) + \chi(X_2) - \chi(X_1 \cap X_2)$.

4. Let X be the space obtained by puncturing a 2-sphere at two distinct points. Prove that $H_0(X) \cong \mathbb{Z}$, $H_1(X) \cong \mathbb{Z}$ and $H_q(X) = 0$ for all $q \geq 2$. (**Hint:** X is homotopy type of a circle.)

5. Consider the torus T as the quotient of the square $I \times I$ by identifying the opposite edges AB with DC and AD with BC respectively (Fig. 6.8). Let $q\colon I \times I \to T$ denote the quotient map. Prove that

   (i) The sets $q(AB)$ and $q(AD)$ are singular 1-simplexes in T which are cycles.

   (ii) The sets $q(ADB)$ and $q(ACD)$ are singular 2-simplexes whose difference is a boundary.

   (iii) If X is a circle lying in the interior of $I \times I$, then show that $q(X)$ is a singular one simplex in T which is a boundary.

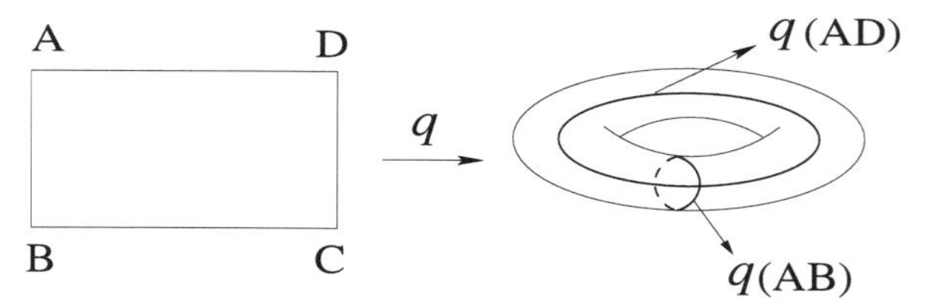

Fig. 6.8: The two nonzero cycles in torus T.

6. Compute the singular homology of the following spaces:

   (a) The Topologist's Sine Curve, i.e., $X = \{(x, y) \in \mathbb{R}^2 : y = \sin 1/x,\ x > 0\} \cup \{(0, y) : |y| \leq 1\}$. (**Hint:** Proposition 6.2.4).

   (b) The Warsaw Circle $Y = \{(x, y) \in \mathbb{R}^2 : y = \sin 1/x, 0 < x < \frac{1}{3\pi}\} \cup A \cup B$, where A is an arc joining $(\frac{1}{3\pi}, 0)$ to $(0, 0)$ lying in the lower plane and $B = \{(0, y) \mid |y| \leq 1\}$ (Fig. 6.9).

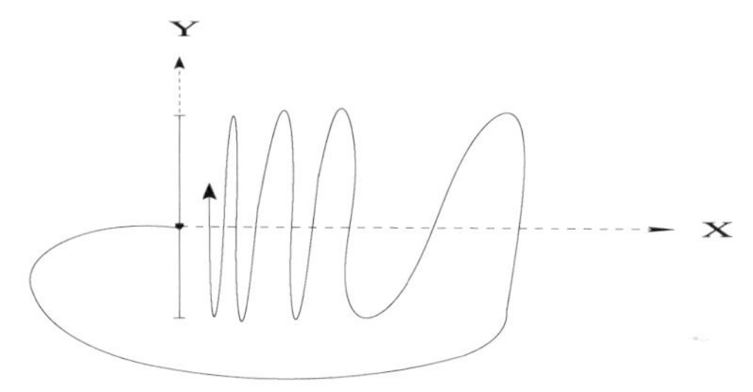

Fig. 6.9: The Warsaw Circle

7. Show that the boundary ∂M of the Möbius band M is not a retract of M. Give a map f from M to the middle circle C of M which is a deformation retraction, i.e., there is a homotopy starting from the identity map of M and ending with the retraction f of M onto C.

8. Let Δ^q denote the q-simplex and for $0 \leq m < q$, let $\Delta^q(m)$ denote the m-skeleton of Δ^q. Compute the singular homology groups of $\Delta^q(m)$ (see Ex. 8, Section 4.4).

9. Let X be the quotient space obtained from the triangle ABC by identifying AB with BC, BC with CA and CA with AB by linear homeomorphisms. Compute the singular homology groups of the space X (see Ex. 7, Section 4.4).

10. Use the above exercise to show that given any $m \geq 1$ and any $n > 0$, there is a compact polyhedron X such that the reduced singular homology $\tilde{H}_q(X) = \mathbb{Z}_m$ when $q = n$, and is 0, otherwise.

11. Use the preceding exercise to prove that given any finite sequence $\{G_0, G_1, \ldots, G_n\}$ of finitely generated abelian groups in which G_0 is free, there exists a compact polyhedron X such that the reduced singular homology groups $\tilde{H}_q(X) \cong G_q$ for all $q \geq 0$.

12. Let X be the space of a 2-sphere with a circle touching the sphere at a point (see Fig. 6.10). Compute the singular homology of X.

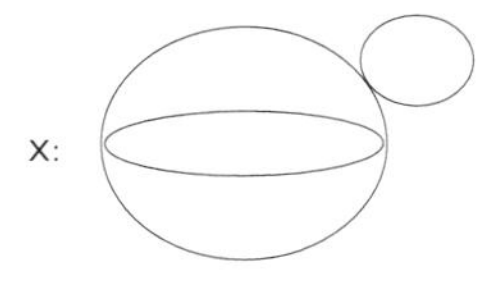

Fig. 6.10: A 2-sphere with a tangent circle

13. Compute the singular homology of a two-holed torus, i.e., a 2-sphere with two handles (see Fig. 6.11) or the connected sum of two tori.

14. Prove that for any k with $0 < k < n$:

Fig. 6.11: A two-holed (double) torus

(i) $\mathbb{S}^k \subset \mathbb{S}^n$ cannot be a retract of $\mathbb{S}^n$.

(ii) $\mathbb{RP}^k \subset \mathbb{RP}^n$ cannot be a retract of $\mathbb{RP}^n$.

(iii) $\mathbb{CP}^k \subset \mathbb{CP}^n$ cannot be a retract of $\mathbb{CP}^n$.

Chapter 7

Appendix

7.1 Basic Algebra – a Review

Here we present some basic definitions and a few results from elementary algebra and homological algebra which are required in introducing and discussing the concepts of algebraic topology. We assume that the reader is familiar with the fundamentals of groups and modules from where we recall only essential portions for completeness.

7.1.1 Groups and Homomorphisms

A set G with an associative binary operation is called a **semigroup**. A semigroup G is said to be a **group** if there is a two-sided identity element e in G, and every element $a \in G$ has an inverse $a^{-1} \in G$. It is readily verified that identity element of a group G is unique, and that inverse of each element in G is also unique. We will, for convenience only, denote the product $a.b$ of two elements of G just by ab. A group G is called **abelian** if $ab = ba$ for each $a, b \in G$.

A map $f\colon G \to G'$ from a group G to another group G' is said to be a **homomorphism** if it preserves binary operations in G and G', i.e., $f(ab) = f(a)f(b)$ for all $a, b \in G$. One can easily prove that f preserves the identity element and also the inverse of each element, i.e., $f(e) = e'$, where $e \in G$ and $e' \in G'$ are identity elements and $(f(a))^{-1} = f(a^{-1})$ for each $a \in G$. A bijective homomorphism is called an **isomorphism**. A homomorphism $f\colon G \to G$ from a group G to itself is called an **endomorphism** and a bijective endomorphism is said to be an **automorphism**.

A subset H of a group G, which is closed under the binary operation of G, is called a **subgroup** of G if H is a group with respect to this induced binary operation. A group G is said to be **finitely generated** if there exists a finite subset $S = \{g_1, g_2, \ldots, g_k\}$ of G such that every element of G can be written as a product of powers of $g_i \in S$, i.e., for each $a \in G$, there exist integers $n_1, n_2, \ldots, n_k$ and a permutation σ on the set $\{1, 2, \ldots, k\}$ such that

$$a = g_{\sigma(1)}^{n_1} \, g_{\sigma(2)}^{n_2} \, \cdots \, g_{\sigma(k)}^{n_k}.$$

We point out that when the group G is abelian, the above expression for $a \in G$ is usually written as $a = n_1 g_{\sigma(1)} + n_2 g_{\sigma(2)} + \ldots + n_k g_{\sigma(k)}$. If S generates G, we write $G = [g_1, \ldots, g_k]$. A group G generated by a single element is said to be a **cyclic** group. In other words, a group G is cyclic if there exists a $g \in G$ such that $G = [g]$. It is well-known that a cyclic group G is isomorphic to either the additive group $\mathbb{Z}$ of integers or to the additive group $\mathbb{Z}_m$ of residue classes modulo m for some $m \in \mathbb{N}$.

A subgroup N of a group G is called a **normal** subgroup if $gng^{-1} \in N$ for each $g \in G$ and for each $n \in N$. This is equivalent to say that $gNg^{-1} \subseteq N$ for each $g \in G$. If N is normal in G then, for each $g \in G$, left cosets and right cosets of N in G are identical, i.e., $gN = Ng$ for all $g \in G$. Let N be normal in G and let G/N denote the set of all cosets (right or left – it is the same thing) of N in G. Then we can define a product in the set G/N by setting $(g_1N)(g_2N) = (g_1g_2)N$. Note that this product in G/N is, in fact, induced by the product of G and is well-defined since N is normal. Then one can easily verify that, with this operation, the set G/N becomes a group with $eN = N$ as the identity element. This group is called a **factor group** (or quotient of G by N) of G by the normal subgroup N. The following is a basic result connecting normal subgroups of G and kernels of homomorphisms emanating from G:

Theorem 7.1.1. (Fundamental Theorem of Homomorphism). *Let $f\colon G \to G'$ be a surjective homomorphism and $K = \ker f$. Then K is a normal subgroup of G and $G/K \cong G'$.*

The map $f\colon \mathbb{R} \to \mathbb{S}^1$, where $\mathbb{S}^1$ is the unit circle in the complex plane, defined by $f(x) = e^{2\pi i x}$, is a surjective homomorphism with group $\mathbb{Z}$ of integers as its kernel. Hence, by the Fundamental Theorem of Homomorphism, $\mathbb{R}/\mathbb{Z} \cong \mathbb{S}^1$. As another interesting example, let $GL(n, \mathbb{R})$ denote the multiplicative group of all $n \times n$ nonsingular real matrices, $n \geq 1$. Then the map $f\colon GL(n, \mathbb{R}) \to \mathbb{R}^* = \mathbb{R} - \{0\}$ (the multiplicative group of nonzero reals), defined by $f(A) = \det A$, is a surjective homomorphism whose kernel is the group $SL(n, \mathbb{R})$ of real matrices with determinant 1. Hence, by the Fundamental Theorem of Homomorphism,

$$GL(n, \mathbb{R})/SL(n, \mathbb{R}) \cong \mathbb{R}^*.$$

Exercises

1. Let $n \geq 1$ and $GL(n, \mathbb{R})$ be the set of all $n \times n$ non-singular matrices over reals. Prove that $GL(n, \mathbb{R})$ is a group with respect to multiplication of matrices. Identify the group $GL(1, \mathbb{R})$.

2. Show that the additive group $\mathbb{Q}$ of rationals is not finitely generated.

3. Let G be a semigroup. Suppose it has a left identity, and every element of G has a left inverse in G. Then show that G is a group. (A similar result also holds assuming right identity and right inverses.)

7.1.2 Direct Product and Direct Sum

Let G_1, G_2 be two groups and consider the cartesian product $G_1 \times G_2$ of the two sets G_1 and G_2. Let us define a binary operation in $G_1 \times G_2$ as follows:

$$(a_1, b_1) \cdot (a_2, b_2) = (a_1 a_2, b_1 b_2).$$

Then $G_1 \times G_2$ is a group with (e_1, e_2), where $e_1 \in G_1$, $e_2 \in G_2$ are identity elements, as the identity element. The resulting group is called the **direct product** of G_1 and G_2. We have projection maps $p_1 \colon G_1 \times G_2 \to G_1$, $p_2 \colon G_1 \times G_2 \to G_2$ defined by $p_1(a,b) = a$, $p_2(a,b) = b$ for $(a,b) \in G_1 \times G_2$ which are surjective group homomorphisms. We have also the inclusion maps $i_1 \colon G_1 \to G_1 \times G_2$, $i_2 \colon G_2 \to G_1 \times G_2$ defined by $i_1(a) = (a, e_2)$, $i_2(b) = (e_1, b)$. These are injective homomorphisms. One can easily see that all these homomorphisms satisfy the following properties: $p_1 \circ i_1 = I_{G_1}$, $p_2 \circ i_2 = I_{G_2}$, $\ker p_1 = \{e_1\} \times G_2$, $\ker p_2 = G_1 \times \{e_2\}$. It follows that $\ker p_1 = \operatorname{Im} i_2$, $\ker p_2 = \operatorname{Im} i_1$ and $G_2 \cong (G_1 \times G_2)/\operatorname{Im} i_1$, $G_1 \cong (G_1 \times G_2)/\operatorname{Im} i_2$. One can clearly extend all of the above and define the direct product $G_1 \times \cdots \times G_n$ of a finite number of groups. The product here will also be coordinate-wise. Projections $p_k \colon \prod_{i=1}^n G_i \to G_k$, defined by $p_k(a_1, a_2, \ldots, a_n) = a_k$, and the inclusions $i_k \colon G_k \to \prod_1^n G_i$ defined by $i_k(a_k) = (e_1, \ldots, a_k, \ldots, e_n)$ (the element a_k at the kth place and identities elsewhere), are homomorphisms and these will also satisfy appropriate analogous conditions.

Let $\{G_i \mid i \in I\}$ be a family of groups. Recall that an element $f \in \prod_{i \in I} G_i$ is just a map $f \colon I \to \bigcup_{i \in I} G_i$ such that $f(i) \in G_i$ for all $i \in I$. Therefore, one can define a product in $\prod G_i$ by putting for all $i \in I$,

$$(f \cdot g)(i) = f(i) \cdot g(i),$$

where $f, g \in \prod G_i$. Note that this is again a coordinate-wise product and, if we write the element $f \in \prod G_i$ as $f = (f(i))_{i \in I}$ as an I-tuple, then the above product can be expressed as

$$(f(i)) \cdot (g(i)) = (f(i) \cdot g(i)).$$

The new group $\prod_{i \in I} G_i$ with the above operation is called the **direct product** of the groups $\{G_i \mid i \in I\}$. In this general situation also we have projection and inclusion homomorphisms which satisfy the analogous properties stated earlier for finite products.

Now, we come to the important concept of direct sum of a given family $\{G_i \mid i \in I\}$ of groups and we assume that $e_i \in G_i$ denotes the identity element of G_i. For this we consider the following subset, denoted by $\bigoplus_{i \in I} G_i$, of the product $\prod_{i \in I} G_i$;

$$\oplus G_i = \{(g_i) \in \prod G_i \mid g_i = e_i \in G_i \text{ for all } i \text{ except finitely many indices}\}.$$

Note that $\oplus_{i \in I} G_i$ defined above is a subgroup of the direct product. This subgroup is called the **direct sum** of the family $\{G_i \mid i \in I\}$ of given groups. It must also be noted that if the indexing set I is finite, then the direct sum $\oplus G_i$ coincides with direct product. Another important observation is that if elements of each group G_k, $k \in I$, are identified with their images under the inclusion maps $i_k \colon G_k \to \prod_{k \in I} G_k$, i.e., $g_k \in G_k$ is treated as the element $(\ldots, e_l, g_k, e_m, \ldots) \in \prod_{G_k}$ where $e_l \in G_l$ is the identity element of G_l for all $l \neq k$, then an arbitrary element $g \in \bigoplus G_k$ can be uniquely written as

$$g = g_{k_1} g_{k_2} \cdots g_{k_p},$$

where $g_{k_i} \in G_{k_i}$ are the nonidentity components of the element g. Using the additive notation for each G_k and for $\prod G_k$, the above expression is written as:

$$g = g_{k_1} + g_{k_2} + \ldots + g_{k_p}, \quad g_{k_i} \in G_{k_i}.$$

The case of a finite product $\prod_{k=1}^{n} G_k$ of n groups $G_1, G_2, \ldots, G_n$ $(n \geq 2)$ requires some further discussion: for each $k = 1, 2, \ldots, n$, we have the inclusion maps $i_k \colon G_k \to \prod_1^n G_k$ defined by $i_k(g) = (e_1, \ldots, g, \ldots, e_n)$, where g is at the kth place. Let us denote the image set $i_k(G_k)$ as $\bar{G}_k$. Then it can be proved that these image sets satisfy the following conditions:

(i) each $\bar{G}_k$ is a normal subgroup of $\prod G_k$.

(ii) the product group $\prod_1^n G_k$ can be written as $\prod_1^n G_k = \bar{G}_1 \cdot \bar{G}_2 \cdots \bar{G}_n$, i.e., the union of subgroups $\bar{G}_k$ generate the group $\prod G_k$.

(iii) for each k, $\bar{G}_k \cap [\bar{G}_1 \cup \cdots \cup \bar{G}_{k-1} \cup \bar{G}_{k+1} \cup \cdots \cup \bar{G}_n] = \{e\}$, the identity element of $\prod G_k$, i.e., each $\bar{G}_k$ intersects the subgroup generated by the rest of them only in the identity element.

It is interesting to observe that the converse of above result is also true. To be precise, suppose a group G has n subgroups $H_1, H_2, \ldots, H_n$ such that these subgroups satisfy the three conditions stated above for $\bar{G}_k$. Then it can be proved that G is isomorphic to the direct product $H_1 \times \cdots \times H_n$ of these subgroups. We come across the above situation quite often and the above result then better clarifies the structure of the group G.

7.1.3 The Structure of a Finite Abelian Group

As an important illustration of the foregoing result, let us consider a finite abelian group G of order n. In this case, the structure of G can be clarified completely. We write $n = p_1^{k_1} \cdots p_r^{k_r}$ as a product of positive primes. By Sylow's Theorem, the group G has a p_i-Sylow subgroup P_i of order $p_i^{k_i}$ for all $i = 1, 2, \ldots, r$. The group G being abelian, each P_i is a normal subgroup of G, and clearly, $P_i \cap [P_1 \cup \cdots \cup P_{i-1} \cup P_{i+1} \cup \cdots \cup P_r] = \{e\}$ for each i. Hence, from what we said earlier, G is isomorphic to the direct product of its p_i-Sylow subgroups P_i. The case when all the exponents are 1 is interesting. Since any group of prime order is cyclic, it follows from the above decomposition that an abelian group of order $p_1 p_2 \cdots p_r$, where each p_i is prime, is cyclic. For understanding the general case, let us recall (I.N. Herstein, *Topics in Algebra*, p. 113) that if P is a p-group, p a prime, then P is isomorphic to the direct sum of cyclic groups, viz.,

$$P \cong \mathbb{Z}_{(p^{l_1})} \oplus \mathbb{Z}_{(p^{l_2})} \oplus \ldots \oplus \mathbb{Z}_{(p^{l_m})},$$

where $l_1 \geq l_2 \geq \ldots \geq l_m$ are uniquely determined by the group P; here, $\mathbb{Z}_{(p^l)}$ denotes the cyclic group of order p^l. Uniqueness means if P is also isomorphic to the direct sum

$$\mathbb{Z}_{(p^{r_1})} \oplus \mathbb{Z}_{(p^{r_2})} \oplus \cdots \oplus \mathbb{Z}_{(p^{r_n})}$$

with $r_1 \geq r_2 \geq \cdots \geq r_n$, then $n = m$ and $l_1 = r_1, \ldots, l_n = r_n$. In fact, these numbers $\{l_1, l_2, \ldots, l_n\}$ are complete invariants of the p-group P. Now, let us go back to the abelian group G with its p_i-Sylow subgroups P_i; $i = 1, 2, \ldots, r$. Then we have already noted that

$$G \cong P_1 \oplus \ldots \oplus P_r.$$

Next, we write each $P_i \cong \mathbb{Z}_{{p_i}^{r_1(p_i)}} \oplus \ldots \oplus \mathbb{Z}_{{p_i}^{r_t(p_i)}}$, $r_1 \geq r_2 \geq \ldots \geq r_t \geq 0$, for $i = 1, 2, \ldots, r$; each r_i depends on p_i. If we put

$$d_1 = p_1^{r_1} \cdot p_2^{l_1} \cdot \ \ldots \ \cdot p_r^{s_1}$$

$$d_2 = p_1^{r_2} \cdot p_2^{l_2} \cdot \ \ldots \ \cdot p_r^{s_2}$$

$$\vdots$$

$$d_t = p_1^{r_n} \cdot p_2^{l_n} \cdot \ \ldots \ \cdot p_r^{s_n},$$

where some of these exponents could be zero, then each d_i is divisible by d_{i+1} and the group G itself can be written as a direct sum of cyclic groups uniquely. To see this, we note that $p_1^{r_1}, \ldots, p_r^{s_1}$ are all relatively primes and, since the direct sum of finite cyclic groups of relatively prime orders is again cyclic, we find that

$$G \cong \mathbb{Z}_{d_1} \oplus \mathbb{Z}_{d_2} \oplus \ldots \oplus \mathbb{Z}_{d_t},$$

where d_1 is divisible by d_2, d_2 is divisible by $d_3, \ldots, d_{t-1}$ is divisible by d_t. These numbers $d_1, d_2, \ldots, d_t$, which are uniquely determined by the abelian group G, are called **torsion coefficients** of G. In fact, two finite abelian groups are isomorphic if and only if they have the same torsion coefficients. In this way the structure of a finite abelian group is completely determined.

7.1.4 Free Groups and Free Products

The notion of a free group is basic to the study of fundamental groups discussed in Chapter 2. We prefer to introduce a free group by its universal property. Defining a concept using the universal property is not only quicker but is also efficient in proving other results about the concept itself. However, it has the disadvantage that nothing is clear about the inner structure of the object being defined: in fact, all of that insight is relegated to the existence theorem of the object itself. We have

Definition 7.1.2. *By a free group F on a given set S (called a set of generators) we mean a group F and a function $\phi\colon S \to F$ which satisfies the following universal property: For any group H and any function $\psi\colon S \to H$, there exists a unique group homomorphism $f\colon F \to H$ such that the following diagram is commutative:*

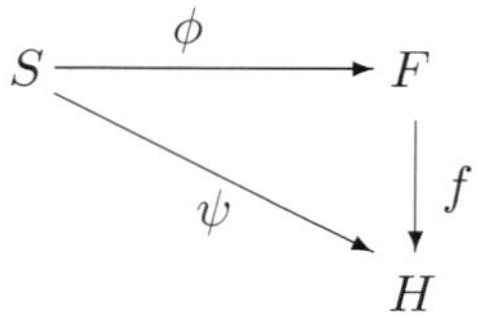

i.e., $f \circ \phi = \psi$.

Let $F = [x]$ be the infinite cyclic group generated by a single element x. In this case, we know that F consists of all powers $x^n, n \in \mathbb{Z}$, of x. Then it is readily seen that F is free group on the set $S = \{x\}$ with inclusion map $i\colon S \to F$ as the map ϕ. On the other hand, the cyclic group $\mathbb{Z}_m$ of order m can never be a free group on any set S. Note that, up to isomorphism, there is a unique free group on a given set S. To see this, let F and F' be two groups on S via the maps $\phi\colon S \to F$ and $\phi'\colon S \to F'$. Then, by the defining universal property, we have a unique homomorphism $f\colon F \to F'$ and a unique homomorphism $g\colon F' \to F$ which make the following two triangles commutative, i.e., $f \circ \phi = \phi'$, $g \circ \phi' = \phi$.

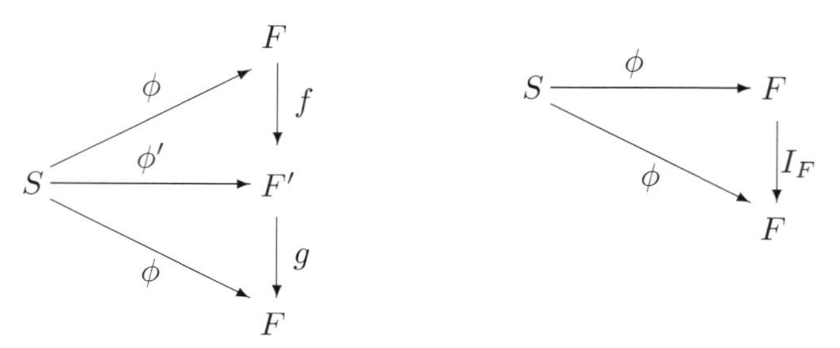

It follows that the homomorphism $g \circ f\colon F \to F$ makes the bigger triangle commutative, i.e., $g \circ f \circ \phi = \phi$. But the identity map $I_F\colon F \to F$ already makes the bigger triangle commutative. Hence, by the uniqueness of the defining homomorphisms, we must have $g \circ f = I_F$. By a similar argument, we find that $g \circ f = I_{F'}$. Hence, it follows that each of the maps f and g is an isomorphism and so $F \cong F'$.

In order to discuss the question of existence of free groups, we need the concept of **free products** of a family $\{G_i \mid i \in I\}$ of groups. This too can be defined by its universal property. We have

Definition 7.1.3. *Let $\{G_i \mid i \in I\}$ be a family of groups. Then a group G together with a family of homomorphisms $\phi_i \colon G_i \to G, i \in I$ is called* **free product** *of the family $\{G_i \mid i \in I\}$ if the following universal property is satisfied: for any group H and any family of homomorphisms $\psi_i \colon G_i \to H$, there exists a unique homomorphism $f \colon G \to H$ such that every triangle*

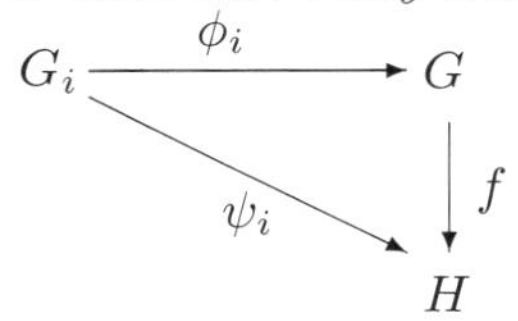

is commutative, i.e., $f \circ \phi_i = \psi_i$ for all $i \in I$.

The uniqueness of the free product of a given family $\{G_i \mid i \in I\}$ of groups follows from the universal property defining the free product. We omit proof of the following existence theorem (see W. Massey [13], p. 98).

Theorem 7.1.4. *Given any collection $\{G_i \mid i \in I\}$ of groups, their free product exists.*

It is also true that each homomorphism $\phi_i \colon G_i \to G$ occurring in the definition of free product is indeed a monomorphism and the free product is generated by the union of their images in G. We denote the free product of groups $G_1, G_2, \ldots, G_n$ by $G_1 * G_2 * \ldots * G_n = \prod^*_{1 \le i \le n} G_i$ whereas the free product of the family $\{G_i \mid i \in I\}$ is denoted by $\prod^*_{i \in I} G_i$.

Example 7.1.5. Let $G_1 = \{1, x_1\}$, $G_2 = \{1, x_2\}$ be two cyclic groups, each of order 2. Then their free product consists of the following elements

$$x_1,\ x_1x_2,\ x_1x_2x_1,\ \ldots \quad \text{or}, \quad x_2,\ x_2x_1,\ x_2x_1x_2,\ \ldots.$$

The elements x_1x_2 and x_2x_1 are different and each one is of infinite order. Note that the direct product $G_1 \times G_2$ of the two groups consists of only 4 elements whereas the free product $G_1 * G_2$ consists of infinite number of elements, each of infinite order.

Now, we come back to the question of existence of a free group on a given set S. Let us index the set of elements of S as $S = \{x_i \mid i \in I\}$ and let $F_i = [x_i]$ be the free group on one generator x_i – this would be the infinite cyclic group generated by x_i. Now, if $F = \prod^*_{i \in I} F_i$ denotes the free product of these groups F_i, then it can be seen that F is indeed a free group on the set S. This proves that given any set S, there always exists a free group on S and S is a set of generators for F. As an example, let $S = \{x_1, x_2\}$ have just two elements. Then the free group $F[S]$ on S will have elements of the type

$$x_1^{n_1} x_2^{n_2} x_1^{n_3} x_2^{n_4} \dots x_1^{n_k} x_2^{n_k+1} \quad \text{or,} \quad x_2^{m_1} x_1^{m_2} x_2^{m_3} x_1^{m_4} \dots x_2^{m_l} x_1^{m_l+1},$$

where n_i and m_i are nonzero integers (compare with the preceding example). These elements are called reduced "words" in symbols x_1 and x_2. The crucial point is that there are no relations among the powers of x_1 and x_2, i.e., none of the elements written above ever equals identity element. Also, note that the free group on a set S having just one element is abelian whereas the free group on a set S having more than one element is never abelian. Furthermore, it follows from the universal property that two free groups on the sets S_1 and S_2 are isomorphic if and only if S_1 and S_2 have the same cardinality.

Let G be a group with S as a set of generators. Consider the free group F on the set S with respect to a map $\phi\colon S \to F$ and let $\psi\colon S \to G$, where G is seen as a group, be the inclusion map. Then, by the universal property of F, there exists a homomorphism $f\colon F \to G$ such that $f \circ \phi = \psi$. This means f is onto and so, by the Fundamental Theorem of Homomorphism, $G \cong F/\ker f$. This proves the useful fact that **every group G is isomorphic to a quotient of a free group**.

When we consider a group G as the quotient of a free group F, as explained above, the elements of $\ker f$ in F are finite products of powers of elements of S (a reduced word). But when the same element is seen in the group G via the inclusion map ψ then each of them reduces to an identity of G. Each of the elements in the $\ker f$ is called a **relation** defining the group G. If the group G is described as the quotient of a free group as above, we say that G is given in terms of **generators** and **relations**. The elements of S are generators and the elements of $\ker f$ are **relations** for G. It may be observed that if r_1, r_2 are relations in G, then their product, their inverses, their conjugates in G are all relations for G. A set R of relations which generates the $\ker f$ as the smallest normal subgroup containing $\ker f$ is called a complete set of relations for G. A group G is completely described by a set S of generators and complete set R of relations.

A group G is said to be **finitely presented** if the sets S and R describing G as above are finite. We write this as $G = \langle S : R \rangle$.

Example 7.1.6. 1. A cyclic group $\mathbb{Z}_n$ of order n has a presentation $\langle x : x^n \rangle$.

2. A free abelian group of rank two, $\mathbb{Z} \times \mathbb{Z}$, has a presentation $\langle x, y : xyx^{-1}y^{-1} \rangle$.

3. The dihedral group D_{2n} having $2n$ elements has a presentation $\langle a, b : a^n, b^2, (ab)^2 \rangle$.

Free Abelian Groups

The concept of a free abelian group is needed in defining various kinds of homology groups. The definition uses the universal property analogous to the one already given for free groups. We have

Definition 7.1.7. *Let S be a set. By a free abelian group F on the set S we mean an abelian group F together with a map $\phi\colon S \to F$ which has the following universal property: For any abelian group A and for any map $\psi\colon S \to A$, there exists a unique homomorphism $f\colon F \to A$ such that the triangle*

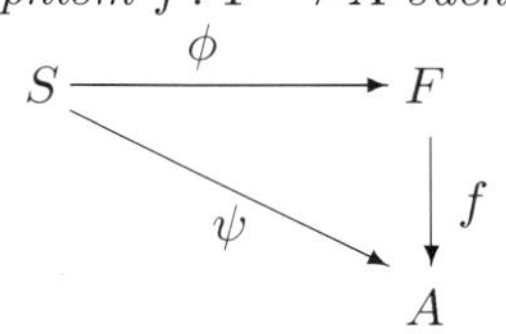

is commutative, i.e., $f \circ \phi = \psi$.

The question of uniqueness of the free abelian group F on S is answered exactly as in the case of free groups discussed earlier. However, the existence question in the case of free abelian group is answered by the concept of direct sum instead of free product: Let $S = \{x_i \mid i \in I\}$ be the given set and let $F_i = \mathbb{Z}x_i$ be the infinite cyclic group on the generators $x_i, i \in I$. Then the direct sum $F = \bigoplus_{i\in I} F_i$, with the map $\phi\colon S \to F$ as the obvious inclusion map, is the free abelian group on S. As pointed out in the definition of direct sum, every element $x \in F$ can be expressed (in additive notation) uniquely:

$$x = \sum n_i x_i, \qquad x_i \in S,\ n_i \in \mathbb{Z},$$

as a finite sum. This explains the easy structure of a free abelian group on a set S of generators of F.

It is easily shown that every abelian group is isomorphic to a quotient of a free abelian group. If F is free abelian on the set S, then the cardinality of the set S is called **rank** of F. It is then readily verified that two free abelian groups are isomorphic if and only if both have the same rank.

Exercises

1. Let $G = \mathbb{Z}_p$ be a cyclic group of order p, p a prime. Show that the group $\mathrm{Aut}(G) \cong \mathbb{Z}_{p-1}$.
2. Let $\mathbb{Q} = (\mathbb{Q}, +)$ be the additive group of rationals. Prove that $\mathrm{Aut}(\mathbb{Q}) \cong \mathbb{Q}^*$, the multiplicative group of non-zero rational numbers. Hence conclude that the only nontrivial finite subgroup of $\mathrm{Aut}(\mathbb{Q})$ is $\mathbb{Z}_2$.

3. Show that a subgroup of a free group is free (This is the important Nielsen-Schreier Theorem).

4. Show that a subgroup H of a free abelian group F is free abelian. Also, prove that rank $H \leq$ rank F (This result need not be true in the case of free group, Ex 3).

5. Let $G = \mathbb{Z}_m$ and $G' = \mathbb{Z}_n$ $(m, n > 1)$. Show that the free product $G * G'$ is not finite and the only elements of finite order in $G * G'$ are the elements of G, G' or their conjugates.

7.1.5 Modules and their Direct Sum

Let R be a commutative ring with unity element $1 \neq 0$. An abelian group M with a multiplication $R \times M \to M$ (called scalar multiplication) is said to be an R-**module** if the following conditions are satisfied:

(i) $(r_1 + r_2) \cdot m = r_1 \cdot m + r_2 \cdot m$

(ii) $r \cdot (m_1 + m_2) = r \cdot m_1 + r \cdot m_2$

(iii) $(r_1 r_2) \cdot m = r_1 \cdot (r_2 \cdot m)$

(iv) $1 \cdot m = m$

for each $r_1,\ r_2,\ r \in R$ and $m_1,\ m_2,\ m \in M$. If $R = F$ is a field, then a module over R is just a vector space over F. An abelian group A is, clearly, a $\mathbb{Z}$-module. An ideal I of a ring R is an R-module. In particular, a ring R is a module over itself. Submodules and quotient modules are defined analogous to subgroups and quotient groups.

If M, M' are two modules over R, then a homomorphism $f\colon M \to M'$ of abelian groups is said to be an R-**homomorphism** if $f(r \cdot m) = r \cdot f(m)$ for all $r \in R, m \in M$, i.e., f preserves scalar multiplication. Isomorphisms, endomorphisms, automorphisms of modules are defined analogous to the corresponding notions in groups. The fundamental theorem of homomorphism for R-modules is also valid.

Let $\{M_\alpha \mid \alpha \in I\}$ be a family of R-modules. Consider the product $\prod_{\alpha \in I} M_\alpha$ as the direct product of abelian groups and define scalar multiplication in this product as $r \cdot (m_\alpha) = (r \cdot m_\alpha)$. Then $\prod_{\alpha \in I} M_\alpha$ is an R-module and is called the **direct product** of the given family of R-modules. Now, let us consider a subset, denoted by $\bigoplus_{\alpha \in I} M_\alpha$ or $\Sigma_{\alpha \in I} M_\alpha$, of this product module defined by

$$\bigoplus M_\alpha = \{(x_\alpha) \in \textstyle\prod_{\alpha \in I} M_\alpha \mid x_\alpha = 0 \text{ for all but finitely many indices } \alpha \in I\}.$$

Then it is evident that $\bigoplus M_\alpha$ is a submodule of the product module $\prod_{\alpha \in I} M_\alpha$. This module is called the **direct sum** of the given family $\{M_\alpha \mid \alpha \in I\}$ of R-modules. If the indexing set I is finite, then note that the direct sum

$\bigoplus M_\alpha$ coincides with the direct product $\prod M_\alpha$. A module M is said to be **free** if $M \cong \oplus_{\alpha \in I} M_\alpha$, where for each $\alpha \in I$, $M_\alpha \cong R$ as R-module. It is not difficult to prove that any R-module M is the quotient of a free R-module.

A module M over R is said to be **finitely generated** if there exists a finite set $X = \{x_1, x_2, \ldots, x_n\}$ of elements of M such that every element $x \in M$ can be written as a linear combination of elements of X, i.e., given $m \in M$ there are elements $r_1, r_2, \ldots, r_n$ of R such that $m = r_1 x_1 + \ldots + r_n x_n$. A module M is said to be **cyclic** if it is generated by a single element $x \in M$, i.e., $M = R \cdot x = (x)$. This generalizes the concept of cyclic groups. A finite set $X = \{x_1, x_2, \ldots, x_n\}$ of elements of M is said to be a **basis** for M if every element of M can be expressed as a linear combination of elements $x_1, x_2, \ldots, x_n$ in a **unique** way. The set $B = \{e_1, \ldots, e_n\}$ of elements of $R^n = R \oplus \ldots \oplus R$, where $e_i = (0, \ldots, 1, \ldots, 0)$ (1 at the ith position) for $i = 1, 2, \ldots, n$, is a basis of the R-module R^n. A finitely generated R-module can be expressed as a quotient of the free R-module $M = R^n$ for some $n \geq 1$. An R-module, even though it is finitely generated, may not have a basis, e.g., the abelian group $\mathbb{Z}_m (m > 1)$ is cyclic over $\mathbb{Z}$ but does not have a basis because the element $\bar{2} = 2 \cdot \bar{1} = (m+2).\bar{1}$ can be expressed in different ways.

Let M be an R-module, where R is a domain. An element $m \in M$ is called a **torsion element** of M if there exists an $r (\neq 0) \in R$ such that $r \cdot m = 0$. The set of all torsion elements of M form a submodule of M and is called the **torsion** submodule of M. An R-module M is said to be **torsion free** if torsion $M = \{0\}$. Note that a torsion free module need not be free: the abelian group $\mathbb{Q}$ of rationals is torsion free over $\mathbb{Z}$, but is not free over $\mathbb{Z}$. What about a finitely generated torsion free module? A module over a field R, i.e., a vector space is always free and has a basis, finite or infinite.

Let $M_1, M_2, \ldots, M_n$ be submodules of an R-module M. Suppose these modules satisfy the condition that every element of M can be uniquely expressed as

$$m = m_1 + m_2 + \cdots + m_n,$$

where $m_i \in M_i$, $i = 1, 2, \ldots, n$. Then one can easily prove that $M \cong M_1 \oplus M_2 \oplus \cdots \oplus M_n$. The following fundamental result is very important. For a proof of this one can see (See N. Jacobson [11] p. 187).

Theorem 7.1.8. (Structure Theorem for finitely generated Modules). *Let R be a PID (Principal Ideal Domain). Then a finitely generated R-module M is isomorphic to the direct sum of a finite number of cyclic R-modules.*

Structure Theorem for Finitely Generated Abelian Groups

As a corollary to the above theorem, it follows that a finitely generated abelian group A can be expressed as a direct sum of finite number of cyclic groups. More precisely, for such an A, there exist integers $n \geq 0$, $k \geq 0$ and $r_1, r_2, \ldots, r_k$ such that

$$A \cong \overbrace{\mathbb{Z} \oplus \cdots \oplus \mathbb{Z}}^{n \text{ copies}} \oplus \mathbb{Z}_{r_1} \oplus \mathbb{Z}_{r_2} \oplus \cdots \oplus \mathbb{Z}_{r_k},$$

where r_1 is divisible by r_2, r_2 is divisible by r_3, etc. If A has such a decomposition then the number n is called the **rank** or **Betti number** of A and the elements of the set $\{r_1, r_2, \ldots, r_k\}$ are called the **torsion coefficients** of A. In fact, the rank and the torsion coefficients of A completely characterize the finitely generated abelian groups A. The subgroup $T = \mathbb{Z}_{r_1} \oplus \cdots \oplus \mathbb{Z}_{n_k}$ of A is called the **torsion subgroup** of A. Note that the factor group A/T is free abelian group and $A \cong (A/T) \oplus T$. The factor group A/T is called the **free part** of A. It is now clear that the two finitely generated abelian groups A and A' are isomorphic if and only if their free parts as well as their torsion parts are isomorphic. This will happen if and only if their ranks and torsion coefficients are the same.

7.2 Categories and Functors

The concept of a category, functor and several accompanying ideas were discovered and formalized by S. Eilenberg and S. Mac Lane around the year 1942. They have provided a powerful language to mathematics in achieving unity of concepts and expression. These have proved to be of immense importance in stating and formulating diverse ideas involving definitions and results, which appeared very different from each other, into a nice whole. Some of these have even brought out new and deeper interpretations of known results which were earlier hidden underneath. Using this new language, dozens of known theorems can be stated and proved as a single result. Here, we will give only the definition of categories and functors and a few examples. The interested reader is referred to (Homology : S. Mac Lane [17]) for a fuller account.

Categories

A **category** $\mathcal{C}$ consists of three items:

(a) a class of **objects**, to be denoted by letters $A, B, C, \ldots$, etc., which can be thought of as sets with some additional structures on them,

(b) for each pair (A, B) of objects in $\mathcal{C}$, a set $\text{Mor}(A, B)$, called the set of **morphisms** $f\colon A \to B$, which may be thought of as maps preserving additional structures, and

(c) a product $\text{Mor}(A, B) \times \text{Mor}(B, C) \to \text{Mor}(A, C)$ for each triple (A, B, C) of objects in $\mathcal{C}$, called **composition** and taking the pair (f, g) to $g \circ f \in \text{Hom}(A, C)$.

These are required to satisfy the following two axioms:

(i) For any three morphisms $f\colon A \to B$, $g\colon B \to C$ and $h\colon C \to D$, we have

$$h \circ (g \circ f) = (h \circ g) \circ f,$$

i.e., the composition of morphisms is associative.

(ii) For each object A in $\mathcal{C}$, there exists a morphism $I_A\colon A \to A$, called **identity morphism**, which has the property that for all $f\colon A \to B$ and all $g\colon C \to A$, we have

$$f \circ I_A = f, \qquad I_A \circ g = g.$$

Some examples of categories are given below:

(1) If we take objects as topological spaces, morphisms as continuous maps and the product to be composition of continuous maps, then we get a category, called the category of topological spaces and continuous maps.

(2) Let us take groups as objects, homomorphisms from one group to another as morphisms and the composition of homomorphisms to be the composition of morphisms. Then we have the category of groups and group homomorphisms.

(3) The category of sets and maps has objects as the class of all sets, morphisms as just maps and the product as the composition of maps.

(4) If we let the objects be the class of all simplicial complexes, the morphisms to be the set of all simplicial maps and the product to be the composition of simplicial maps, then we have the category of simplicial complexes and simplicial maps.

(5) If we take objects as finite groups, morphisms as group homomorphisms and the product to be the composition of homomorphisms, then again we get a category. This category is in fact a subcategory of the category of all groups and homomorphisms – one can define, in a natural way, a **subcategory** of a category.

There are numerous important categories which we often come across, especially in algebraic topology. The category of all R-modules and module homomorphisms for a given ring R is an important category for homological algebra as well as for commutative algebra. The class of all sets with morphisms as injective maps is a subcategory of the category of sets and maps. Note that in this category, the set $\mathrm{Mor}(A, B)$ between sets A and B are clearly small as compared to the set $\mathrm{Mor}(A, B)$ in the category of sets. Thus, one can see that if $\mathcal{C}$ is a subcategory of $\mathcal{D}$, then the class of objects of $\mathcal{C}$ may be quite small compared to that of $\mathcal{D}$. Also, the $\mathrm{Mor}(A, B)$ between any two objects A and B of $\mathcal{C}$ could be smaller or even equal to that for the category $\mathcal{D}$.

A morphism $f\colon A \to B$ is called an **equivalence** (or isomorphism) if there is a morphism $g\colon B \to A$ such that $g \circ f = I_A$ and $f \circ g = I_B$. Note that an equivalence in the category of sets is just a bijective map; an equivalence in the category of topological spaces is a homeomorphism and an equivalence in the category of R-modules is a module isomorphism.

Functors

A **covariant** functor F from a category $\mathcal{C}$ to a category $\mathcal{D}$ is a pair of functions (both denoted by the same letter F) which maps objects of $\mathcal{C}$ to the objects of $\mathcal{D}$ and, for any pair (A, B) of $\mathcal{C}$, it maps the set $\mathrm{Mor}(A, B)$ to the set $\mathrm{Mor}(F(A), F(B))$ and is required to satisfy the following two conditions:

(i) $F(I_A) = I_{F(A)}$ for every $A \in \mathcal{C}$.

(ii) $F(g \circ f) = F(g) \circ F(f)\colon F(A) \to F(C)$ for any two morphisms $f\colon A \to B$ and $g\colon B \to C$ in the category $\mathcal{C}$.

A **contravariant** functor F from a category $\mathcal{C}$ to the category $\mathcal{D}$ is similarly defined. The only difference is that the contravariant functor maps a morphism $f\colon A \to B$ to a morphism $F(f)\colon F(B) \to F(A)$ in the opposite direction to that of f. Thus, the two conditions in this case will be

(i) $F(I_A) = I_{F(A)}$ for all $A \in \mathcal{C}$.

(ii) $F(g \circ f) = F(f) \circ F(g)\colon F(C) \to F(A)$.

Example 7.2.1. Let $\mathcal{C}$ be the category of finite sets and maps between them as morphisms. Let $\mathcal{D}$ be the category of free abelian groups having a finite bases and morphisms as group homomorphisms between them. For each object X in $\mathcal{C}$, let $F(X)$ be the free abelian group with X as a basis and, for any map $f\colon X \to Y$ in the category $\mathcal{C}$, let $F(f)\colon F(X) \to F(Y)$ be the abelian group homomorphism defined by the map f (any set theoretic map f from a basis of a free abelian group G defines a unique homomorphism from the free abelian group G). Then it can be easily verified that F is a covariant functor from the category $\mathcal{C}$ to the category $\mathcal{D}$.

Example 7.2.2. For each topological space X, let $C(X)$ denote the ring of continuous real-valued functions on X with point-wise operations. For any continuous map $f\colon X \to Y$, we get a ring homomorphism $f^*\colon C(Y) \to C(X)$ defined by $f^*(\alpha) = \alpha \circ f$. Then one readily verifies that we get a contravariant functor from the category of topological spaces and continuous maps to the category of rings and homomorphisms.

Natural Transformations

Let F and G be two covariant functors from a category $\mathcal{C}$ to the category $\mathcal{D}$. Then a map η, which assigns to each object $A \in \mathcal{C}$ a unique morphism $\eta_A : F(A) \to G(A)$, is called a **natural transformation** from F to G, if for each $f : A \to B$ in $\mathcal{C}$, the diagram

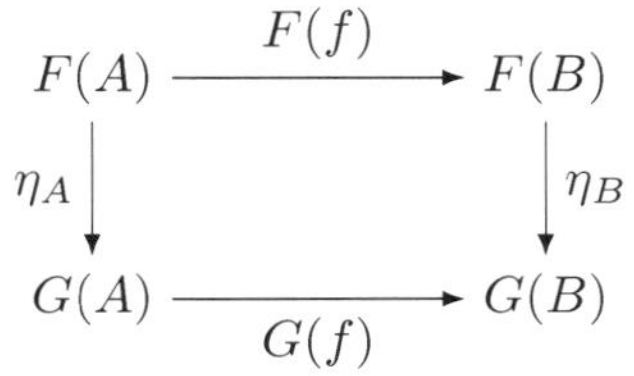

$$\begin{array}{ccc} F(A) & \xrightarrow{F(f)} & F(B) \\ {\scriptstyle \eta_A}\downarrow & & \downarrow{\scriptstyle \eta_B} \\ G(A) & \xrightarrow[G(f)]{} & G(B) \end{array}$$

is commutative. A natural transformation between contravariant functors can be similarly defined. A natural transformation $\eta : F \to G$ is called an **equivalence** if η_A is an equivalence in the category $\mathcal{D}$ for every object A in $\mathcal{C}$.

Example 7.2.3. Let $\mathcal{C}$ be the category of all vector spaces (over a field F) and linear transformations as morphisms. Let S and T be two functors from $\mathcal{C}$ to $\mathcal{C}$ defined by

$$S = I_{\mathcal{C}} \quad and \quad T(V) = V^{**},$$

where V^{**} is the double dual of V. For each linear map $f : V \to W$ in V, let

$$T(f) = f^{**} : V^{**} \to W^{**}$$

be the homomorphism induced in the double dual, i.e.,

$$f^{**}(\phi)(\alpha) = \phi(\alpha \circ f) \qquad \text{for all } \alpha \in W^* \text{ and } \phi \in V^{**}.$$

Now, if for any $V \in \mathcal{C}$, we define $\eta_V : V \to V^{**}$ by

$$\eta_V(x)(\alpha) = \alpha(x),$$

where $x \in V$ and $\alpha \in V^*$, then we get a natural transformation from the functor S to the functor T. This follows since, for any linear map $f : V \to W$, the diagram

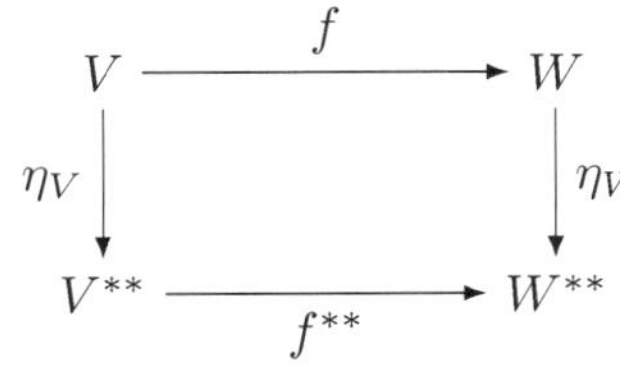

$$\begin{array}{ccc} V & \xrightarrow{f} & W \\ {\scriptstyle \eta_V}\downarrow & & \downarrow{\scriptstyle \eta_W} \\ V^{**} & \xrightarrow[f^{**}]{} & W^{**} \end{array}$$

is easily seen to be commutative. If we take the subcategory $\mathcal{C}'$ of $\mathcal{C}$ having only finite-dimensional vector spaces, then $\eta_V : V \to V^{**}$ is a vector space isomorphism and hence η is, in fact, a **natural equivalence** from the identity functor S to the double dual functor T on $\mathcal{C}'$.

We come across numerous examples of categories, functors and natural transformations when we study different topics of algebraic topology. In fact, it is sometimes said "algebraic topology is the study of a few functors!!".

7.2.1 The Hom(M, N) Functor

Here, we begin with the remark that the material being discussed from this point onward is indeed a portion of elementary Homological Algebra. Let M, N be R-modules and let $\mathrm{Hom}_R(M, N)$ be the set of all R-homomorphisms from M to N. Then we can define an addition $f + g$ and a scalar multiplication $r \cdot f$ in the set $\mathrm{Hom}_R(M, N)$ as follows:

$$(f + g)(m) = f(m) + g(m)$$
$$(rf)(m) = r \cdot f(m)$$

for all $m \in M$ and for all $r \in R$. It is easy to verify that the set $\mathrm{Hom}_R(M, N)$ becomes an R-module w.r.t. the above operations. This module will be denoted by just $\mathrm{Hom}(M, N)$ whenever there is no confusion. Note that if $\alpha\colon M \to M'$ is any R-module homomorphism, then this induces an R-module homomorphism $\alpha^*\colon \mathrm{Hom}(M', N) \to \mathrm{Hom}(M, N)$, in a direction opposite to α, by the formula $\alpha^*(f) = f \circ \alpha$. Similarly, a homomorphism $\beta\colon N \to N''$ induces an R-homomorphism $\beta_*\colon \mathrm{Hom}(M, N) \to \mathrm{Hom}(M, N'')$, in the direction of β, by the formula $\beta_*(f) = \beta \circ f$.

With above definitions, one can easily verify the following properties of induced homomorphisms:

(i) If $\alpha_1\colon M \to M'$ and $\alpha_2\colon M' \to M''$ are any two homomorphisms, then $(\alpha_2 \circ \alpha_1)^* = \alpha_1^* \circ \alpha_2^*$.

(ii) If $I_M\colon M \to M$ is the identity homomorphism, then the induced map $I_M^*\colon \mathrm{Hom}(M, N) \to \mathrm{Hom}(M, N)$ is also the identity map.

Likewise, if $\beta_1\colon N \to N'$ and $\beta_2\colon N' \to N''$ are two homomorphisms, then $(\beta_2 \circ \beta_1)_* = \beta_{2*} \circ \beta_{1*}$ and if $I_N\colon N \to N$ is the identity map, then $(I_N)_*\colon \mathrm{Hom}(M, N) \to \mathrm{Hom}(M, N)$ is also the identity map.

Recall that if $f\colon M \to N$ is an R-homomorphism, then $\ker f = \{x \in M \mid f(x) = 0\}$, $\mathrm{Im}\, f = f(M)$ and $\mathrm{coker}\, f = N/\mathrm{Im}\, f$ are all R-modules. Using the Fundamental Theorem of Homomorphism for R-modules, one can easily prove the following facts:

(i) If $L \supseteq M \supseteq N$ are R-modules, then $(L/N)/(M/N) \cong L/M$.

(ii) If M_1, M_2 are submodules of M, then $(M_1 + M_2)/M_1 \cong M_2/(M_1 \cap M_2)$ (Noether Isomorphism Theorem).

7.2.2 Exact Sequences

Let us consider the following sequence (finite or infinite) of R-modules and R-homomorphisms

$$\ldots \longrightarrow M_{n+1} \xrightarrow{f_{n+1}} M_n \xrightarrow{f_n} M_{n-1} \longrightarrow \ldots.$$

We say that this sequence is **exact** at M_n if $\text{Im } f_{n+1} = \ker f_n$. It is said to be **exact** if it is exact at M_n for each n. One can easily verify that

(i) A homomorphism $f\colon M' \to M$ is 1-1 if and only if the sequence $0 \to M' \xrightarrow{f} M$ is exact.

(ii) A homomorphism $f\colon M \to M''$ is onto if and only if the sequence $M \xrightarrow{f} M'' \to 0$ is exact.

(iii) A sequence $0 \to M' \xrightarrow{\alpha} M \xrightarrow{\beta} M'' \to 0$ is exact if and only if α is 1-1, β is onto and $\text{Im } \alpha = \ker \beta$, i.e., $M'' \cong M/\alpha(M')$.

An exact sequence $0 \to M' \xrightarrow{\alpha} M \xrightarrow{\beta} M'' \to 0$ is said to be **split exact** if there exists a map $\gamma : M'' \to M$ such that $\beta\gamma = I_{M''}$. In such a case $M \cong M' \oplus M''$. This, of course, already implies that $M'' \cong M/\alpha(M')$ and is indeed stronger than (iii). A sequence of the kind (iii) above, consisting of five terms is called a **short exact sequence**.

The Hom functor is left exact

Concerning the exactness of the contravariant functor $\text{Hom}(.\,, N)$, we have

Proposition 7.2.4. *Let*

$$M' \xrightarrow{\alpha} M \xrightarrow{\beta} M'' \to 0 \qquad (*)$$

be a sequence of R-modules and homomorphisms. Then it is exact if and only if for any R-module N, the induced sequence

$$0 \to \text{Hom}(M'', N) \xrightarrow{\beta^*} \text{Hom}(M, N) \xrightarrow{\alpha^*} \text{Hom}(M', N) \quad (**)$$

is exact.

Proof. Suppose $(*)$ is exact. To show that $(**)$ is exact, we must show that β^* is 1-1 and $\text{Im } \beta^* = \ker \alpha^*$. Suppose $\beta^*(\alpha) = 0$, where $\alpha\colon M'' \to N$ is a homomorphism. Then we claim that $\alpha = 0$. If $\alpha \neq 0$, then note that there exists an element $x'' \in M''$ such that $\alpha(x'') \neq 0$. But, since β is onto, there exists an element $x \in M$ such that $x'' = \beta(x)$ which means $\beta^*(\alpha)(x) = (\alpha \circ \beta)(x) = \alpha(x'') \neq 0$, i.e., $\beta^*(\alpha) \neq 0$, a contradiction. Thus, β^* is 1-1. Next, note that $\alpha^* \circ \beta^* = (\beta \circ \alpha)^* = 0^* = 0$. Hence $\text{Im } \beta^* \subseteq \ker \alpha^*$. Conversely, suppose $f \in \ker \alpha^*$, i.e., $\alpha^*(f) = f \circ \alpha = 0$. Then f maps $\alpha(M'')$ into zero which means f maps $\ker \beta = \alpha(M')$ to zero. Then f defines a unique homomorphism

$g\colon M'' \to N$ satisfying the condition that $g \circ \beta = f$, i.e., $f = \beta^*(g)$. Thus, $\ker \alpha^* \subseteq \operatorname{Im} \beta^*$.

Conversely, suppose the sequence $(**)$ is an exact sequence for each R-module N. Then we will prove that the sequence $(*)$ is exact. Assume that the map β is not an onto map. This means $M''/\beta(M) \neq 0$. Hence the quotient map $\nu\colon M'' \to M''/\beta(M)$ is nonzero. Since β^* is 1-1, taking $N = M''/\beta(M)$, we find that $\beta^*(\nu) \neq 0$. But clearly $\nu\beta = 0$, a contradiction. Next, take $N = M''$ and consider the identity map $I_{M''} \in \operatorname{Hom}(M'', M'')$. By the exactness of $(**)$ for this choice of N, we must have $(\alpha^* \circ \beta^*)(I_{M''}) = 0$. But this means $(\beta \circ \alpha)^*(I_{M''}) = I_{M''} \circ (\beta \circ \alpha) = \beta \circ \alpha = 0$. This shows that $\operatorname{Im} \alpha \subseteq \ker \beta$. Conversely, take $N = M/\operatorname{Im} \alpha$ and let $\phi\colon M \to M/\operatorname{Im} \alpha$ be the quotient map. Then $\alpha^*(\phi) = \phi \circ \alpha = 0$ which implies that there exists a $\psi\colon M'' \to M/\operatorname{Im} \alpha$ such that $\phi = \psi \circ \beta$. It follows that $\ker \beta \subseteq \ker \phi = \operatorname{Im} \alpha$. Thus, $(*)$ is exact. ■

A similar result stated below is also true for the covariant functor $\operatorname{Hom}(M, \,.)$. A proof analogous to the above is left as an exercise.

Theorem 7.2.5. *Let*

$$0 \to N' \xrightarrow{f} N \xrightarrow{g} N''$$

be a sequence of R-modules. Then it is exact if and only if for any R-module M, the induced sequence

$$0 \to \operatorname{Hom}(M, N') \xrightarrow{f_*} \operatorname{Hom}(M, N) \xrightarrow{g_*} \operatorname{Hom}(M, N'')$$

is exact.

The basic fact contained in the above two propositions is frequently expressed by saying that the functor $\operatorname{Hom}_R(.,.)$ is **left exact** in each variable. The following example shows that the functor $\operatorname{Hom}(., N)$ is not exact on the right, i.e., we have examples of injective homomorphisms of modules $\alpha\colon M' \to M$ but the corresponding map $\alpha_*\colon \operatorname{Hom}(M, N) \to \operatorname{Hom}(M', N)$ are not injective. It is the important fact which gives rise to the functors $\operatorname{Ext}^1(.,.), \operatorname{Ext}^2(.,.)$ etc.

Example 7.2.6. Note that $\operatorname{Hom}(\mathbb{Z}, \mathbb{Z}_n) \cong \mathbb{Z}_n, n \geq 2$. Let $f\colon \mathbb{Z} \to \mathbb{Z}$ be the 1-1 map defined by $f(x) = 2x$. Then the induced map $f_*\colon \operatorname{Hom}(\mathbb{Z}, \mathbb{Z}_2) \to \operatorname{Hom}(\mathbb{Z}, \mathbb{Z}_2)$ is indeed the zero map.

In dealing with homomorphisms in homology and cohomology groups, one frequently needs the following simple result:

Lemma 7.2.7. (Five Lemma). *Let*

$$\begin{array}{ccccccccc} 0 & \longrightarrow & M' & \xrightarrow{\alpha} & M & \xrightarrow{\beta} & M'' & \longrightarrow & 0 \\ & & \downarrow f & & \downarrow g & & \downarrow h & & \\ 0 & \longrightarrow & N' & \xrightarrow{\gamma} & N & \xrightarrow{\delta} & N'' & \longrightarrow & 0 \end{array}$$

be a commutative diagram of R-modules and homomorphisms in which the two rows are exact. If the two vertical maps f and h are isomorphisms, then g is also an isomorphism.

Proof. The map g is 1-1. For, suppose $g(m) = 0$. This means $h\beta(m) = \beta' g(m) = 0$. Since h is 1-1, we find that $\beta(m) = 0$, i.e., $m \in \ker \beta$. The exactness of first row implies that there exists an $m' \in M'$ such that $\alpha(m') = m$. Thus, $\alpha'(f(m')) = g\alpha(m') = 0$. Since α' is 1-1, we see that $f(m') = 0$. Since f is 1-1, we find that $m' = 0$. Therefore, $m = \alpha(m') = 0$.

To show that g is onto, let $n \in N$. Then $\beta'(n) \in N''$. Since h is onto, there is an $m'' \in M''$ such that $h(m'') = \beta'(n)$. But, since β is onto, there is an $m \in M$ such that $m'' = \beta(m)$, i.e., $h(\beta(m)) = \beta'(n) = \beta' g(m)$. This means $\beta'(g(m) - n) = 0$. Therefore, $g(m) - n = \alpha'(n')$ for some $n' \in N'$. Since f is onto, there is an $m' \in M'$ such that $n' = f(m')$. Thus, $g(m) - n = \alpha' f(m') = g\alpha(m')$ which implies $g(m - \alpha(m')) = n$. ■

Exercises

1. For any R-module M, show that $\text{Hom}(R, M) \cong M$.
2. Let A be a finite abelian group. Prove that $\text{Hom}(A, \mathbb{Z}) = 0$.
3. Identify the following modules: $\text{Hom}_{\mathbb{Z}}(\mathbb{Z}_m, \mathbb{Z}_m)$, $\text{Hom}_{\mathbb{Z}}(\mathbb{Q}, \mathbb{Q})$ and $\text{Hom}_{\mathbb{Q}}(\mathbb{Q}, \mathbb{Q})$.
4. Suppose

$$0 \to M' \overset{i_1}{\rightleftharpoons} M \overset{p_2}{\rightleftharpoons} M'' \to 0$$

is a direct sum diagram, i.e., $p_1 \circ i_1 = I_{M'}$, $p_2 \circ i_2 = I_{M''}$, $i_1 \circ p_1 + i_2 \circ p_2 = I_M$. Then show that, for any module P, the diagram induced by $Hom(\ , P)$, viz.,

$$0 \to \text{Hom}(M'', P) \overset{p_2^*}{\rightleftharpoons} \text{Hom}(M, P) \overset{i_1^*}{\rightleftharpoons} \text{Hom}(M', P) \to 0$$

is also a direct sum diagram. State and prove a similar result for the functor $\text{Hom}(P,\ )$ also.

5. Let

$$\begin{array}{ccccccccc}
A_1 & \xrightarrow{f_1} & A_2 & \xrightarrow{f_2} & A_3 & \xrightarrow{f_3} & A_4 & \xrightarrow{f_4} & A_5 \\
\downarrow \alpha & & \downarrow \beta & & \downarrow \gamma & & \downarrow \delta & & \downarrow \epsilon \\
B_1 & \xrightarrow{g_1} & B_2 & \xrightarrow{g_2} & B_3 & \xrightarrow{g_3} & B_4 & \xrightarrow{g_4} & B_5
\end{array}$$

be a commutative diagram of R-modules and R-homomorphisms with exact rows. If β and δ are isomorphisms, α is onto and ϵ is 1-1, then show that γ is also an isomorphism (This is a generalization of five lemma).

6. A short exact sequence $0 \to A \xrightarrow{f} B \xrightarrow{g} C \to 0$ of R-modules is said to **split** if there exists a homomorphism $k\colon C \to B$ such that $gk = I_C$. Prove that the following two conditions are equivalent:

   (i) The above exact sequence splits.

   (ii) There exists a homomorphism $l\colon B \to A$ such that $l \circ f = I_A$.

   In either of the above cases, show that $B \cong A \oplus C$.

7. Given two sequences $A \xrightarrow{f_1} A_n \xrightarrow{\alpha} A' \to 0$ and $0 \to A' \xrightarrow{\beta} A_{n-1} \xrightarrow{f_2} A''$, let us define a map $f\colon A_n \to A_{n-1}$ by putting $f = \beta \circ \alpha$. Show that the sequence $A \xrightarrow{f_1} A_n \xrightarrow{f} A_{n-1} \xrightarrow{f_2} A''$ is exact if and only if the two given sequences are exact. Conclude that a long exact sequence can be broken into a sequence of short exact sequences and vice-versa.

7.2.3 The Tensor Product of Modules and Homomorphisms

Tensor product of R-modules is crucial in explaining homology and cohomology with general coefficients as well as for introducing "product" in cohomology groups of a topological space. We continue to assume that R is a commutative ring with unity element $1 \neq 0$. We define tensor product of two R-modules, not by the universal property, as we did in the case of free groups and free products, but by directly showing its existence.

Let M and N be two R-modules. Consider the free R-module C on the set $M \times N$. Thus, the elements of this free module are finite linear combinations of elements of $M \times N$ with coefficients in R, i.e., the elements are of the form $\sum_{i=1}^{n} r_i(x_i, y_i)$, where $r_i \in R$ and $(x_i, y_i) \in M \times N$. Let D be the submodule of C generated by all elements of C of the type:

$$(x + x', y) - (x, y) - (x', y),$$

$$(x, y + y') - (x, y) - (x, y'),$$

$$(rx, y) - r(x, y),$$
$$(x, ry) - r(x, y).$$

The quotient module C/D, denoted by $M \otimes_R N$, is called the **tensor product** of M and N. Observe that if $x \otimes y$ denotes the image of $(x, y) \in M \times N$ under the quotient map $C \to D$ then, in the tensor product $M \otimes N$, the following relations must be true:

$$(x + x') \otimes y = x \otimes y + x' \otimes y,$$

$$x \otimes (y + y') = x \otimes y + x \otimes y',$$

$$(rx) \otimes y = r(x \otimes y) = x \otimes (ry).$$

Furthermore, the set $\{x \otimes y \mid (x, y) \in M \times N\}$ is, clearly, a set of generators for the tensor product $M \otimes_R N$.

Let M, N, P be three R-modules. Recall that a map $f \colon M \times N \to P$ is said to be R-**bilinear** if for all $y \in N$, the map $x \mapsto f(x, y)$ from M to P and for all $x \in M$ the map $y \to f(x, y)$ from N to P are R-linear maps. We make the following observations:

(1) The map $M \times N \to M \otimes_R N$, taking $(x, y) \to x \otimes y$, is an R-bilinear map – this follows from the relations mentioned above.

(2) If $f \colon M \times N \to P$ is any R-bilinear map, where P is an R-module, then f can be extended by linearity to a module homomorphism f (denoted by the same symbol) from the free R-module C to P which maps generators of D into zero. Hence f induces a well-defined R-homomorphism $f' \colon M \otimes N = C/D \to P$ such that $f'(x \otimes y) = f(x, y)$. In fact, the map f', so defined by f, will be unique.

From the above two properties, we find that the tensor product $M \otimes_R N$ of two R-modules M and N satisfies the following universal property:

Universal Property: Tensor product of two R-modules M and N is an R-module $M \otimes_R N$ along with an R-bilinear map $g \colon M \times N \to M \otimes N$ such that whenever there is an R-bilinear map $f \colon M \times N \to P$, where P is any R-module, then there exists a unique R-homomorphism $f' \colon M \otimes N \to P$ making the following triangle commutative:

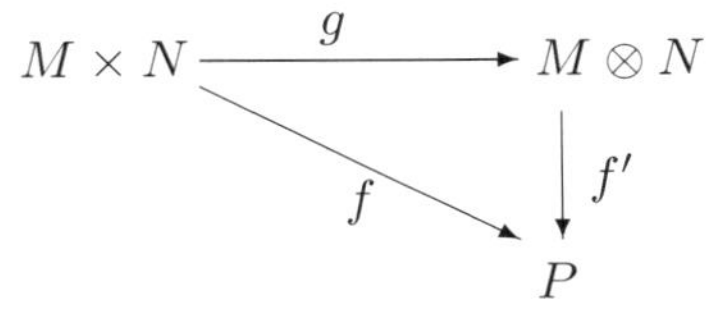

As usual, it is the above universal property of tensor product, rather than its definition, which is used in proving results about tensor products. In fact, the construction used in defining tensor product was needed to show the existence of tensor product: Once that is done, one will hardly use the construction again and the universal property will suffice. A few basic results illustrating the point just made are included in the following proposition:

Proposition 7.2.8. *Let M, N, P be three R-modules. Then there exist unique isomorphisms given below:*

(i) $M \otimes N \cong N \otimes M$.

(ii) $(M \otimes N) \otimes P \cong M \otimes (N \otimes P)$.

(iii) $R \otimes M \cong M$.

(iv) $(M \oplus N) \otimes P \cong M \otimes P \oplus N \otimes P$.

Proof. In each case, we will use the universal property of tensor product to prove the existence of a homomorphism and then we will find a homomorphism using the same or some other property which will be its inverse.

(i) In this case, consider the map $f\colon M \times N \to N \otimes M$ taking $(x, y) \to y \otimes x$, i.e., $f(x, y) = y \otimes x$. Then observe that

$$\begin{aligned} f(x + x', y) &= y \otimes (x + x') \\ &= y \otimes x + y \otimes x' \\ &= f(x, y) + f(x', y). \end{aligned}$$

Similarly, we prove that $f(x, y + y') = f(x, y) + f(x, y')$, $f(rx, y) = rf(x, y) = f(x, ry)$ for each $r \in R$. Thus, f is indeed a bilinear map. Hence f defines a unique homomorphism $f' : M \otimes N \to N \otimes M$ satisfying $f(x \otimes y) = y \otimes x$. Note that in proving bilinearity of f, we used the bilinearity of tensor product in $N \otimes M$. Using the same procedure, we find a homomorphism $h'\colon N \otimes M \to M \otimes N$ such that $h'(y \otimes x) = x \otimes y$. Thus, it is now evident that $h' \circ f'(x \otimes y) = x \otimes y$ and, similarly, $f' \circ h'(y \otimes x) = y \otimes x$. Then uniqueness of these homomorphisms will yield that each of $h' \circ f'$ and $f' \circ h'$ is the identity map, i.e., $f'\colon M \otimes N \to N \otimes M$ is an isomorphism.

(ii) This requires more work. The whole idea is use the universal property of tensor product and find a homomorphism from $(M \otimes N) \otimes P$ to $M \otimes (N \otimes P)$ taking $(x \otimes y) \otimes z$ to $x \otimes (y \otimes z)$ for each $x \in M$, $y \in N$, $z \in P$. Then, one also finds a homomorphism from $M \otimes (N \otimes P) \to (M \otimes N) \otimes P$ taking $x \otimes (y \otimes z)$ to $(x \otimes y) \otimes z$. We leave the details as an exercise. Once this has been done, it is then clear that each of the above two homomorphisms is an isomorphism.

(iii) The map $f\colon R \times M \to M$ defined by $f(r, x) = r \cdot x$ is, clearly, bilinear and hence defines an R-homomorphism $f'\colon R \otimes M \to M$ such that $f'(r \otimes x) = r \cdot x$. The map $g\colon M \to R \otimes M$, defined by $g(x) = 1 \otimes x$, is a homomorphism which is the inverse of f' because

$$gf'(r \otimes x) = g(r \cdot x) = 1 \otimes rx = r \otimes x,$$

showing that gf' is identity. Similarly, $f'g$ is clearly the identity map. Hence f' is an isomorphism.

(iv) This one is also straightforward. We find a homomorphism $(M \oplus N) \otimes P \to M \otimes P \oplus N \otimes P$ taking $(x, y) \otimes z$ to $(x \otimes z, y \otimes z)$. Conversely, we find a homomorphism from $M \otimes P$ to $(M \oplus N) \otimes P$ taking $x \otimes z$ to $(x, 0) \otimes z$ and also find a homomorphism taking $y \otimes z$ to $(0, y) \otimes z$. Then we use the definition of direct sum of R-modules to find a homomorphism $M \otimes P \oplus N \otimes P \to (M \oplus N) \otimes P$ taking $(x \otimes z, y \otimes z)$ to $(x, 0) \otimes z + (0, y) \otimes z = (x, y) \otimes z$. The two homomorphisms are clearly inverse of each other. ■

Suppose we have homomorphisms $f\colon M \to M'$ and $g\colon N \to N'$. Then we can use the universal property of tensor products to define the **tensor product of two homomorphism**, viz., $f \otimes g\colon M \otimes N \to M' \otimes N'$. For this, just consider the map $M \times N \to M' \otimes N'$ taking (x, y) to $f(x) \otimes g(y)$. It is readily verified that this map is R-bilinear and so we have unique homomorphism, denoted by $f \otimes g$, from $M \otimes N \to M' \otimes N'$ defined by $(f \otimes g)(x \otimes y) = f(x) \otimes g(y)$.

The tensor product of homomorphisms preserves composites also, viz., if $f'\colon M' \to M''$ and $g'\colon N' \to N''$ are further two homomorphisms, then it is true that $(f' \otimes g') \circ (f \otimes g) = (f' \circ f) \otimes (g' \circ g)$. To verify this, just notice that both the homomorphisms agree on the generators $x \otimes y \in M \otimes N$.

Example 7.2.9. It will follow from the next proposition that tensor product of two epimorphisms is an epimorphism. It is, however, not true that the tensor product of two monomorphisms is again a monomorphism. To see an example of this, consider the map $f\colon \mathbb{Z} \to \mathbb{Z}$ defined by $f(x) = 2x$. If we tensor by $\mathbb{Z}_2$, we get the induced map $f \otimes 1\colon \mathbb{Z} \otimes \mathbb{Z}_2 \to \mathbb{Z} \otimes \mathbb{Z}_2$, and this map is not 1-1 because $1 \otimes 1 \in \mathbb{Z} \otimes \mathbb{Z}_2$ is a generator of $\mathbb{Z} \otimes \mathbb{Z}_2 \cong \mathbb{Z}_2$ which goes to $(f \otimes 1)(1 \otimes 1) = 2 \otimes 1 = 2(1 \otimes 1) = 1 \otimes 2 = 0$.

Tensor Product is Right Exact

Let

$$0 \longrightarrow M' \xrightarrow{f} M \xrightarrow{g} M'' \longrightarrow 0 \quad (*)$$

be an exact sequence of R-modules. If we tensor the above sequence by an R-module N, then we get an induced sequence

$$0 \longrightarrow M' \otimes N \xrightarrow{f \otimes 1} M \otimes N \xrightarrow{g \otimes 1} M'' \otimes N \longrightarrow 0 \quad (**).$$

As we have seen in the earlier example, even though f is a monomorphism, $f \otimes 1$ need not be a 1-1 map, i.e., the sequence $(**)$ need not be exact on left. However, the induced sequence is exact at all other terms as the following result shows!

Proposition 7.2.10. *If*

$$M' \xrightarrow{f} M \xrightarrow{g} M'' \longrightarrow 0$$

is an exact sequence of R-modules then, for any R-module N, the induced sequence

$$M' \otimes N \xrightarrow{f \otimes 1} M \otimes N \xrightarrow{g \otimes 1} M'' \otimes N \longrightarrow 0$$

is exact.

Proof. Clearly, $g \otimes 1$ is onto. For, if $m'' \otimes n$ is a generator of $M'' \otimes N$ then, since g is onto, we can find an $m \in M$ such that $(g \otimes 1)(m \otimes n) = g(m) \otimes n = m'' \otimes n$.

Next, since $g \circ f = 0$, we find that $(g \otimes 1) \circ (f \otimes 1) = (g \circ f) \otimes 1 = 0$ which means $\text{Im}\,(f \otimes 1) \subseteq \ker(g \otimes 1)$.

For converse, let $\nu \colon M \otimes N \to (M \otimes N)/\text{Im}\,(f \otimes 1) = L$, say, be the quotient map. By the inclusion just stated, we have an obvious homomorphism $\mu \colon L \to M'' \otimes N$. On the other hand, for each $(m'', n) \in M'' \otimes N$, we can choose an element $m \in M$ such that $g(m) = m''$. Then the element $\nu(m \otimes n) \in L$ is independent of the choice of m and this defines a bilinear map $(m'', n) \to \nu(m \otimes n)$. Therefore, we have a homomorphism $\omega \colon M'' \otimes N \to L$ which is just the inverse of μ. Hence $M'' \otimes N \cong L$ and we find that $\text{Im}\,(f \otimes 1) = \ker(g \otimes 1)$. ■

The fact that tensoring is not exact on the left is very important because it is this observation which gives rise to the functors $\text{Tor}^1(.,.)$, $\text{Tor}^2(.,.)$ etc. We must remark that if the sequence $(*)$ above is split exact, then tensoring by a module will make $(**)$ exact on the left also. For example, if M'' is free, then $(**)$ will be exact everywhere. It is interesting to observe that the two functors, viz., Hom and Tensor product are closely related to one another. We have

Proposition 7.2.11. *For any R-modules M, N and P, there is a canonical isomorphism*

$$\text{Hom}(M \otimes N, P) \cong \text{Hom}(M, \text{Hom}(N, P)).$$

Proof. We know that any bilinear map $f \colon M \times N \to P$ uniquely defines a homomorphism $f' \colon M \otimes N \to P$ which, in fact, gives a 1-1 correspondence between the set of all above bilinear maps and the set $\text{Hom}(M \otimes N, P)$. On the other hand, given a bilinear map $f \colon M \times N \to P$, we have a linear map $f_x \colon N \to P$ for each $x \in M$ defined by $f_x(y) = f(x, y)$. Therefore, f yields a map $\phi \colon M \to \text{Hom}(N, P)$ defined by $f(x) = f_x$. Note that for any $y \in N$,

$$f_{x+x'}(y) = f(x + x', y) = f(x, y) + f(x', y) = (f_x + f_{x'})(y).$$

Similarly, $f_{rx} = r \cdot f_x$. Thus, the map ϕ is indeed a linear map. Conversely, given a linear map $\phi \colon M \to \text{Hom}(N, P)$, we can define a bilinear map $g \colon M \times N \to P$ by $g(x, y) = \phi(x)(y)$. This establishes a 1-1 correspondence between the set of all bilinear maps $M \times N \to P$ and the set $\mathrm{Hom}(M, \mathrm{Hom}(N, P))$. Combining the two 1-1 correspondences, we find that there is 1-1 correspondence between the two sets $\mathrm{Hom}(M \times N, P)$ and $\mathrm{Hom}(M, \mathrm{Hom}(N, P))$. To conclude the proof, we just observe that the above 1-1 correspondence $\eta\colon \mathrm{Hom}(M \otimes N, P) \to \mathrm{Hom}(M, \mathrm{Hom}(N, P))$ is defined by the formula

$$[(\eta f')(x)](y) = f'(x \otimes y)$$

for $f' \in \mathrm{Hom}(M \otimes N, P)$ and this η is, in fact, a linear map. ■

Exercises

1. If m, n are relatively prime, then show that $\mathbb{Z}_m \otimes \mathbb{Z}_n = 0$. More generally, for any two positive integers m and n, prove that $\mathbb{Z}_m \otimes \mathbb{Z}_n \cong \mathbb{Z}_d$, where $d = (m, n)$ is the gcd of m and n.
2. Use Proposition 7.2.10 to prove that the tensor product of two epimorphisms is an epimorphism. Give two distinct examples to show that the tensor product of two monomorphisms need not be a monomorphism.
3. Compute $\mathbb{Z}_m \otimes_{\mathbb{Z}} \mathbb{Q}$, $\mathbb{Q} \otimes_{\mathbb{Z}} \mathbb{Q}$ and $\mathbb{Q} \otimes_{\mathbb{Q}} \mathbb{Q}$.
4. Let G be a finite abelian group. Prove that $G \otimes_{\mathbb{Z}} \mathbb{Q} = 0$.

Ext and Tor Functors

Suppose A is an abelian group, i.e., A is a $\mathbb{Z}$-module. We can generalise the results that follow when A is any R-module, R a PID. We say that A is **injective** if the following diagram of abelian groups, with exact row, can always be completed to make it a commutative diagram:

$$\begin{array}{ccccc} 0 & \longrightarrow & G & \xrightarrow{i} & G' \\ & & {\scriptstyle h}\downarrow & \swarrow {\scriptstyle h'} & \\ & & A & & \end{array}$$

In other words, the abelian group A is injective if any homomorphism $h\colon G \to A$ can always be extended to a homomorphism $h'\colon G' \to A$. Similarly, we say that an abelian group B is **projective** if the following diagram, with exact row, can be completed to a commutative diagram

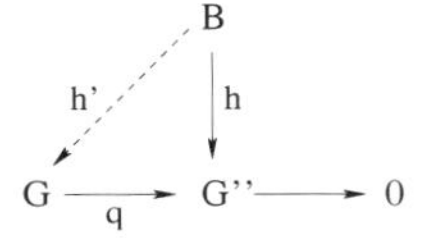

It is well-known that a $\mathbb{Z}$-module A can always be embedded into an injective $\mathbb{Z}$-module. Similarly, a $\mathbb{Z}$-module B can always be written as the quotient of a projective module. It is also true that a quotient of an injective $\mathbb{Z}$-module is again injective and a submodule of a projective $\mathbb{Z}$-module is again projective.

Now let us define $\mathrm{Ext}(A, B)$ for any two abelian groups A and B (i.e., for $\mathbb{Z}$-modules) as follows:

Definition 7.2.12. *Embed B into an injective $\mathbb{Z}$-module I_0 and let I_1 be the quotient module. Thus we have an exact sequence*

$$0 \longrightarrow B \xrightarrow{i} I_0 \xrightarrow{q} I_1 \longrightarrow 0$$

of abelian groups. Apply the covariant functor $\mathrm{Hom}(A, .)$ *on the above exact sequence to get an exact sequence*

$$0 \longrightarrow \mathrm{Hom}(A, B) \xrightarrow{i_*} \mathrm{Hom}(A, I_0) \xrightarrow{q_*} \mathrm{Hom}(A, I_1).$$

Note that q_* need not be onto. We define

$$Ext(A, B) = Coker\, q_*$$

Thus $\mathrm{Ext}(A, B)$ is an abelian group, which makes the sequence

$$0 \longrightarrow \mathrm{Hom}(A, B) \longrightarrow \mathrm{Hom}(A, I_0) \longrightarrow \mathrm{Hom}(A, I_1) \longrightarrow \mathrm{Ext}(A, B) \longrightarrow 0$$

exact. It can be proved that $\mathrm{Ext}(A, B)$, defined as above, is really independent of the choice of the injective group I_0. In case, P is a projective group (a free abelian group is always projective), then the induced sequence

$$0 \longrightarrow \mathrm{Hom}(P, B) \longrightarrow \mathrm{Hom}(P, I_0) \longrightarrow \mathrm{Hom}(P, I_1) \longrightarrow 0$$

is always exact, by the definition of projective group. Therefore, it follows that if A is projective, then

$$\mathrm{Ext}(A, B) = 0$$

for any group B.

We can also define $\mathrm{Ext}(A, B)$ as follows: Write down A as the quotient of a projective group P_0 and let P_1 be the kernel. Then we have a short exact sequence

$$0 \longrightarrow P_1 \xrightarrow{q_1} P_0 \xrightarrow{q_0} A \longrightarrow 0$$

of abelian groups. We now apply the contravariant functor $\mathrm{Hom}(., B)$ to the above sequence. Then we know that the induced sequence

$$0 \longrightarrow \mathrm{Hom}(A, B) \xrightarrow{q_0^*} \mathrm{Hom}(P_0, B) \xrightarrow{q_1^*} \mathrm{Hom}(P_1, B)$$

is exact and the map q_1^* need not be onto. We define

$$\text{Ext}(A, B) = \text{Coker } q_1^*.$$

In other words, $\text{Ext}(A, B)$ is the abelian group which makes the following sequence exact :

$$0 \longrightarrow \text{Hom}(A, B) \xrightarrow{q_0^*} \text{Hom}(P_0, B) \xrightarrow{q_1^*} \text{Hom}(P_1, B) \longrightarrow \text{Ext}(A, B) \longrightarrow 0.$$

Again it can be proved that this definition of $\text{Ext}(A, B)$ is independent of the choice of the projective group P_0 and is, in fact, equivalent to the one given earlier using injective groups.

If we regard $C\colon 0 \to B \to I_0 \to I_1 \to 0$ as an injective resolution of B, and consider the induced cochain complex

$$C^*\colon 0 \to \text{Hom}(A, B) \to \text{Hom}(A, I_0) \to \text{Hom}(A, I_1) \to 0$$

then, by definition, $H^0(C^*) \cong \text{Hom}(A, B)$ and $H^1(C^*) \cong \text{Ext}(A, B)$. This interpretation yields the following interesting result.

Proposition 7.2.13. *Let* $0 \to A' \to A \to A" \to 0$ *be an exact sequence of abelian groups. Then for any abelian group* B *there is an induced exact sequence of abelian groups*

$$0 \longrightarrow \text{Hom}(A'', B) \longrightarrow \text{Hom}(A, B) \longrightarrow \text{Hom}(A', B) \longrightarrow \text{Ext}(A'', B)$$
$$\longrightarrow \text{Ext}(A, B) \longrightarrow \text{Ext}(A', B) \longrightarrow 0.$$

Proof : Let $0 \to B \to I_0 \to I_1 \to 0$ be an injective resolution of B. Consider the following commutative diagram in which the horizontal rows are exact (because I_0, I_1 both are injective)

$$\begin{array}{ccccccccc}
 & & 0 & & 0 & & 0 & & \\
 & & \downarrow & & \downarrow & & \downarrow & & \\
0 & \to & \text{Hom}(A'', B) & \to & \text{Hom}(A, B) & \to & \text{Hom}(A', B) & \to & 0 \\
 & & \downarrow & & \downarrow & & \downarrow & & \\
0 & \to & \text{Hom}(A'', I_0) & \to & \text{Hom}(A, I_0) & \to & \text{Hom}(A', I_0) & \to & 0 \\
 & & \downarrow & & \downarrow & & \downarrow & & \\
0 & \to & \text{Hom}(A'', I_1) & \to & \text{Hom}(A, I_1) & \to & \text{Hom}(A', I_1) & \to & 0 \\
 & & \downarrow & & \downarrow & & \downarrow & & \\
 & & 0 & & 0 & & 0 & &
\end{array}$$

Applying the homology exact sequence theorem (ker-coker theorem) to the two bottom horizontal rows of the above diagram (see later in this section) and interpreting the Hom and Ext terms as the kernel and cokernel of vertical maps, the result follows immediately. ■

Using projective resolution of the first variable in $\mathrm{Ext}(A, B)$, one can easily derive the following analogous result for the second variable:

Proposition 7.2.14. *Let $0 \to B' \to B \to B'' \to 0$ be an exact sequence of abelian groups. Then for any abelian group A, there is the following exact sequence of abelian groups*

$$0 \to \mathrm{Hom}(A, B') \to \mathrm{Hom}(A, B) \to \mathrm{Hom}(A, B'') \to \mathrm{Ext}(A, B') \to \mathrm{Ext}(A, B) \to \mathrm{Ext}(A, B'') \to 0.$$

Examples

(i) $\mathrm{Ext}(\mathbb{Z}, A)$=0 for each abelian group A. This follows since $\mathbb{Z}$ is free (and hence projective).

(ii) $\mathrm{Ext}(\mathbb{Z}_n, B) \cong B/nB$. In particular $\mathrm{Ext}(\mathbb{Z}_n, \mathbb{Z}) \cong \mathbb{Z}_n$.

Observe that $0 \to \mathbb{Z} \xrightarrow{n} \mathbb{Z} \to \mathbb{Z}/n\mathbb{Z} \to 0$ is a free resolution of $\mathbb{Z}/n\mathbb{Z}$. Hence the following sequence is exact

$$0 \to \mathrm{Hom}(\mathbb{Z}/n\mathbb{Z}, B) \to \mathrm{Hom}(\mathbb{Z}, B) \to \mathrm{Hom}(\mathbb{Z}, B) \to \mathrm{Ext}(\mathbb{Z}/n\mathbb{Z}, B) \to 0.$$

Since $\mathrm{Hom}(\mathbb{Z}, B) \cong B$ and the middle map is just multiplication by n, $\mathrm{Ext}(\mathbb{Z}/n\mathbb{Z}, B) \cong B/nB$.

(iii) $\mathrm{Ext}(\mathbb{Z}_n, \mathbb{Z}_m) = \mathbb{Z}_d$, where d =gcd(n, m).

Consider the exact sequence

$$0 \to \mathbb{Z}_m \to \mathbb{Q}/\mathbb{Z} \xrightarrow{m} \mathbb{Q}/\mathbb{Z} \to 0,$$

where the map on the right is the multiplication by m. Then we get the following exact sequence

$$0 \to \mathrm{Hom}(\mathbb{Z}_n, \mathbb{Z}_m) \to \mathrm{Hom}(\mathbb{Z}_n, \mathbb{Q}/\mathbb{Z}) \xrightarrow{m_*} \mathrm{Hom}(\mathbb{Z}_n, \mathbb{Q}/\mathbb{Z}) \to \mathrm{Ext}(\mathbb{Z}_n, \mathbb{Z}_m) \to 0$$

in which one checks easily that m_* is also multiplication by m.

Since $\mathrm{Hom}(\mathbb{Z}_n, \mathbb{Q}/\mathbb{Z}) \cong \mathbb{Z}_n$, we see that the groups at both ends should be cyclic. Exactness of the sequence yields that both end groups must also be of the same order, say d. Now we claim that ker $m_* = d$ =gcd(n, m). Let $n = dp$ and $m = dq$ where p, q are relatively prime. Then

$$r \in \ker m_* \Leftrightarrow n|rm \Leftrightarrow dp|rdq \Leftrightarrow p|rq \Leftrightarrow p|r.$$

Therefore $r = p$ works and is the smallest such positive integer. This proves our assertion. Hence $\mathrm{Ext}(\mathbb{Z}_n, \mathbb{Z}_m) \cong \mathbb{Z}_d \cong \mathrm{Ext}(\mathbb{Z}_m, \mathbb{Z}_n)$, where d =gcd(n, m).

Tor Functor

To define $\text{Tor}(A, B)$ for any two abelian groups A, B, let $0 \to P_1 \xrightarrow{i} P_0 \to A \to 0$ be a projective resolution of A. Then the sequence

$$0 \to P_1 \otimes B \xrightarrow{i \otimes B} P_0 \otimes B \to A \otimes B \to 0$$

is exact everywhere except possibly on the left. We define $A * B$ (A torsion B) or $\text{Tor}(A, B)$ to be just the kernel of the homomorphism $i \otimes B$. In other words, $A * B$ is an abelian group which makes the following sequence exact

$$0 \to A * B \to P_1 \otimes B \to P_0 \otimes B \to A \otimes B \to 0.$$

The torsion product $A * B$ is also symmetric in a natural way. Furthermore, given any exact sequence $0 \to A' \to A \to A'' \to 0$ of abelian groups, there is an induced long exact sequence for any abelian group B viz.,

$$0 \to A' * B \to A * B \to A'' * B \to A' \otimes B \to A \otimes B \to A'' \otimes B \to 0.$$

Now one can easily verify the following facts using the same techniques which were adopted earlier for computing the $\text{Ext}(A, B)$.

Examples

(i) $\text{Tor}(A, \mathbb{Z}) = 0 \;\forall$ abelian group A.

(ii) $\text{Tor}(A, \mathbb{Z}_n) \cong \ker\{n \colon A \to A\}$, where n is the multiplication by n.

(iii) $\text{Tor}(\mathbb{Z}_n, \mathbb{Z}_m) = \mathbb{Z}_d = \text{Tor}(\mathbb{Z}_m, \mathbb{Z}_n)$, where $d =$gcd (n, m). To see the last fact, note that the exact sequence

$$0 \to \mathbb{Z} \xrightarrow{n} \mathbb{Z} \to \mathbb{Z}_n \to 0$$

induces an exact sequence

$$0 \to \text{Tor}(\mathbb{Z}, \mathbb{Z}_m) \to \text{Tor}(\mathbb{Z}, \mathbb{Z}_m) \to \text{Tor}(\mathbb{Z}_n, \mathbb{Z}_m) \to$$

$$\mathbb{Z} \otimes \mathbb{Z}_m \to \mathbb{Z} \otimes \mathbb{Z}_m \to \mathbb{Z}_n \otimes \mathbb{Z}_m \to 0.$$

Since $\mathbb{Z}$ is free, this reduces to

$$0 \to \text{Tor}(\mathbb{Z}_n, \mathbb{Z}_m) \to \mathbb{Z}_m \xrightarrow{n_*} \mathbb{Z}_m \to \mathbb{Z}_n \otimes \mathbb{Z}_m \to 0$$

where n_* is the multiplication by n. This means both of the end groups are cyclic and are of the same order, say d. Then, as before, $d =$gcd (n, m).

Exercises

1. An abelian group A is said to be **divisible** if given any a and an $n(\neq 0) \in \mathbb{N}$, there exists a $b \in A$ such that $nb = a$. Prove that A is divisible if and only if A is injective as a $\mathbb{Z}$-module.
2. Prove that the quotient of an injective abelian group is again injective.
3. Show that a subgroup of a projective abelian group is again projective.
4. Prove that for a given abelian group A, $\text{Ext}(A, B) = 0$ for all B if and only if A is projective.
5. Prove that for a given B, $\text{Ext}(A, B) = 0$ for all A if and only if B is injective.
6. For an abelian group A, show that $\text{Tor}(\mathbb{Z}_n, A) = \{a \in A \mid na = 0\}$.
7. Prove that A is torsion free if and only if $\text{Tor}(A, B) = 0$ for every B.

7.2.4 Chain Complexes and Homology

A sequence $M = \{(M_n, \partial_n), n \in \mathbb{Z}\}$, of R-modules M_n with R-homomorphisms $\partial_n \colon M_n \to M_{n-1}$, called **boundary homomorphisms**, viz.,

$$M\colon \ldots \longrightarrow M_{n+1} \xrightarrow{\partial_{n+1}} M_n \xrightarrow{\partial_n} M_{n-1} \longrightarrow \ldots,$$

is called a **chain-complex** if $\partial_n \circ \partial_{n+1} = 0$ for all $n \in \mathbb{Z}$. The chain complexes of practical importance which we would come across will have all the negative terms zero – these are called **positive** chain complexes.

Let $Z_n = \ker \partial_n$ and $B_n = \text{Im}\ \partial_{n+1}$. Then Z_n and B_n are submodules of M_n. The elements of Z_n are called ***n*-cycles** and the elements of B_n are are called ***n*-boundaries**. The elements of M_n are called ***n*-chains**. Since $\partial_n \circ \partial_{n+1} = 0$, we find that $B_n \subseteq Z_n$ for each n. The quotient module Z_n/B_n, denoted by $H_n(M)$, is called the ***n*-dimensional homology module** of the chain complex M_*. The elements of $H_n = Z_n/B_n$ are called **homology classes** and are denoted by $\{z\}$, where $z \in Z_n$.

If we take a sequence $M = \{(M^n, \delta^n)\}$, viz.,

$$M\colon \ldots \longrightarrow M^{n-1} \xrightarrow{\delta^{n-1}} M^n \xrightarrow{\delta^n} M^{n+1} \longrightarrow \ldots$$

of R-modules and R-homomorphisms such that $\delta^n \circ \delta^{n-1} = 0$ for each $n \in \mathbb{Z}$, then M is called a **cochain complex** of R-modules. Occasionally, a chain complex M is denoted by M_* and a cochain complex M by M^*. The indices in (M^n, δ^n) are superscripts rather than subscripts just to distinguish it from a chain complexe M – it is true that except for the position of indices, there is no conceptual difference between a chain complex and a cochain complex. We call the elements of $Z^n = \ker \delta^n$ **cocycles**, elements of $B^n = \operatorname{Im} \delta^{n-1}$ **coboundaries**, elements of M^n **cochains** and the elements of the quotient group $H^n(M) = Z^n/B^n$ **cohomology classes**. The quotient group itself is called **n-dimensional cohomology module** of the cochain complex M. Note that if the homology groups $H_n(M) = 0$, then the chain complex becomes exact at M_n. Thus, the homology groups of a chain complex M measure its deviation from exactness of the sequence.

Let M and N be two chain complexes of R-modules. Then a sequence $\{f_n : M_n \to N_n\}$ of R-homomorphisms is called a **chain map** from M to N if these homomorphisms commute with boundary homomorphisms, i.e., the square

$$\begin{array}{ccccccc} \longrightarrow & M_n & \xrightarrow{\partial_n} & M_{n-1} & \longrightarrow \\ & \downarrow f_n & & \downarrow f_{n-1} & \\ \longrightarrow & N_n & \xrightarrow{\partial'_n} & N_{n-1} & \longrightarrow \end{array}$$

is commutative for each $n \in \mathbb{Z}$, i.e., $f_{n-1} \circ \partial_n = \partial'_n \circ f_n$ for all $n \in \mathbb{Z}$. The definition of a chain isomorphism (or chain equivalence) is now evident. We readily verify that f_n maps cycles into cycles and boundaries into boundaries and, therefore, it defines a homomorphism $(f_n)_* : H_n(M) \to H_n(N)$, $n \in \mathbb{Z}$ by the formula $(f_n)_*\{z\} = \{f_n(z)\}$, where $\{z\} \in H_n(M)$. These are called **induced homomorphisms in homology**. If there is no confusion, we write $(f_n)_*$ just as $f_* : H_n(M) \to H_n(N)$. Let us observe that these induced homomorphisms are functorial, viz.,

(i) If $f : M \to N$ and $g : N \to P$ are two chain maps, then their composite $g \circ f$ is a chain map and

$$(g \circ f)_* = g_* \circ f_* : H_n(M) \to H_n(P).$$

(ii) If $I : M \to M$ is the identity chain map, then the induced homomorphism $I_* : H_n(M) \to H_n(M)$ is also the identity map.

A chain complex M is said to be **acyclic** if $H_n(M) = 0$ for all $n \in \mathbb{Z}$, i.e., all the homology groups of M are trivial. When we dualize each of the notions defined above, we get cochain map, cochain isomorphisms, etc. Next, we have

Definition 7.2.15. *Let $f, g : M \to N$ be two chain maps. We say that f is* **chain homotopic** *to g, written as $f \approx g$, if there is a sequence $\{D_n : M_n \to N_{n+1}\}$ of homomorphisms such that for each $n \in \mathbb{Z}$,*

$$f_n - g_n = \partial_{n+1} D_n + D_{n-1} \partial_n.$$

We now observe that if we consider the set of all chain maps from M to N, then the relation of chain-homotopy is an equivalence relation in that set. The following result is noteworthy:

Proposition 7.2.16. *If two chain maps $f,\ g\colon M \to N$ are chain homotopic, then the homomorphism in homology induced by them are identical, i.e., for all $n \in \mathbb{Z}$,*

$$f_* = g_* \colon H_n(M) \to H_n(N).$$

Proof. Let $\{D_n \colon M_n \to M_{n+1}\}$ be a chain homotopy from f to g and $\{z\} \in H_n(M)$. This means $\partial_n(z) = 0$. Since $f_n - g_n = \partial_{n+1}D_n + D_{n-1}\partial_n$, we find that $f_n(z) - g_n(z) = \partial_{n+1}D_n(z)$ is a boundary. Hence $\{f_n(z)\} = \{g_n(z)\}$, i.e., $f_*(\{z\}) = g_*(\{z\})$. ■

Similarly, one can define a cochain-homotopy between two cochain maps. Then it follows from the above proposition that any two cochain homotopic maps induce identical maps in cohomology.

Let us recall the Hom and Tensor functors discussed earlier. Note that if

$$M_* \colon \ \ldots \to M_{n+1} \to M_n \to M_{n-1} \to \ldots$$

is a chain complex, and if for any R-module N we apply the Hom(, N) functor to this chain complex, then we get the induced sequence $\mathrm{Hom}(M_*, N)$:

$$\ldots \leftarrow \mathrm{Hom}(M_{n+1}, N) \leftarrow \mathrm{Hom}(M_n, N) \leftarrow \mathrm{Hom}(M_{n-1}, N) \leftarrow \ldots$$

of R-modules going in opposite direction. In other words, we get a cochain complex $\mathrm{Hom}(M_*, N)$. Here, we must point out that if the chain complex M_* of R-modules is acyclic, i.e., $H_n(M_*) = 0$ for all $n \in \mathbb{Z}$ and we take the induced cochain complex $\mathrm{Hom}(M_*, N)$, it is not necessary that this cochain complex will also be acyclic - this is clear because Hom(. , N) is only left exact, it is not an exact functor. This observation is used in defining the concept of higher order Hom(,) functors called Ext^n(. ,) functors. Similarly, when we tensor the chain complex M_*, which is acyclic, by another R-module N, it is not necessary that the resulting complex $M_* \otimes N$ would be acyclic. This phenomenon gives rise to higher order tensor products called Tor_n(. ,) functors (see S. MacLane [17] for all the details). The special case when $N = R$ is of frequent occurrence in algebraic topology. In this case, the homology of the cochain complex $\mathrm{Hom}(M_*, R)$ is called the **cohomology** of the chain complex M_*. Now, it is pertinent to ask: What is the relationship between the homology groups of the chain complex M_* and the homology groups of the cochain complex $\mathrm{Hom}(M_*, R)$? In other words, how is the homology of a chain complex M_* related with its cohomology? The answer to this question is given by the following theorem known as the **Universal Coefficient Theorem for Cohomology.**

Theorem 7.2.17. *Let K be a chain complex of free R-modules and G be an R-module, where R is a PID (principal ideal domain). Then, with $H^n(K,G) = H_n(\mathrm{Hom}(K,G))$, there exists a split exact sequence*

$$0 \to \mathrm{Ext}(H_{n-1}(K), G) \xrightarrow{\beta} H^n(K,G) \xrightarrow{\alpha} \mathrm{Hom}(H_n(K), G) \to 0.$$

The homomorphism β is natural with respect to the chain map as well as with respect to the coefficient module, but the splitting is not natural with respect to the chain map.

Another significant situation arises when we take the tensor product of a given chain complex M_* by an R-module N. We get the new chain complex, viz., $M_* \otimes N$. Again, the question is: How is the homology of the chain complex M_* related to the homology of the new chain complex $M_* \otimes N$? The answer is known as the "Universal Coefficient Theorem for Homology" (See Spanier, E.H. [18] for this and the preceding result).

Theorem 7.2.18. (Universal Coefficient Theorem for Homology). *Let K be a chain complex of free R-modules, R a PID, and let G be an R-module. Then there exists a split exact sequence*

$$0 \to H_n(K) \otimes G \xrightarrow{\alpha} H_n(K \otimes G) \xrightarrow{\beta} \mathrm{Tor}^1(H_{n-1}(K), G) \to 0,$$

which is natural with respect to the chain complex K and coefficient module G, but the splitting is not natural with respect to the chain maps.

7.2.5 Tensor Product of two Chain Complexes

Let (K, ∂') and (L, ∂'') be two chain complexes of R-modules which are positive, i.e., $K_n = 0 = L_n$ for all negative integers n. The tensor product of K and L is a new chain complex, denoted by $K \otimes L$, whose nth term is given by the module $(K \otimes L)_n = \sum_{p+q=n} K_p \otimes L_q$ and the boundary operator

$$\partial_n \colon (K \otimes L)_n \to (K \otimes L)_{n-1}$$

is defined using ∂', ∂'' as follows: for any generating element $a \otimes b \in K_p \otimes L_q$,

$$\partial_n(a \otimes b) = \partial' a \otimes b + (-1)^{\deg a} a \otimes \partial'' b.$$

Here, $\deg a$ means degree of the element a, where $\deg a = p$ if $a \in K_p$. We have

Proposition 7.2.19. *The homomorphism $\partial \colon (K \otimes L)_n \to (K \otimes L)_{n-1}$ satisfies the property $\partial \circ \partial = 0$, i.e., $(K \otimes L, \partial)$ is a chain complex.*

Proof. Note that $(K \otimes L)_n$ consists of the direct sum of $(n+1)$ terms of the type $K_p \otimes L_q$, where $p + q = n$. If $a \otimes b$ is a generator of $K_p \otimes L_q$, then

$$\begin{aligned}
\partial \circ \partial(a \otimes b) &= \partial(\partial' a \otimes b + (-1)^{\deg a} a \otimes \partial'' b) \\
&= \partial'(\partial' a) \otimes b + (-1)^{\deg \partial' a}(\partial' a) \otimes \partial' b + (-1)^{\deg a} \partial' a \otimes \partial' b \\
&\quad + (-1)^{\deg a + \deg a} a \otimes \partial''(\partial'' b) \\
&= 0,
\end{aligned}$$

since $\deg a$ and $\deg(\partial' a)$ differ by 1, and this proves the assertion. ■

One can similarly define the tensor product of two positive cochain complexes. Once again, it may now be asked as to how the homology groups of the new chain complex $K \otimes L$ are related to the homologies of K and L. There is a well-known answer to this question. It is known as the "Künneth Formula", which provides an exact sequence connecting the homology of $K \otimes L$ in terms of the tensor product and torsion product of homology groups of K and L. The Künneth formula is extremely useful in computing the homology (respectively, cohomology) of the product topological space in terms of the homology (respectively, cohomology) groups of the individual spaces (see E.H. Spanier [18]).

Theorem 7.2.20. (Künneth Formula for Homology). *Let K and L be two chain complexes of R-modules, at least one of which is a chain complex of free modules and R is a PID. Then there exists a split exact sequence*

$$0 \to \sum_{i+j=n} H_i(K) \otimes H_j(L) \xrightarrow{\alpha} H_n(K \otimes L)$$

$$\xrightarrow{\beta} \sum_{i+j=n-1} \mathrm{Tor}_1(H_i(K), H_j(L)) \to 0.$$

The homomorphisms α and β are natural with respect to the chain maps, but the splitting is not natural.

At this point, it is worth mentioning that the tensor product respects chain homotopies between chain maps. To explain this, let us denote the chain homotopy between chain maps f and f' as $D\colon f \cong f'$. In this notation, we have

Proposition 7.2.21. *Let $f_1,\ f_2\colon K \to L$ be two chain-homotopic chain maps and $g_1,\ g_2\colon K' \to L'$ be another chain-homotopic chain maps. Then the chain maps $f_1 \otimes g_1,\ f_2 \otimes g_2\colon K \otimes K' \to L \otimes L'$ are also chain-homotopic.*

Proof. Let $S\colon f_1 \cong f_2$ and $T\colon g_1 \cong g_2$ be two chain homotopies. Then we can define a chain homotopy $H\colon f_1 \otimes g_1 \cong f_2 \otimes g_2$ by the formula $H = S \otimes g_1 + f_2 \otimes T$, i.e., we define

$$H(x \otimes y) = S(x) \otimes g_1(y) + (-1)^{\deg x} f_2(x) \otimes T(y).$$

In fact, $S \otimes 1$ gives a chain homotopy $K \otimes L \to K' \otimes L$ from $f_1 \otimes 1$ to $f_2 \otimes 1$, and $1 \otimes T$ gives a homotopy between $1 \otimes g_1 \cong 1 \otimes g_2 \colon K' \otimes L \to K' \otimes L'$. Composing these homotopies yields the desired result. ■

7.2.6 Exact Homology Sequence Theorem

The next result, called **Exact Homology Sequence Theorem**, which we are going to state and discuss, is sometimes known as the **Fundamental Theorem of Homological Algebra**. It is indeed a very powerful result which has numerous applications in the development of all kinds of homology groups and is of frequent occurrence elsewhere also. We will not prove the result as the proof is not only technical but also requires lengthy verifications. We will, however, explain the crucial steps involved in the statement of the theorem.

Let C', C and C'' be three chain complexes of R-modules and suppose there are chain maps $f \colon C' \to C$ and $g \colon C \to C''$ which yield a short exact sequence

$$0 \longrightarrow C' \xrightarrow{f} C \xrightarrow{g} C'' \longrightarrow 0$$

of chain complexes. Since f and g are chain maps, they induce homomorphisms

$$H_n(C') \xrightarrow{f_*} H_n(C) \xrightarrow{g_*} H_n(C'')$$

for each $n \in \mathbb{Z}$. The Exact Homology Sequence Theorem asserts the existence of a sequence of homomorphisms $\delta \colon H_n(C'') \to H_{n-1}(C')$, $n \in \mathbb{Z}$, such that the resulting long sequence

$$\to H_n(C') \xrightarrow{f_*} H_n(C) \xrightarrow{g_*} H_n(C'') \xrightarrow{\delta} H_{n-1}(C')$$

$$\xrightarrow{f_*} H_{n-1}(C) \xrightarrow{g_*} H_{n-1}(C'') \to,$$

connecting the homology groups of C', C and C'', is exact. The homomorphism $\delta \colon H_n(C'') \to H_{n-1}(C')$ which connects the nth level to the $(n-1)$th level of homology groups is called the **connecting homomorphism**. The crucial aspect of this homomorphism is that it is natural in the given short exact sequence of chain complexes, i.e., if we have two short exact sequences

$$E\colon \quad 0 \to C' \to C \to C'' \to 0$$

and

$$F\colon \quad 0 \to D' \to D \to D'' \to 0$$

which are related by chain maps $\alpha \colon C' \to D'$, $\beta \colon C \to D$, $\gamma \colon C'' \to D''$ making various squares commutative, then the connecting homomorphism

$\delta_E\colon H_n(C'') \to H_{n-1}(C')$ and $\delta_F\colon H_n(D'') \to H_{n-1}(D')$ make the resulting squares in the diagram

$$\begin{array}{ccccccc} \longrightarrow & H_n(C'') & \xrightarrow{\delta_E} & H_{n-1}(C') & \longrightarrow \\ & \downarrow{\gamma_*} & & \downarrow{\alpha_*} & \\ \longrightarrow & H_n(D'') & \xrightarrow{\delta_F} & H_{n-1}(D') & \longrightarrow \end{array}$$

commutative for each $n \in \mathbb{Z}$. We can now state

Theorem 7.2.22. *Let*

$$0 \to C' \xrightarrow{f} C \xrightarrow{g} C'' \to 0$$

be an exact sequence of chain complexes of R-modules. Then the long sequence induced in their homology, viz.,

$$\cdots \to H_n(C) \xrightarrow{g_*} H_n(C'') \xrightarrow{\delta} H_{n-1}(C') \xrightarrow{f_*} H_{n-1}(C) \to \cdots$$

is exact, where δ is the connecting homomorphism.

We can easily indicate the definition of δ as follows: take a homology class $\{z''\} \in H_n(C'')$. Using ontoness of g, find a $c \in C_n$ such that $g_n(c) = z''$. Then $\partial(c) \in C_{n-1}$. It can be shown that $\partial c \in \ker g_{n-1} = \text{Im } f_{n-1}$. Hence we can find a $z \in C'_{n-1}$ such that $f_{n-1}(z) = \partial c$. Then it can be proved that z' is also a cycle of C' and we define $\delta\{z''\} = \{z'\}$, i.e., $\delta\{z''\} = \{f_{n-1}\partial g_n^{-1}\}$. The technical details are needed to show that δ is well defined in the sense that the definition is independent of all the choices which we have made in going from $z'' \in C''_n$ to $z' \in C'_{n-1}$. Once it is proved that δ is a well-defined homomorphism, then the exactness of the long sequence is proved by "diagram chase". That completes the proof of the long exact homology sequence theorem – the fundamental theorem of homological algebra (see G.E. Bredon [2] p. 178–180 for details).

Exercises

1. Let $f,\ g\colon C \to C'$; $f',\ g'\colon C' \to C''$ be chain maps and $\cong$ denote chain homotopy. Prove that $f \cong g$ and $f' \cong g'$ implies $f' \circ f \cong g' \circ g$.

2. Let

   $$0 \to C' \to C \to C'' \to 0$$

   be an exact sequence of chain complexes. If any two of these are acyclic, then show that the third one is also acyclic.

3. Let

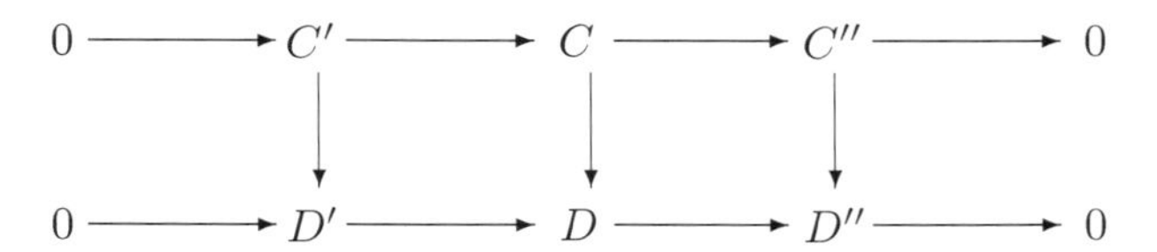

be a commutative diagram of chain complexes in which the two rows are exact. Show that if any two of the vertical chain maps induce isomorphisms in homology then so does the remaining third chain map.

4. Let $f\colon C \to C'$ be a chain map between two nonnegative chain complexes of free abelian groups such that the induced homomorphism $f_*\colon H_q(C) \to H_q(C')$ is an isomorphism for all $q \geq 0$. Then prove that the chain map f is a chain equivalence.

7.3 Topological Transformation Groups

The concepts, e.g., a topological group acting on a topological space continuously, a Lie group acting on a smooth manifold smoothly, a discrete group acting on a simplicial complex simplicially, etc., are all similar and can be defined in an analogous manner with obvious changes. Here, we deal with only continuous actions as they provide the best examples of topological spaces in the guise of orbit spaces. Before we define it formally, let us see a very general concept which is always present in the background of everything mentioned above.

Group Acting on a Set

Let X be a set and G be group. A map $G \times X \to X$, $(g, x) \mapsto g.x$, is called an **action** of G on the set X if the two conditions stated below are satisfied for all $g_1, g_2 \in G$ and $x \in X$:

(i) $(g_1.g_2).x = g_1.(g_2.x)$,

(ii) $e.x = x$.

Here, $e \in G$ is the identity element of G. Fix a point $x \in X$ and consider the set $G_x = \{g \in G \mid g.x = x\}$. Then G_x is a subgroup of G and is called the **isotropy subgroup** of the point $x \in X$. Next, the subset $G(x) = \{g.x \mid g \in G\}$ of X is called the **orbit** of x in X. One can easily verify that any two orbits of X are either identical or disjoint. Hence these orbits decompose the set X into mutually disjoint orbits of X. One can easily prove that, for any $x \in X$, the map $g \to g.x$ induces a 1-1, onto map $f\colon G/G_x \to G(x)$ defined by $f(g.G_x) = g.x$. Thus, the set of all cosets of the isotropy subgroup G_x of x in G is in bijection with the set of all elements in the orbit $G(x)$. Now, if we choose exactly one x_i from each orbit $G(x_i), i \in I$, then since orbits $\{G(x_i) \mid i \in I\}$ provide a partition of X, we find an important formula, viz.,

$$|X| = \sum_{i \in I} |G(x_i)|,$$

where $|X|$ denotes the number of elements in the set X. It may also be noted that if x, y are in the same orbit, then the isotropy subgroups G_x and G_y are conjugate in G, i.e., $G_x = gG_yg^{-1}$ for some $g \in G$. The converse of this is also true, viz., if H is conjugate to an isotropy subgroup $G(x)$, say, $H = gG_xg^{-1}$, then there exists a point $y \in G(x)$, viz., $y = gx$ such that $G_y = H$.

We now illustrate some interesting actions of a group G on different sets.

Example 7.3.1. Let us consider the group G acting on the set G which is the underlying set of the group itself defined by $g.x = gxg^{-1}$, where the right-hand side is the conjugate of x by the element g.

In the above example, we readily see that the isotropy subgroup G_a of an element $a \in G$ is simply the normalizer $N(a)$ of $\{a\}$ in G. Thus, $a \in Z(G)$, the centre of the group if and only if $G(a) = \{a\}$ or $G_a = G$. The orbit of an element $a \in G$ is just the conjugacy class of the element a. It follows that if G is a finite group then $|G| = \sum |G/N(a)| = \sum |G|/|N(a)|$. Now, the important formula mentioned earlier gives us the following equation, called **class equation** of the finite group G,

$$\begin{aligned} |G| &= \sum |G/N(a)| \\ &= |Z(G)| + \sum_{N(a) \neq G} \frac{|G|}{|N(a)|}. \end{aligned}$$

There is yet another useful case of a group acting on a set:

Example 7.3.2. Let us consider the set $L(G)$ of all subgroups of G. Consider the group G acting on the set $L(G)$ under the action defined by $g.H := gHg^{-1}$. Here, we should note that the point H of $L(G)$ is a fixed point under the above action if and only if H is normal in G. The orbit of H is just the set of all subgroups of G which are conjugate to H, i.e., the conjugacy class of H. The isotropy subgroup of a subgroup H of G is simply the normalizer $N(H)$ of H in G. Hence it follows from the general facts about group actions that $|G/N(H)|$ must be equal to the number of distinct conjugacy classes of H in G. This concept of conjugacy classes of subgroups of a group G plays a fundamental role in the classification of covering projections.

7.3.1 Topological Transformation Groups

Now, we will discuss the important notion of a "topological transformation group". Let G be a topological group, i.e., G is a group with a topology on it so that the group operations $G \times G \to G$ and $G \to G$, defined by $(g_1, g_2) \mapsto g_1.g_2$ and $g \mapsto g^{-1}$, respectively, are continuous. Let X be a topological space. Then a **continuous** map $G \times X \to X$, denoted by $(g, x) \mapsto g.x$, is said to be a **topological action** (or just an action) of G on X if the following two conditions (same as mentioned earlier) are satisfied:

(i) $(g_1.g_2).x = g_1(g_2.x)$

(ii) $e.x = x$

for every g_1, $g_2 \in G$ and for all $x \in X$. Here, $e \in G$ is the identity element.

The pair (G, X) along with the given action is called a **topological transformation group**. If we change any one of the three items, viz., G, X or the action, then we get a different transformation group. If, in the definition of a transformation group, we forget the topologies from the space X, the group G and the action, then we revert back to the notion of "a group G acting on a set X" discussed earlier.

Recall the definitions of **isotropy group** and **orbit** given earlier. Note that if the group G is compact and the space X is Hausdorff, then the orbits will be closed sets of X. In fact, in this case, the coset space G/G_x will be homeomorphic to the orbit $G(x)$. The action of G on X is said to be **free** if the isotropy group of each point is trivial, i.e., $G_x = \{e\}$ for all $x \in X$. The set $F(G, X) = \{x \in X \mid G_x = G\}$ is called the **fixed-point set** of the transformation group (G, X) and is also denoted by X^G. We have already noted that any two orbits in X are either identical or disjoint. The set of all distinct orbits of X, denoted by X/G, with the quotient topology induced from X, is called the **orbit space** of the transformation group. We will come across many spaces which are simply the orbit spaces of a space X under a suitable action of a topological group G.

Before giving examples, let us remark that if a topological group G acts on a space X then, for each $g \in G$, we can define a map

$$f_g \colon X \to X$$

by setting

$$f_g(x) = g.x.$$

One can, using conditions (i) and (ii) above, easily verify that for any pair g_1, $g_2 \in G$, $f_{g_1.g_2} = f_{g_1} \circ f_{g_2}$ and $f_e = I_X$, where $e \in G$ is the identity element of G. We also note that, since the action is continuous, each $f_g \colon X \to X$ is a continuous map. Now, taking $g_1 = g$ and $g_2 = g^{-1}$, we find that

$$f_g \circ f_{g^{-1}} = f_{g.g^{-1}} = f_e = I_X.$$

Similarly, we have

$$f_{g^{-1}} \circ f_g = I_X.$$

This shows that each of the maps $f_g\colon X \to X$ is indeed a homeomorphism. Furthermore, the map $\eta\colon G \to \mathrm{Homeo}(X)$ defined by $\eta(g) = f_g$ is a homomorphism of groups. We say that the given action of G on X is **effective** if η is 1-1, i.e., each $g(\neq e)$ yields a nonidentity homeomorphism of X. In case the action is effective, we can clearly identify every element $g \in G$ with a homeomorphism $x \mapsto g.x$ of X and thus, G can be viewed as a group of homeomorphism of the space of X. With this viewpoint, the definitions of isotropy groups, orbits, free actions, etc., have nice geometric flavour, e.g., the action of G on X is free if and only if each $g(\neq e) \in G$ moves every point of X.

Conversely, given a group G of homeomorphism of X, we can always define an action of G on X by putting $g.x = g(x)$. The only question which is not immediately obvious is: what is the topology on G? Since $G \subseteq \mathrm{Homeo}(X)$, any topology on $\mathrm{Homeo}(X)$ (for instance, the compact open topology), will induce a topology on G. But, then the question will be whether or not the map $G \times X \to X$ defined above will be continuous. It is well known that if X is a locally compact Hausdorff space then the compact open topology does give a continuous action of the desired type (see Dugundji [5] p. 259).

An important case, which occurs frequently, is the case when we take the discrete topology on a group G of homeomorphisms of a space X. In this case, the map $G \times X \to X$, defined above, is always continuous and we have a nice action of G on X. In the special case of a finite group G, the topology on G is always assumed to be discrete; we even do not mention this explicitly. The finite group of homeomorphisms of a space X have been studied and used extensively in the study of fixed-point sets.

Example 7.3.3 (Real Projective Space)**.** Let $\mathbb{S}^n$ be the standard n-sphere, $n \geq 1$. Let $A\colon \mathbb{S}^n \to \mathbb{S}^n$ be the antipodal map, i.e., $A(x) = -x$. Then A^2 is identity. Thus, $G = \{I, A\}$ is a group of homeomorphisms of $\mathbb{S}^n$ and so it acts on $\mathbb{S}^n$. The action is free and the orbit space $\mathbb{S}^n/G$, denoted by $\mathbb{RP}^n$, is called the **real projective space**. The space $\mathbb{RP}^n$ is a compact, Hausdorff, connected manifold of dimension n.

Example 7.3.4 (Complex Projective Space)**.** Let $\mathbb{S}^1 \subset \mathbb{C}$ be the circle group. Then it is easily seen that $\mathbb{S}^1$ is a topological group. Let

$$\mathbb{S}^{2n+1} = \{(z_0, z_1, \ldots, z_n) \in \mathbb{C}^n \colon \sum_{i=1}^{n} |z_i|^2 = 1\},$$

be the $(2n+1)$-dimensional unit sphere. Then $\mathbb{S}^1$ acts on $\mathbb{S}^{2n+1}$ continuously under the action defined by

$$z.(z_0, z_1, \ldots, z_n) = (zz_0, zz_1, \ldots, zz_n).$$

This again is a free action. The orbit space $\mathbb{S}^{2n+1}/\mathbb{S}^1$, denoted by $\mathbb{CP}^n$, is called the **complex projective space**. This space is a compact, Hausdorff, connected manifold of real dimension $2n$.

Example 7.3.5 (Torus)**.** Let $\mathbb{R}$ be the real line and define a map $h\colon \mathbb{R} \to \mathbb{R}$ by $h(x) = x+1$. Then h is a homeomorphism and for each integer n, $h^n : \mathbb{R} \to \mathbb{R}$ is also a homeomorphism. These are just translations by integer amounts. Thus, we get the infinite cyclic group $\mathbb{Z} = [h]$ generated by the homeomorphism h of $\mathbb{R}$. Giving $\mathbb{Z}$ the discrete topology, we find that $\mathbb{Z}$ acts as a group of homeomorphisms on $\mathbb{R}$. Again, the action is free and the orbit space, denoted by $\mathbb{R}/\mathbb{Z}$, can be seen to be homeomorphic to the circle group $\mathbb{S}^1$.

We can take the product action of the discrete group $\mathbb{Z} \times \mathbb{Z}$ on $\mathbb{R} \times \mathbb{R}$ by considering two homeomorphisms h and k of $\mathbb{R}$ and defining the action of (h, k) on $\mathbb{R} \times \mathbb{R}$ by putting

$$(h, k)(x, y) = (x+1, y+1).$$

Then

$$(h^m, k^n)(x, y) = (x+m, y+n)$$

for every pair of integers (m, n). Again, the action is free and the orbit space is $\mathbb{S}^1 \times \mathbb{S}^1$, which is the **2-torus**. One can similarly obtain n**-torus** as an orbit space of $\mathbb{R}^n$ for all $n \geq 2$ under the action of $\mathbb{Z}^n$.

Example 7.3.6 (Klein bottle)**.** A group may act on a space in many different ways. To see a simple example of this situation, consider the torus T as a subspace of $\mathbb{R}^3$ given by rotating the circle $(x-3)^2 + z^2 = 1$ about z-axis. Define maps f, g and h as follows:

$$\begin{aligned} f(x, y, z) &= (x, -y, -z), \\ g(x, y, z) &= (-x, -y, z), \\ h(x, y, z) &= (-x, -y, -z). \end{aligned}$$

Each of these is a homeomorphism of T of order 2. Hence, the group $G = \mathbb{Z}_2 = \{1, a\}$ can act on T in three distinct ways. The interesting point to be verified and observed is that orbit spaces corresponding to all the three actions defined by homeomorphisms f, g, h on T are distinct. In the first case, it is homeomorphic to the sphere; in the second case, it is homeomorphic to the torus itself and in the last case, it is homeomorphic to the Klein bottle.

Example 7.3.7 (Lens Spaces). Let us take the 3-sphere

$$\mathbb{S}^3 = \{(x_1, x_2, x_3, x_4) \in \mathbb{R}^4 \mid \sum_{i=1}^{4} {x_i}^2 = 1\}.$$

We can also describe this sphere in terms of pairs of complex numbers $z_1 = x_1 + \iota x_2, z_2 = x_3 + \iota x_4$ as

$$\mathbb{S}^3 = \{(z_1, z_2) \mid |z_1|^2 + |z_2|^2 = 1\}.$$

Let us take two positive integers p and q which are relatively prime. Now, we consider the complex number $e^{\frac{2\pi\iota q}{p}}$, whose modulus is clearly 1. Let us define a map $h : \mathbb{S}^3 \to \mathbb{S}^3$ by

$$h(z_1, z_2) = (z_1.e^{\frac{2\pi\iota}{p}}, z_2.e^{\frac{2\pi\iota q}{p}}).$$

Then, clearly, h is a continuous map. Furthermore, since h^p is identity on $\mathbb{S}^3$, h is in fact a homeomorphism of $\mathbb{S}^3$ of order p. It follows, as before, that the group $\mathbb{Z}_p = [h]$ acts on $\mathbb{S}^3$. Again, the action is free and so the orbit of every point has p distinct points. The orbit space $\mathbb{S}^3/\mathbb{Z}_p$, denoted by $L(p, q)$, is called the **lens space**. The question is: how to visualize the lens space? Since $\mathbb{S}^3$ itself cannot be easily visualized, the quotient space $L(p, q)$ is still worse to imagine. Note that the special case $L(2, 1)$ is evidently $\mathbb{RP}^3$, which contains $\mathbb{RP}^2$. We have already pointed out that $\mathbb{RP}^2$ cannot be embedded in $\mathbb{R}^3$. Thus, we should forget about visualizing the lens space and must rather study it on the basis of its definition only. Being the orbit space of $\mathbb{S}^3$ under a free action, these lens spaces have several interesting properties. One question which crops up right now is the following: The lens space $L(p, q)$ clearly depends on the pair (p, q) of relatively prime integers. Then, is it true that $L(p, q)$ are all distinct spaces for different values of pairs (p, q)? Note that if (p, q_1) and (p, q_2) are two such pairs so that $q_1 - q_2 = kp$, then $e^{\frac{2\pi\iota q_1}{p}} = e^{\frac{2\pi\iota q_2}{p}}$ and hence, the two actions of $\mathbb{Z}_p$ on $\mathbb{S}^3$ with respect to the pairs (p, q_1) and (p, q_2) are same. This means that $L(p, q_1) = L(p, q_2)$. In fact, it is well-known that $L(p, q_1) \cong L(p, q_2)$ if and only if $q_1 \equiv \pm q_2 (\text{mod } p)$ or $q_1 q_2 \equiv \pm 1 (\text{mod } p)$ (see [16] p. 242).

Now, suppose a topological group G acts continuously on a space X. Fix a point $x \in X$ and consider the orbit $G(x)$ of x. If G_x is the isotropy group of the point x, then we can define a map

$$f : G/G_x \to G(x)$$

simply by putting

$$f(gG_x) = g.x.$$

This is well defined since $g_1 G_x = g_2 G_x$ implies that $(g_1)^{-1}.g_2 \in G_x$, i.e., $((g_1)^{-1}.g_2).x = x$, i.e., $g_1.x = g_2.x$. Clearly, the map f is continuous and is a bijection. However, the map f need not be a homeomorphism. (see one of the exercises later). In case the group G is compact and X is Hausdorff, then, clearly, f is a homeomorphism. In other words, for compact groups G acting on a Hausdorff space X, the orbits are simply homeomorphic to the coset spaces of the type G/H, where $H \subset G$ is a closed subgroup of G. An action of G on X is said to be **transitive** if given any two points $x, y \in X$, there exists a $g \in G$ such that $g.x = y$. This really means that the space X is just a single orbit and so X is homeomorphic to the coset space G/H for some closed subgroup H of G. Such spaces are known as "homogeneous spaces".

In our study of the covering projections and fundamental groups, we come across the following special kind of group actions.

Definition 7.3.8. *A group G of homeomorphisms of a space X is said to act on X* **properly discontinuously** *if each point $x \in X$ has a neighbourhood U such that $U \cap g(U) \neq \emptyset$ implies that $g = e$, where $e \in G$ is the identity element of G.*

One can easily verify that any action of a finite group on a Hausdorff space is properly discontinuous. Thus, except for the linear actions (see following exercises) of $GL(n, \mathbb{R})$, $O(n)$ and the irrational flow mentioned later in Exercises, the rest of the examples are properly discontinuous.

Exercises

1. Suppose a connected topological group G acts continuously on a space X. Prove that X is connected if and only if the orbit space X/G is connected.

2. Find an action of $\mathbb{Z}$ on $\mathbb{R} \times [0, 1]$ such that the orbit space is homeomorphic to the Möbius band.

   If a noncompact group G acts on a topological space X, the orbit $G(x)$ of a point $x \in X$ need not be homeomorphic to the quotient space G/G_x. Even the orbit space X/G may have unexpected topology. The following exercise provides a good example to show these pathological situations.

3. **(Irrational Flow)** Let α be a fixed irrational number. We define an action of the additive group $\mathbb{R}$ of real numbers on the torus $T = \mathbb{S}^1 \times \mathbb{S}^1$ using α as follows: Let $(e^{2\pi i x}, e^{2\pi i y})$ be an arbitrary point of T, and for $t \in \mathbb{R}$, put
$$t.(e^{2\pi i x}, e^{2\pi i y}) = (e^{2\pi i(x+t)}, e^{2\pi i(y+\alpha t)}).$$
   Prove that the above is a free action of $\mathbb{R}$ on T and that every orbit in T is a proper subset of T which is dense in T. Also show that the orbit space $T/\mathbb{R}$ has the indiscrete topology.

(**Hint:** Imagine $\mathbb{S}^1$ as the quotient space $\mathbb{R}/\mathbb{Z}$ so that the torus $T = \mathbb{S}^1 \times \mathbb{S}^1$ can be thought of as the unit square $I \times I$ in the plane $\mathbb{R} \times \mathbb{R}$, where the opposite sides of the square have been identified. Now, chase the orbit of the point $(1/2, 0)$ in the square to conclude that the orbit cannot be locally connected whereas $\mathbb{R}$ is locally connected).

4. An $n \times n$ real matrix can be thought of as a point in the Euclidean space $\mathbb{R}^{n^2}$. Prove that the set $GL(n, \mathbb{R})$ of all nonsingular $n \times n$ real matrices with the Euclidean subspace topology from $\mathbb{R}^{n^2}$ forms a topological group with respect to multiplication of matrices. Show that this group acts on the Euclidean space $\mathbb{R}^n$ by multiplication, i.e., $A \cdot X$ is the product of $n \times n$ matrix $A \in GL(n, \mathbb{R})$ with the $n \times 1$ column vector $X \in \mathbb{R}^n$.

5. Prove that the group $O(n)$ of all $n \times n$ orthogonal real matrices acts on the unit disk $\mathbb{D}^n$ as well as on the unit sphere $\mathbb{S}^n$ under the action of preceding exercise. Show that this action is transitive on $\mathbb{S}^n$. Hence or otherwise, deduce that $O(n)/O(n-1) \cong \mathbb{S}^{n-1}$.

References

[1] Agoston, M.K., *Algebraic Topology: A First Course*, Marcel Dekker, 1967.

[2] Bredon, G.E., *Topology and Geometry*, Springer-Verlag, 1997.

[3] Chinn, W.G., and Steenrod, N., *First Concepts of Topology*, Random House, 1966.

[4] Dieudonné J., *A History of Algebraic and Differential Topology (1900–1960)*, Birkhäuser, 1989.

[5] Dugundji, J., *Topology*, Allyn and Bacon, 1966.

[6] Eilenberg, S. and Steenrod, N., *Foundations of Algebraic Topology*, Princeton University Press, 1952.

[7] Greenberg, M. and Harper, J., *Algebraic Topology- A First Course*, Benjamin/Cummings, 1981.

[8] Herstein, I.N., *Topics in Algebra*, John Wiley and Sons, 1975.

[9] Hilton, P.J. and Wylie, S., *Homology Theory*, Cambridge University Press, 1960.

[10] Hocking, J. and Young, G.S., *Topology*, Addison Wesley, 1961.

[11] Jacobson, N., *Basic Algebra I*, W.H. Freeman and Co., 1985.

[12] Lundell, A. and Weingram, S., *The Topology of CW-Complexes*, Van Nostrand Reinhold, 1969.

[13] Massey, W.S., *Algebraic Topology: An Introduction*, Springer-Verlag, 1967.

[14] Maunder, C.F., *Algebraic Topology*, Van Nostrand Reinhold, 1970.

[15] Munkres, J.R., *Topology: A First Course*, Prentice-Hall, 1975.

[16] Munkres, J.R., *Elements of Algebraic Topology*, Addison Wesley, 1984.

[17] MacLane, S., *Homology*, Academic Press, Springer-Verlag, 1963.

[18] Spanier, E.H., *Algebraic Toplogy*, McGraw-Hill, 1966.

[19] Seifert, H., and Threlfall, W., *A Textbook of Topology* (Translation from German Edition of 1947), Academic Press, 1980.

[20] Whitehead, G.W., *Elements of Homotopy Theory*, Springer-Verlag, 1978.

Index

Texts and Readings in Mathematics

1. R. B. Bapat: Linear Algebra and Linear Models (3/E)
2. Rajendra Bhatia: Fourier Series (2/E)
3. C.Musili: Representations of Finite Groups
4. Henry Helson: Linear Algebra (2/E)
5. Donald Sarason: Complex Function Theory (2/E)
6. M. G. Nadkarni: Basic Ergodic Theory (3/E)
7. Henry Helson: Harmonic Analysis (2/E)
8. K. Chandrasekharan: A Course on Integration Theory
9. K. Chandrasekharan: A Course on Topological Groups
10. Rajendra Bhatia(ed.): Analysis, Geometry and Probability
11. K. R. Davidson: C* -Algebras by Example
12. Meenaxi Bhattacharjee et al.: Notes on Infinite Permutation Groups
13. V. S. Sunder: Functional Analysis - Spectral Theory
14. V. S. Varadarajan: Algebra in Ancient and Modern Times
15. M. G. Nadkarni: Spectral Theory of Dynamical Systems
16. A. Borel: Semi-Simple Groups and Symmetric Spaces
17. Matilde Marcoli: Seiberg Witten Gauge Theory
18. Albrecht Bottcher:Toeplitz Matrices, Asymptotic Linear Algebra and Functional Analysis
19. A. Ramachandra Rao and P Bhimasankaram: Linear Algebra (2/E)
20. C. Musili: Algebraic Geomtery for Beginners
21. A. R. Rajwade: Convex Polyhedra with Regularity Conditions and Hilbert's Third Problem
22. S. Kumaresen: A Course in Differential Geometry and Lie Groups
23. Stef Tijs: Introduction to Game Theory
24. B. Sury: The Congruence Subgroup Problem - An Elementary Approach Aimed at Applications
25. Rajendra Bhatia (ed.): Connected at Infinity - A Selection of Mathematics by Indians
26. Kalyan Mukherjea: Differential Calculas in Normed Linear Spaces (2/E)
27. Satya Deo: Algebraic Topology - A Primer (2/E)
28. S. Kesavan: Nonlinear Functional Analysis - A First Course
29. Sandor Szabo: Topics in Factorization of Abelian Groups
30. S. Kumaresan and G.Santhanam: An Expedition to Geometry
31. David Mumford: Lectures on Curves on an Algebraic Surface (Reprint)
32. John. W Milnor and James D Stasheff: Characteristic Classes(Reprint)
33. K.R. Parthasarathy: Introduction to Probability and Measure
34. Amiya Mukherjee: Topics in Differential Topology
35. K.R. Parthasarathy: Mathematical Foundation of Quantum Mechanics (Corrected Reprint)
36. K. B. Athreya and S.N.Lahiri: Measure Theory
37. Terence Tao: Analysis - I (3/E)
38. Terence Tao: Analysis - II (3/E)

39. Wolfram Decker and Christoph Lossen: Computing in Algebraic Geometry
40. A. Goswami and B.V.Rao: A Course in Applied Stochastic Processes
41. K. B. Athreya and S.N.Lahiri: Probability Theory
42. A. R. Rajwade and A.K. Bhandari: Surprises and Counterexamples in Real Function Theory
43. Gene H. Golub and Charles F. Van Loan: Matrix Computations (Reprint of the 4/E)
44. Rajendra Bhatia: Positive Definite Matrices
45. K.R. Parthasarathy: Coding Theorems of Classical and Quantum Information Theory (2/E)
46. C.S. Seshadri: Introduction to the Theory of Standard Monomials (2/E)
47. Alain Connes and Matilde Marcolli: Noncommutative Geometry, Quantum Fields and Motives
48. Vivek S. Borkar: Stochastic Approximation - A Dynamical Systems Viewpoint
49. B.J. Venkatachala: Inequalities - An Approach Through Problems (2/E)
50. Rajendra Bhatia: Notes on Functional Analysis
51. A. Clebsch: Jacobi's Lectures on Dynamics (2/E)
52. S. Kesavan: Functional Analysis
53. V.Lakshmibai and Justin Brown: Flag Varieties - An Interplay of Geometry, Combinatorics and Representation Theory
54. S. Ramasubramanian: Lectures on Insurance Models
55. Sebastian M. Cioaba and M. Ram Murty: A First Course in Graph Theory and Combinatorics
56. Bamdad R. Yahaghi: Iranian Mathematics Competitions 1973-2007
57. Aloke Dey: Incomplete Block Designs
58. R.B. Bapat: Graphs and Matrices (2/E)
59. Hermann Weyl: Algebraic Theory of Numbers(Reprint)
60. C L Siegel: Transcendental Numbers(Reprint)
61. Steven J. Miller and RaminTakloo-Bighash: An Invitation to Modern Number Theory (Reprint)
62. John Milnor: Dynamics in One Complex Variable (3/E)
63. R. P. Pakshirajan: Probability Theory: A Foundational Course
64. Sharad S. Sane: Combinatorial Techniques
65. Hermann Weyl: The Classical Groups-Their Invariants and Representations (Reprint)
66. John Milnor: Morse Theory (Reprint)
67. Rajendra Bhatia(Ed.): Connected at Infinity II- A Selection of Mathematics by Indians
68. Donald Passman: A Course in Ring Theory (Reprint)
69. Amiya Mukherjee: Atiyah-Singer Index Theorem- An Introduction
70. Fumio Hiai and Denes Petz: Introduction to Matrix Analysis and Applications
71. V. S. Sunder: Operators on Hilbert Space
72. Amiya Mukherjee: Differential Topology
73. David Mumford and Tadao Oda: Algebraic Geometry II
74. Kalyan B. Sinha and Sachi Srivastava: Theory of Semigroups and Applications